NEUTRON INTERFEROMETRY

Neutron Interferometry

Lessons in Experimental Quantum Mechanics, Wave-Particle Duality, and Entanglement

Second Edition

Helmut Rauch and Samuel A. Werner

OXFORD
UNIVERSITY PRESS

Great Clarendon Street, Oxford, OX2 6DP,
United Kingdom

Oxford University Press is a department of the University of Oxford.
It furthers the University's objective of excellence in research, scholarship,
and education by publishing worldwide. Oxford is a registered trade mark of
Oxford University Press in the UK and in certain other countries

First published 2000
Second Edition published in 2015
First published in paperback 2017

Impression: 1

Published in the United States of America by Oxford University Press
198 Madison Avenue, New York, NY 10016, United States of America

British Library Cataloguing in Publication Data
Data available

Library of Congress Cataloging in Publication Data
Data available

ISBN 978–0–19–871251–0 (Hbk.)
ISBN 978–0–19–880981–4 (Pbk.)

Printed and bound by
CPI Group (UK) Ltd, Croydon, CR0 4YY

Preface to the Second Edition

In this new edition we have retained the overall structure of the original book. It now consists of 12 chapters. We have shifted some material from later chapters to the first chapter. In particular, we have described gravitationally induced quantum interference (COW experiment) and early experiments on spin rotation, spinors, and spin-related phases in Chapter 1. These results have gotten broad attention in recent years and are now described in various standard physics textbooks. This change in format was done with a view of the close connection of these experiments with the atom interferometry field and with the modern ideas on entanglement now permeating all of the modern quantum mechanical literature.

Neutron interferometry is a mature technique in experimental physics. The use of the perfect silicon crystal interferometer to accurately measure scattering lengths of many isotopes is given in Chapter 3. Very accurate measurements of the neutron scattering lengths of the proton, deuteron, and He-3 have been carried out in recent years. These results confine the parameters in various theoretical models of few-body nuclear physics.

Chapter 4 is retitled "Coherence and Decoherence," so as to include many new topics connected to the current emphasis in certain areas of physics related to how coherent quantum systems decohere and evolve into systems described by classical Lagrangian and Hamiltonian mechanics. The quantum concepts of entanglement, contexuality, squeezing, Wigner functions, interactions of quantum systems with the environment, post selection experiments, measures of distinguishability, intensity–intensity correlations, and Bell's inequalities are the general topics of this expanded chapter. "Contexuality and Kochen–Specker Phenomena" are discussed in a new Chapter 7.

Our original Chapter 8 dealing with forthcoming and more speculative experiments is now Chapter 10. Some of those experiments have advanced considerably over the intervening 12 years since the publication of the 1st edition. Experiments related to searches for non-Newtonian gravity at short distances have been pursued in various confined geometries, such as transmission through narrow slits. Related ideas, called "bouncing ball" experiments, have successfully observed quantized states of ultra-cold neutrons in the Earth's gravity. An experiment designed to observe the Anandan acceleration has not yet been attempted. The observation of relativistic proper time effects, as suggested in Chapter 8, occurs at the level of 10^{-9}. Suggested observation of such effects in atom beam interferometry, related to a gravitational redshift, has generated interesting, but controversial papers. Perhaps there is a way to pursue such effects with neutron interferometry utilizing the Pendellösung interference fringes, as discussed in this chapter on "Gravitational, Inertial, and Motional Effects." A feasible time-dependent Fizeau experiment is described to study new motional effects and for the cooling of neutron beams.

Various new techniques and experiments of phase contrast radiography, and phase contrast interferometry and microscopy are discussed in Chapter 9 on "Solid State Physics Applications."

The final Chapter 12 deals with epistemological questions connected with basic quantum experiments with massive and composed systems. This has started a new era of experimental quantum physics.

This new edition has a vastly expanded and detailed index. We want to thank our many collaborators and students mentioned in the first edition and those joining our groups in the meantime, particularly R. Cappelletti, Y. Hasegawa, M. Huber, H. Lemmel, D. Pushin, K. Schoen, S. Sponar, G. Sulyok, T.E. Wietfeldt who were involved in various experiments and discussions included in this book. Continuous interest of D.M. Greenberger (New York), A.G. Klein (Melbourne) and W. Schleich (Ulm) is gratefully acknowledged as well. From many sides we got help in identifying and eliminating various typographical errors in the 1st edition.

Helmut Rauch, Vienna, Austria
Samuel Werner, Gaithersburg, MD, USA
December 2014

Preface to the First Edition

This book is the result of an ongoing scientific journey that the two of us began a quarter-century ago on opposite sides of the Atlantic Ocean. The experimental observation of interference between coherently split, well-separated beams of matter waves, in this case neutron de Broglie waves, is now central to the fabric of quantum physics. The revolution in this field was brought about by the wonderfully simple, didactically exquisite, and beautifully stable perfect silicon crystal interferometer, invented by Ulrich Bonse and Michael Hart for X-rays in 1964, and first applied to neutrons in Vienna in 1974 in a cooperative work together with U. Bonse and W. Treimer. It is topologically identical to the Mach–Zehnder interferometer of classical optics. Over the past 25 years, more than 40 neutron interferometry experiments, having an impact on fundamental quantum and neutron physics, have been carried out with these devices. In recent years interferometers based upon cold and ultra-cold neutrons, and also the Larmor- and Ramsey-type interferometers, have been successfully employed. A description of the instrumentation, analysis of the results, and the theoretical motivation and interpretation of these experiments are the main subjects of this book.

The similarities of classical optics and neutron optics are easy to understand and anticipate, since the time-dependent Schrödinger equation is formally equivalent to the Helmholtz scalar wave equation, which accounts for the behavior of light waves (aside from polarization effects). However, since for a thermal neutron (energy $\approx$ 20 meV) having a wavelength about 2 Å moves with a velocity $\approx$ 2000 m/s, the time-of-flight across a silicon interferometer is typically $\approx$ 100 μs. The combination of these long transit times, short wavelengths, and rather long data collection periods requires more attention to vibration, thermal, and environmental isolation of the interferometer than the typical optical interferometer. Chapter 2 is devoted to a discussion of the necessary instrument design to accomplish the required level of stability.

Since the neutron experiences all four fundamental forces of nature (gravitation, electromagnetic, and weak and strong interactions), the landscape for neutron interferometry investigations has proven to be quite broad, extending beyond wave-optical phenomena to encompass the fundamental particle attributes of the neutron itself and the nuclei, with which it interacts strongly, and the electrons and their fields, with which it interacts more weakly. The fundamental dual nature of thermal and cold neutrons—sometimes a particle (when detected) and sometimes a wave (when traversing the interferometer)—is wonderfully manifested by the highly non-local effects observed in neutron interferometry. Chapters 3 and 4 are devoted to a comprehensive discussion of the various neutron interactions, especially the strong interactions and the measurement of the nuclear–nuclear scattering lengths by interferometry, and the intricacies of the mutual coherence properties of the wave packets traversing the two legs of the interferometer. Chapters 5, 6, and 7 provide detailed discussions of fundamental quantum interference phenomena induced by neutron spinor rotation, topology, and geometry (Aharonov–Bohm effects and Berry's phase), gravity, acceleration, rotation (Sagnac effect), and translational motion (Fizeau effects).

Non-relativistic quantum mechanics, based upon the Schrödinger equation, is generally accepted to be a linear theory involving wave functions, which are complex numbers. However, a quantum theory which is neither linear nor restricted to complex numbers is conceivable. Chapter 8 is devoted to a review of neutron interferometry experiments which search for these "speculative" prospects. Some experiments of this type, looking for non-linear effects, quaternion wave functions, AB effects not involving the electromagnetic field, and non-ergodic effects, have already been carried out.

Applications of neutron interferometry in solid state physics have just begun. Interesting experiments on the density fluctuations due to inhomogeneities caused by hydrogen in metals and the density of polymeric overlayers have already been done. These, along with other potentially significant ideas, such as phase-contrast tomography and Fourier spectroscopy, are discussed in Chapter 9. A full understanding of the Bragg reflection process in the perfect silicon crystals of our neutron interferometers requires a detailed description of the dynamical theory of diffraction. Chapter 10 provides the expert with the necessary theory. An understanding of this theory is not necessary for most of the experiments described in this book. The final chapter is devoted to interpretational questions related to quantum mechanics, and specifically to the quantum interference of matter waves. The reader will find the book by Varley Sears, *Neutron Optics*, to be an excellent companion to this book. We refer to it often.

Finally, we would like to thank our many colleagues and students located in many places throughout the world for their very valuable cooperation and for providing us with many figures and original data during the writing of this book over the past four years. With the hope of not omitting any of them, we simply mention them by name: B.E. Allman, M. Arif, G. Badurek, E. Balcar, W. Bauspiess, U. Bonse, A. Cimmino, R. Clothier, R. Colella, C.F. Eagen, G. Eder, G.L. Greene, Y. Hasegawa, A.I. Ioffe, D.L. Jacobsen, H. Kaiser, A.G. Klein, G. Kroupa, W.-T. Lee, K.C. Littrell, P.D. Mannheim, B. Mashhoon, M. Namiki, G.I. Opat, A.W. Overhauser, S. Pascazio, D. Petrascheck, J.-L. Staudenmann, E. Seidl, M. Suda, J. Summhammer, W. Treimer, D. Tuppinger, A. Wilfing, M. Zawisky, and A. Zeilinger.

A large amount of work has been financially supported by the Austrian Science Foundation (FWF) and the National Science Foundation of US (NSF), which is gratefully acknowledged. The hospitality of the Institute Laue–Langevin (ILL) in Greboble, the National Institute of Standards and Technology (NIST) in Gaithersburg, and our respective home institutions, the Atominstitut in Vienna, and the University of Missouri Research reactor (MURR) in Columbia, where many of the experiments have been performed, is gratefully acknowledged as well.

The manuscript was patiently typed by Evi Haberl, who saw it through many restructurings, revisions, and changes. To her we offer our sincerest thanks. Also, to Ilse Futterer, who gave technical assistance.

Vienna, Austria, and Columbia, Missouri, USA H. R.
September 1999 S. A. W.

Contents

1 Introduction

1.1 Neutron Optics and the Analogy with Light Optics

Optical phenomena are directly connected to wave phenomena, which are known for many kinds of radiation. Elsasser (1936) first suggested that the motion of neutrons could be determined by quantum mechanics, shortly after the discovery of the neutron by Chadwick (1932). His suggestion that as waves they would be diffracted by crystalline matter was soon verified experimentally by Halban and Preiswerk (1936) and by Mitchell and Powers (1936). The diffraction and scattering of thermal neutrons ($E \sim kT \approx 0.025$ eV) having a de Broglie wavelength comparable to the interatomic spacings of atoms in solids has become one of the most powerful modern techniques in the study of the dynamic and static structure of condensed matter. The word "optics," when applied to the propagation of neutron radiation and its modifications by various potentials or fields of force and by material objects, is by no means used here metaphorically. The extremely close mathematical analogies between the propagation of light as described by Maxwell's equations and the propagation of low-energy neutrons as described by the Schrödinger equation suggest that most phenomena occurring for light will have their analogs in neutron optics.

Neutrons experience interactions with their environment via all four forces of nature: strong nuclear force, the electromagnetic, the weak and the gravitational force, whose strengths are related as $1{:}10^{-2}{:}10^{-7}{:}10^{-39}$. Neutrons have a spin of $1/2$ and are therefore fermions, while photons are bosons. Thus, the breadth and richness of neutron optical phenomena are a consequence of their wave nature, their full interaction with the four forces of nature, and their mathematical similarities to light optics. The observation of interference phenomena of neutron matter waves having a wavelength of a few ångstroms, but with coherence features extending over macroscopic distances in the perfect silicon crystal interferometer as described in this book, represents a compendium of exquisite and didactic experimental elucidations of quantum mechanics.

Since the invention of the laser, the field of quantum optics has experienced a resurgence with some spectacular discoveries related to squeezed states, multiple photon effects, entanglement, cooling of atoms by laser light, and the Bose–Einstein condensation (Born and Wolf 1975, Walls and Milburn 1994, Anderson et al. 1995, Mandel and Wolf 1995, Berman 1997, Haroche and Raimond 2006). It can be anticipated that more of these developments will suggest new and related neutron optical experiments. Due to the differences in particle statistics for neutrons and photons, and the fact that neutrons have a rest mass while the mass of the photon is zero, future developments in these two fields will be complementary, borrowing experimental techniques and theoretical frameworks from each other. Neutrons are massive and composite (quark-structure) systems which exhibit a well-known level structure within a magnetic field where transitions can

Neutron Interferometry. Second Edition. Helmut Rauch and Samuel A. Werner.

Table 1.1 *Properties of the Neutron (Taylor 1990).*

Particle properties	Connection	Wave properties
$m = 1.674928(1) \times 10^{-27}$ kg		$\lambda_c = \frac{h}{m.c} =$ 1.319695(20) $\times 10^{-15}$ m
$s = \frac{1}{2}\hbar$	de Broglie	(thermal neutrons: $\lambda = 1.8$ Å, $v = 2200$ m/s)
$\mu = -9.6491783(18) \times 10^{-27}$ J/T	$\lambda_B = \frac{h}{mv}$	$\lambda_B = \frac{h}{m.v} = 1.8 \times 10^{-10}$ m
$\tau = 887(2)$ s		$\Delta_c = \frac{1}{2\delta k} \cong 10^{-8}$ m
$R = 0.7$ fm	Schrödinger	$\Delta_p = v.\Delta t \cong 10^{-2}$ m
$\alpha = 12.0\ (2.5) \times 10^{-4}$ fm^3	$\mathcal{H}\psi(r,t) = i\hbar\frac{\delta\psi(r,t)}{\delta t}$	$\Delta_d = v.\tau = 1.942(5) \times 10^6$ m
u–d–d = quark structure		$0 \leq \chi \leq 2\pi\ (4\pi)$
m = mass, s = spin, μ = magnetic moment, τ = β-decay lifetime, R = (magnetic) confinement radius, α = electric polarizability; all other quantities like electric charge, magnetic monopole, and electric dipole moment are compatible with 0.		λ_c = Compton wavelength, λ_B = de Broglie wavelength, Δ_c = coherence length, Δ_p = packet length, Δk = momentum width, Δt = chopper opening time, v = group velocity, χ = phase.

be induced by several means. Several typical particle and wave properties of neutrons are given in Tables 1.1 and 1.2.

The more recent achievements of atom and molecular interferometry show that there is apparently no natural upper mass limit to which interferometry can be developed, but the experimental requirements become increasingly more stringent. The status and several perspectives of atom and molecular interferometry can be found in a book edited by Berman (1997) and in a review article by Cronin et al. (2009). Interferometry with Bose–Einstein condensates (Shin et al. 2004, Wang et al. 2005) and superfluid Helium-4 (Hoskinson et al. 2006, Sato and Packard 2012) has been developed as well.

The development of quantum mechanics took an important new direction with the complimentarity hypothesis of particle motion and wave behavior, most clearly displayed in the de Broglie relation (de Broglie 1923, 1925, 1926)

$$p = mv = h/\lambda = \hbar k \tag{1.1}$$

where the momentum p of a particle of mass m and group velocity v is related to a matter wave with wavelength λ. Matter wave fields $\Psi(r,t)$ are described by the Schrödinger equation (Schrödinger 1926).

$$\mathcal{H}\Psi(r,t) = \left(-\frac{\hbar^2}{2m}\nabla^2 + V(r,t)\right)\Psi(r,t) = i\hbar\frac{\partial\psi(r,t)}{\partial t} \tag{1.2}$$

Table 1.2 *Useful Conversion Factors in Neutron Physics.*

Kinetic energy	$E = \frac{\hbar^2 k^2}{2m} = \frac{h^2}{2m\lambda^2} = hf = \frac{1}{2}mv^2 = \frac{1}{2}m\left(\frac{d}{t}\right)^2 \hat{=} k_B T$
	$E\ [\text{meV}] = 2.0723k^2 = \frac{81.81}{\lambda^2} = 4.136f = 5.2267 \times 10^{-6}v^2$
	$= 5.2267 \times 10^6 \frac{1}{(t/d)^2} = 0.086173T$
	$\lambda\ [\text{Å}], f[\text{TH}_Z], v\ [m/s], k[\text{Å}^{-1}], \frac{t}{d}\left[\frac{\mu\text{s}}{\text{m}}\right], T\ [\text{K}]$
Velocity	$v\ [\text{m/s}] = 3956/\lambda\ [\text{Å}]$
Rest mass energy	$E_0 = m_0 c^2 = 939.6\ \text{MeV}$
Zeeman energy (for $B = 1$ tesla)	$E = -\mu B = 60.311\ \text{neV}$
Larmor precession frequency (for $B = 1$ tesla)	$\omega_L = -\frac{2\mu}{\hbar} B = -\gamma B = 1.833 \times 10^8 \text{rad/s}$

and electromagnetic wave fields are given by the wave equation of classical optics

$$\nabla^2 \psi(\boldsymbol{r}, t) - \frac{1}{c^2} \frac{\partial^2 \psi(\boldsymbol{r}, t)}{\partial t^2} = 0. \tag{1.3}$$

These are linear equations which can be solved in free space using the plane wave Ansatz

$$\Psi_k(\boldsymbol{r}, t) = a_k\, e^{i(\boldsymbol{k}\cdot\boldsymbol{r} - \omega_k t)} = \Psi(\boldsymbol{r}) e^{-i\omega_k t}, \tag{1.4}$$

which gives in both cases the well-known Helmholtz equation

$$\nabla^2 \psi(\boldsymbol{r}) + k^2 \psi(\boldsymbol{r}) = 0, \tag{1.5}$$

with the dispersion relations

$$k^2 = \frac{2mE}{\hbar^2} \text{(matter-waves)}, \tag{1.6a}$$

and

$$k^2 = \frac{E^2}{\hbar^2 c^2} \text{(e. m. waves)}, \tag{1.6b}$$

where the energy is related to the frequency of the wave by $E = \hbar\omega$. The velocity of wave propagation for electromagnetic waves in free space is always equal to the velocity of light, whereas for matter waves it is determined by the de Broglie relation (Eq. 1.1).

One should differentiate between two velocities for all wave motions: the group (v_g) and phase velocity (v_{ph}). The group velocity is defined as

$$v_g = \frac{d\omega}{dk} = \frac{\hbar k}{m}, \tag{1.7}$$

which coincide with the de Broglie relation (Eq.1.1) when the total (kinetic) energy is written as $\hbar\omega = \frac{\hbar^2 k^2}{2m}$. The phase velocity follows from its definition as

$$v_{\mathrm{ph}} = \frac{\omega}{k} = \frac{v_{\mathrm{g}}}{2}, \tag{1.8}$$

which is in contradiction with the correct relativistic result. One should start with the relativistic relation connecting the energy E to the momentum p. That is,

$$E^2 = p^2c^2 + \left(m_0c^2\right)^2,$$

or

$$\omega^2 = c^2k^2 + \frac{m_0^2c^4}{\hbar^2}. \tag{1.9}$$

This gives a phase velocity of (de Broglie 1923)

$$v_{\mathrm{ph}} = \frac{\omega}{k} = \frac{c^2}{v_{\mathrm{g}}}, \tag{1.10}$$

which shows that the phase velocity of a matter wave always exceeds the velocity of light (e.g., Rogalski and Palmer 1999, Zettili 2001).

The physical significance of the phase velocity is still under discussion and the difference between the two values has not yet been resolved (e.g., Zettili 2001, Dunningham and Vedral 2011). On the other hand the particle rest mass m_0 determines the Compton wavelength $\lambda_{\mathrm{C}} = h/m_0c$ which can be used to define a Compton frequency

$$\omega_{\mathrm{C}} = \frac{m_0c^2}{\hbar}. \tag{1.11}$$

This gives for the neutron a frequency of $f_{\mathrm{Compton}} = m_0c^2/h = 2.27 \times 10^{23}$ Hz. Several authors have proposed to use this quantity as an internal clock in an atom-beam interference experiment (Müller et al. 2010, Hohensee et al. 2011). This idea resides in the spirit of de Broglie (1923, 1924), but there is also criticism and skepticism regarding this interpretation of the physical reality of the Compton frequency (see Sections 8.4 and 8.5.6, Sinha and Samuel 2011, Wolf et al. 2011, Greenberger et al. 2012). There is a relation between the Compton frequency and de Broglie wavelength, which reads in the non-relativistic limit as $\lambda_{\mathrm{C}} = \lambda_{\mathrm{dB}}v/c$ and which suggests a physical meaning to the Compton frequency as well. A more detailed description of this view is given in Section 8.5.6.

When stationary situations are described, the time-independent Schrödinger equation (and the Helmholtz equation) can be used for matter waves (and electromagnetic radiation). They have the same mathematical structure and therefore the expected diffraction phenomena must be analogous, aside from the different dispersion relations and the different kinds of interactions. Equation (1.5) is a linear differential equation and, therefore, linear combinations (superpositions) of solutions are also solutions. For time-dependent phenomena, differences are expected because the time derivatives in the Maxwell and Schrödinger wave equations are different, i.e., of first order and second order, respectively. The spreading of matter wave packets may be seen as a most

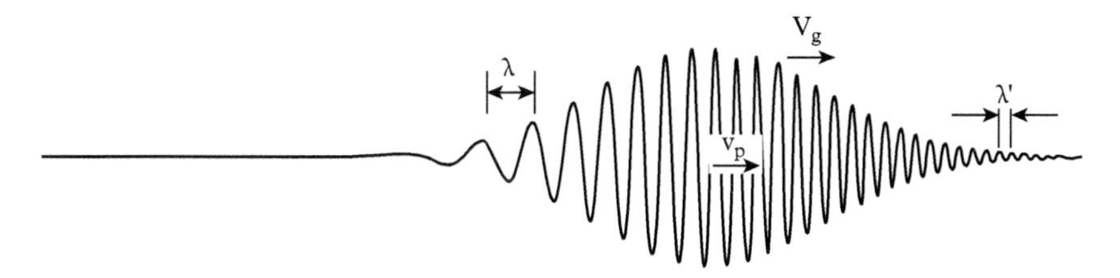

Figure 1.1 *Real part of a neutron wave packet traveling in the x-direction with group velocity v_g and phase velocity v_p. The imaginary part of the wave packet is complementary. The diagram is drawn for $\Delta k/k = 0.1$. Thanks to Robert Dimeo, NIST*

well-known phenomenon of this kind (see, e.g., Messiah 1965, Cohen-Tannoudji et al. 1992), but diffraction in time which occurs for matter waves but not for electromagnetic waves in vacuum may be mentioned as another example where marked differences appear (Moshinsky 1952, Brukner and Zeilinger 1997).

The pervasive conceptual idea of a wave packet provides us with a common language for describing many of the experiments discussed in this book. We show in Fig. 1.1 a snapshot of such a packet of waves moving with velocity v_g along the x-axis. Because of the quadratic dispersion law (Eq. 1.9) the packet spreads in time according to the well-known quantum mechanics rule

$$[\delta x(t)]^2 = [\delta x(0)]^2 + \left[\frac{\hbar t}{2m\,\delta x(0)}\right]^2. \tag{1.12}$$

See the quantum mechanics books by Schiff (1955), Messiah (1965), and Sakurai (1994), for example. This spreading occurs in time-dependent situations only because Eq. (1.12) is based on the time-dependent Schrödinger equation (1.2). The dispersion requires the wavelength of the internal oscillations to be shorter on the front end than on the tail end of the wave packet. The planes of constant phase are moving with the velocity v_{ph} which is much greater than the velocity of light, and the frequency of the internal oscillations is very large, in the neutron case the Compton frequency $f_{\mathrm{Compton}} = m_0c^2/h = 2.27 \times 10^{23}$ Hz. Thus, the question remains: what is it that is oscillating and moving with such high velocities? Is the phase velocity real? To what extent is it playing a role in the neutron interferometry experiments discussed in detail in this book?

The recent and continuing controversy regarding the possible observability of a gravitational redshift in atom interferometry is intimately related to the phase velocity idea. Louis de Broglie (1923) recognized these conceptual problems from the very start when he used the words "associated fictitious wave" in describing the moving electron. Contributions to this ongoing controversy come from Müller et al. (2010), Wolf et al. (2011), and Greenberger et al. (2012).

For dispersive phase shifts caused by forces, such as those arising from the mean optical nuclear potential or from magnetic fields, there will be a change in the group velocity v_g. However, for non-dispersive phase shifts arising from Aharonov–Bohm-type potentials, there will be a change of the phase velocity v_{ph}, but no change of the packet group velocity.

Lamb (1995) has shown that such wave fields are dynamically equivalent to a system of quantum-mechanical harmonic oscillators exhibiting a coherent state behavior. Since reflection, refraction, diffraction, and interference are consequences of the stationary equations, the complete array of wave optical phenomena known from the wave nature of light is also expected to occur for neutron matter waves. For an overview of light optical phenomena see the classical book by Born and Wolf (1975). Diffraction of neutrons by macroscopic objects, such as edges, slits, and zone plates, and the refraction of neutrons by wedges and lenses have been demonstrated in a full range of experiments that parallel those of classical optics (Sears 1989, Kaiser and Rauch 1999, Utsuro and Ignatovich 2010). In addition to the standard measurement of scattered intensities, neutron interferometry also provides experimental access to the phases of neutron wave functions as they are perturbed and modified by gravity, inertia, motion, magnetic and electric fields, topology, and material media (Bonse and Rauch 1979, Werner and Klein 1986, Badurek et al. 1988a, Rauch 2004).

The close mathematical analogy between the theory of neutron optics and the optics of electromagnetic waves allows one to formally connect the two types of fields through the index of refraction $n(\boldsymbol{r})$. For a neutron moving adiabatically while experiencing a spatially dependent potential $V(\boldsymbol{r})$, the wave function is described by an energy eigenstate ($E = \hbar\omega$)

$$\Psi(\boldsymbol{r}, t) = \psi(\boldsymbol{r})\, e^{-i\omega t}, \tag{1.13}$$

where $\psi(\boldsymbol{r})$ satisfies the time-independent Schrödinger equation

$$-\frac{\hbar^2}{2m}\nabla^2\psi(\boldsymbol{r}) + V(\boldsymbol{r})\psi(\boldsymbol{r}) = E\,\psi(\boldsymbol{r}), \tag{1.14}$$

which is a Helmholtz scalar wave equation, as shown above. Including the additional interaction one gets

$$\nabla^2\psi(\boldsymbol{r}) + K^2(\boldsymbol{r})\psi(\boldsymbol{r}) = 0. \tag{1.15}$$

The spatially dependent wave vector $K(\boldsymbol{r})$ in the region of the potential is given by

$$K^2(\boldsymbol{r}) = \frac{2m}{\hbar^2}[E - V(\boldsymbol{r})]. \tag{1.16}$$

It is therefore natural to define the spatially dependent index of refraction as the ratio of this wave vector $K(\boldsymbol{r})$ to the free space wave vector k such that (Halpern et al. 1941, Sears 1989)

$$n(\boldsymbol{r}) \equiv \frac{K(\boldsymbol{r})}{k} = \left[1 - \frac{V(\boldsymbol{r})}{E}\right]^{1/2}. \tag{1.17}$$

Strictly speaking, $n(\boldsymbol{r})$ is a tensor since some media or regions of space where the neutron interacts with its environment will in general be anisotropic. Furthermore, for regions of space containing media that absorb or scatter the neutron incoherently, the index of refraction will be complex. This aspect of the index of refraction is discussed in detail in Chapter 3.

Neutron optics differs from light-, X-ray-, and electron-optical phenomena in a number of significant ways, but most importantly because the neutron's interaction with matter is dominated by the strong neutron–nuclear interaction, which can be described by the point-like Fermi pseudopotential (Fermi 1936), for each nucleus at site $\boldsymbol{r}_j$, such that

$$V_{\text{nuc}}(\boldsymbol{r}) = \sum_j \frac{2\pi\hbar}{m}\, b_c\delta(\boldsymbol{r} - \boldsymbol{r}_j). \tag{1.18}$$

This pseudopotential approximation is generally valid because the range of the strong nuclear force is roughly the nuclear radius R, which is much smaller than the de Broglie wavelength λ for thermal neutrons. This neutron–nuclear interaction is characterized by the single parameter b_c, called the scattering length. The values are typically in the range of –5 to +10 fm, and they are in most cases positive.

In magnetic materials, the neutron interacts with the magnetic induction field $\boldsymbol{B}$ via its magnetic dipole moment $\boldsymbol{\mu}$, which is related to the Pauli spin operator $\boldsymbol{\sigma}$ (Eqs. 3.44–3.46, Halpern and Johnson 1939):

$$V_{\text{mag}}(\boldsymbol{r}) = -\boldsymbol{\mu} \cdot \boldsymbol{B}(\boldsymbol{r}) = -\mu\boldsymbol{\sigma} \cdot \boldsymbol{B}(\boldsymbol{r}). \tag{1.19}$$

The magnetic interaction acts on the spinor part of the neutron wave function. The so-called 4π-symmetry of spinors has been verified by neutron interferometry (Rauch et al. 1975, Werner et al. 1975, Klein and Opat 1976; see Chapter 5). It should be pointed out that a coupling (entanglement) exists between the spinor part and the momentum part of the wave function (Stern–Gerlach effect). The strengths of V_{mag} and V_{nuc} are comparable for magnetic materials such as the ferromagnetic metals Fe, Ni, and Co. The corresponding magnetic scattering lengths, generally called p, are dependent upon the scattering vector $\boldsymbol{q}$ due to the magnetic shape of the atoms. Magnetism in materials generally arises from unpaired electron spins, as discussed in Section 3.2.

The neutron feels the gravitational attraction of the Earth through the local Newtonian potential

$$V_{\text{grav}}(\boldsymbol{r}) = m\boldsymbol{g} \cdot \boldsymbol{r}\,, \tag{1.20}$$

where $\boldsymbol{g}$ is the gravitational field. The neutron follows a parabolic trajectory in free space, and falls according to Newton's laws of motion. This classical trajectory was first observed by McReynolds (1951) and later more precisely by Dabbs et al. (1965). The free fall of neutrons under gravity has led to experiments by Koester (1976) that establish the equivalence of the gravitational and inertial masses of the neutron to an accuracy of 3 parts in 10^4. It has also given rise to the gravity spectrometer, an important method for measuring the index of refraction of materials. The phase shift accompanying the bending of the neutron rays by gravity can only be observed by interferometry as was first done by Colella, Overhauser, and Werner (1975).

Most of the experiments are based on perfect silicon crystal interferometers where the perfect arrangement of the atoms within the monolithic perfect crystal provides a wide angle coherent beam splitting and superposition. Such an interferometer was first tested by Rauch, Treimer, and Bonse in 1974. Figure 1.2 shows the principle of this technique and the first results obtained at a small 250-kW research reactor in Vienna. This technique has been used for many experiments exploiting basic phenomena of quantum physics. These experiments are described in Chapters 5–8. Two of them are also shown here to demonstrate the capabilities of this technique. First, there is the observation of a gravity-induced phase when the interferometer is tilted about the horizontal beam line causing a relative gravitational difference between the two beams. Second, there is the verification of the 4π-symmetry of spinor wave functions.

The neutron interferometry experiments verifying the gravitational and spinor symmetry effect have become part of many textbooks. They are discussed in many quantum mechanics textbooks (e.g., Ballentine 1990, Sakurai 1994, Dubbers and Stöckmann 2013) and have entered the broader physics literature in a lively manner, such as in a *Physics Today* article by Snow (2013). Figure 1.3 shows these results which stimulated many other quantum optics investigations with neutrons and have become a motivation for the development of matter wave interferometry in general reaching to antimatter-antiproton interferometry (Aghion et al. 2014).

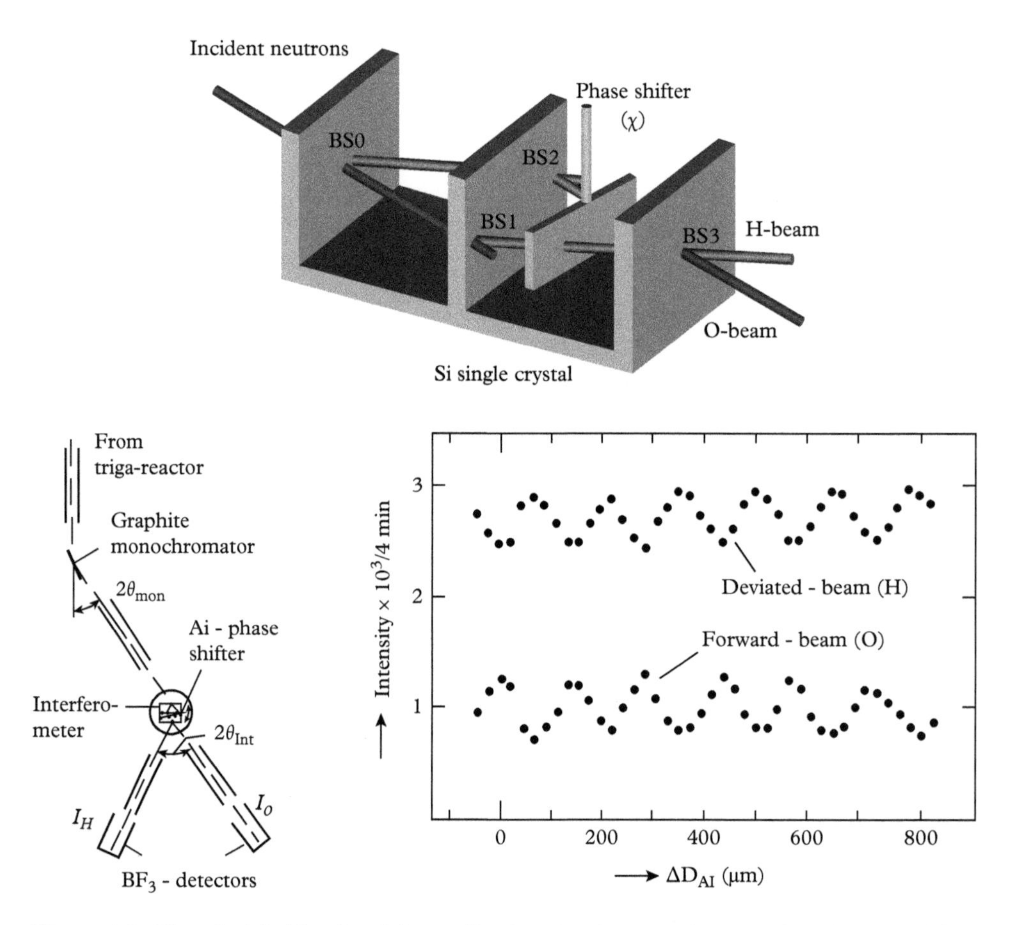

Figure 1.2 *The principle (above) and first realization of perfect crystal neutron interferometry (below) (Rauch et al. 1974). The optical path difference is obtained by rotating a flat aluminum plate subtending both beams around a vertical axis*

The neutron also interacts weakly with the electric field $\boldsymbol{E}(\boldsymbol{r})$ surrounding nuclei and electrons in matter via the spin–orbit coupling, namely

$$V_{\text{spin-orbit}}(\boldsymbol{r}, \boldsymbol{p}) = \frac{1}{mc}\boldsymbol{\mu} \cdot (\boldsymbol{E} \times \boldsymbol{p}), \tag{1.21}$$

where $\boldsymbol{p}$ is the neutron's canonical momentum. This velocity-dependent term in the Hamiltonian leads to Schwinger (1948) scattering. It was first observed for thermal neutrons by Shull (1963, 1967) in a polarized beam experiment on vanadium metal. In 1984, Aharonov and Casher proposed that a beam of neutrons moving around a line of charge will experience a topological quantum phase shift, which was subsequently observed by Cimmino et al. (1989) by neutron interferometry. This experiment is discussed in Section 6.1. The interaction giving rise to this effect is the spin–orbit coupling.

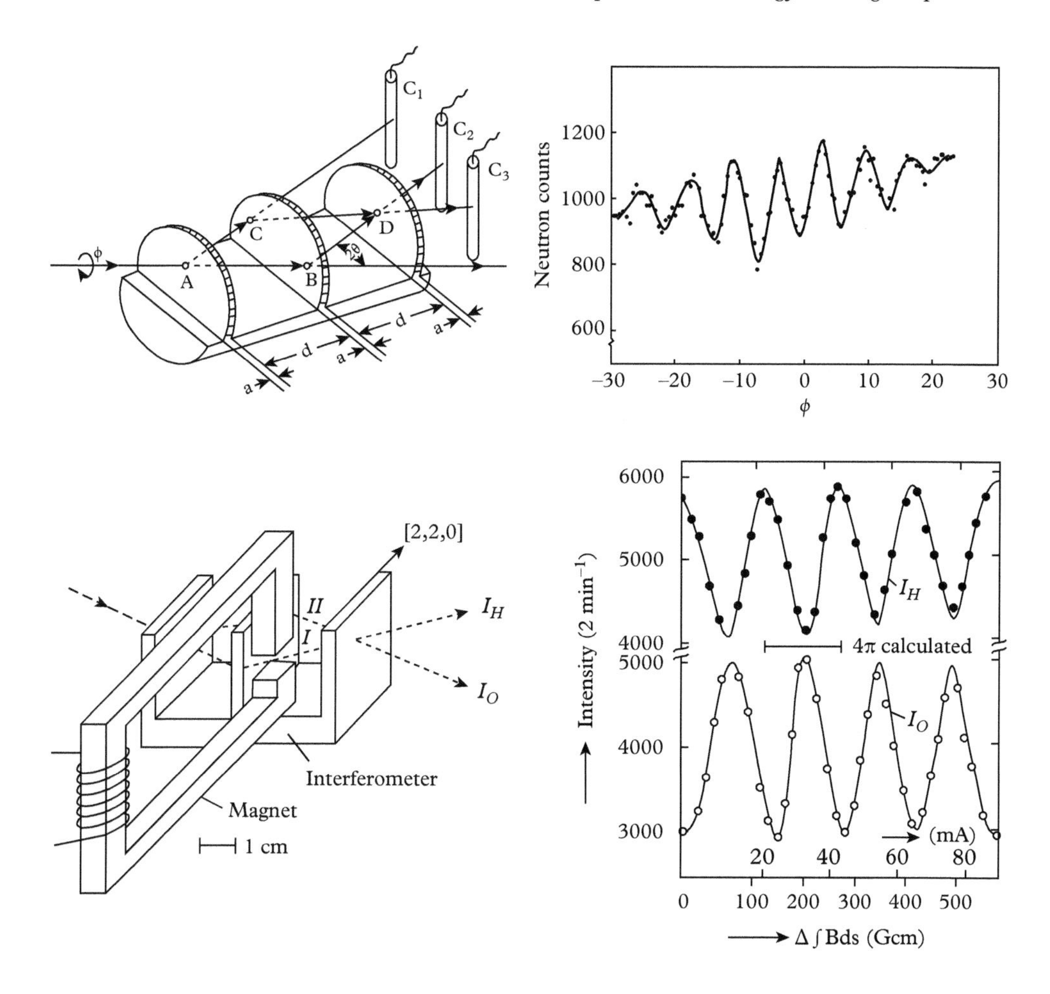

Figure 1.3 *(Above) Observed gravitationally induced phase shift (COW experiment; Colella, Overhauser, Werner 1975); (below) the spinor 4π-symmetry experiment (Rauch et al. 1975)*

The collective interaction of a neutron wave with a large assembly of nuclei leads to elastic Bragg scattering, inelastic scattering by the elementary excitations in condensed matter (phonons and magnons), and small-angle scattering by density fluctuations, for example in polymers and alloys. As alluded to earlier, these neutron scattering techniques have become indispensable in solid state physics, biology, chemistry, and materials science. Besides these diffraction and spectroscopic methods, neutron optics has evolved as a complementary field where the collective interaction of neutron waves with macroscopic objects is of central interest. In this regime, the phenomena depend upon the index of refraction n and the dimensions of these objects. The refraction of a neutron beam at a vacuum–solid interface is governed by Snell's Law

$$n \hat{=} \frac{K}{k} = \frac{v}{v_0} = \frac{\sin \varphi_0}{\sin \varphi}, \text{ (neutrons)} \tag{1.22}$$

where the incident ray of wave vector $\boldsymbol{k}$ makes an angle φ_0 with the surface normal, and the refracted ray inside the medium of wave vector $\boldsymbol{k}$ makes an angle φ with the surface normal. For light, the wave vector $\boldsymbol{k}$ inside the medium is inversely proportional to the photon's velocity v, so that

$$n \hat{=} \frac{K}{k} = \frac{v_0}{v} = \frac{c}{v} = \frac{\sin \varphi_0}{\sin \varphi}. \text{ (light)} \tag{1.23}$$

One notes that the index of refraction for neutrons is defined in terms of the ratio of velocities for neutrons, which is inverse to that for light. Snell discovered this law empirically for light in 1621, but died before publishing it. Apparently it first appeared in the published literature in Descartes' *Dioptrique* (1637), and was subsequently extensively used by Isaac Newton (1730) in his investigations of the diffraction of light. Historical aspects of Snell's law are recounted in the book by Silverman (1997).

For a non-magnetic medium, averaging the Fermi pseudopotential (Eq. 1.18) over a macroscopic volume gives us an expression for the effective optical potential (e.g., Sears 1989)

$$V_{\text{optical}} = \frac{2\pi\hbar^2}{m} b_c N, \tag{1.24}$$

where N is the atom density and b_c the neutron–nuclear scattering length. Any absorption or nuclear reaction effect is described by an imaginary term of the interaction potential (e.g., Blatt and Weisskopf 1952); that is, the scattering length b_c becomes complex: $b_c = b' - ib''$. This yields according to expressions (1.17) and (3.16) a complex index of refraction (Goldberger and Seitz 1947, Sears 1989)

$$n = 1 - \frac{\lambda^2 N}{2\pi} \sqrt{b_c^2 - \left(\frac{\sigma_r}{2\lambda}\right)^2} + i\frac{\sigma_r N \lambda}{4\pi}, \tag{1.25}$$

thus accounting for absorption (σ_a) and incoherent scattering (σ_{incoh}) processes (where the total reaction cross-section per atom is $\sigma_r = \sigma_a + \sigma_{\text{incoh}}$). The imaginary part is in most cases small and, therefore, one can often use the relation

$$n = 1 - \lambda^2 N b_c / 2\pi , \tag{1.26}$$

where we have used the de Broglie relation in the expression for the neutron's kinetic energy ($p^2/2m$). The optical potential is typically of order 10^{-7} eV; thus, for thermal neutrons one has $1-n \sim 1-10^{-5}$. A small number of isotopes (less than 5% of all isotopes) have a negative coherent scattering length; examples are ^{1}H (–3.74 fm), ^{48}Ti (–5.84 fm), and ^{55}Mn (–3.73 fm). Consequently, the neutron optical potential of most materials is positive (repulsive) such that the index of refraction is slightly less than unity. This fact leads to total external reflection of neutrons at surfaces at grazing angles θ less than a critical angle θ_c, which occurs when the neutron's kinetic energy related to the momentum perpendicular to the surface ($p^2 \sin^2\theta/2m$) is less than V_{optical}, such that θ_c is given by

$$\sin \theta_c = \lambda (Nb/\pi)^{1/2}. \tag{1.27}$$

This critical angle is generally less than 0.3° in the thermal neutron energy range. For example, $\theta_c/\lambda = 0.097°/\text{Å}$ for Ni, an element quite often used for coating the glass surfaces in neutron guides. For long wavelength neutrons (ultra-cold neutrons) the critical angle increases and reaches 90° at

$\lambda_c = (\pi/Nb)^{1/2}$, which corresponds to neutrons with wavelengths greater than about 700 Å. These ultra-cold neutrons are totally reflected at all angles of incidence. This result has opened up a field of research on neutrons where they are confined within bottles (Steyerl 1977, Gollub and Pendlebury 1979) or within magnetic traps (Paul and Trinks 1978).

The index of refraction for X-rays, like neutrons, is generally slightly less than unity, while for photons in the optical region of the electromagnetic spectrum the index of refraction for insulators, like SiO_2, is greater than unity, typically in the range 1.2–1.7. It may appear somewhat surprising that the neutron optical potential is generally repulsive, since the individual nuclei always provide an attractive potential well for neutrons. The origin of this fact is strictly quantum mechanical, having to do with nuclear size effect resonances and the requirements on the neutron wave function continuity at the nuclear surface. For a detailed derivation and explanation of this effect the reader is referred to Peshkin and Ringo (1971) or to Klein and Werner (1983) and Byrne (1994). This is discussed in Chapter 3.

Neutron optical phenomena are fundamentally connected to the coherence properties of the wave function $\Psi(\boldsymbol{r},t)$, which describes the neutron in a certain beam by means of a wave packet. This linear superposition of plane waves with amplitude $a(\boldsymbol{k} - \boldsymbol{k}_0)$ is a general solution of the Schrödinger equation (1.2), namely

$$\Psi_{k_0}(\boldsymbol{r}, t) = \frac{1}{(2\pi)^{3/2}} \int a(\boldsymbol{k} - \boldsymbol{k}_0) e^{i(\boldsymbol{k}\cdot\boldsymbol{r} - \omega_k t)}\, d^3\boldsymbol{k}. \tag{1.28}$$

The group or particle velocity of the neutron beam is given by $v_g = \hbar k_0/m$, where k_0 is the central wave vector of the wave packet and the frequency is given by the quadratic dispersion relation $\omega_k \cong \hbar k^2/2m_0 + m_0c^2/\hbar$ in free space (Eq. 1.9), which, for pulsed beams, causes distinct differences compared to electromagnetic wave phenomena, where all Fourier components of a pulse travel with the velocity of light, c, and $\omega_k = ck$ (Eq. 1.6b). The quadratic dispersion causes the well-known, time-dependent spreading of the neutron wave packet (Section 4.5.5). The amplitude function $a(\boldsymbol{k} - \boldsymbol{k}_0)$ is determined by the experimental constraints applied to the beam by collimation, monochromatization, polarization, apertures, etc. The autocorrelation function of the wave function is defined as (Glauber 1963, Walls and Milburn 1994, Mandel and Wolf 1995)

$$\Gamma(\boldsymbol{\Delta}, \tau) = \langle \psi(0,0) \cdot \psi(\boldsymbol{\Delta}, \tau) \rangle. \tag{1.29}$$

This function describes the spatial and temporal coherence at two space–time points separated by $\boldsymbol{\Delta} = \boldsymbol{r} - \boldsymbol{r}'$ and $\tau = t - t'$. The spatial coherence lengths are given by characteristic lengths $\boldsymbol{\Delta}^c$ of the function $\Gamma(\boldsymbol{\Delta},0)$, and therefore the spatial extent δx_i of $\Gamma(\boldsymbol{r},t)$, which in turn depends upon the widths δk_i of the amplitude function $a(\boldsymbol{k} - \boldsymbol{k}_0)$. These widths must satisfy the uncertainty relation

$$\delta k_i\, \delta x_i = \delta k_i \Delta_i^c \geq 1/2. \tag{1.30}$$

These coherence phenomena are discussed in Chapter 4.

1.2 The Quantum Phase Shift of Matter Waves

The curious dual nature of the neutron, sometimes a particle, sometimes a wave, is wonderfully manifested in the various non-local interference effects observed in neutron interferometry experiments. The point-by-point motion of particles in space–time as described by relativity considerations and the seemingly incompatible non-local quantum mechanical phenomena are brought into close juxtaposition by neutron interference experiments, for example, by spinor rotation in a magnetic field (Rauch et al. 1975, Werner et al. 1975, Klein and Opat 1976), by the Earth's gravity and rotation (Colella et al. 1975, Staudenmann et al. 1980, Werner et al. 1988), by Aharonov–Bohm (1959) type topological potentials (Cimmino et al. 1989, Allman et al. 1992, Wagh et al. 1997), and by quantum contextuality phenomena (Hasegawa et al. 2003).

We begin here with a general discussion of the quantum phase $\Phi(x, t)$ of a matter wave evolving in space and time according to the Feynman–Dirac path integral along the trajectories defined by classical mechanics (Dirac 1945, Feynman 1948, Feynman et al. 1965, Goldstein 1980). This is identical to the Wentzel–Kramers–Brillouin (WKB) approximation and usually called the eikonal approximation in classical optics used for slowly varying potentials. The phase $\Phi(x, t)$ of the wave function $\Psi(x, t)$ should be regarded as a scalar field extending throughout the apparatus, in our case a neutron interferometer, which includes the slits, phase shifting interactions, and detectors. The perfect Si-crystal neutron interferometer is geometrically identical to the classical optics Mach–Zehnder interferometer, and topologically equivalent to a ring as shown in Fig. 1.4. At some point A on the ring an incident wave Ψ_0 is brought into the ring and split coherently into two parts; one propagates clockwise on path I and the other counter-clockwise on path II around the ring. After interacting with a potential $V(x', v', t')$ which depends upon position x' along the trajectories, the time t' and occasionally also on the neutron's velocity v', these two waves are mixed and allowed to interfere in a small region of space surrounding point B, such that an exit beam is formed which is a linear superposition of the two wave functions Ψ_{I} and Ψ_{II} traversing the two paths. We will see that differences in phase are the only measurable quantities. Here one faces the same situation as in the case of position, momentum, angular momentum, or time measurements in classical and quantum mechanics where also only relative observables are measurable, which are always defined relative to a system of reference.

The phase accumulated on either path is a line integral over the Lagrangian $\mathcal{L}$ in space–time given by (Feynman et al. 1965, Opat 1995)

$$\Phi(x, t) = \frac{1}{\hbar}\int \mathcal{L}dt'. \tag{1.31}$$

The Lagrangian $\mathcal{L}$ is related to the Hamiltonian $\mathcal{H}$ by a Legendre transformation

$$\mathcal{L} = p \cdot v - \mathcal{H}, \tag{1.32}$$

where p is the canonical momentum of the neutron and v is the classical (group) velocity, i.e., $v = ds/dt$. Thus, Eq. (1.31) gives the phase at the detector at position x as a function of time t, namely

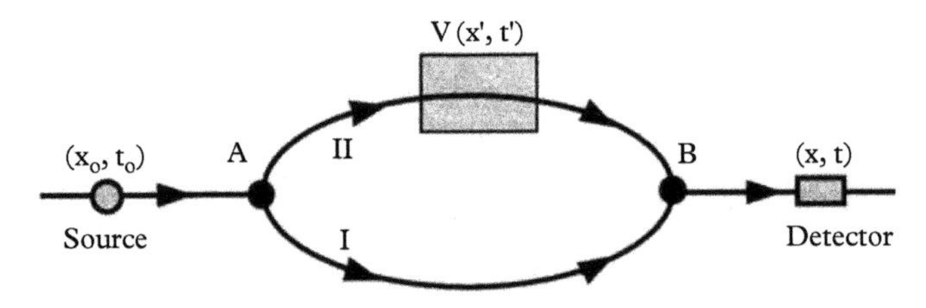

Figure 1.4 *General scheme of an interferometer experiment*

$$\Phi(x,t) = \frac{1}{\hbar}\int_{x_0}^{x} \boldsymbol{p}\cdot d\boldsymbol{s} - \frac{1}{\hbar}\int_{t_0}^{t} \mathcal{H}\, dt' = \int_{x_0}^{x} \boldsymbol{k}\cdot d\boldsymbol{s} - \int_{t_0}^{t} \omega\, dt', \tag{1.33}$$

where the wave vector $k = 2\pi/\lambda$ corresponding to the de Broglie wavelength λ, and ω is the frequency related to the total energy of the wave at any point (x', t') along the trajectories.

We must evaluate the line integrals in Eq. (1.33) along each of the paths in Fig. 1.4, namely

$$\Phi_{\mathrm{I}}(x,t) = \frac{1}{\hbar}\int_{x_0}^{x} \boldsymbol{p}_{\mathrm{I}}\cdot d\boldsymbol{s} - \frac{1}{\hbar}\int_{t_0}^{t} \mathcal{H}_{\mathrm{I}}\, dt', \tag{1.34a}$$

and

$$\Phi_{\mathrm{II}}(x,t) = \frac{1}{\hbar}\int_{x_0}^{x} \boldsymbol{p}_{\mathrm{II}}\cdot d\boldsymbol{s} - \frac{1}{\hbar}\int_{t_0}^{t} \mathcal{H}_{\mathrm{II}}\, dt'. \tag{1.34b}$$

Thus, the last terms cancel due to energy conservation and the phase difference for the waves traversing the two paths is

$$\Delta\Phi(x,t) = \Phi_{\mathrm{II}}(x,t) - \Phi_{\mathrm{I}}(x,t) = \frac{1}{\hbar}\int \boldsymbol{p}_{\mathrm{II}}.d\boldsymbol{s} - \frac{1}{\hbar}\int \boldsymbol{p}_{\mathrm{I}}.d\boldsymbol{s}, \tag{1.35}$$

which is a path integral around the ring in Fig. 1.4. In neutron interferometry, it is the *phase shift* $\Delta\Phi_{\mathrm{V}}(x,t)$ caused by the potential $V(x', v', t')$ that is of physical interest, namely

$$\Delta\Phi_V = \Delta\Phi - \Delta\Phi_0\,, \tag{1.36}$$

where $\Delta\Phi_0$ is the "empty" interferometer phase shift, that is, when $\mathrm{V}(x', v', t')$ is everywhere zero (or spatially constant). This *potential-dependent phase shift* is the line integral along the *classical trajectories* of the neutron, which is the "golden rule" of small perturbations (e.g., Opat 1995, Greenberger et al. 2012). A stationary phase shifter is represented by a potential $V(x)$ which changes the momenta due to energy conservation:

$$\frac{(p+\delta p)^2}{2m} + V(x) = E \Rightarrow v.\delta p \approx -V(x) \tag{1.37}$$

and

$$\Delta\Phi_V = \frac{1}{\hbar}\oint \delta p.ds = -\frac{1}{\hbar}\oint V(x)dt. \tag{1.38}$$

This is the basic equation for most interferometer measurements.

The canonical momentum $\boldsymbol{p}$ must be used in evaluating this potential-dependent phase shift to account for gauge invariant potentials. In general it consists of two parts

$$\boldsymbol{p} = \boldsymbol{p}_{\mathrm{kinetic}} + \boldsymbol{p}_{\mathrm{hidden}}\,, \tag{1.39}$$

where the kinetic momentum is given by the product of the neutron's mass and velocity $\boldsymbol{v}$,

$$\boldsymbol{p}_{\text{kinetic}} = m\boldsymbol{v}. \tag{1.40}$$

The hidden part of the momentum arises in situations where the potential is velocity-dependent. In the Aharonov–Casher (AC) effect experiment, the velocity-dependent potential comes from the spin–orbit coupling of the neutron's motion to the electric field $\boldsymbol{E}$ generated by a line of charge, where it is found that

$$\boldsymbol{p}_{\text{hidden}} = \frac{1}{c}\,\boldsymbol{\mu} \times \boldsymbol{E}. \tag{1.41}$$

This will be discussed in detail in Section 6.1. In the Sagnac effect experiments the hidden momentum is

$$\boldsymbol{p}_{\text{hidden}} = m\boldsymbol{\omega} \times \boldsymbol{r}, \tag{1.42}$$

and comes from the state of rotation of the frame of the interferometer, that is, its frequency of angular rotation $\boldsymbol{\omega}$. These non-inertial frame experiments are described in Sections 8.3 and 8.5.

There are three general situations to consider:

1. The potential depends only upon position, and is independent of time t and velocity $\boldsymbol{v}$, that is:

$$V = V(\boldsymbol{x}), \tag{1.43}$$

Since the Hamiltonian $\mathcal{H}$ is time-independent, the neutron's total energy E is a constant of the motion. The force is said to be conservative and the neutron decelerates when entering the region $\mathcal{R}$ where V is non-zero and accelerates when leaving $\mathcal{R}$ (and vice versa when the potential changes sign). There is no hidden momentum in this case and the phase shift depends only upon the action of the kinetic momentum

$$\Delta\Phi_V = \frac{1}{\hbar}\int_R \Delta\boldsymbol{p}_{\text{kinetic}} \cdot \mathrm{d}\boldsymbol{s}. \tag{1.44}$$

This is the situation that applies for a neutron traversing of slab of material and interacting with many nuclei which creates an effective optical potential as discussed in more detail in Chapter 3 (Eq. 3.11). It also applies to the gravitationally induced quantum interference experiments (Section 8.1), and to spin precession in a time-independent magnetic field $\boldsymbol{B}$ (Section 5.3).

2. Zero-force situations when $\mathcal{H}$ is independent of time:

The phase shift $\Delta\Phi_V$ may depend explicitly upon the geometry and topology. In the AC effect geometry, the neutron's acceleration due to the electric field $\mathbf{E}$ of a line charge can be shown to be

$$\boldsymbol{a} = -\frac{1}{mc}(\boldsymbol{\mu} \cdot \nabla)\,(\boldsymbol{v} \times \mathbf{E}). \tag{1.45}$$

This is worked out explicitly in Section 6.1. If the axis of quantization is along the line charge, call this the z-direction, and we take $\boldsymbol{\mu} = \mu\hat{z}$, while $\boldsymbol{E} = \boldsymbol{E}(x,y)$ as is the case for a line charge, we see that the acceleration $\boldsymbol{a} = 0$. Thus, the phase shift depends only upon the hidden momentum, namely

$$\Delta\Phi_V = \frac{1}{\hbar}\oint \Delta \boldsymbol{p}_{\text{hidden}} d\boldsymbol{s}. \tag{1.46}$$

The energy E and the kinetic momentum $\Delta \boldsymbol{p}_{\text{kinetic}}$ are both constants of the motion.

3. Zero-force situations where the potential depends only upon the intermediate time t', but not upon position:

The experiments in this case require synchronized time-of-flight analysis since

$$\Delta V = \Delta V(t') = \Delta \mathcal{H}(t'), \tag{1.47}$$

and the phase shift is

$$\Delta\Phi_V(t) = -\frac{1}{\hbar}\int_{\text{pulse}} \Delta \mathcal{H}(t')dt', \tag{1.48}$$

where the time structure of Hamiltonian determines the phase shift and integration range, indicated by the label "pulse" on the integral. This is the situation appropriate to the neutron version of the scalar AB effect discussed in Section 6.2. The neutron feels no force and the kinetic momentum $m\boldsymbol{v}$ is a constant of the motion but the total energy E is not. The more general case is one in which the neutron enters a region of space $\mathcal{R}$ where the potential is changing with time while it experiences a spatial gradient of the potential, thus being accelerated by a position-dependent, time-varying force.

A compilation of various interactions (conservative and non-conservative) and their related phase shifts are listed up in Table 1.3. Table 1.4 gives a listing of neutron interferometry experiments carried out over the past three decades, along with a chapter and section number notation where they are discussed in this book.

Neutron optics continues to be an important field of research, which combines the unique particle properties of neutrons and fundamental wave optical phenomena. The status of the whole field has been summarized in proceedings of workshops (Bonse and Rauch 1979, Badurek et al. 1988a, Kawano et al. 1996), in various review articles (for example, Klein and Werner 1983, Werner and Klein 1986, Rauch 1986, Wagh and Rakhecha 1996, Kaiser and Rauch 1999), and in the books of Sears (1989) and of Utsuro and Ignatovich (2010). Here, we focus on neutron interferometry and its relation to fundamental quantum physics problems, its applications in solid state physics and for precise scattering length measurements. The wave–particle dualism of quantum mechanics can now be discussed on a more profound basis with considerably more epistemological and pedagogic aspects. More recently it has been shown that neutron interferometry is a branch of quantum optics, where quantum statistics, squeezing, and the quantum mechanical measuring process can be studied for massive particles. The two-level system which the neutron experiences inside a magnetic field provides the basis for distinct state manipulation and for Bordé–Ramsey state interferometry.

The question how the (quantum) system interacts with, and how it can be separated from, the (classical) measuring apparatus plays an essential role in the discussions throughout this book. It will be shown how this somewhat arbitrary borderline between the quantum and classical system

Table 1.3 *Neutron Interferometric-Measured Phase Shifts.*

Interaction	Potential	Phase Shift	Reference
nuclear	$\frac{2\pi\hbar^2}{m} b_c \delta(\boldsymbol{r})$	$-N b_c \lambda D$	Rauch et al. (1974)
magnetic	$-\boldsymbol{\mu} \cdot \boldsymbol{B}(\boldsymbol{r})$	$\pm \frac{\mu B m \lambda D}{2\pi\hbar^2}$	Rauch et al. (1975)
gravitation	$m\boldsymbol{g} \cdot \boldsymbol{r}$	$\frac{m^2 g \lambda A \sin\alpha}{2\pi h^2}$	Collela et al. (1975)
Coriolis	$-\hbar\boldsymbol{\omega} \cdot (\boldsymbol{r} \times \boldsymbol{k})$	$\frac{2m}{\hbar} \omega_e \cdot A$	Werner et al. (1979)
Aharonov–Casher (Schwinger)	$-\boldsymbol{\mu} \cdot (\boldsymbol{v} \times \boldsymbol{E})/c$	$\pm \frac{2\mu}{\hbar c} E \cdot D$	Cimmino et al. (1989)
scalar Aharonov–Bohm	$-\boldsymbol{\mu} \cdot \boldsymbol{B}(t)$	$\pm \frac{\mu B T}{\hbar}$	Allman et al. (1992)
magnetic Josephson	$-\boldsymbol{\mu} \cdot \boldsymbol{B}(t)$	$\pm\, \omega t$	Badurek et al. (1986)
Fizeau	—	$-N b_c \lambda D \left(\frac{w_x}{v_x - w_x}\right)$	Klein et al. (1981a)
geometry (Berry)	—	$\Omega/2$	Wagh et al. (1997)

B = magnetic field strength, $\boldsymbol{g}$ = gravitational acceleration, A = normal area enclosed in the coherent beams, α = angle between the horizontal and the area A, $\boldsymbol{\omega}_e = 0.727 \times 10^{-4}\ \mathrm{s}^{-1}$ = angular rotation velocity of the Earth, E = electric field, $\hbar\omega$ = energy transfer due to the time-dependent field $B(t)$, T = time during which the constant field B is switched on, w_x, v_x = velocity components of the phase shifter and the neutrons perpendicular to the moving surface of the phase shifter, Ω = solid angle subtended by a closed circuit in parameter space.

depends on the historical debate on whether quantum mechanics is a complete description of the physical reality. The uncertainty principle implies that, unlike classical mechanics, the quantum mechanical backreaction can never be negligible. In certain cases this backreaction can be described by an induced topological phase (Aharonov et al. 1998). It should be kept in mind that experiments cannot decide by themselves between different interpretations of the quantum formalism. The reason is that it is not the interpretation of the theory that predicts the result of the experiments, it is the theory itself. Quantum theory is a part of modern quantum field theories which are mainly based upon superstring theories (e.g., Gibbin 1998). In this respect neutron interference experiments also contribute to our understanding of these more general views of nature.

The demonstration of quantum interference of neutron waves has led to a wealth of experimental and theoretical work concerned with the understanding of quantum interference of massive particles. The complementarity between particle and wave behavior is embedded in the broader complementarity between observables, whose simultaneous precise knowledge is impossible due to basic uncertainty relations. Neutron interferometry is a proper tool for such investigations due to the fact that neutrons carry well-defined particle and wave properties (Table 1.1). The relevance of neutrons in cosmology and particle physics has been summarized by Dubbers and Schmidt (2011).

The problems of locality and non-locality of quantum separability are inherently involved in any interpretation (e.g., Bell 1965, Wigner 1970, d'Espargnat 1979, Wotters and Zurek 1979,

Table 1.4 *Neutron Interferometry Experiments (1974–2014).*

- First test of perfect Si-crystal interferometer with neutrons: Vienna (1974)
- Observation of gravitationally induced quantum interference: Michigan, Missouri (1975, 1980, 1988, 1993, 1996)
- Observation of the change of sign of the wave function of a fermion due to precession of 360° in a magnetic field: Michigan, Vienna–Grenoble (1975, 1978)
- Observation of the effect of the Earth's rotation on the quantum mechanical phase of the neutron (Sagnac effect): Missouri (1980)
- Measurement of the energy-dependent scattering length of Sm-149 in the vicinity of a thermal nuclear resonance: Missouri (1982)
- Charge dependence of the four-body nuclear interaction in n-^{3}He versus n-^{3}H: Vienna-Grenoble (1979, 1985)
- Search for nonlinear terms in the Schrödinger equation: MIT (1981)
- Search for the Aharonov–Bohm effect for neutrons with a magnetized single crystal of Fe inside interferometer: MIT (1981)
- Measurement of the longitudinal coherence length of a neutron beam: Missouri (1983)
- Observation of the coherent superposition of spin states ("Wigner Phenomenon") with both static and RF spin flippers: Vienna–Grenoble (1983, 1984)
- Neutron interferometric search for quaternions in quantum mechanics: Missouri (1984)
- Sagnac effect using a laboratory turntable-shows phase shift due to rotation is linear in ω: MIT (1984)
- Observation of acceleration-induced quantum interference: Dortmund–Grenoble (1984)
- Experiment on the null Fizeau effect (stationary boundaries) for thermal neutrons in moving matter: Missouri–Melbourne (1985)
- Observation of the neutron Fizeau effect with moving boundaries of moving matter: Dortmund–Grenoble (1985)
- Double-RF coil experiment–analogue of the magnetic Josephson experiment: Vienna–Grenoble (1986)
- Precision measurement of the bound-coherent neutron scattering lengths of U-235, U-238, V, Eu, Gd, Th, Kr, H, D, Si, Bi, etc.: Vienna–Grenoble, Missouri (1975–93)
- Observation of a motion-induced phase shift of neutron de Broglie waves passing through matter near a nuclear resonance (Sm-149): Missouri–Melbourne (1988)
- Observation of stochastic versus deterministic absorption of the neutron wave function: Vienna-Grenoble (1984, 1987, 1990)
- Observation of the topological Aharonov–Casher phase shift: Missouri–Melbourne (1989)
- Test of possible non-ergodic memory effects: Vienna-Grenoble (1989)
- Observation of the effects of spectral filtering in neutron interferometry: Missouri–Vienna (1991)
- Counting statistics experiments—particle number/phase uncertainty: Vienna (1990, 1992)
- Observation of the neutron phase echo effect: Missouri–Vienna (1991)
- Coherence effects in time-of-flight neutron interferometry: Missouri–Vienna (1992)
- Observation of the scalar Aharonov–Bohm effect: Missouri–Melbourne (1992, 1993)
- Spectral modulation and squeezed states in neutron interferometry: Missouri–Vienna (1994)
- Observation of multiphoton exchange amplitudes by interferometry: Vienna–Missouri (1995)
- Observation of the topological phase by coupled loop interferometers: Vienna–Berlin (1996)
- Experimental separation of geometric (Berry) and dynamical phases by neutron interferometry: Bombay–Missouri–Vienna (1997)

(*continued*)

Table 1.4 *(continued)*

- Light-induced interferometer: Geestacht–Vienna (1997)
- Verification of quantum contextuality in matter wave quantum optics: Grenoble–Vienna (2003)
- Polarized He-3 scattering length: NIST, ILL (2006–13)
- *n-p* and *n-d* scattering lengths: NIST, Missouri, Indiana (2003)
- Non-defocusing, non-dispersive phase shifters: Vienna, Grenoble, Mumbai (2010–13)
- Kochen–Specker and weak measurements: Vienna, Grenoble, Nagoya (2013–)
- Observation of the Cheshire Cat effect for neutron matter waves: Vienna, Grenoble, Orange (2014)

Pitowsky 1982, Selleri 1990, Schleich 2001). Non-locality means in this connection that the results of measurements performed at distant parts with distant local pieces of apparatus are statistically correlated in a way that indicates the existence of a link between the measurement events, and this correlation persists even when these distant measurement events are separated by a Minkowskian space-like distance. It will be shown that even if a quantum entity (say a neutron) is in a wave-like situation it can be influenced without destroying its wave properties, giving rise to interference in a sense that one could only imagine for a particle-like entity. This influence is not a measurement of localization, but the apparatus defines several widely separated regions of space where the neutron can be influenced as a whole by a local device that acts at a given instant of time in any one of these regions. A heuristical approach which focuses on the description of the experiments and of their interference results will be given. In connection with post-selection measurements which are treated in Section 4.5 we conclude that locality should be considered in phase space rather than in ordinary space (Schleich and Wheeler 1987, Rauch 1993a, Schleich 2001, Suda 2005). Interference phenomena can be transformed between different parameter spaces, e.g., between ordinary and momentum space. Different features of a quantum entity can be transported along different paths through an interferometer (Denkmayr et al. 2014).

The interpretation of neutron interferometry results is closely related to the well-known situation in ordinary double-slit interference experiments, "which has in it the heart of quantum mechanics; in reality it contains the *only* mystery of the theory" (Feynman et al. 1965, Ghose 2009). The experimental results are described by the laws of standard quantum mechanics. However, it should be emphasized that the classical method of defining the location of the particle (in beam I or in beam II) is inconsistent with quantum mechanics, since this definition presumes the existence of a hidden variable which determines particle location. When an interference pattern exists it is even misleading to describe the situation with the particle picture in the sense that the neutron has chosen one of both possible beam paths. It is fair to say that there are many aspects to this mystery, perhaps even "more than one mystery" (Silverman 1997).

While the whole topic of neutron interferometry is described by the wave picture of quantum mechanics, the particle trajectories appear only in the de Broglie–Bohm interpretation of quantum mechanics where a quantum potential is introduced to guide the neutrons properly (Bohm and Vigier 1984, Dewdney 1985, Dürr 2001, Sanz and Miret-Artès 2012). This topic will be addressed again in the last chapter of this book. However, the main purpose of this book is to provide a comprehensive elucidation of the experimental facts obtained by neutron interferometry.

The many-fold possibilities of neutron interferometry stimulated the development of matter-wave interferometry for atoms and molecules. Both transmission gratings and standing light wave gratings have been tested as proper beam splitter devices (Keith et al. 1988, Martin et al. 1988,

Cronin et al. 2009). A special challenge for the realization of atom interferometry is the possibility to store and to recover information in the form of atomic excitations. Carnal and Mlynek (1991) tested a double-slit interferometer for He atoms and achieved a beam separation of 8 μm. The slit width was 1 μm and the He atoms had a mean velocity of 500 m/s. Keith et al. (1991) developed a grating interferometer for sodium atoms with a beam separation of 27 μm. Riehle et al. (1991) and Kasevich and Chu (1991) tested interferometers with laser-cooled calcium and sodium atoms which are coherently split and superposed by proper stimulated Raman transitions. These Ramsey–Bordé-type interferometers are analogous to neutron zero-field spin-echo systems discussed in Section 2.4 (Fig. 2.23), and they are operational with and without spatial beam separation. Rather comprehensive reviews have been written by Adams et al. (1994), Berman (1997), and Cronin et al. (2009). Chapman et al. (1995) tested successfully an interferometer for Na_2 molecules which is based on three nanofabricated diffraction gratings which constitute a Mach–Zehnder interferometer. The beam separation was 38 μm and the observed contrast was on the order of 50%. This kind of interferometer has been used to measure the real and imaginary parts of the forward scattering amplitude for atoms and molecules (Chapman et al. 1995, Schmiedmayer et al. 1995, Berman 1997). Transmission gratings have also been used by Schoellkopf and Toennis (1994, 1996) to observe the mass selective diffraction of He clusters with up to 26 He atoms. Rare gas atomic beam diffraction and diffraction phenomena of CH_3F and CHF_3 molecules have also been observed. The diffraction and coherent splitting of Bose–Einstein condensates from optically induced lattices has been reported (Kozuma et al. 1999) and the diffraction of fullerenes from material grating has been reported by Arndt et al. (1999) and Hackermüller et al. (2003). This shows that diffraction and interference phenomena can be observed for rather large and fragile objects as well. New perspectives for matter-wave interferometry arise when further experimental methods become available and more complex objects can be used as quantum objects. The wide spectrum of new matter wave interferometry experiments may bring additional technological applications of quantum phenomena, enhancing the daily experience of the subtleties of nature by everyone. From a fundamental physics perspective new instruments and possibilities continue to influence the debate about quantum mechanics. This has been addressed, e.g., by Bromberg (2008), Snow 2013, Klepp et al. 2014.

1.3 Basic Neutron Diffraction Phenomena

The diffraction of light from macroscopic objects is best described by the coherent superposition of Huygens waves, which gives the Fraunhofer limit for the case of small phase shifts between the diffracted waves and the Fresnel limit for large phase shifts. These phenomena also apply to neutron optics and are described by the Helmholtz–Kirchhoff theorem which reduces to the Kirchhoff formula if the wavelength of the neutron is smaller than the dimensions of the object (D) and the distance of the source to the object (L_1) and also the distance between the object and the point of observation (L_2). Crystal diffraction is a special case of the Fraunhofer limit, with a linear phase dependence (James 1950) when

$$\frac{D}{2}\left(\frac{1}{L_1}+\frac{1}{L_2}\right) << \lambda/D. \tag{1.49}$$

In all other cases the diffraction integral becomes more complicated and depends explicitly on the distance between the object and the points of observation. Diffraction is described in $\boldsymbol{k}$-space and depends on the momentum transfer $\hbar\Delta\boldsymbol{k}$ which correlates with the spatial interaction distances Δx by the Heisenberg uncertainty relation (1.30). In the Fraunhofer limit Δx equals D, while in the Fresnel limit Δx becomes the Fresnel length defined by

$$\ell = \left(\frac{\lambda\, L_1 L_2}{L_1 + L_2}\right)^{1/2}, \tag{1.50}$$

which in the thermal neutron case is generally of order 50 μm. Therefore, $D << \ell$ and $D >> \ell$ are alternative definitions of the Fraunhofer and Fresnel regimes.

Neutron diffraction phenomena are described by the solutions of the Schrödinger equation using appropriate boundary conditions. Due to the same structure of the wave equations, these effects are equivalent to those of classical optics and many features of light optics can be adapted to the neutron case (Born and Wolf 1975, Sears 1989, Kaiser and Rauch 1999). Fraunhofer diffraction appears as long as the plane wave approximation holds and Fresnel diffraction appears if the curvature of the wave front has to be considered in the calculation of the path differences. These limiting cases are characterized by the condition $D \ll \ell$ for the Fraunhofer regime and $D \gg \ell$ for the Fresnel regime, where D denotes the dimension of the slit (aperture) and ℓ the Fresnel length (coherence length) as defined in Eq. (1.49). The finite dimensions of the entrance (d_1) and the detector slit (d_2) influence the coherence properties, as well. Visibility of an interference pattern can be expected only if $Dd_1/L_1 \leq \lambda/2$ and $Dd_2/L_2 \leq \lambda/2$. Related diffraction measurements from narrow slits (10–100 μm) have been performed and interpreted with wave front division interferometry by Mayer-Leibnitz and Springer (1962), by Landkammer (1966), by Friedrich and Heintz (1978), and by Zeilinger et al. (1982). Shull (1969) realized the Fraunhofer limit by using a non-dispersive perfect-crystal double-crystal arrangement, allowing him to experimentally deduce a transverse coherence length of 21 μm, which is in rough agreement with the interferometer result described in Section 4.2.2. In both cases, the dynamical diffraction from a perfect crystal determines the transverse coherence length. The broadening of the zero-order slit diffraction peak is given according to the Fraunhofer formula ($\Delta\varepsilon = 0.888\ \lambda/d$, where d is the slit width). The outcome of the classical experiments of double-slit diffraction (Zeilinger et al. 1982), the diffraction at an absorbing wire (Gähler et al. 1981), and the focusing effects from a cylindrical zone plate (Klein et al. 1981) are shown in Fig. 1.5 together with the calculated diffraction pattern where the finite resolution has been taken into account (Tumulka et al. 2007). In the neutron case, gratings can be designed as phase gratings (see Fig. 2.14) and zone plates which act rather by phase reversal instead of absorption and, thereby, achieve greater transmission efficiencies (Kearney et al. 1980). Such zone plates have also been used instead of a biprism to establish neutron interferometry by division of the wave front (Klein et al. 1981).

These experiments were performed with a 10-m-long optical bench (Fig. 1.5; Gähler et al. 1980) at a neutron wavelength of about 18.4 Å. The other characteristic dimensions were as follows: center-to-center distance between the two-slit openings, 126 μm; diameter of the absorbing wire, 100 μm; total width of the cylindrical zone plate, 2 mm; number of zones, 200. All the results are in excellent agreement with the calculated intensity profiles if the related resolution function is convoluted with the ideal two-slit diffraction curve (Zeilinger et al. 1988). The agreement with calculation is very good, allowing an upper limit of a nonlinear term of the Schrödinger equation to be extracted (Gähler et al. 1981; Section 10.1) from this experiment. The double-slit diffraction pattern has also been observed with a perfect crystal amplification camera, where the diffraction pattern from rather macroscopic slits (~0.2 mm) has been projected onto a scale of about 25 mm (Section 11.3; Lacroix et al.1999).

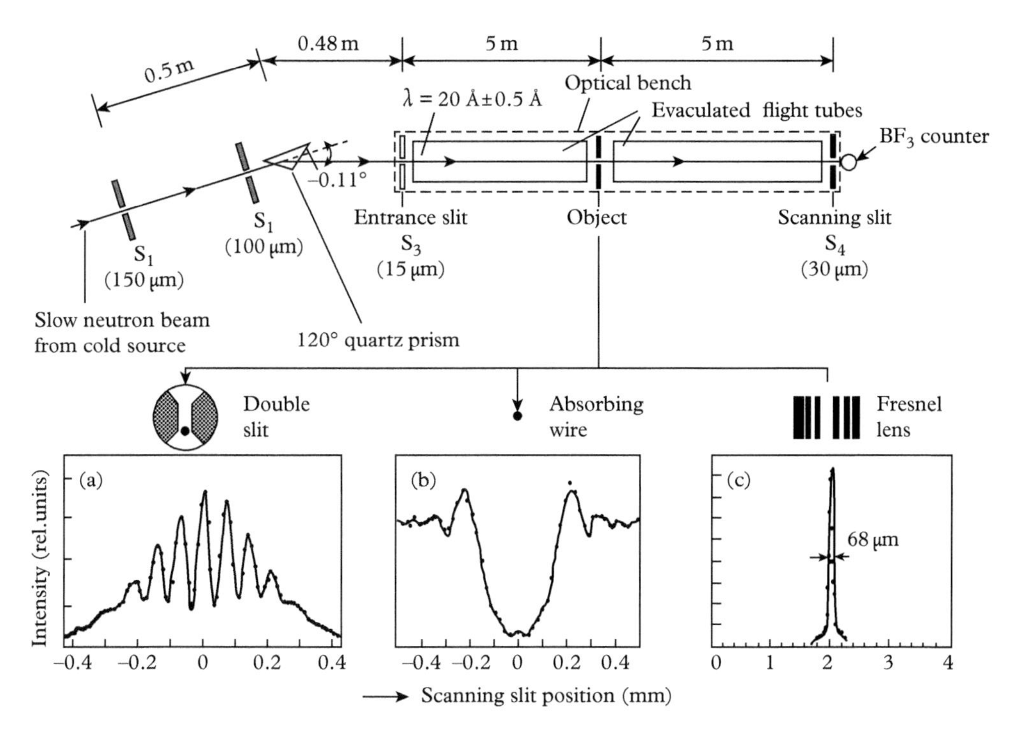

Figure 1.5 *Sketch of the optical bench used for various diffraction experiments form macroscopic objects: (a) double-slit diffraction (Zeilinger et al. 1982), (b) diffraction from an absorbing wire (Gaehler et al. 1991), and (c) the diffraction from a Fresnel zone plate (Klein et al. 1981b). (b) reproduced with permission from Gaehler et al. 1991. Copyright 1991, American Association of Physics Teachers.*

The diffraction of thermal and cold neutrons at ruled gratings showed that multiple beam interferences from structured macroscopic objects can be observed (Kurz and Rauch 1969, Graf et al. 1979). Such lattices have been used in combination with mirror reflections by Ioffe et al. (1985) to develop a grating interferometer for 3.15 Å neutrons which has aberration-free features (Ioffe 1988; see Fig. 2.1). Transmission phase gratings for very cold neutrons ($\lambda \cong 102$ Å) have been tested by Gruber et al. (1989) and combined to form a Mach–Zehnder interferometer device with rather large beam paths. In this case, intensity can be concentrated into certain diffraction orders. The lattice constant was 2 μm and the step height for the first and third grating produced a phase shift difference of $\pi/2$ and the middle one a phase shift difference of π which provides optimal neutron economy of the interferometer. Diffraction of ultra-cold neutrons ($\lambda \sim 1500$ Å) at blazed gratings indicated a coherence length greater than 100 μm (Scheckenhofer and Steyerl 1977). Diffraction from multilayer mirrors and supermirrors has become a standard tool for advanced neutron beam tailoring (e.g., Mezei and Dagleish 1977, Majkrzak and Passel 1985). Reflectometry from various surface structures and small-angle scattering measurements from various precipitates, magnetic domains, or magnetic flux lines in superconductors bridge the field of coherent neutron optics to important solid state physics applications (Chapter 9). In all cases, diffraction peaks become visible if the typical momentum transfer ($\Delta k \cong \pi/d$) becomes larger than the momentum resolution of the beam. In this case a new state—a new entity—which carries information about the diffracting object is created. Reflections from nano-structured substances show resonance, tunneling, and surface wave effects (Maaza 2011). These aspects will be discussed in more detail in Section 4.3 in connection to an experiment where an absorbing lattice is rotated within one beam of a crystal interferometer.

Even much higher resolution can be achieved when multiple Laue-rocking curves are used where the reflecting lattice planes d_{hkl} of a perfect crystal of thickness t act as a kind of monochromator, so that d_{hkl}/t instead of D/L defines the collimation (see Section 11.3, Bonse et al. 1977). With such a system the diffraction of thermal neutrons from slits having a width of several millimeters has been observed (Rauch et al. 1983a, Fig. 11.10). From the structure of these rocking curves, scattering lengths for finite momentum transfer can be extracted as has been done for X-rays (Bonse and Teworte 1980).

The collective interaction of the neutron with magnetic fields is an important topic of neutron optics. The development of neutron spin-echo spectroscopy uses this collective interaction to achieve a very high energy resolution ($\sim$1 neV) by marking each neutron with its own clock in the form of its Larmor rotation angle (Mezei 1972, 1980a). The collective interaction with oscillating fields appears most clearly in the case of resonance fields where single neutron wave–field interactions and dressed neutron phenomena have been studied (Muskat et al. 1987). Investigations concerning the intrinsic 4π-symmetry of fermions (Rauch et al. 1975, Werner et al. 1975), discussed in Chapter 5, and the observation of the geometric Berry phase are other examples of important quantum measurements (Bitter and Dubbers 1987, Richardson 1988, Wagh et al. 1997). Neutron interferometry experiments related to the topological and the geometric quantum phases are described in Chapter 6 and those dealing with quantum contextuality in Chapter 7.

All neutron experiments performed up till now belong to the field of self-interference where at a given time only one neutron—if at all—is within the interferometer and the next one is not yet released from the fission (or spallation) process in the neutron source. This can be quantified by the mean numbers of neutrons N_{Ph} (degeneracy parameter) within the phase space volume, which is defined as $\Delta x\ \Delta y\ \Delta z\ \Delta k_x\ \Delta k_y\ \Delta k_z = (1/2)^3$. For advanced neutron sources N_{Ph} is on the order of 10^{-14}, whereas for thermal light, synchrotron and electron sources N_{ph} reach values of 10^{-3}; but lasers produce phase space densities up to 10^{14} (Mandel_and Wolf 1965, Boffi and Caglotti 1966, Maier-Leibnitz 1966a). The degeneracy parameter of a beam can also be understood as the mean number of particles passing through a coherence area normal to the beam in an interval of time called the longitudinal coherence time. According to the quantum Liouville theorem the phase space density ρ obeys the quantum Liouville equation (von Neumann equation)

$$i\hbar\frac{\delta\rho}{\delta t} = [\mathcal{H}, \rho], \tag{1.51}$$

where $[\mathcal{H},\rho]$ denotes the commutator of the Hamiltonian $\mathcal{H}$ with ρ. The phase space density cannot be changed by any conservative force acting on the system. Self-interference means that the neutron interferes with itself, supporting Dirac's statement that in ordinary light optics the photon interferes with itself (Dirac 1930).

Modern neutron scattering instrumentation has benefitted greatly from advances in neutron optical components. Total reflection inside neutron guides is used to transfer the luminosity existing near to the neutron source to many instruments placed in low background environments far from the source (Christ and Springer 1962, Maier-Leibnitz and Springer 1963, Willis and Carlile 2009). Multilayers with variable spacing consisting of materials with different indices of refraction can enhance the region of total reflection by a factor of about 3 (Mezei and Dagleish 1977). Neutron microscopy, with corrections for gravitationally induced chromatic aberration, has been developed and has reached a magnification of 50 by using an achromatic two-mirror arrangement and gravity focusing (Schütz et al. 1980, Arzumanov et al. 1984, Herrmann et al. 1985). Charged particle physics has benefitted substantially by the invention of phase space cooling (e.g., van der Meer 1985), but no similar methods are known for neutrons. However, various bunching systems, particularly for pulsed beams, are feasible by mechanical and electromagnetic means (Maier-Leibnitz 1966b, Buras and Kjems 1973, Steyerl 1975, Rauch 1985, Mayer et al. 2009).

2
Neutron Interferometers and Apparatus

Among the various areas of neutron optical research, neutron interferometry has provided some of the most challenging and spectacular developments (Bromberg 2008, Snow 2013). In most cases, separated coherent beams which are produced either by wave-front division (Young's type interferometer) or by amplitude division (Mach–Zehnder-type interferometer) are used. These beams are subsequently coherently superposed after passing through regions of space where the neutron wave function is modified in phase and amplitude by various interactions: nuclear, magnetic, electromagnetic, gravitational, or geometry and topology (as in Aharonov–Bohm-type experiments). Various types of neutron interferometers that have been tested are shown in Fig. 2.1. The first neutron interferometer was based upon wave-front division and biprism deflection, but it had several constraints for interferometric applications mainly due to the small beam separation (100 μm) (Maier-Leibnitz and Springer 1962). Nevertheless, this method gained further interest for diffraction experiments from macroscopic objects and for the detection of weak interaction effects where the long flight paths (10 m) enhance the sensitivity for the measurements of these effects (Gaehler et al. 1980, Klein et al. 1981a). The perfect crystal interferometer (Rauch, Treimer, and Bonse 1974) provides widely separated coherent beams and has become a standard method for advanced neutron optical investigations. The development of diffraction and phase-grating interferometers extended the Mach–Zehnder interferometer method to very slow neutrons (Ioffe et al. 1985, Gruber et al. 1989). The superposition of spin-"up" and spin-"down" states in the longitudinal direction provides the basis for Larmor and Ramsey interferometry where no lateral beam splitting is necessary. The well-known spin-echo systems (Mezei 1972) and zero-field spin-echo systems (Gaehler and Golub 1987, Dubbers et al. 1989) are examples of such interferometers, which also play an important role in atom interferometry (Bordé 1989, Kasevich and Chu 1991, Riehle et al. 1991, Parazzoli et al. 2012). Ramsey (1993) has addressed the complementarity of neutron two-path and spin-rotation interferences. These various types of interferometers will be described in some detail in the following sections and they are schematically shown in Fig. 2.1.

2.1 The Perfect Si-Crystal Interferometer

A great resurgence of activity in neutron optics occurred after the development of the perfect crystal interferometer by Rauch et al. (1974), which was first tested at a rather small (250-kW) TRIGA-reactor in Vienna. The first results and the setup are shown in Fig. 1.2. The results

Neutron Interferometry. Second Edition. Helmut Rauch and Samuel A. Werner.

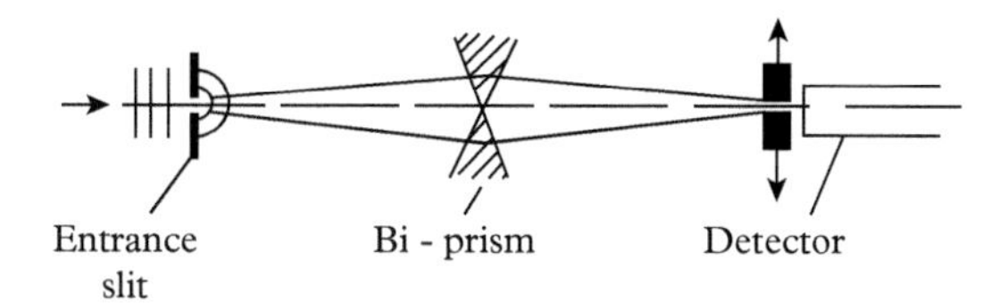

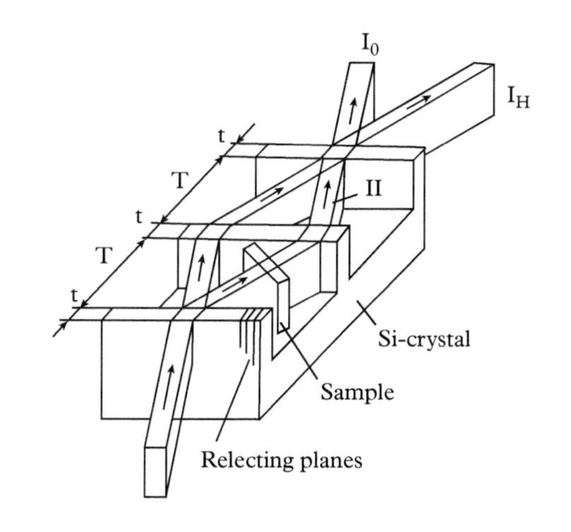

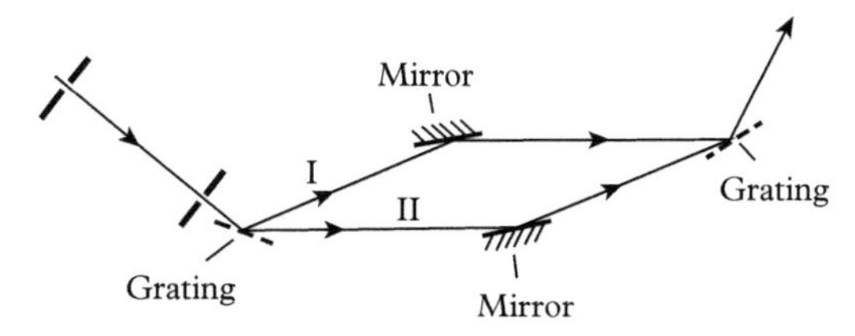

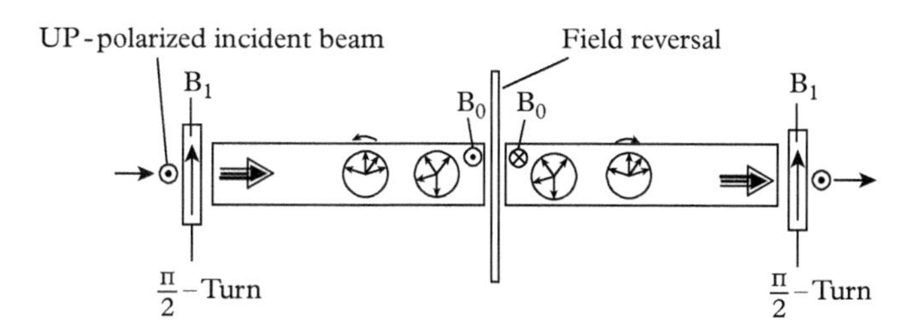

Figure 2.1 *Sketch of various neutron interferometer types. From top to bottom: wave front division interferometer based on single-slit diffraction, amplitude division interferometer based on perfect crystal or on grating diffraction, spin-echo interferometer based on spin-state superposition*

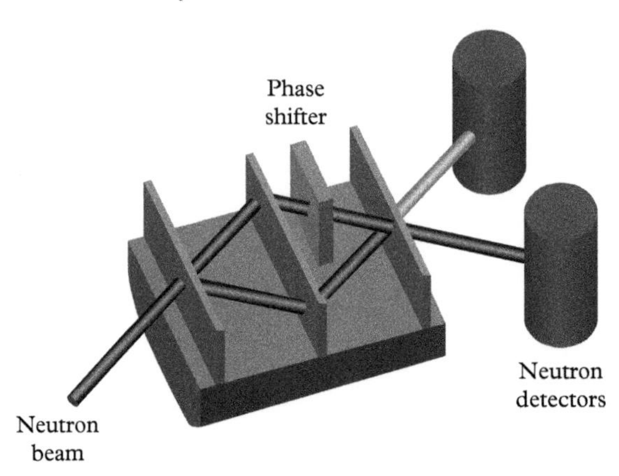

Figure 2.2 *Sketch of a monolithic perfect crystal interferometer. An optical path difference is obtained by rotating a flat aluminum plate, subtending both beams around a vertical axis*

indicate the opposite intensity variation (as a function of phase shift) of the forward and deviated beam (Fig. 1.2). The oscillations are 180° out of phase with each other. A schematic view of such an interferometer is shown in Fig. 2.2.

Its operation is based on the perfect arrangement of the lattice planes which causes a coherent division of amplitudes by dynamical Bragg diffraction from perfect crystals of silicon as was first implemented for X-rays (Bonse and Hart 1965a). In the standard version, a monolithic triple-plate system in the Laue (L) transmission geometry which provides a wide beam separation ($\lesssim$5 cm) and a non-dispersive response to the incident neutron beam is used. This LLL device permits the use of large-beam cross-sections providing good intensity conditions (Fig. 2.1). It is geometrically analogous to the well-known Mach–Zehnder interferometer of light optics (Zehnder 1891, Mach 1892). Various configurations of this interferometer have been suggested for neutrons and to some extent implemented, such as the skew symmetric device shown in Fig. 2.3, left. Some designs have only two crystal plates, analogous to the Rayleigh (1896)-type interferometer (Fig. 2.3, middle), while others have four (Fig. 2.3, right) and even more plates (Heinrich et al. 1988, Suda et al. 2004). The essential feature of all these devices is that the reflecting lattice planes are arranged undisturbed throughout the whole crystal with a precision comparable to the lattice parameter. This is most easily provided by a monolithic design of such interferometers. Details on the necessary silicon crystal perfection and the interferometer fabrication, including machining with diamond cutting tools and the subsequent chemical etching, are given at the end of this section. When the interferometer crystal has the required perfection and is cut and etched properly, neutron interferometry can be introduced rather easily in any neutron laboratory. We point out also that perfect crystal interferometry opens many possibilities for educational purposes, because it exhibits the fundamentals of quantum mechanics in a very direct and obvious way. The perfect Si crystal interferometer is extremely useful for fundamental neutron physics studies and has proven to be a marvelous didactic laboratory for probing and elucidating the basic quantum mechanical principles of nature.

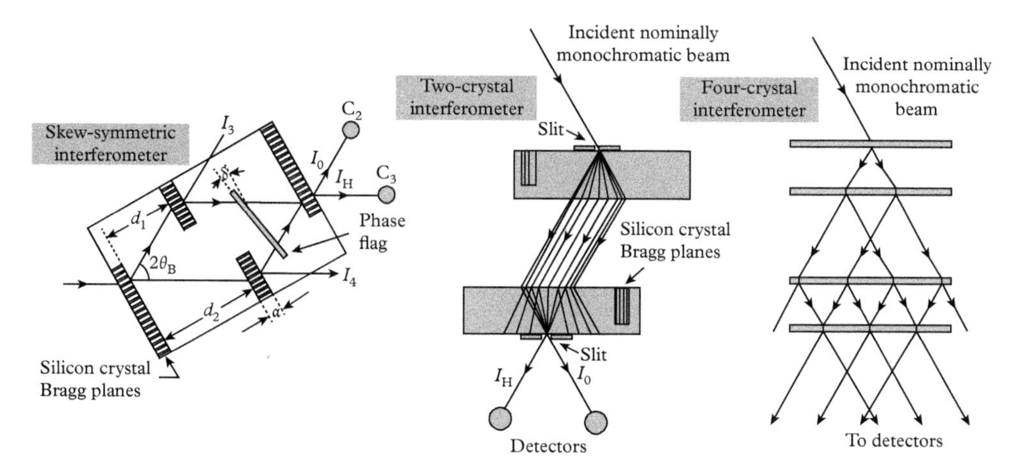

Figure 2.3 *Various types of perfect crystal interferometers: skew symmetric (left), two-plate (middle), and four-plate interferometer (right)*

For most applications the standard triple Laue-case interferometer (Fig. 2.4, top) is generally the best configuration. It follows from symmetry considerations that the amplitude and the phase of the wave function in the forward direction (0-beam) behind the empty interferometer are composed of equal parts coming from both beams traversing paths I and II. The wave on path I arrives in the detector in the "0-beam" after having made a transmission (t) in the first crystal, a reflection (r) is the second crystal, and another reflection (r) in the third crystal. On path II the sequence is rrt. From symmetry, it follows that these two waves are equal in phase and amplitude. The phases (χ_1 and χ_2) can become different when the neutron experiences an interaction along the two beam paths which are different. Thus, the intensity in the 0-beam is given by

$$I_0 = |\psi_{\mathrm{I}} + \psi_{\mathrm{II}}|^2 = \left|\mathrm{trr}\ \psi_0\, e^{\mathrm{i}\chi_1} + \mathrm{rrt}\ \psi_0\, e^{\mathrm{i}\chi_2}\right|^2, \tag{2.1}$$

where ψ_0 is the incident wave function. Setting $\Delta\chi = \chi_2 - \chi_1$ to be the phase difference, one sees that the intensity pattern for the 0-beam interferogram displays complete modulation in this ideal case:

$$I_0(\Delta\chi) = A[1 + \cos\ \Delta\chi]. \tag{2.2}$$

The constant A is given as $A = |\psi_0|^2|\mathrm{r}|^4|\mathrm{t}|^2$.

By similar reasoning one finds that the intensity in the H-beam is

$$I_{\mathrm{H}}(\Delta\chi) = \left|\mathrm{trt}\ \psi_0\, e^{\mathrm{i}\chi_1} + \mathrm{rrr}\ \psi_0\, e^{\mathrm{i}\chi_2}\right|^2 = B - A\cos\Delta\chi, \tag{2.3}$$

where $B = |\psi_0|^2[|\mathrm{t}|^4|\mathrm{r}|^2 + |\mathrm{r}|^6]$. The minus sign in front of the cosine term arises because there is an odd number of reflections, r, for each component contributing to I_{H}. From particle conservation it also follows that

$$I_0 + I_{\mathrm{H}} = \mathrm{constant}. \tag{2.4}$$

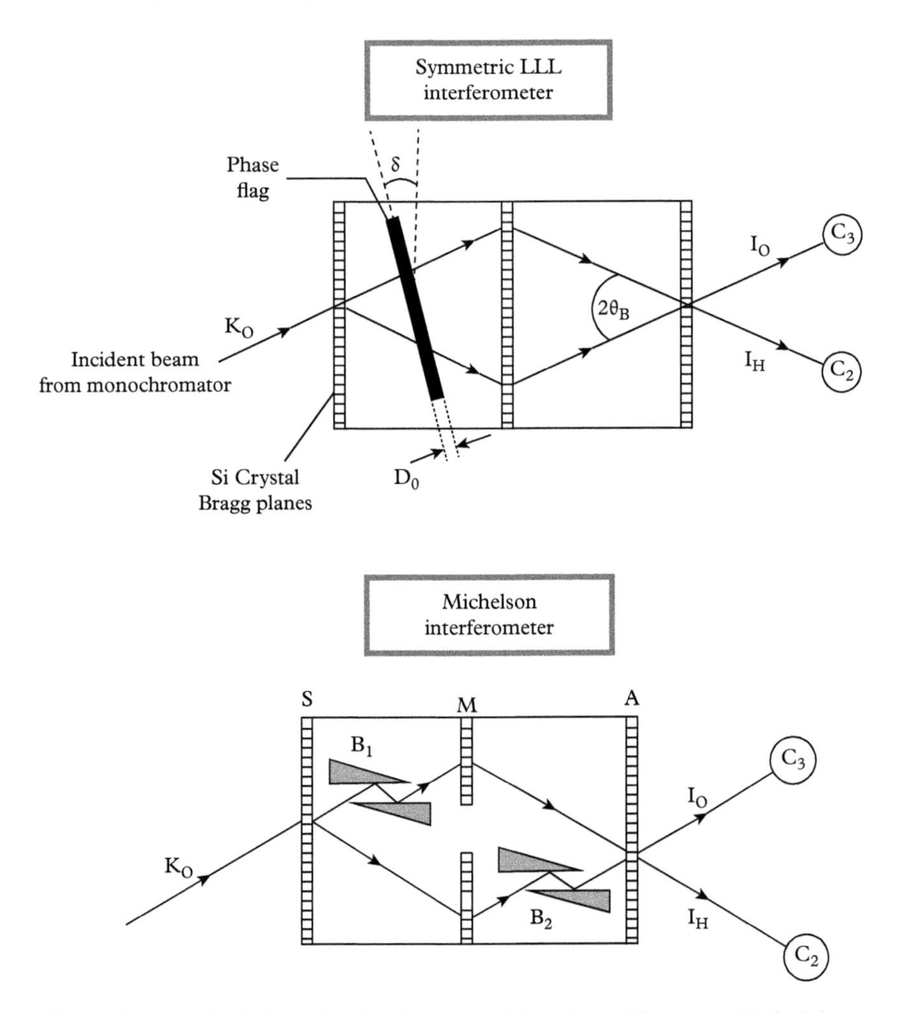

Figure 2.4 *Standard three-plate interferometer with a phase shifter rotated in both beam paths (above) and with a perfect crystal optical path length controller*

The explicit dynamical diffraction calculation of the transmission and reflection coefficients, t and r, is carried out in Chapter 11 on the basis of plane wave dynamical diffraction theory. Here it should be mentioned that r and t strongly depend on the deviation ($|\boldsymbol{k}-\boldsymbol{k}_B|$) of the different momentum components $\boldsymbol{k}$ and $\boldsymbol{k}_B$ fulfilling exactly the Bragg condition. Thus, in cases where $r=0$ or $t=0$ a path labeling exists. The formulas show that these wave components do not contribute to the interference pattern, but they contribute to the non-interfering part of the H-beam and to the beams leaving the interferometer at the second interferometer plate (see Fig. 2.3). Whether individual $\boldsymbol{k}$-components can or cannot be associated with individual neutrons having the related momentum is an epistemological question and will be discussed in Chapter 12. The most important feature here is that the wave functions can be calculated because the whole interaction region between the neutrons and the perfect crystal is known up to an accuracy better than the wavelength of the neutrons. The sum of the 0-beam intensity and the H-beam intensity is a constant as it must be, since Si has essentially zero absorption for thermal neutrons. Thus, the neutron intensity is swapped back and forth between the 0-beam and the H-beam detectors as the phase difference $\Delta\chi$ is varied (Bonse and Graeff 1977, Rauch and Petrascheck 1978, Lemmel 2013). A general discussion of the calculation of the quantum phase shift $\Delta\Phi$ according to the Feynman–Dirac path integral formalism along the classical trajectories was given in Section 1.2. Due to various imperfections of the apparatus (the interferometer, especially) and of the neutron beam (finite monochromaticity), the intensity oscillations are somewhat damped compared to the predicted behavior (Eq. 2.2):

$$I_{\text{meas}} = A' + B' \cos(\Delta\chi + \phi_0), \tag{2.5}$$

where A', B', and ϕ_0 are characteristic parameters for each experimental setup.

From the point of view of using the change of wave vector due to the optical potential provided by the collection of nuclei in a slab of matter of thickness D_0 in one of the sub-beams of the interferometer, we find that the phase shift is

$$\Delta\chi_{\text{nuc}} = (n-1)\,k\,D_0 \;=\; -\lambda N b_c D_0, \tag{2.6}$$

where we have used Eq. (1.26) for the index of refraction n. A direct and simple technique to produce an interferogram is to rotate the slab about an axis perpendicular to the interferometer through angles δ, such that the neutron path length within the slab is $D_0/\cos(\theta_B + \delta)$. If the slab extends across both beams as shown in Fig. 2.4 (top), the path length difference $\Delta D(\delta)$ of the neutron passing through the slab on path II minus that on path I is given by

$$\Delta D(\delta) = \left(\frac{1}{\cos(\theta_B + \delta)} - \frac{1}{\cos(\theta_B - \delta)}\right)\cdot D_0. \tag{2.7}$$

For small δ, this function is nearly linear in δ. Consequently, an interferogram obtained by rotating the slab will be approximately a sinusoid in δ. Figure 2.5 shows a typical scan taken with a symmetric LLL-interferometer. A rather high contrast in the 0-beam and particle conservation between the 0- and the H-beams is observed. To provide an idea of the size of the phase shift for thermal neutrons traversing a slab of matter, suppose that the neutrons of wavelength 2 Å traverse a 1-cm-thick aluminum slab at normal incidence. The phase shift is $\Delta\chi = 420$ rad $\approx 150{,}000°$. It is clear from this calculation that neutron interferometry provides a very sensitive method for accurately measuring the neutron coherent scattering lengths b_c. Results and discussion of such experiments are given in the next chapter. Macroscopic optical path differences can be achieved in a Michelson (1881)-type interferometer (Fig. 2.4) where Laue and Bragg case diffraction from perfect crystals are used (Appel and Bonse 1991). This idea has not yet been utilized in neutron interferometry. The application of phase shifts in the transverse and vertical direction will be discussed in Section 4.2. Combinations of Laue and Bragg diffraction interferometers, two-, three-, and multiple-plate interferometers, and monolithic- and polylithic-type interferometers are discussed in the literature and have been partly tested for X-rays and neutrons (Zeilinger et al. 1983, Bonse et al. 1994, Massa et al. 2010). The largest perfect crystal interferometers tested have an enclosed area up to 114 cm^2 (Zawisky et al. 2010, Springer et al. 2011a) and interferometers with an enclosed area of 174 cm^2 are in a testing phase. Monolithic interferometers seem to be feasible up to an enclosed area of about 500 cm^2; beyond that, polylithic interferometers may become superior but much more challenging than monolithic ones.

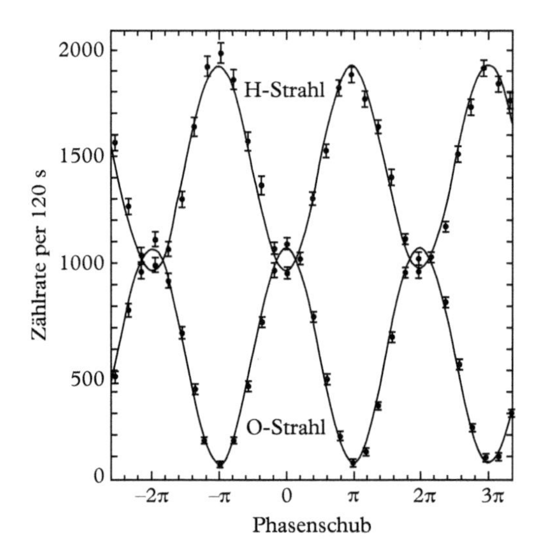

Figure 2.5 *High contrast interference pattern obtained with a perfect silicon crystal interferometer. Courtesy of M. Arif and D.L. Jacobson, NIST 1997*

Pushin et al. (2009) presented calculations and later on test measurements (Pushin et al. 2011) which show that a four-plate interferometer with a double-space distance between the second and third plates (Fig. 2.3, right) may be very robust against low-frequency environmental disturbances. Figure 2.6 shows the influence of 8-Hz vibrations on the contrast of a standard triple plate and a dephasing-free four-plate interferometer. Vibrations are specifically disturbing in the case of neutron interferometry since the time-of-flight through the interferometer (about 50 μs) becomes comparable to the typical time scale of such vibrations. Similar problems exist in atom interferometry and therefore some related estimates can be used for neutron setups as well. The most disturbing frequencies lie in the range between 5 and 200 Hz and are caused by excitations of the setup and the seismic noise spectrum (Jacquey et al. 2006).

As discussed earlier, an interferometer crystal is a monolithic device consisting of two, three, or more perfect crystal blades cut from a large perfect dislocation-free Si-crystal ingot perpendicular to a set of strongly reflecting Bragg planes (typically 220). Figure 2.7 shows a photograph of a symmetric LLL interferometer, a skew symmetric LLL interferometer, and multi-blade Laue-geometry devices. Due to the requirements of the modern semiconductor industry, there is a very-large-scale production of perfect silicon crystals. Only high purity, float-zone silicon crystals can be used for the fabrication of neutron interferometers. Crystals grown by the Czochralski technique (pulling with a seed from the melt) contain too much oxygen, which creates local strain and density fluctuations; therefore, they are not suitable for neutron interferometers. The most critical impurity component appearing also in float-zone production is carbon due to its unavoidable inhomogeneous distribution. This produces slight lattice distortion which destroys the parallelism of the lattice planes. The carbon distribution can be investigated by infrared absorption spectroscopy. This permits a proper selection of appropriate crystal ingots. Crystals purchased from various companies worldwide have been used to fabricate interferometers. In recent years, crystals

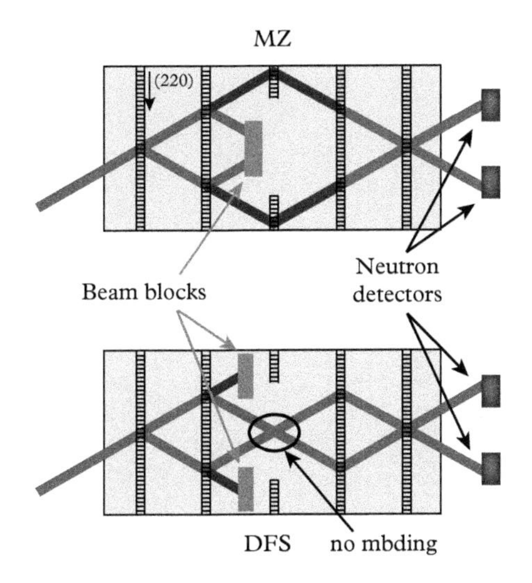

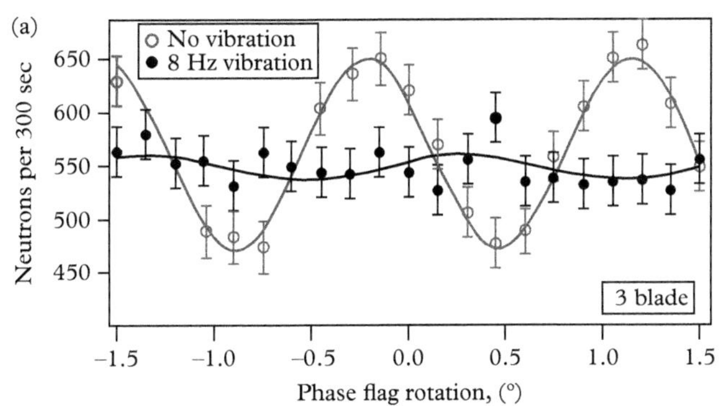

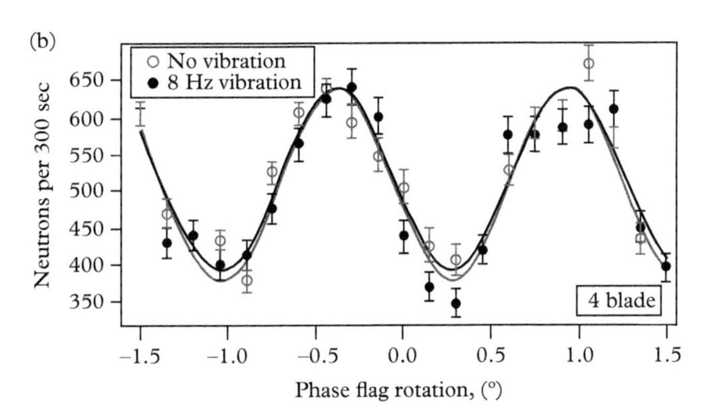

Figure 2.6 *Five-plate interferometer used as a triple-plate (above) and a vibration-compensated four-plate interferometer (below). Contrast under the influence of 8-Hz vibrations of a triple-plate interferometer (a) and a dephasing-compensated four plate interferometer (b). Reprinted with permission from Pushin et al. 2011, copyright 2011 by the American Physical Society.*

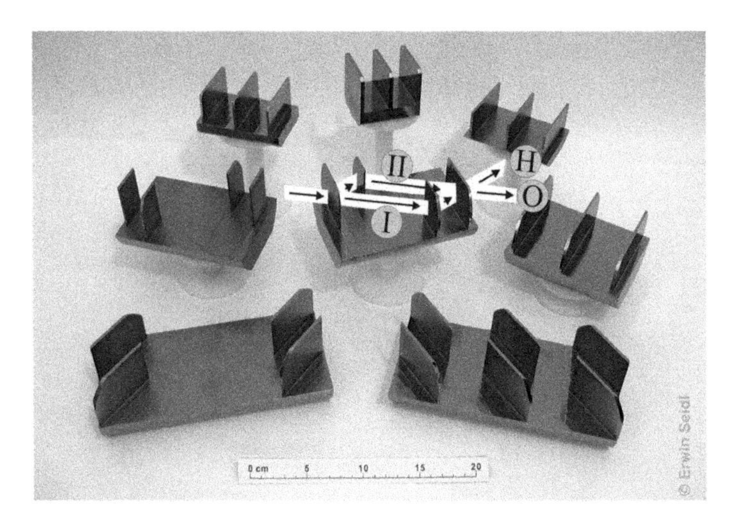

Figure 2.7 *Photograph of an assembly of several perfect crystal interferometers at the Atominstitut, TU-Vienna. Courtesy E. Seidl*

obtained from Wacker Siltronic, GmbH (Germany) have been found to be of the highest quality for interferometer fabrication. Although the price has increased in recent years, a 10-cm-diameter ingot, adequate for the interferometers shown in Fig. 2.7 costs about 10,000 Euros.

Machining the interferometer out from a monolithic perfect single crystal ensures that the respective crystal lattice planes in each blade are aligned to a tolerance of fractions of a lattice constant. This can be tested by observing the Moiré pattern, which appear when the lattice planes are slightly non-parallel, where a deviation angle α causes a bright/dark pattern with a rather macroscopic distance of D given by lattice distance divided by α (Hart and Bonse 1970, Bartscher and Bonse 1998, Amidor 1999). In order to fabricate the interferometer from a perfect Si crystal, the bulk material between the blades is first machined away in the top part of the crystal, leaving the upright blades attached to the rather thick ($\sim$1 cm) base or backbone of the device. The machining of the blades is performed using diamond grinding wheels (200 to 600 grit), mounted on a high-precision spindle on a stable, heavy milling-type machine fitted with optical encoders to assure that the surfaces of the blades are parallel to within about 1 μm over their entire area. The distances between the blades and their thickness must be the same to an accuracy of about 2 μm. This requirement follows from the constraint that the geometry of an interferometer must be accurate in comparison with the Pendellösung lengths, which are on the order of 50 μm for most silicon reflections (Eq. 11.54). The surface damage created by the machining must be etched away with acid, while still maintaining the dimensional accuracy of the original machining operation. Typically this is done in steps, etching away several micrometers of Si at each stage, and then checking the interferometer contrast and intensity with neutrons and/or X-rays. Etching is usually performed by a mixture of hydrofluoric acid and nitric acid. A mixture of one part hydrofluoric acid in 50 parts of nitric acid produces an etching rate of some 0.2 μm/s. Optical flatness of the blades has been checked in some laboratories with optical interferometer techniques. Experimentally, it is found that the surface damage strains the bulk of the crystal inhomogeneously. This strain increases the neutron reflectivity of the crystal above its perfect crystal value. Thus, as the interferometer is etched with acid in steps, the sum of the 0- and H-beam intensities

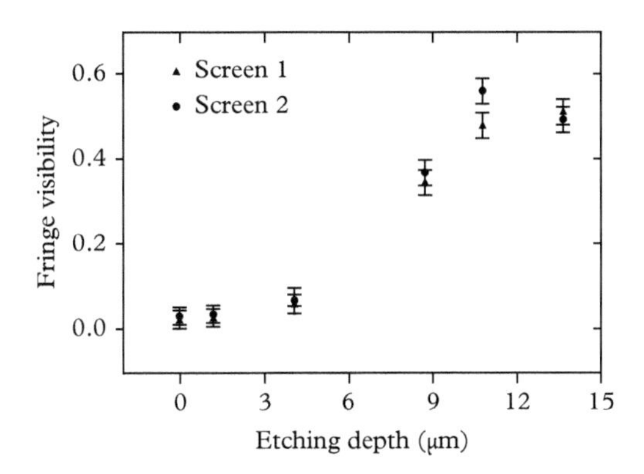

Figure 2.8 *Variation of the contrast of a perfect crystal interferometer as a function of the etching depth*

decreases, while the contrast or fringe visibility increases. We show in Fig. 2.8 this behavior of the intensity and contrast as the amount of Si material removed by etching increases for two new interferometers recently put into operation at NIST and at ILL. Some additional details of the machining and etching of interferometers can be found in articles of Treimer et al. (1967) and Zawisky et al. (2010).

The crystal's own weight results in very small internal strains. This can rotate the crystal plates by fractions of a lattice constant, which causes inhomogeneous phases across the reflected beam, thereby forming a Moiré pattern. In order to keep this effect small, a glass plate or glass balls between the goniometer and the interferometer can reduce thermal strains due to various thermal expansions and a thin housing around the interferometer can reduce thermal gradients. Adding an additional weight on the order of several grams at special parts of the base of the interferometer can balance a pre-existing Moiré strain pattern and increase the contrast of the overall interference pattern.

The various position-dependent interference properties can be visualized by scanning across the beam cross-section. As shown in Fig. 2.9 there is a marked variation of the local contrast and especially a strong variation of the internal phase (ϕ_0 in Eq. 2.5) over the beam cross-section of 6×4 cm (Lemmel 2000). This indicates that all measured quantities of an interference pattern are average values over distinct parts in ordinary real space and in momentum space. A slight non-parallelism of the lattice planes and slight internal strains cause such inhomogeneous Moiré pattern in the outgoing beams.

The most sensitive quantity of an interference pattern is the internal phase ϕ_0 in Eq. (2.5). The internal phase shift ϕ_0 may vary due to temperature fluctuations, small drifts of the crystal orientation, or humidity content of the air. These things are often limiting factors on how accurately phase shifts can be measured. Figure 2.10 shows such a variation for an overnight run. This limits the accuracy of phase measurements and explains why a sample-in/sample-out method has been developed, which can reduce this error contribution up to a factor of 10 (Section 3.1.2). These measurements are also examples of post-selection experiments that will be discussed in Section 4.5.1.

In order to achieve high contrast it is recommended to avoid narrow slits (<3 mm) within the interferometer since single-slit diffraction broadens the beam, which changes the reflectivity from the crystal plates that follow (see Fig. 11.9).

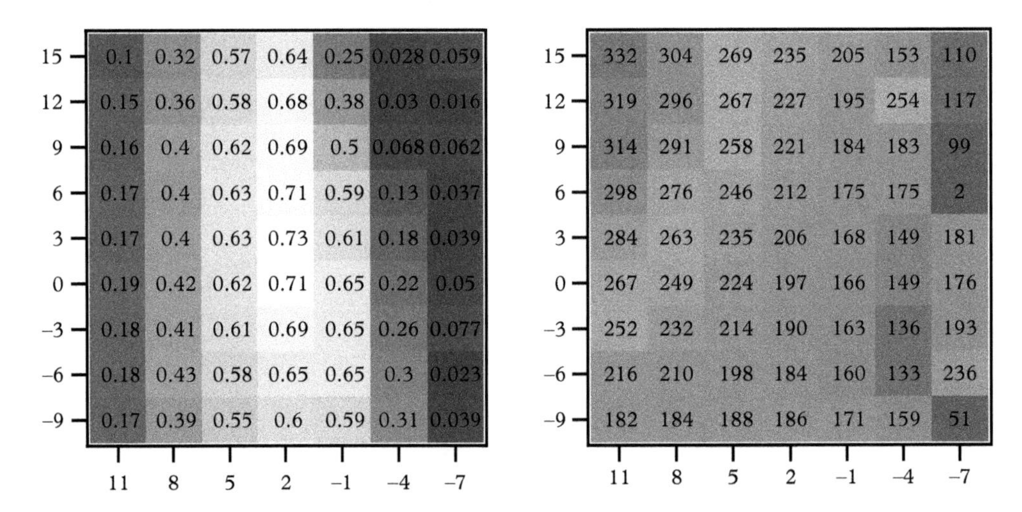

Figure 2.9 *Measured local contrast (left) and local internal phase (right) of a typical perfect crystal interferometer. Courtesy of H. Lemmel ILL (2010)*

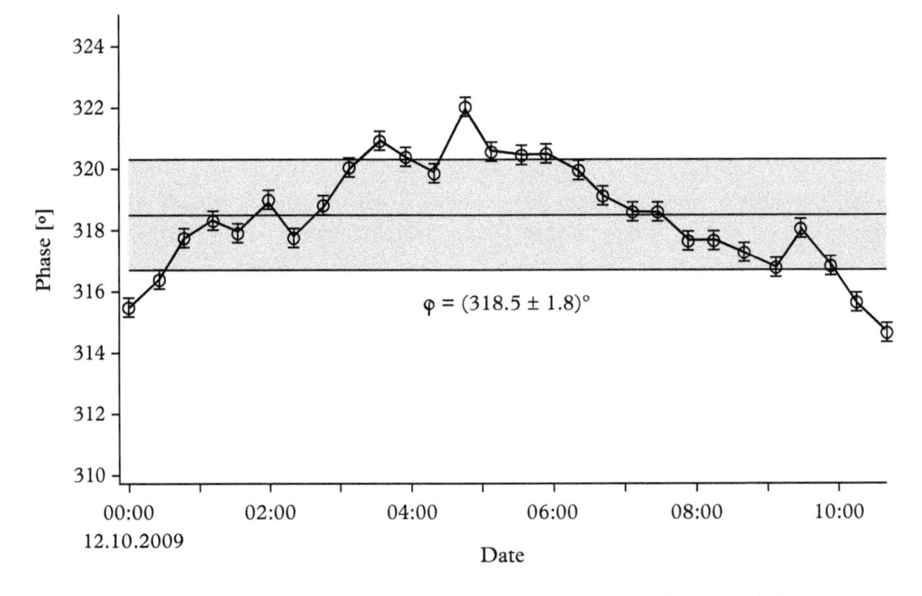

Figure 2.10 *Variation of the internal interferometer phase for a typical overnight run*

2.2 Perfect Crystal Interferometer Setups

Most experimental techniques in physics evolve in somewhat different ways in various laboratories. In addition to certain space constraints in the neutron laboratories (beam halls, reactor halls), the available neutron beams also have dictated a somewhat different design strategy of neutron interferometry instrumentation. It was recognized very early that the perfect silicon crystal neutron interferometer is very sensitive to various deleterious environmental effects, in particular vibrations, microphonics, and thermal gradients. It was clear that the interferometer needed to be vibration-isolated from the beam hall floor and would need to be placed within some serious environmental isolation enclosure. Perfect crystal neutron interferometers have now been used routinely in Austria, the Czech Republic, France, Germany, India, Japan, and the USA at neutron sources ranging from 250 kW up to 51 MW. Here we give some examples of rather advanced setups.

2.2.1 The ILL Setup

The original S18 station at the Institut Laue Langevin (Fig. 2.11) was installed as a joint project between the Atominstitut in Vienna and the Universität Dortmund on the H25 curved thermal guide in the guide hall at the 57-MW high flux (HFR) in Grenoble. In the usual mode of operation

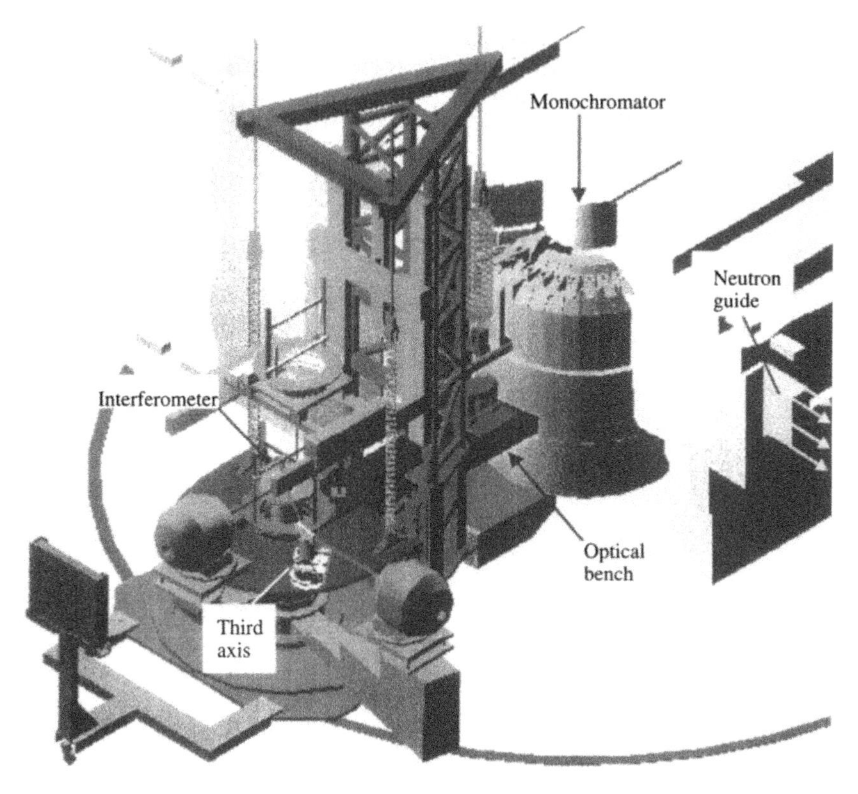

Figure 2.11 *Sketch of the ILL interferometer setup*

the 5-cm-high perfect Si monochromator crystal extracts a beam of adjustable wavelength dependent upon the $2\theta_M$ angle which is used in the non-dispersive (parallel) setting with the neutron interferometer (Bauspiess et al. 1978). This means that for every incident ray reflected by the Si monochromator there is a corresponding outgoing ray in the H-beam that is precisely parallel to it. The resulting rocking curve of the interferometer is therefore very narrow, with FWHM less than about 3 arcsec. The monochromator crystal and the interferometer crystal with its associated goniometer are mounted on a steel bench which serves as a rigid "optical table." This table is suspended with a set of three springs from a steel structure that overhangs the apparatus, thereby providing the necessary vibration isolation. The interferometer crystal itself rests on three sapphire balls which are glued to a silicon crystal mounting plate. Although the alignment procedure required to find the Bragg condition of the interferometer using the perfect Si crystal monochromator is tedious, and keeping it aligned requires substantial stability, it provides for a very low background situation. It also facilitates the rather straightforward production of a polarized beam by simply inserting a wedge-shaped region of magnetic field (of about 1 T) between the monochromator and the interferometer crystals (Badurek et al. 1979). This region is birefringent, deflecting the spin-up and spin-down components in opposite senses through small angles (a few arcseconds). Aligning the interferometer with the spin-up component allows only spin-up neutrons to traverse the interferometer. Microphonic disturbances, air currents, and thermal gradients are mitigated by enclosing the entire setup in a sound-proof hut. An additional bicrystal monochromator could be inserted to narrow the wavelength band to $\Delta\lambda/\lambda_0 = 10^{-3}$ without

Figure 2.12 *Photo of the ILL-S-18 neutron interferometer setup*

deflecting the beam off the axis of the optical bench (Bauspiess et al. 1977). An X-ray tube is incorporated into the system for calibration and stabilization purposes. The neutrons in the two interfering beams are counted with well-shielded ^{3}He gas proportional detectors, which are essentially "black" to thermal neutrons. The nuclear reaction in the ^{3}He detectors is $n + {}^3\mathrm{He} \rightarrow {}^1\mathrm{H} + {}^3\mathrm{H}$, where the 770 keV of energy liberated in the reaction is shared between the proton and the triton. This ionizing energy is very large, thus providing a perfect discrimination in detecting neutrons against background compared with other forms of radiation. Since the shielding around these detectors is rather heavy, they are mounted on separate arms that rotate around the interferometer, and are independent of the vibration isolation system. The actual Bragg angle is measured from the angle difference between the non-dispersive (antiparallel) and the dispersive (parallel) position of the monochromator–interferometer arrangement. The wavelength spread is obtained from the width of the Bragg peak in the dispersive position. Typical counting rates in the I_0 interfering beam for an entrance aperture of 1 mm $\times$ 10 mm are 5000 counts/min. The entire setup has been upgraded several times during refurbishments of the high flux reactor in Grenoble. A third axis has been added behind the interferometer (see Fig. 2.11) and a supermirror guide in front of the interferometer setup gives a further increase in intensity. An integrated Bonse–Hart (1965b) neutron small-angle scattering camera provides possibilities for the investigation of sample inhomogeneities on the order of a micrometer. The first test measurements gave an intensity behind the interferometer or small-angle scattering camera of about 7000 cm^{-2}s^{-1} (Kroupa et al. 2000). A photo of the central part of the ILL interferometer setup is shown in Fig. 2.12.

2.2.2 The MURR Setup

There were two neutron interferometry stations at the 10-MW MURR reactor in Columbia, Missouri, in the time frame 1975–2000. We give a brief description here of the beam port B setup, which is shown in Fig. 2.13. A thermal neutron beam was brought out of the reactor through a helium-filled beam tube (15 cm diameter at the source end) and monochromated by a double-crystal monochromator assembly. The distance from the reactor core to the first monochromator was about 4 m. The double-crystal monochromator assembly used a 10-cm-high flat copper (220) crystal at the first position and a 7-element, 10-cm-high, focusing copper crystal assembly at the second position. The incident beam direction was fixed along the local north–south axis of the Earth, so that the nominally monochromatic beam ($\Delta\lambda/\lambda \approx 0.5\%$) incident upon the interferometer was also directed along the north–south direction. Adjusting the Bragg angles of double-crystal assembly allowed a variable wavelength beam to be used. The interferometer crystal is ordinarily mounted in a V-shaped aluminum cradle, and held semi-rigidly to it with double-sticky-back tape and soft felt strips. The outgoing interfering beams were detected with 1.2-cm-diameter, 20-atm ^{3}He gas-filled proportional detectors which were positioned close to the interferometer crystal and inside an aluminum box about 25 cm on a side. This aluminum box provided an isothermal enclosure for the interferometer. It was bolted to a platform which could be rotated about the incident beam direction, a degree of freedom important for the gravitationally induced quantum interference experiments. This assembly was mounted inside of a heavy Benelex-70 box, which in turn rested upon a vibration isolation pad consisting of four Firestone pneumatic "tires." A plexiglass enclosure surrounded this Benelex-70 box. This triple-box strategy provides the necessary vibration, microphonic, and thermal isolation of the neutron interferometer from its environment in the reactor hall. Only a small fraction, less than 1/1000, of the neutrons incident upon the interferometer satisfy the perfect crystal Si (220) Bragg condition. The rocking curve width of the interferometer was about $\frac{1}{2}^\circ$, corresponding to the mosaic spreads of the copper monochromator crystals. This arrangement has the advantage of eliminating the alignment and stability

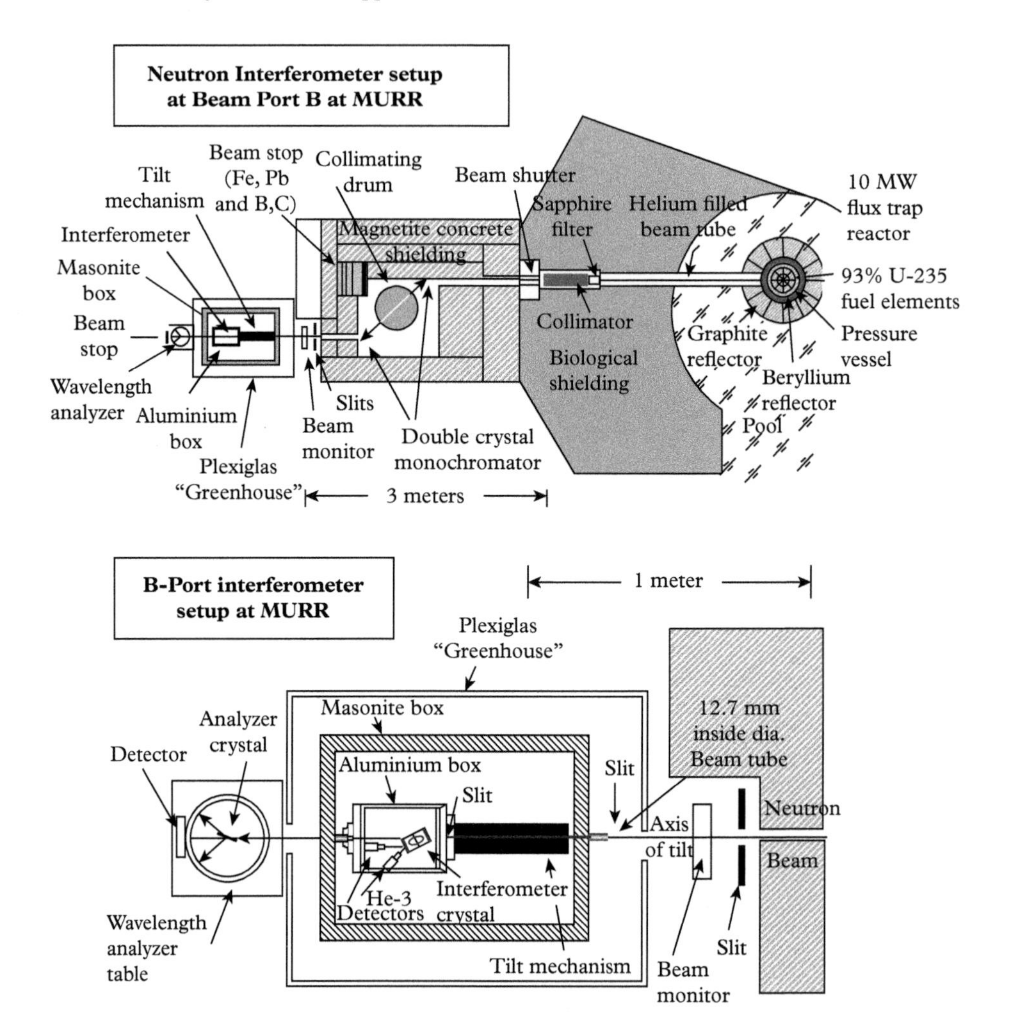

Figure 2.13 *Schematic diagrams of interferometer setups at the beam part B at MURR*

difficulties of the ILL instrument. It has the disadvantage of allowing many neutrons not utilized in the interferometer to be in the region of the detectors, thus providing a potential source of excess background. However, since the detectors are small and well shielded this has proven not to be a serious drawback. It is clear that the technique to produce polarized neutrons with a magnetic wedge cannot be used here with such an angularly dispersive incident beam. The counting rate in the interfering H-beam was of order 1500 counts/min with a 1×10 mm entrant slit. The background counting rate was typically a factor of 100 less than this signal. A more detailed description can be found in an article by Allman et al. (1998).

2.2.3 The NIST Setup

The experience gained with the ILL and MURR has been used for the design and construction of a new instrument at the 20-MW NBSR reactor at NIST in Gaithersburg, Maryland. A schematic

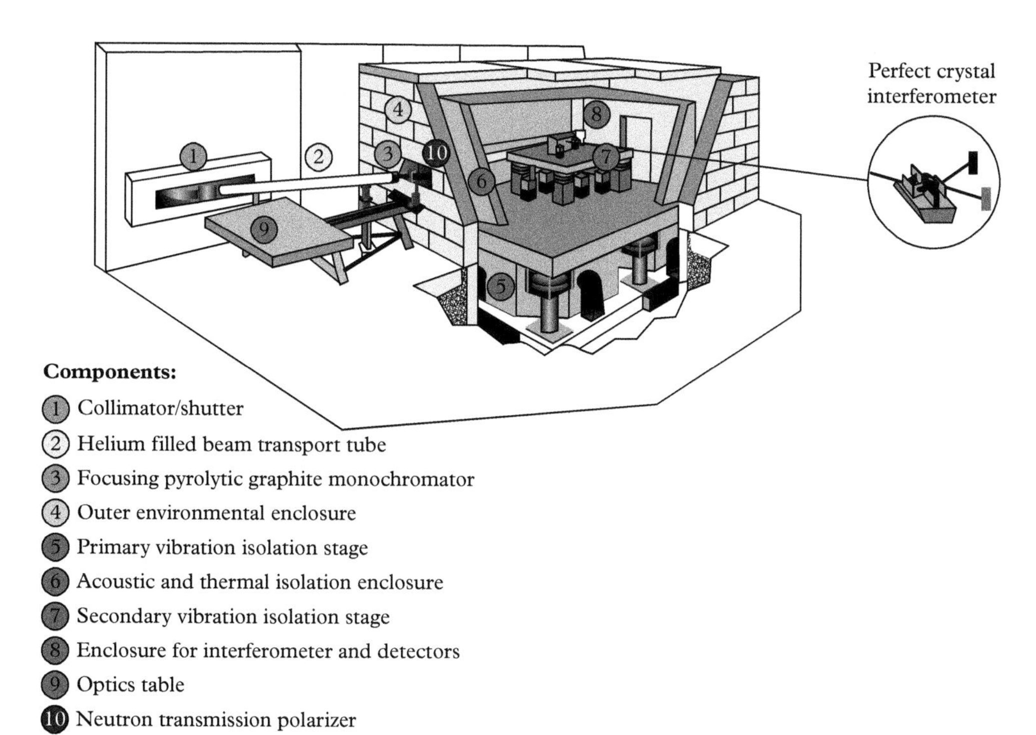

Figure 2.14 *Schematic diagram of the NIST interferometer setup*

diagram of this setup is shown in Fig. 2.14. It utilizes a combination of the experimental strategies of the ILL and MURR instruments. The instrument is positioned in the guide hall on guide NG7. The significant advance in technology involves the use of a three-stage vibration isolation system (see Arif et al. 1997). The lower stage was designed into the building during the planning stage of the guide hall construction. This level of sophistication in the vibration isolation is anticipated to be necessary for large-scale (1-m) separated component interferometers of the future. A pyrolytic graphite double monochromator with a vertical focusing option provides the possibility of directing neutrons of different energies onto the interferometer without changing the position of the interferometer table. Figure 2.5 shows a typical interferogram with a high contrast taken with a 2×8 mm entrance slit.

Early in 2013 a second interferometer setup was installed at NIST using a fixed angle pyrolytic graphite monochromator and 4.5-Å neutrons from the same cold neutron beam guide as the original setup.

2.3 Interferometers Based upon Cold and Ultra-Cold Neutrons

When the neutron wavelength becomes much larger than the interatomic spacings in crystals, Bragg diffraction no longer exists, giving rise to a sharp Bragg cutoff. However, such long wavelength neutrons can be diffracted from macroscopically structured materials like slits, gratings, multilayers, or artificially produced lattices.

The first attempt to make a single-slit interferometer was made by Maier-Leibnitz and Springer (1962) and later on further developed at the ILL in Grenoble by Gähler et al. (1980). Single-slit diffraction and biprism deflection have been combined to form a wave-front division interferometer (Fig. 2.1). Single- and double-slit diffraction patterns are analogous to the related light and electron diffraction pattern (Zeilinger et al. 1981, 1988). The predicted single-slit diffraction pattern, in the Fraunhofer (plane wave) limit (e.g., Born and Wolf 1975, Cowley 1981, Sears 1989, Kaiser and Rauch 1999), can be written as a differential scattering cross-section, namely

$$\frac{d\sigma}{d\Omega} = \left| D \frac{\sin\left(\frac{\pi D}{\lambda} \sin \Theta\right)}{\frac{\pi D}{\lambda} \sin \Theta} \right|^2, \tag{2.8}$$

where D denotes the slit width and Θ the deflection angle. Rather narrow slits (<100 μm) and a high angular resolution must be used to separate the diffraction peaks from the much more intense zero-order peak. According to the wavelength used (3–100 Å), the deflection angles are rather small except for ultra-cold neutrons where unfortunately the source intensity becomes rather small. The intensity of the diffraction peaks strongly decreases with increasing diffraction order which makes this method less attractive for interferometer applications.

Double-slit systems with a slit separation T show more pronounced diffraction peaks (see Section 1.3, Fig. 1.5, left) which can be understood from the well-known plane wave diffraction formula. It is the diffraction pattern of a single slit modulated by the slit distance function, namely

$$\frac{d\sigma}{d\Omega} = \left| 2D \frac{\sin\left(\frac{\pi D}{\lambda} \sin \Theta\right)}{\frac{\pi D}{\lambda} \sin \Theta} \cos\left(\frac{\pi T}{\lambda} \sin \Theta\right) \right|^2. \tag{2.9}$$

The constraints due to rather narrow slits and high collimation reduce the intensity considerably. To compensate for the low intensity a rather broad wavelength band $\delta\lambda$ must be used which limits the visibility of the interference pattern to rather low orders (see Section 4.2). A double-slit interferometer has been used by Klein et al. (1976, 1981a) for the observation of the sign change of the neutron spinor wave function in the case of a 2π-precession (Section 5.1) and of the neutron Fizeau effect (Section 8.5).

A more effective use of the available neutron intensity becomes possible if gratings are used as optical components. The coherent diffraction of neutrons from various gratings has been observed by Kurz and Rauch (1969), Graf et al. (1979), and Scheckenhofer and Steyerl (1977). In the case of surface refraction the deflection angles follow from the relation

$$\begin{aligned} \cos\Theta_m - \cos\Theta_0 &= \frac{m\lambda}{a} \\ \cos\xi_m - \cos\Theta_0 &= \frac{m\lambda}{a} \end{aligned}, \tag{2.10}$$

where a is the grating constant, n the index of refraction of the material (Eq. 1.26), and $m = 0$, ± 1, ± 2, etc. See Fig. 2.15a. The intensity depends on the ratio of the reflecting (b) and the

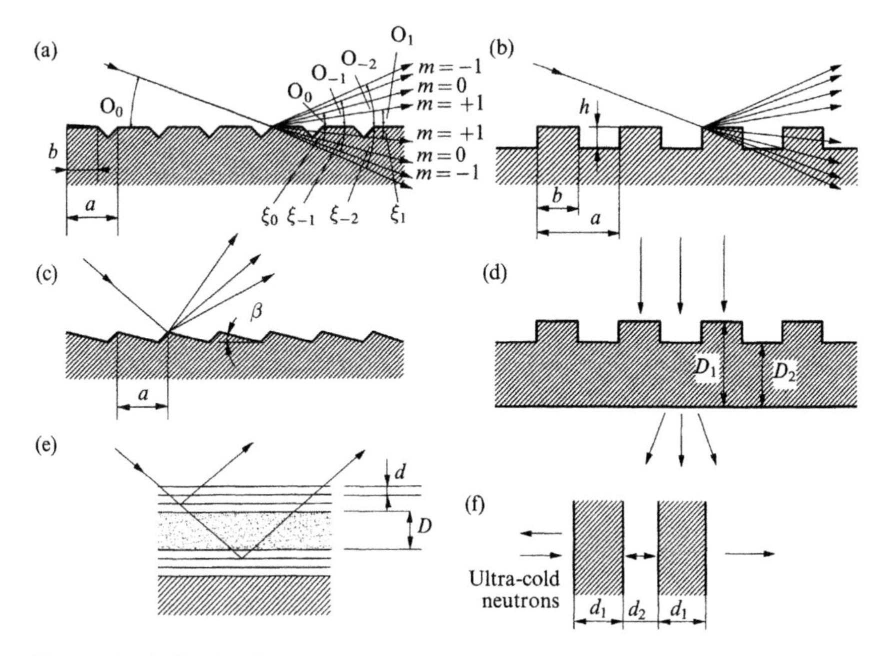

Figure 2.15 *Sketch of various diffraction gratings used for neutron interferometry, (a, b) ruled gratings, (c) blazed grating, (d) phase grating, (e) layered structure grating, (f) Fizeau-type resonator*

non-reflecting (a, b) parts of the grating (Sprague et al. 1955). When the angle of incidence Θ_0 is smaller than the critical angle Θ_c (Eq. 1.27) one gets

$$R_m = \left[\frac{\sin\left(\pi m \dfrac{a-b}{a}\right)}{\pi m}\right]^2 \left[\frac{\sin\Theta_0 - n\sin\xi_m}{\sin\Theta_m + n\sin\xi_m}\right]^2. \tag{2.11}$$

For reasonable parameters, the reflectivities for $|m| > 0$ diffraction intensities are on the order of 1% compared to the total reflection part ($m = 0$). The theoretical values of the reflectivity have been approached for the symmetric grating ($b = a/2$) by Ioffe et al. (1981). The intensity can be more concentrated into certain diffraction orders when blazed (echelette) gratings are used (Fig. 2.15c; Wood 1910, 1912; Steyerl et al. 1988). For ultra-cold neutrons and a grating blazed for first-order diffraction, reflectivities of 14 and 8% for first- and zero-order diffraction are obtained.

Ioffe et al. (1985) succeeded in the construction of a ruled grating interferometer (see Fig. 2.1). They used flat step diffraction gratings for the beam splitter and combiner with a lattice constant of 21 μm and a profile depth of 0.1 μm and total reflecting mirrors. For 3.15 Å neutrons a beam separation of about 1 mm has been achieved. The interference in that interferometer is produced by the superposition of a twice diffracted beam with diffraction orders $m = 1$ and $m = -1$. The observed contrast was 23%.

Phase gratings are other effective optical components for long wavelength interferometers (Fig. 2.15d). Such gratings are generally used in the transmission geometry, which permits a larger beam cross-section to be used. The optimal step heights are determined by calculating the phase shift difference between the thicker and thinner parts of the grating (see Eq. 2.6). In most cases phase gratings for phase shifts of multiples of π and $\pi/2$ are used, which then determines the step heights

$$\Delta h = h(\pi) - h(\pi/2) = \frac{\pi}{N b_c \lambda}. \tag{2.12}$$

For the plane wave case (Fraunhofer limit) the diffraction cross-section formula is (e.g., Eder et al. 1991)

$$\frac{d\sigma}{d\Omega} = |Na\, f(\Theta)\; F(\Theta)|^2, \tag{2.13}$$

where N is the number of slits and $f(\Theta)$ the ideal grating diffraction function which becomes for $N \rightarrow \infty$

$$f(\Theta) = \delta(\Theta - m\lambda/a). \tag{2.14}$$

$|F(\Theta)|^2$ represents the grating form factor which is the Fourier transform of the (phase) transmission function. The zero-order diffraction vanishes for $h = h(\pi)$ and equal widths of the steps and the grooves, where the first-order contribution is maximized (40%). For an interferometer application, the first-order diffraction peaks are needed as a beam splitter, beam recombiner, and a beam deflector. A successful test of such an interferometer was achieved at the ultra-cold neutron facility at the ILL reactor in Grenoble (Gruber et al. 1989). The lattice constant of the microstructured grating on SiO_2 was $a = 2$ μm, and the step heights were 0. 75 and 0.35 μm, respectively. For a flight path of 50 cm and very long wavelength neutrons (102 Å) a beam separation of about 1.2 cm was achieved. The critical adjustment of all lattices to within one lattice constant was done on a vibration-isolated optical bench using a parallel laser interferometer. A sketch of the experimental arrangement and typical results is shown in Fig. 2.16. The contrast amounts to about 53% and the total measurement time was about 7 h at the high flux reactor at Grenoble (Weber and Zeilinger 1997, van der Zouw et al. 2000). Work is now going on to extend such interferometers to path lengths of 4 m, which makes such interferometers very sensitive to many small interaction effects. Various aberration-free arrangements are reported in the literature (Ioffe 1986).

When such interferometers—perfect crystal interferometers as well—are moved perpendicularly to the interferometer axis with velocity v_m, the neutron velocity in the moving frame changes to $v' = v + v_{mot}$. This introduces a phase shift

$$\Delta\chi_{mot} = (2\pi L/d).\tan(v/v_{mot}). \tag{2.15}$$

This provides the basis for interference in velocity space and for new types of high-resolution neutron spectrometers (Ioffe 1997).

Another interferometer for long wavelength neutrons is based upon multilayer diffraction and has been tested by a Japanese group for 12.6-Å neutrons (Funahashi et al. 1996). It has features similar to those of the Jamin interferometer in classical optics (Born and Wolf 1975) and consists of two pairs of multilayer mirrors where each pair consists of two multilayer mirrors which are separated by a thicker intermediate monolayer (Figs. 2.15e and 2.17). The first pair splits the incident beam into two coherent parts whose phase difference is

$$\Delta\chi = 2\pi D/d, \tag{2.16}$$

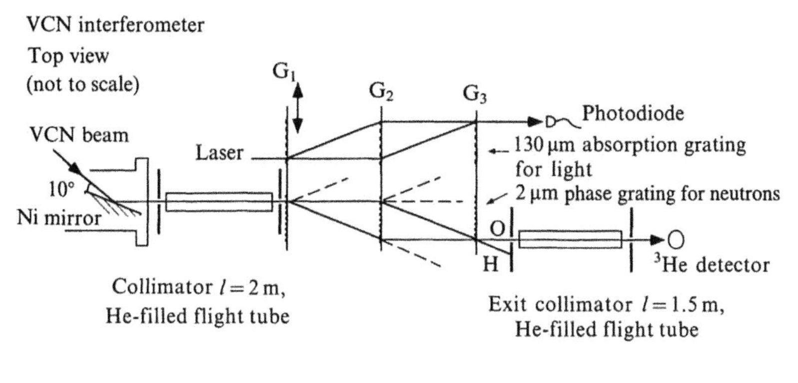

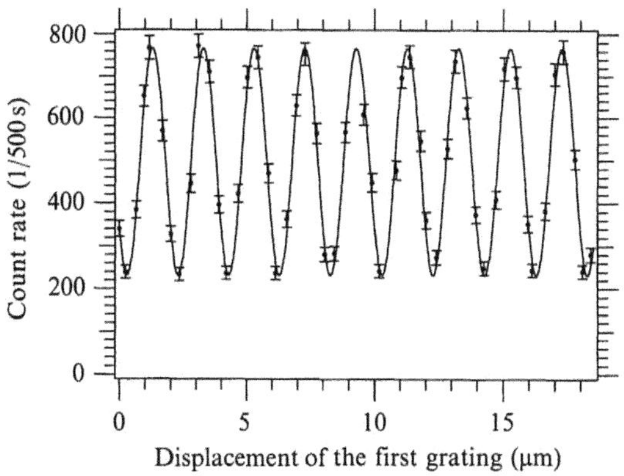

Figure 2.16 *Sketch and typical results of the ILL phase grating interferometers. Courtesy of M. Weber and A. Zeilinger, Innsbruck (1997)*

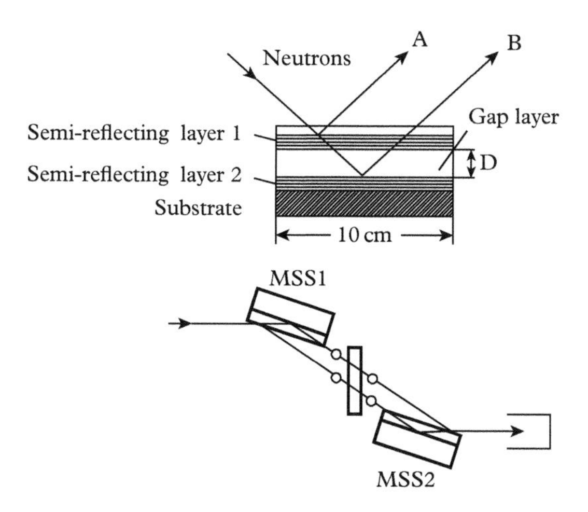

Figure 2.17 *Principle of the structured multilayer Jamin-type interferometer. Reprinted with permission from Funahashi et al. 1996, copyright 1996 by the American Physical Society.*

where D is the effective thickness of the monolayer and d the lattice constant of the multilayer. It is assumed that the monolayer has an index of refraction $n \sim 1$. Germanium–titanium bi-layers with a lattice constant of d = 360 Å and an intermediate germanium monolayer with an effective thickness D = 1 μm have been used.

Fringes of the Brewster type in close analogy to light optics were observed when the second pair of multilayer mirrors was rotated. The area enclosed by the two coherent beams is rather small, which reduces the sensitivity in many cases, but the intensity situation is rather reasonable. The intensity calculation of multilayer diffraction is generally based upon a matrix approach and dynamical diffraction theory (e.g., Stepanov et al. 1995). Resonance-enhanced standing neutron waves have been used to observe the simultaneously emitted capture gamma radiation from an absorbing Gd-157 layer (Zhang et al. 1994). A combination of two structured multilayers was used to build a Mach–Zehnder-type interferometer based upon multilayer mirror diffraction (Fig. 2.16; Ebisawa et al. 1994, 1998; Funahashi et al. 1996). A unit for its own—two multilayers and a gap layer between—turned out to be an effective mini-interferometer with nearly no beam path separation but a considerable phase shift between the interfering beams. More details are discussed in connection with Larmor interferometers (Section 2.4, Eq. 2.37). Such Fabry–Perot systems are the subject of a related review by Maaza and Hamidi (2012).

Other artificially produced lattices which have been tested, or which are capable of interferometer applications, are based upon light-induced lattices in polymers and on flux line lattices in superconductors. Light-induced lattices in polymers result from the small density fluctuations ($\Delta\rho/\rho \cong 10^{-4} - 10^{-3}$) caused by a light intensity-dependent polymerization in standing laser light fields (Rupp et al. 1990; Matull et al. 1990, 1991). The reflectivity of a lattice with a sinusoidal density variation can be written as

$$R_1 = A\sin^2(\lambda b_c \, \Delta N \, d/2), \tag{2.17}$$

assuming no correction for extinction effects. A is the usual attenuation factor accounting for absorption and incoherent scattering, d denotes the thickness of the artificial lattice crystal, and ΔN is the maximum density fluctuation. Reflectivity values up to 60% have been reported (Pruner et al. 2006). Such holographic lattices have been put together to form a Mach–Zehnder interferometer for 15 Å neutrons (Schellhorn et al. 1997, Pruner et al. 2006). Fringe visibilities up to 20% have been achieved with a LLL-geometry interferometer having a lattice constant of $a = 7980$ Å, a slab thickness of 2.5 mm, and a slab separation $L = 30$ mm (Fig. 2.18). Due to polymerization such lattices can be fixed inside various polymers and used as a static density variation phase lattice. The separation of the interfering beams (0- and H-beam) from the parasitic beams is often not possible due to the rather small Bragg angles involved. Therefore, the measured intensity oscillations belong to the composed R- and S-beams where the contrast is reduced due to these parasitic beams. This shows a general difficulty of interferometry based on artificially produced lattices when they are not used for very long wavelength neutrons. The separation of the interfering beams from the parasitic ones requires the use of rather narrow slits.

For cold neutrons the newly developed optical beam splitter based on holographically structured nanoparticle polymer composites may become powerful new elements in neutron optics since reflectivities up to 50% for rather thin splitters (~200 μm) have been reported (Fally et al. 2010, Klepp et al. 2012). This technique can be used for cold neutron interferometers.

Another alternative for long wavelength interferometry may be provided by the flux line lattice of superconductors in the mixed state. In this case the lattice constant can be varied by an external magnetic field as (Cribier et al. 1964, Weber et al. 1973)

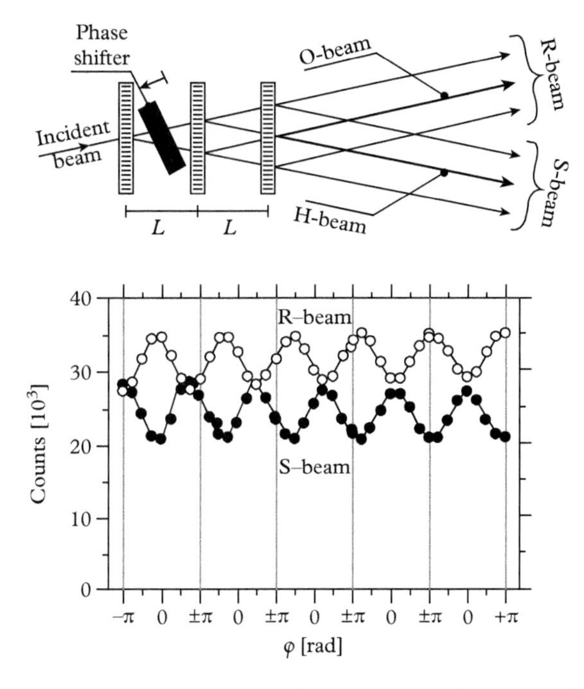

Figure 2.18 *Sketch and typical results of a light induced polymer neutron interferometer. Reprinted from Schellhorn et al. 1997, copyright 1997, with permission from Elsevier.*

$$a = \sqrt{\frac{2\phi_0}{\sqrt{3}\,B_0}}, \tag{2.18}$$

where ϕ_0 denotes the elementary flux quantum ($\phi_0 = h/2|e|$). Very regular flux line lattices have been observed, which provides at least the possibility of use in neutron interferometry.

For ultra-cold neutrons the optical potential (Eq. 1.24) becomes comparable to the kinetic energy of the neutrons ($\leq$ 100 neV) such that they are totally reflected at all angles of incidence. Thus, semi-transparent barriers and double and multiple barrier resonators can be built in close analogy to the light optical Fabry–Perot interferometers (Fig. 2.15f; Fabry and Perot 1899, Born and Wolf 1975). The reflectivity R_S and the transmittance T_S of such a system is described by the Airy relation

$$R_S = 1 - T_S - \frac{T^2}{T^2 + 4R\sin^2(kd_2 + \varphi_0)}. \tag{2.19}$$

T is the transmission coefficient of a single barrier of height V and width d_1 given by the standard formula

$$T = \left\{1 + \frac{V^2}{4E(V-E)}\,\sinh \kappa d_1\right\}^{-1}, \tag{2.20}$$

where κ is given by $\kappa^2 = 2m(V - E)/\hbar^2$. Resonance transmission of the double-well barrier occurs if the incident wave vector fulfills the relation

$$k = k_m = \frac{m\pi - \varphi_0}{d_2}. \tag{2.21}$$

This corresponds to a strongly enhanced wave field between the potential wells. This can be understood as a size effect resonance where there is a multiple (N) back-and-forth reflection of neutrons between the potential wells, where

$$N \cong \frac{d_1 + d_2}{d_2}\frac{\sqrt{R}}{T}. \tag{2.22}$$

Related experiments for Cu–Al–Cu barriers with $d_1 = 240$ Å and $d_2 = 860$ Å and another one with $d_1 = 180$ Å and $d_2 = 1670$ Å have verified the expected behavior for $m = 1$ and $m = 2$ resonators (Steinhauser et al. 1980). Multiple barriers which act as coupled resonators (Cu–Al–Cu–Al–Cu) have also been investigated (Steyerl et al. 1981). The inherent problem of the rather low intensity of ultra-cold neutrons limits the application of this technique. Maaza et al. (1996) measured the trapping time of neutrons in a Ni–V–Ni Fabry–Perot thin film resonator in grazing angle neutron reflectometry.

2.4 Larmor and Ramsey Interferometers

Beam polarization and its rotation within a region of a magnetic field $\boldsymbol{B}$ result from the interference of the spin-up and the spin-down states. We will see that the classical picture of Larmor precession of a magnetic dipole moment vector of a point-like particle around a magnetic field is inadequate to account simultaneously for the Stern–Gerlach effect and for the coupling of kinematical and spin variables (Mezei 1980b, 1988). Here, we deal with interferences in the forward direction without spatial beam splitting, but splitting in momentum space. All characteristic interference effects will be seen to occur. For the calculation, the Schrödinger equation (1.2) must be used in its two-component form (Pauli equation)

$$\mathcal{H}\Psi = \left[-\frac{\hbar^2}{2m}\nabla^2 - \mu\boldsymbol{\sigma}\cdot\boldsymbol{B}(\boldsymbol{r},t)\right]\Psi(\boldsymbol{r},t) = i\hbar\frac{\partial\Psi(\boldsymbol{r},t)}{\partial t} \tag{2.23}$$

with

$$\Psi = \begin{pmatrix}\Psi_+\\ \Psi_-\end{pmatrix} = \alpha|+> + \beta|->= f_+(\boldsymbol{r},t)\cos\frac{\Theta}{2}|+> + f_-(\boldsymbol{r},t)e^{i\phi}\sin\frac{\Theta}{2}|->.$$

Here (Θ, ϕ) are the polar angles of the spin vector of a point-like particle and the axis of quantization is chosen along the direction of the magnetic field. The vector $\boldsymbol{\sigma} = (\sigma_x, \sigma_y, \sigma_z)$ is composed of the 2×2 Pauli spin matrices (see Eqs. 3.44–3.46) which couple all components of the spin to the magnetic field $\boldsymbol{B}$. $f_\pm$ are the space-time-dependent wave functions.

The beam polarization is defined as

$$\boldsymbol{P} = \mathrm{Tr}(\rho\boldsymbol{\sigma}) = <\psi|\boldsymbol{\sigma}|\psi>, \tag{2.24}$$

where ρ denotes the density matrix. The degree of polarization is usually denoted by $|\boldsymbol{P}| \le 1$.

The equation of motion of an observable A is (e.g., Messiah 1965)

$$\frac{\mathrm{d}}{\mathrm{d}t} < \underline{A} > = \left\langle \frac{\partial A}{\partial t} \right\rangle + \frac{1}{\hbar} \langle [\mathcal{H}, A] \rangle. \tag{2.25}$$

For a stationary magnetic field the Hamiltonian $\mathcal{H}$ is time independent ($\partial \underline{\mathcal{H}}/\partial t = 0$), and the total energy is conserved, but the kinetic energy and potential energy may be spatially dependent (Zeeman effect). In the case of a purely time-dependent magnetic field ($\partial \underline{\mathcal{H}}/\partial t \neq 0$) the total energy changes but the kinetic energy remains unchanged if the field is spatially uniform ($[\underline{\nabla}, \underline{\mathcal{H}}] = 0$). The motion of the spin components in a static field follows from the Heisenberg equation

$$\frac{\mathrm{d}}{\mathrm{d}t} <\boldsymbol{\sigma}> = \frac{\mu}{\mathrm{i}\hbar} \langle [\boldsymbol{\sigma}, \boldsymbol{\sigma}.\boldsymbol{B}] \rangle. \tag{2.26}$$

This gives the well-known Bloch equation for the neutron Larmor precession according to the commutation relation of the Pauli spin matrices (Eq. 3.46) (Halpern and Holstein 1941, Williams 1988); that is,

$$\frac{\mathrm{d}}{\mathrm{d}t} <\boldsymbol{\sigma}> = \frac{2\mu}{\hbar} \langle \boldsymbol{\sigma} \times \boldsymbol{B} \rangle = \gamma \langle \boldsymbol{\sigma} \times \boldsymbol{B} \rangle = \gamma (\boldsymbol{P} \times \boldsymbol{B}). \tag{2.27}$$

This is formally the equation of motion of a classical magnetic dipole moment inside a magnetic field region. In this case the polarization vector $\boldsymbol{P}$ precesses about the magnetic field $\boldsymbol{B}$ with the Larmor frequency ($\omega_{\mathrm{L}} = |2\mu B/\hbar|$) as shown in Fig. 2.19. When a neutron traveling in the y-direction enters a region of magnetic field $\boldsymbol{B} = B\hat{z}$, of length L with polarization $\boldsymbol{P}$ initially along $\hat{x}$, we have when it leaves the magnetic field

$$\boldsymbol{P} = \begin{pmatrix} \cos \dfrac{2\mu B}{\hbar} \dfrac{L}{v} \\ \sin \dfrac{2\mu B}{\hbar} \dfrac{L}{v} \\ 0 \end{pmatrix}. \tag{2.28}$$

This shows that a complete spin-reversal occurs when $2|\mu|BL/\hbar v = \pi$. This idea is used routinely for polarization (spin) rotations in DC-spin flipper coils (Mezei 1972, Badurek et al. 1974, Williams 1988).

For neutrons having an initial, arbitrary polarization $\mathbf{P}_0$, the total energy is conserved, but the kinetic energy is split into two parts (Zeeman splitting)

$$\frac{\hbar^2 k^2}{2m} = \frac{\hbar^2 k_{\pm}^2}{2m} \pm |\mu| B, \tag{2.29}$$

which shows the coupling of the momentum and spin variables. For motion along the y-axis, this then causes a small shift of the wave amplitudes and creates a relative phase shift between the spin-up and spin-down wave packets whose wave functions become

$$f_{\pm}(y, t) = \frac{1}{(2\pi)^{1/2}} \int a_{\pm}(k \mp \Delta k) e^{\mathrm{i}[(k \mp \Delta k) y - \omega_k t]} \mathrm{d}k, \tag{2.30}$$

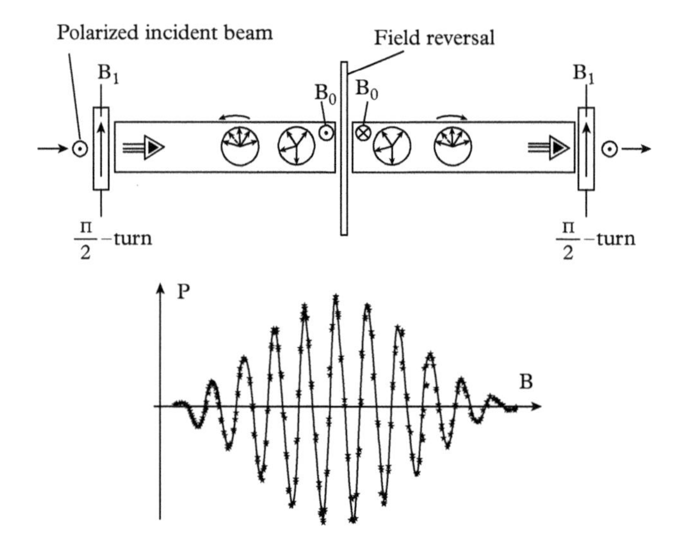

Figure 2.19 *Sketch of the spin-echo spectrometer (interferometer) and the retrieval of the polarization behind the second precession field. Reprinted from Mezei 1972, 1988, with permission from Elsevier.*

where

$$\Delta k \cong m|\mu|B/\hbar^2 k\,.$$

One notices that the whole wave function in Eq. (2.23) describes a momentum-spin entangled quantum state, which cannot be factorized into a product state. In terms of quantum optics it can be written as

$$\Psi = |\alpha(k_+)>|\uparrow> + |\beta(k_-)>|\downarrow>\,. \tag{2.31}$$

This entanglement will be discussed and used in more detail in Chapters 5 and 7.

In most cases the Zeeman shift Δk of the wave vector is much smaller than the neutron momentum widths δk_i of the momentum distribution functions $g_\pm(k) = |a_\pm(k)|^2$ and, therefore, $|a_\pm(k \pm \Delta k/2)| \cong |a_\pm(k)|$. In this case one gets from Eq. (2.24) the components of the polarization vector $\boldsymbol{P}$ after the neutron has traversed a distance y:

$$P_x(y) = \int g(k)\sin\Theta(k)\cos\phi(k,y)\,\mathrm{d}\,k$$

$$P_y(y) = \int g(k)\sin\Theta(k)\sin\phi(k,y)\,\mathrm{d}\,k \tag{2.32}$$

$$P_z(y) = \int g(k)\cos\Theta(k)\,\mathrm{d}\,k,$$

where $g(k) = g_+(k) + g_-(k)$. Here, the momentum distribution functions $g_+(k)$ and $g_-(k)$ determine the fixed polar angle Θ, namely that

$$\cos\Theta(k) = [|g_+(k)|^2 - |g_-(k)|^2]/|g(k)|^2$$

$$\sin\Theta(k) = 2\,|a_+(k)||a_-(k)|\,/g(k). \tag{2.33}$$

The accumulated phase difference between the spin-up and spin-down states is

$$\phi(k,y) = 2y\Delta k, \tag{2.34}$$

where Δk is given by Eq. (2.30). This corresponds to a precession of the polarization vector around the field axis (z) where the angle of rotation is given by (Larmor angle)

$$\phi(y,k) = \phi(y,\lambda) = \gamma\int_0^y B(y')dy'/v = \gamma B\, m\,\lambda\, y/h, \tag{2.35}$$

where the gyromagnetic ratio $\gamma = 2\mu/\hbar$ is used. When the neutron polarization at the entrance into the field lies in the x-direction (i.e., perpendicular to the field) the degree of polarization in the x-direction behind a field region of length L becomes reduced by (Mezei 1972, 1980; Gaehler et al. 1996)

$$P_x(B,L) = \int \tilde{g}(\lambda)\cos(\gamma B\, L\, m\,\lambda/h)\mathrm{d}\,\lambda, \tag{2.36}$$

where the spectral distribution function for wavelength follows from the relation $\tilde{g}(\lambda)\mathrm{d}\lambda = g(k)dk$.

This loss of polarization at large BL values is caused by the dispersive action of the Larmor rotation, and causes the polarization vectors for different wavelengths to point in different directions within the (x,y)-plane, which results in a vanishing overall polarization of the beam. The polarization directions are correlated to the related wavelengths and full polarization can still be restored (Fig. 2.19). This situation is analogous to a spatially split beam interferometer experiment when a magnetic field along one beam path is varied (see Section 3.2.1). The contrast, i.e., the coherence function (Section 5.1), of the interference pattern appearing in an interferometer experiment shows the same behavior even though these measurements are made with unpolarized neutrons.

The revival of full contrast can be achieved when a second magnetic field is applied in the opposite direction, or when all individual spins are inverted between two fields in the same direction. These ideas lead to the well-known spin-echo arrangements (Fig. 2.19; Mezei 1972, 1980), which are analogous to phase-echo arrangements discussed in Section 4.2.4. In this case BL in Eq. (2.36) then denotes the difference between the two integrated field values ($BL = B_1L_1 - B_2L_2$) and the polarization revival matching point occurs when $B_1L_1 = B_2L_2$. This preserves the high sensitivity of the Larmor precession method and reduces considerably the polarization reduction effect caused by the wavelength spread of the beam. The fine-tuning of the echo condition can be achieved by changing the optical path length due to a material inserted in one part of the spin-echo arrangement (Eq. 2.6; Hino et al. 1995). This fine-tuning of spin-echo spectrometers can be used to measure the tunneling time of neutrons through thin magnetic films by using the Larmor precession of the neutrons as a clock (Achiwa et al. 1996, Hino et al. 1998).

It follows from Eq. (2.30) that the spin-up and the spin-down wave packet components move inside the precession field with slightly different velocities, thus separating the components by $\Delta x = 2|\mu|BL/mv^2$ (i.e., Schrödinger cat-like states are created—Section 4.5.2). They arrive at the same position behind a precession field of length L with a time delay

$$\Delta t = \frac{2|\mu|BL}{mv^3}, \tag{2.37}$$

which defines the spin-echo time, a characteristic time scale (~10–100 ns) where dynamical variations of the system can be measured (Gaehler et al. 1996). In all these cases, the coupling of the spin space (Larmor precession) and the momentum space (Zeeman splitting) becomes obvious. This time delay also provides the basis for Larmor clocks with time resolutions on the order of 10^{-10} s, as tested by Frank et al. (2001).

Equation (2.36) shows clearly that the polarization measured in a certain direction is the Fourier transform of the wavelength distribution function. This is shown in Fig. 2.19 for a beam with a rather broad wavelength distribution ($\Delta\lambda/\lambda_0 \cong 18\%$). When the energy of the neutrons is changed between the two precession fields, due to inelastic scattering processes within a sample, the spin-echo method provides a practicable way of achieving high sensitive neutron Fourier spectroscopy (Mezei 1980). Instruments based upon this idea are installed at many advanced neutron sources. High resolution is achieved at high Larmor precession orders which require rather strong and very homogeneous precession fields ($\Delta B/B_0 \cong 10^{-5}$).

The spin-echo systems described above cause a longitudinal shift of the wave packets, which results in the momentum-spin entangled quantum state given by Eq. (2.31). An alternative method where the two packets become shifted normal to the beam direction has been proposed by Rekveldt (1996) and realized by Bouwman et al. (2004). They use tilted magnetic fields produced by DC coils or magnetized foils (Fig. 2.20). In the upper case a longitudinal separation and a transverse

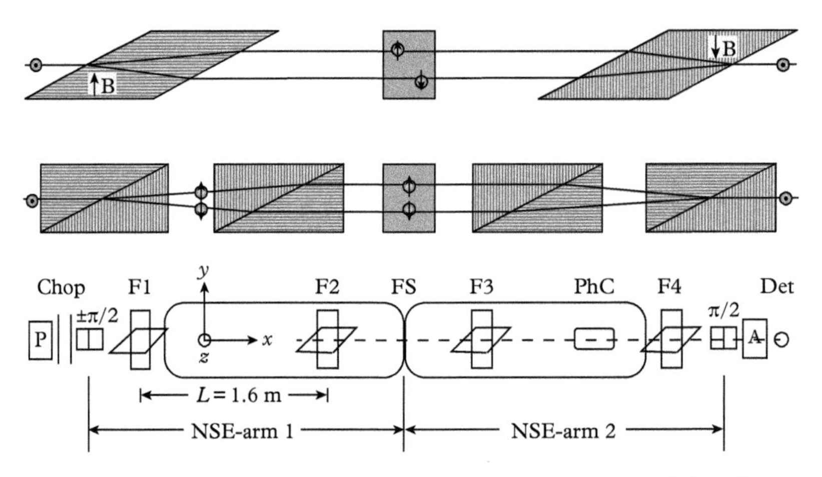

Figure 2.20 *Spin-echo arrangement for a transverse separation of the coherent states. In both setups (above and below) the polarization of the incident neutrons is perpendicular to the magnetic field direction. In the lower case a spin flipper is installed along the diagonal lines. Reprinted from Bouwman et al. 2008, copyright 2008, with permission from Elsevier.*

separation occur. The transverse separation Δz is determined by the deflection angles at the entrance into the tilted fields having a tilt angle Θ, a length (L), and a magnetic field strength (B), namely

$$\Delta z = \frac{2\mu m\lambda^2 BL}{h^2}\tan\Theta. \tag{2.38}$$

When each unit consists of oppositely oriented magnetic fields as shown in Fig. 2.20 the separation is $\Delta z' = \Delta z/\cos\Theta$. This can be varied between 0.1 and 20 μm. The magnetic fields in the first and second units point in the same direction and those in the third and fourth units in the opposite direction. Along the diagonal line of each unit a DC-spin flipper changes the direction of spin rotation. The transverse spatial separation given in Eq. (2.38) plays the same role as the spin-echo time in the standard spin-echo system. Such systems are used for small-angle scattering experiments where direct information about the spatial structure of a sample in the micrometer range can be obtained (Gähler et al. 1996, Bouwman et al. 2004). Figure 2.21 shows a photo of such an arrangement. A detailed analysis including a sensitivity estimation based on dynamical diffraction theory has been given by Rana Ashkar et al. (2010, 2011).

A low field version of a spin-echo arrangement has been developed and used as a phase spin-echo interferometer (Ebisawa et al. 1996, 1998). In this case, the system satisfies the phase-echo (discussed in Section 4.2.4) and spin-echo condition simultaneously. It is based upon a Jamin-type interferometer (Fig. 2.17; Funahashi et al. 1996) and a spin-echo system. By this method high-resolution and rather compact spin-echo spectrometers become feasible. In this case, a magnetic multilayer mirror on top, followed by a gap layer and a nonmagnetic layer mirror, was used. The top magnetic mirror reflects one spin eigenstate while the non-magnetic mirror reflects the other eigenstate. The presence of the gap layer, however, creates a path length difference for the two eigenstates, which results in a phase shift and, therefore, a quantum precession of the neutron spin. The related phase shift can be calculated from the geometric situation shown in Fig. 2.15e (Ebisawa et al. 1998):

$$\phi = \frac{2\pi D}{d}\left[1 - \frac{Nb_c}{\pi}\left(\frac{\lambda}{\sin\Theta}\right)^2\right]^{1/2}, \tag{2.39}$$

Figure 2.21 *Photo of a spin-echo spectrometer at the research reactor in Delft, Netherlands. Courtesy W. G. Bouwman*

where Θ is the incident angle of the neutrons; D, N, and b_c are the thickness, density, and coherent scattering length of the gap layer, respectively; and d is the lattice constant of the magnetic mirrors. For typical values (d = 100 Å, D = 1 μm Ge) one obtains phase shifts on the order of 600 rad. The equivalence of the related quantum precession with Larmor precession has been demonstrated by measuring spin-echo profiles with small additional magnetic fields. The whole setup is shown in Section 4.2.4 (Fig. 4.11) and in Section 5.5 (Fig. 5.9) in connection with spin-echo and spin-superposition experiments. This equivalence of phase shifts and Larmor rotation also appeared in interferometric measurements of the spin-superposition law (Summhammer et al. 1983).

The broad-band DC-flippers acting as Larmor rotators (Eq. 2.28) can be replaced by narrow band flippers when an alternating meander field of strength B_1 superposed to a guide field B_0 is used (Fig. 2.22). Such a field structure gives rise to a rotating magnetic field in the rest frame

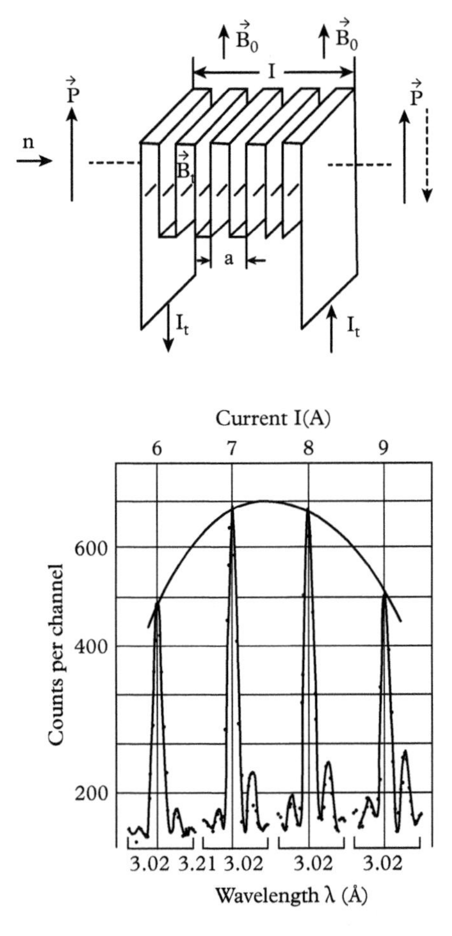

Figure 2.22 *Meander spin flipper with features similar to those of a magnetic wiggler at an electron synchrotron (Drabkin et al. 1988)*

of the moving neutron. This rotating field flips only those neutrons in a rather narrow velocity band given by $\delta v/v_0 \cong 1/2N$, with N the number of meanders. For a complete spin reversal the strength of the meander field must fulfill the amplitude condition $B_1 L = \pi\, h\, v_0/8\mu$ with $L = N{\cdot}a$. The characteristic velocity v_0 is determined from the equivalence of the time-of-flight through one meander and the Larmor precession time, namely $v_0 = 2|\mu|B_0\, a/h$, where a denotes the width of one meander element. This coherent superposition of rotational phases can be used as a neutron monochromator in combination with a polarizer and analyzer mirror, which has the same features as that of a magnetic wiggler (Drabkin et al. 1968, Agamalyan et al. 1988, Badurek et al. 2011), Gösselsberger et al. 2013).

The crucial requirement of strong and homogeneous magnetic fields in spin-echo spectrometers can be circumvented by the so-called zero-field spin-echo method (Gaehler and Golub 1987, 1988), which has several features in common with the well-known double-resonance method developed for molecular beams (Ramsey 1949, 1956). In this case the spin rotation is achieved by a neutron magnetic resonance system (Rabi (1937) flipper). When a rotating field B_1 is added to a region having a static guide field B_0 such that

$$\boldsymbol{B}(t) = \begin{pmatrix} B_1 \cos\omega t \\ B_1 \sin\omega t \\ B_0 \end{pmatrix}, \tag{2.40}$$

the Pauli equation (2.23) can be solved for plane waves (Rabi 1937, Ramsey 1956, Krüger 1980, Golub et al. 1994, Gaehler et al. 1996). For a beam initially polarized in the z-direction the wave function behind the resonance system can be shown to be

$$\psi(y,t) = \frac{1}{(2\pi)^{1/2}} \exp\left[i\left(ky - \frac{\hbar k^2}{2m}t\right)\right] \bullet \begin{pmatrix} e^{-i\varpi t/2} & 0 \\ 0 & e^{i\omega t/2} \end{pmatrix} \bullet \mathrm{M}, \tag{2.41}$$

where the matrix $\mathbf{M}$ is given as

$$\mathbf{M} = \begin{pmatrix} \cos\dfrac{\gamma B_{\mathrm{eff}} t}{2} - i\dfrac{B_0 + \omega/\gamma}{B_{eff}} \sin\dfrac{\gamma B_{\mathrm{eff}} t}{2} & i\dfrac{B_1}{B_{\mathrm{eff}}} \sin\dfrac{\gamma B_{\mathrm{eff}} t}{2} \\ i\dfrac{B_1}{B_{\mathrm{eff}}} \sin\dfrac{\gamma B_{\mathrm{eff}} t}{2} & \cos\dfrac{\gamma B_{\mathrm{eff}} t}{2} + i\dfrac{B_0 + \omega/\gamma}{} B_{\mathrm{eff}} \sin\dfrac{\gamma t}{2B_{eff}} \end{pmatrix}$$

and B_{eff} is

$$B_{\mathrm{eff}} = \sqrt{\left(B_0 + \frac{\omega}{\gamma}\right)^2 + B_1^2}.$$

For a spin which starts at $t = 0$ in the spin-up state $\psi_\uparrow(0,0)$ the probability to reach the spin-down state at time t is given as

$$P_{1,2} = \left|\psi_\downarrow(t)\right|^2 = \frac{(\gamma B_1)^2 \sin^2\left[\frac{\alpha}{2}\sqrt{(\omega_0 - \omega)^2 + (\gamma B_1)^2}\right]}{(\omega_0 - \omega)^2 + (\gamma B_1)^2}, \tag{2.42}$$

where

$$\alpha(t) = \gamma t\, B_{eff}.$$

For strong guide fields ($B >> B_1$) the following conditions for a complete spin-reversal are obtained, at time $t = \tau = \ell/v$, namely that

$$\omega = \omega_r = -\gamma B_0 \tag{2.43}$$

and

$$-\gamma B_1 \tau = (2n+1)\pi.$$

For strong oscillating fields ($B_0 \cong B_1$) the Bloch–Siegert (1940) shift of the resonance conditions needs to be included.

Such a neutron magnetic resonance system (Rabi flipper) represents a time-dependent interaction so that in addition to the spin rotation an energy exchange occurs according to the Zeeman energy $\hbar\omega_r = 2\mu B_2$ (Alefeld et al. 1981b). In a zero-field spin-echo experiment a $\pi/2$ spin rotation is applied at the beginning, a π-rotation at the middle, and an additional $\pi/2$ spin rotation at the end (Fig. 2.23). The phases of the flipper fields are synchronized. Thus, a spin-up $|\uparrow>$ and a spin-down state $|\downarrow>$ with slightly different energies interfere. This results in a Larmor rotation in the (x, y)-plane even in a zero magnetic field. This effect has been observed in a dedicated split beam interference experiment (Badurek et al. 1983b; Section 5.3). Equation (2.31) can be written as

$$\psi = \alpha\,|+z> + \beta e^{i\omega_r t}\,|-z>, \tag{2.44}$$

which gives for $|\alpha|^2 = |\beta|^2$ a polarization

$$\boldsymbol{P} = \begin{pmatrix} \cos\omega_r t \\ \sin\omega_r t \\ 0 \end{pmatrix}. \tag{2.45}$$

The neutrons also accumulate a velocity-dependent phase (because $t = \ell/v$) in the field-free regions between the first $\pi/2$ and the π-flipper (length ℓ_1) and the π-flipper and the second $\pi/2$-flipper (length ℓ_2). When the phases accumulated in the guide fields in the vicinity of the flippers match each other, the spin-echo condition is

$$\omega_r t_1 - \omega_r t_2 = \frac{\mu B_0}{\hbar v}(\ell_1 - \ell_2) = 0. \tag{2.46}$$

The confinement of the rather strong guide fields B_0 needed in the flipper regions and the transfer of the neutron polarization into and out of that region cause some experimental problems. On the other hand, the zero-field spin-echo method has distinct advantages compared to the DC-spin-echo method discussed earlier. Higher order spin-echo signals resulting in higher resolution can be realized by long field-free regions, and a bootstrap method which increases the energy difference by a multiple spin-flip device can be used (Gaehler and Golub 1988, Gaehler et al. 1992). A prototype zero-field spin-echo spectrometer was tested by Dubbers et al. (1989). Figure 2.23 shows typical spin-echo signals for various widths of the neutron wavelength band. Tilted fields which also cause tilted field-free regions have been tested for spectroscopic applications in the micro-electron volt resolution regime (Keller et al. 1998). The spin-echo signal and the wavelength spectrum are Fourier related to each other as in the case of the DC-spin-echo method (Eq. 2.36). The complementarity of split-beam interferences and Larmor interferences is striking as it has been addressed by Ramsey (1993) and Lamoreaux (1992). Pursuing the comparison between diffraction in ordinary and spin space one notices that the original Pauli

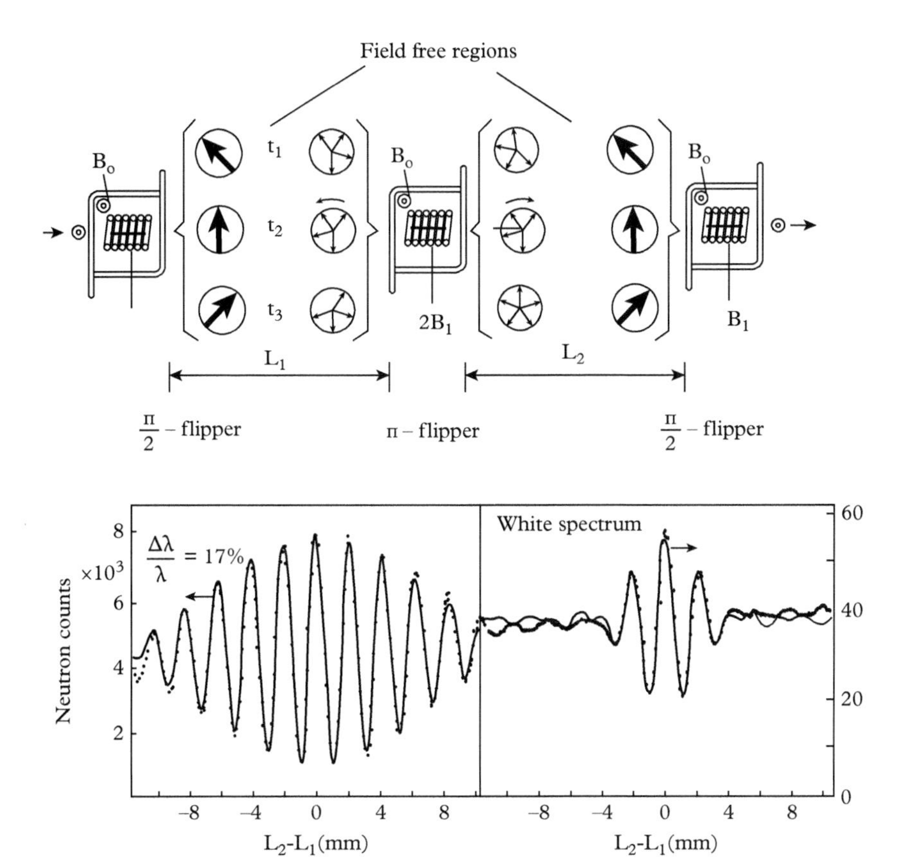

Figure 2.23 *Sketch of the zero-field spin-echo system showing the equivalence to the Ramsey-type atom interferometers and the retrieved polarization at the end of the second field-free region for two different wavelength spreads of the neutrons (below; Gaehler and Golub 1988). Reprinted from Dubbers et al. 1989, Copyright 1989, with permission from Elsevier.*

single-coil resonance technique resembles the optical single-slit interference experiments, whereas the more sophisticated Ramsey separated resonance coil method is the analogue of the double-slit interference experiment (Fig. 1.5).

When four flippers are used and their flipping probability is kept between 0 and 1 many more interference phenomena can be observed, as is known in Bordè–Ramsey atom interferometry (Bordè 1989, Kasevich and Chu 1991, Riele et al. 1991). For the neutron case such interferometers have been tested (Grigoriev et al. 2000, Mulder et al. 2000, Kraan et al. 2010) and discussed in connection with the measurement of higher order correlation functions by Grigoriev et al. (2004).

A comparison of standard Mach–Zehnder (spatial) and Larmor (time) interferometers has been given by Felber et al. (1999). More details about multi-coil resonance systems are given in Section 6.7.

3

Neutron Interactions and the Coherent Scattering Lengths

Scattering lengths are the relevant quantities for the description of the neutron–nuclear interaction at low energies and they are of substantial interest for the understanding of the basic nucleon–nucleon interaction. In this connection they are directly connected to the elementary hadron–hadron interaction, namely neutron–proton, neutron–neutron, and proton–proton scattering and to the problems of charge dependence and charge symmetry of nuclear forces (Henley 1966, Slaus et al. 1989, Gardestig 2009). Quantum chromodynamics is the fundamental theory of hadronic structure and interaction but a rigorous calculation of low-energy neutron–nuclear data is still not possible. The observed interactions between hadrons are influenced by direct and indirect electromagnetic effects due to Coulomb interaction, magnetic interaction, effects due to the finite charge and magnetic moment distribution and due to the mass differences between charged and neutral pions producing various nuclear forces. Various aspects of fundamental properties of the neutron and its interaction are summarized in the books of Byrne (1994) and Alexandrov (1992) and by a review of Dubbers and Schmidt (2011). In the low-energy regime the scattering lengths are the only parameters necessary to describe the neutron–nuclear interaction, whereas at higher energies the effective range and the shape parameter must be included to properly account for the scattering cross-section data. As neutron physics moved more and more toward condensed matter research, the precise values for scattering lengths of all the elements and many isotopes became quite important, because they determine the strength of the elastic and inelastic scattering signal from structurally complex samples. A collection of all measured neutron scattering length data and recommended values for the elements and most of the isotopes is given in reviews by Koester et al. (1991), Sears (1992), and Rauch and Waschkowski (1993).

3.1 Nuclear Interaction

3.1.1 General Relations

The range of the neutron–nucleus interaction is much smaller than the wavelength of thermal neutrons and therefore the scattering is isotropic within the center-of-mass system. This justifies the use of the Fermi pseudopotential as introduced in Chapter 1, Eq. (1.18) to describe it as a point-like interaction within the context of the Born approximation (Fermi 1936). For a single unbound nucleus we have

$$V = \frac{2\pi\hbar^2}{\mu} a\delta(r), \tag{3.1}$$

where $\mu = mM/(m + M)$ is the reduced mass between the neutron of mass m and the nucleus of mass M. The scattering length a for a free (unbound) nucleus is related to the forward scattering amplitude $f_0 = -a$. In the case of spherical scattering a is given in terms of the s-wave phase shift δ_0 or equivalently by the logarithmic derivative of the wave function at the nuclear surface R (e.g., Blatt and Weisskopf 1952), that is

$$a = -\lim_{k'\to 0}\left(\frac{\sin\delta_0}{k'}\right) \cong \frac{-\delta_0}{k'}, \tag{3.2a}$$

or, more precisely,

$$-\frac{1}{a} = \left(\frac{1}{\psi}\frac{\mathrm{d}\psi}{\mathrm{d}r}\right)_{r=R} = k' ctg\,\delta_0. \tag{3.2b}$$

Here k' is the wave vector in the center-of-mass system which is related to the wave vector and wavelength in the laboratory system as $k = 2\pi/\lambda = k'm/\mu$. For an infinitely repulsive potential the scattering length becomes equal to the hard core radius. For any other repulsive potential the wave function in the interior of the nucleus decreases exponentially giving a positive derivative at the nuclear surface and a scattering length $0 < a < R$. In the case of an attractive potential permitting a bound state the wave function inside the nuclear volume will be oscillatory and exponentially decreasing outside giving a negative derivative at the nuclear surface and a scattering length $a > R$. For an attractive potential without a bound state the wave function is oscillatory for both regions with a positive curvature at R which results in a scattering length $a < 0$. The behavior can be quite different for the two possible spin states $I + 1/2$ and $I - 1/2$ between the nucleus of spin I and the neutron of spin $1/2$ due to the spin dependence of nuclear forces.

In terms of the Breit–Wigner formalism for each spin state a separation into a potential and a resonance scattering length can be written in the form (Breit and Wigner 1936, Koester 1977, Mughabghab et al. 1981)

$$a_\pm = R' + \sum_j \frac{1}{2k_j'} \frac{\Gamma_{nj}}{\left[(E' - E_j) + i\Gamma_{j/2}\right]}, \tag{3.3}$$

where Γ_{nj} and Γ_j are the neutron and the total widths at the resonance energy E_j, and R' is the potential scattering radius. The summation must be taken over all resonances with the same spin state, including those below the binding energy ($E_j < 0$). This equation can be rewritten in terms of real and imaginary parts as

$$a_\pm = R' + \sum_j \frac{\Gamma_{nj}(E' - E_j)}{2k_j'\left[(E' - E_j)^2 + \Gamma_j^2/4\right]} - i\sum_j \frac{\Gamma_{nj}\Gamma_j}{4k_j'\left[(E' - E_j)^2 + \Gamma_j^2/4\right]}, \tag{3.4}$$

so that, in general, $a_\pm$ is a complex number $a_r - ia_i$. The imaginary part is related to the absorption cross-section σ_a by the optical theorem (Bacon 1975, Felcher et al. 1975)

$$a_i = \frac{k'\sigma_a}{4\pi} = \frac{\sigma_a}{2\lambda'}, \tag{3.5}$$

which becomes energy independent for $1/v$ absorbers. For most nuclei a_i is much smaller than a_r. Often the neutron resonances are located far away from thermal energies and the relation $E_j >> \Gamma_j$ is fulfilled. This simplifies the expression for the scattering length to

$$a_\pm = R' - \sum_j \frac{\Gamma_{nj}}{2k'_j E_j} = R' - 2.277 \times 10^3 \left(\frac{m+M}{M}\right) \cdot \sum_j \frac{\Gamma^0_{nj}}{E_j}, \tag{3.6}$$

where the reduced neutron width $\Gamma^0_{nj} = \Gamma_{nj}(1eV/E_j)^{1/2}$ is introduced. The potential scattering radius R' can be calculated by means of Eq. (3.2) when δ_0 is replaced by the real part of the optical model s-wave phase shift and it can be determined experimentally by the analysis of the neutron–nuclei cross-section in the resonance region (Feshbach et al. 1954, Seth et al. 1958, Pineo et al. 1974). Figure 3.1 shows recommended R' values in comparison with optical model calculations and with a mean nuclear radius $R = 1.35A^{1/3}$ fm (Mughabghab et al. 1981). Although a large number of nuclear level parameters are known for most nuclei, it is not possible to calculate the resonance contribution from these values because often the correct spin assignments and resonances below the binding energy are often unknown.

A spin-dependent scattering length operator $\underline{\hat{a}}$ can be defined to account for the different scattering lengths belonging to the interaction channels $I + 1/2$ and $I - 1/2$, respectively, that is

$$\underline{\hat{a}} = \frac{I+1}{2I+1} a_+ + \frac{I}{2I+1} a_- + \frac{2(a_+ - a_-)}{2I+1} \boldsymbol{I} \cdot \boldsymbol{s}_n, \tag{3.7}$$

where $\boldsymbol{s}_n$ is the neutron spin operator. The statistical weight factors are $g_+ = (I+1)/(2I+1)$ and $g_- = I/(2I+1)$ for unpolarized nuclei. For a partially polarized collection of nuclei, we can define the degree of polarization by $f_n = \langle I \rangle / I$. Thus, for neutrons with polarization, $\boldsymbol{P}_n = 2\langle \boldsymbol{s}_n \rangle$ the statistical weight factors are changed to

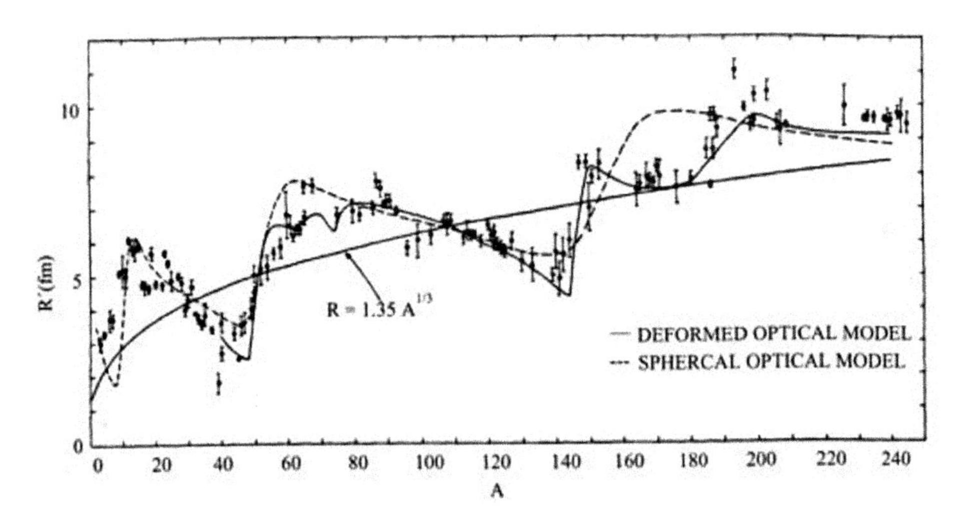

Figure 3.1 *Variation of the potential scattering radius with the mass number and the results of various model calculations (Mughabghab 1984)*

$$g_+ = \frac{I+1}{2I+1}\left(1 + \frac{I}{I+1} f_n \cdot \boldsymbol{P}_n\right), \tag{3.8}$$

and

$$g_- = \frac{I}{2I+1}(1 - f_n \cdot \boldsymbol{P}_n).$$

These relations can also be used to evaluate the scattering lengths and cross-sections for the two spin states of an unpolarized beam (e.g., Williams 1988).

For neutron optical phenomena, the neutron must be viewed as interacting with many nuclei simultaneously. From the uncertainty relations $\Delta x \Delta k \geq ½$ and the fact that in neutron optics the momentum transfer Δk is zero or at least very small, one sees that the region of interaction Δx and the number of interacting nuclei become very large. This justifies averaging the neutron interaction over many nuclei with different scattering lengths. In Chapter 4 we will relate Δx to the coherence length of the beam. Because there is no energy exchange with the assembly, the nuclei appear to be fixed within the target and therefore a bound scattering length b should be defined. The "strongly" bound neutron–nucleus scattering length is defined by

$$b = a\,\frac{m+M}{M} = a \cdot m/\mu\,. \tag{3.9}$$

This is also valid in the scalar and operator forms of Eq. (3.7). There is a discussion in the literature about the meaning of strongly bound nuclei which was needed in the derivation of Eq. (3.9). Strongly bound means that the nucleus is embedded elastically to its surrounding nuclei, permitting a purely elastic interaction of the neutron with the medium. That is, the nucleus has a definite time-averaged position. This situation is not a nucleus with a mass tending to infinity which would simulate a "rigidly" bound nucleus where the Born approximation becomes violated (Sears 1978). Within the impulse approximation, the interaction of a neutron with a sample containing many nuclei at the positions $\boldsymbol{R}_j$ can be written as a sum of δ-functions, that is

$$V(\boldsymbol{r}) = \frac{2\pi\hbar^2}{m}\sum_j \underline{b}_j \delta(\boldsymbol{r} - \boldsymbol{R}_j). \tag{3.10}$$

This result enables the definition of a mean interaction potential, or *optical potential* for a material

$$< V(\boldsymbol{r}) > = \overline{V} = \frac{2\pi\hbar^2}{m} b_c N. \tag{3.11}$$

Here N is the particle density and $b_c = < \underline{b} >$ is the mean coherent scattering length of the collection of nuclei, which for an unpolarized nuclear system is ($< \vec{I} > = 0$, Eqs. 3.7 and 3.8)

$$b_c = < b > = \frac{I+1}{2I+1} b_+ + \frac{I}{2I+1} b_-. \tag{3.12}$$

This can also be understood as an interaction parameter caused by a mean phase shift $<\delta_0> = -k \cdot <b>$ affecting the neutron wave scattering from the different nuclei. The variances of these quantities $<(\Delta\delta_0)^2> = (< \delta_0{}^2> - < \delta_0>^2)$ define the incoherent scattering length b_{inc}

$$b_{inc}{}^2 = <b^2> - <\overline{b}>^2 = g_+ g_-(b_+ - b_-)^2 = \frac{I(I+1)}{(2I+1)^2}(b_+ - b_-)^2. \tag{3.13}$$

If various isotopes or elements with abundances p_j are present a further averaging procedure must be done, yielding the coherent scattering length of the sample as whole

$$b_c = \sum_j p_j \left(g_{j+} b_{j+} + g_{j-} b_{j-}\right) = \sum_j p_j b_{cj}. \tag{3.14}$$

From Eqs. (3.5) and (3.11) together with the relation between the wave vectors in the laboratory and center-of-mass system $k/k' = A/(A+1)$, it follows that the absorption cross-section of the jth nucleus is given by the imaginary parts of the scattering lengths

$$\sigma_{aj} = \sigma_{aj}^{+} + \sigma_{aj}^{-} = \frac{4\pi}{k} b_{cj} = \frac{4\pi}{k}\left(g_{+} b''_{+}\right)_j + \frac{4\pi}{k}\left(g_{-} b''_{-}\right)_j, \tag{3.15}$$

where A is the mass number of this nucleus and the double prime indicates the imaginary part of the scattering lengths. In many cases, the absorption process is associated with one interaction channel only (e.g., $He^3 : \mathcal{J} = I - 1/2 = 0$).

The index of refraction n is the ratio of the wave vector (K) inside to the wave vector (k) outside a material (or region of a potential). It can most easily be obtained by solving the Schrödinger equation for a neutron moving through a potential step of a height $\overline{V}$ given by the mean interaction potential (Eq. 3.11). This yields a complex index of refraction (Eq. 1.24)

$$n = \frac{K}{k} = \sqrt{1 - \frac{\overline{V}}{E}} \cong 1 - \frac{\lambda^2 N}{2\pi}\sqrt{b_c^2 - \left(\frac{\sigma_r}{2\lambda}\right)^2} + i\frac{\sigma_r N \lambda}{4\pi} = n_r + i n_i, \tag{3.16}$$

which is in agreement with the expression first obtained by Goldberger and Seitz (1947) by adding the imaginary part (from the absorption cross-section) to the scattering cross-section to account for the optical theorem of general scattering theory (Sears 1982a, 1988). Thus, σ_a has been replaced by $\sigma_r = \sigma_a + \sigma_s$, which now also fulfills Lambert's law of beam attenuation

$$\frac{I}{I_0} = \exp[-(\sigma_a + \sigma_s)ND]. \tag{3.17}$$

It can be anticipated that the deviation of the index of refraction from unity becomes purely imaginary for a very strong absorber like Gd-157.

Strictly speaking, the related phase shift χ for a beam traversing a medium is complex and is given by

$$\chi = k(1-n)\,D_{\text{eff}} = \chi' + i\chi'', \tag{3.18}$$

where D_{eff} is the effective path length of the beam in the material medium. The interference pattern in the 0-beam (Chapter 1) in an interferometer experiment is then given by

$$I_0(\chi) = \left|\psi_0^{\text{I}} + \psi_0^{\text{II}}\right|^2 = \frac{I_0(0)}{2} e^{-\chi''}\left(\cosh \chi'' + \cos \chi'\right). \tag{3.19}$$

Coherence effects related to the beam attenuation term ($\chi'' = \sigma_r ND/2$) are discussed in Section 4.3. For thermal neutrons ($E >> \overline{V}$) and low-absorbing materials, the index of refraction and the phase shift simplify to

$$n = 1 - \lambda^2 \frac{N b_c}{2\pi} \quad \text{and} \quad \chi' = -N b_c \lambda D, \tag{3.20}$$

which results in the simpler form of the interferogram

$$I_0(\chi') = \left|\psi_0{}^{\mathrm{I}} + \psi_0{}^{\mathrm{II}}\right|^2 = \frac{I_0(0)}{2}(1 + \cos\chi'). \tag{3.21}$$

The phase shift is real under these conditions.

General scattering theory which accounts for mutual wave interaction of neighboring scattering centers also predicts slight changes of the real part of the phase shift. Predicted correction factors due to the internal structure of the sample material causing such local field effects are on the order of b_c/d, where d is the mean distance between particles (Ekstein 1953, Dietze and Nowak 1981, Nowak 1982, Sears 1982a, Adli and Summerfield 1984). They have not been observed experimentally up till now (see Section 3.4.4).

The coherent scattering length is directly related to the neutron–nucleus interaction only for spinless nuclei. Otherwise, one must extract the spin-dependent scattering length a_+ and a_- (or b_+ and b_-) from the results of two independent measurements. In addition to the measurement of the coherent scattering length by neutron interferometry, such an independent measurement can be the following:

(a) The total scattering cross-section

$$\sigma_s = 4\pi\left(\frac{I+1}{2I+1}b_+^2 + \frac{I}{2I+1}b_-^2\right), \tag{3.22}$$

which is strongly influenced by solid state effects in many cases and, therefore, such measurements must be done with epithermal neutrons (e.g., Koester 1977);

(b) The spin incoherent cross-section (Eq. 3.13) plus a term describing isotope incoherence

$$\sigma_{\mathrm{inc}} = 4\pi b_{\mathrm{inc}}^2 + \sigma_{\mathrm{inc}}^{\mathrm{isot.}}, \tag{3.23}$$

which is also influenced by solid state effects and only applicable for strongly incoherent scattering materials; or thirdly

(c) The coherent interaction with a nuclear polarized sample, where the coherent scattering length depends on the neutron polarization $\boldsymbol{P}_n = 2 < \boldsymbol{s}_n >$ and on the nuclear polarization $\boldsymbol{f} = <\boldsymbol{I}>/I$. From Eq. (3.7) one gets (Lushchikov et al. 1970, Williams 1988)

$$b_c(P_n) = \bar{b}_c + b_{\mathrm{inc}} I \boldsymbol{P}_n \cdot \boldsymbol{f}/2. \tag{3.24}$$

This method is less dependent on solid state effects and well suited for the extraction of spin-dependent scattering lengths (Abragam 1972, Abragam et al. 1973, Glättli et al. 1979). In a way similar to the magnetic interaction (Section 3.2.1) the nuclear coherent scattering length depends on the neutron polarization, thus permitting the definition of a pseudo-magnetic induction which is proportional to b_{inc} and to the nuclear polarization $\boldsymbol{f}$, namely

$$\boldsymbol{B}^* = -\frac{4\pi\hbar^2}{m\gamma}\sqrt{\frac{I}{I+1}}N b_{\mathrm{inc}}\boldsymbol{f}. \tag{3.25}$$

In this case, the polarization vector rotates around this pseudo-magnetic field like it rotates in the case of Larmor precession around the magnetic field $\boldsymbol{B}$ (Baryshevski and Podgoretskii 1965, Glättli and Goldman 1987). The degree of nuclear polarization can often be determined by the spin-dependent absorption cross-section. The pseudo-magnetic field acts like an ordinary magnetic field when Stren–Gerlach-like experiments are considered (Zimmer et al. 2001). The nuclear polarization method is used in connection with Bragg scattering measurements, where the relevant structure factors must be added, and with transmission measurements where the spin dependence of the absorption cross-section must be taken into account in some cases.

3.1.2 Experimental Results

The format of this chapter places the experimental aspects in the foreground. The results obtained by neutron interferometry must compete with the results from many other methods such as transmission measurements, mirror reflection, prism reflection, Bragg diffraction, Christiansen filter, gravity refractometer, etc. (e.g., Koester 1977).

All the measurements with the neutron interferometer are based on the intensity modulation caused by the phase shift of the sample. This modulation is given by Eqs. (3.18) and (3.21), but must be modified due to unavoidable imperfections of any experimental setup (see Section 2.1). In general, an interferogram is of the form

$$I = A + B\cos(\chi + \phi)\,. \tag{3.26}$$

The quantity A is the mean counting rate in one of the interfering beams and includes the non-interfering background; B is the amplitude of the interfering part, χ is the phase shift of a sample whose thickness or whose pressure is varied ($\chi = -Nb_c\lambda D$), and ϕ is composed of an internal phase shift ϕ_0 of the empty interferometer, plus the phase shift of any other phase shifter or field ϕ_p, and it is also reduced by the phase shift of air ϕ_a displaced along the beam path within the interferometer by the sample, i.e., $\phi = \phi_0 + \phi_p - \phi_a$ (for discussion of ϕ_a see Section 3.4). All the quantities χ, ϕ, and B must be determined in such a scattering length measurement, which takes on a variety of experimental procedures. These are indicated for the case of a skew symmetrically cut interferometer crystal in Figs. 2.3 and 3.2. The sample in/out option becomes important to compensate for the drift of the internal phase (see Fig. 2.10).

3.1.2.1 *Measurement of Flat Solid and Liquid Samples*

Samples are often available in a slab-shaped form or they are in a flat container and they can be measured quite conveniently by various methods as indicated in Fig. 3.2. In the conventional method (A) the phase shift can be varied continuously by rotating the sample within the coherent beams by an angle δ as discussed briefly in Chapter 2. Here the phase shift depends upon the difference of the optical path lengths of path I and II, namely (Eq. 2.7)

$$\chi = Nb_c\lambda D_0\left[\frac{1}{\cos(\theta_B - \delta)} - \frac{1}{\cos(\theta_B + \delta)}\right]. \tag{3.27}$$

The first measurements with this method dealt with Sn, Al, V, Bi, and Nb samples (Bauspiess et al. 1976, Rauch et al. 1986). The accuracy for the scattering length, namely $\Delta b_c/b_c$ is typically on the order of several parts in 1000 and is determined by how accurately the period of oscillation can be determined from the interference pattern; but also by the errors of the density (ΔN), of

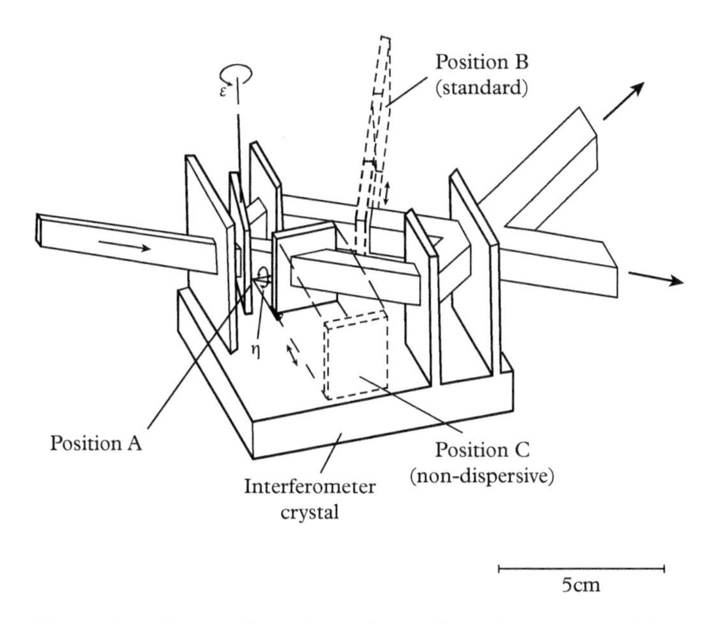

Figure 3.2 *Various dispersive and non-dispersive sample positions within a skew symmetric neutron interferometer*

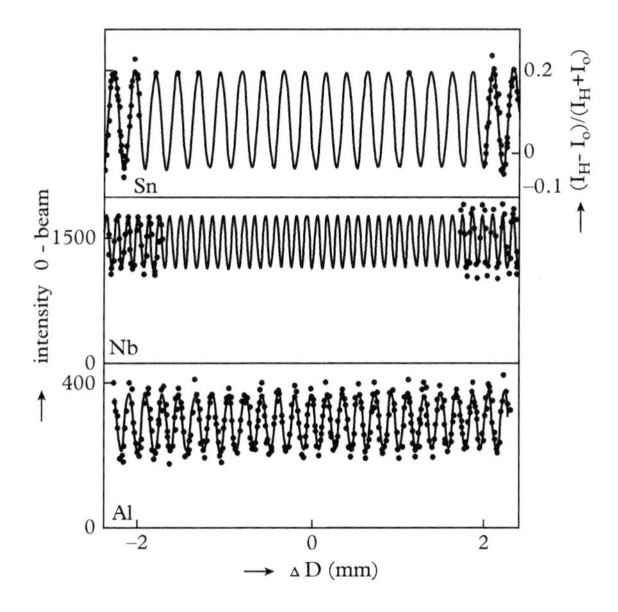

Figure 3.3 *Conventional interferometer scans when a phase shifter is rotated in both beams in continuous steps (below) and when the measurement positions are concentrated to the edges of the interference pattern (above; Bauspiess 1977)*

the sample thickness (ΔD), of the mean wavelength determination ($\Delta\lambda$), and in the sample orientation ($\Delta\delta$). For several experiments the error due to wavelength uncertainty has been reduced by the installation of a novel Ge–Simonochromator system which was stabilized by means of an additional X-ray beam (Section 2.2.1; Bauspiess et al. 1977). An even more economic method has been developed later on by using the normalized counting rate ratio $((I_H - I_0)/(I_H + I_0))$ and by measuring the interferences at high positive and negative orders and interpolating in between (Bauspiess 1977, Bauspiess et al. 1978). Figure 3.3 shows typical results of different measuring methods for various samples (Bauspiess 1977). The phase shift of the air displaced by the sample must be included in precise measurements (see Section 3.4.1). A four-parameter fit according to Eq. (3.26) was applied to get the optimal mean values and their statistical error bars as listed in Table 3.1.

Another measuring method (standard sweep) consists of alternately inserting and removing the sample into one beam (method B in Fig. 3.2). In this case, the interference patterns are taken by an auxiliary phase shifter in position A in Fig. 3.2. The phase shift of the sample (ϕs) is obtained modulo 2π from the interferograms with the sample in and then out of beam path II. In this case the internal phase drift as shown in Fig. 2.10 is to a large degree compensated. The phase shift between the two interferograms is given as

$$\Delta\chi = \chi_{\text{sample}} - \phi_{\text{a}} - 2\pi m, \tag{3.28}$$

where m is an integer to be found from an approximate value of χ_{sample} or a rough measurement using the standard rotation method. One notices that the air correction ϕ_{a} is more important in the sweep measuring method than in the standard rotation method. This method has been adapted from X-ray interferometry (Cusatis and Hart 1975) and it provides the advantage that small samples can be used and that the total phase shift of the sample can be made larger than for the standard sample rotation method. It has been used for many neutron scattering length measurements, where in some cases accuracies better than 1 part in 1000 for the scattering length b_c have been achieved (Hammerschmied et al. 1981; Boeuf_et al. 1982, 1985; Bonse and Kischko 1982, 1985; Freund et al. 1985; Rauch and Tuppinger 1985). This method has also been used to observe the energy dependence of the scattering length due to resonance effects (Word and Werner 1982, Arif et al. 1986, Kaiser et al. 1986). They established a negative resonance level of ^{235}U at −1.4 eV. A typical result is shown in Fig. 3.4.

An interesting method was proposed by Scherm (1981) and was first tested by Rauch et al. (1985). It is a non-dispersive measuring method (position C in Fig. 3.2), where the boundary of the sample is oriented parallel to the reflecting planes of the interferometer crystal. As the path length inside the sample is $D_0/\sin\theta_B$ the whole phase shift becomes independent from the wavelength, namely (see also Section 4.2.2)

$$\chi = -2Nb_c d_{hkl} D_0. \tag{3.29}$$

Here d_{hkl} is the lattice plane spacing of the interferometer crystal. This method permits two variants: the modified sweep method where the sample is moved in and out, and the modified rotation method where the sample is rotated around a horizontal axis. The obvious advantages of this method are the elimination of the necessity of a precise wavelength determination and the use of an incident beam with a broad wavelength spectrum. This method is discussed in more detail in Section 4.2. Figure 3.5 shows an example of such a measurement for a Bi sample in comparison with a conventional (dispersive) measurement (Rauch et al. 1987, Tuppinger 1987). The

Table 3.1 *Interferometrically Measured Bound Coherent Scattering Lengths*

	λ (Å)	b_c (fm)	Method[a]	Reference
H	2.71266(12)	−3.7384(20)	SS	Schoen et al. (2003)
H	1.898(3)	−3.64(3)	PV	Kaiser et al. (1979)
D	2.71266(12)	6.665(4)	SS	Black et al. (2003)
				Schoen et al. (2003)
D	1.898(3)	6.55(8)	PV	Kaiser et al. (1979)
T	1.876(5)	5.1(1)	SS	Hammerschmied et al. (1981)
	1.858(3)	4.792(27)	SS	Rauch et al. (1985)
He	1.898(3)	3.26(3)	PV	Kaiser et al. (1979)
^{3}He	1.910(2)	6.010(20)	SS	Ketter et al. (2006)
	2.71(1)	5.8572(72)	SS	Huffman et al. (2006)
	1.898(3)	5.74(7)	PV	Kaiser et al. (1979)
C	1.8389(6)	6.647(5)	SS	Freund et al. (1985)
^{13}C	1.9233(10)	6.542(3)	SS	Fischer et al. (2008)
N	1.898(3)	9.30(8)	PV	Kaiser et al. (1979)
O	1.898(3)	5.83(5)	PV	Kaiser et al. (1979)
^{17}O	1.9233(10)	5.867(4)	SS	Zeidler et al. (2011)
^{18}O	1.9233(10)	6.009(5)	SS	Zeidler et al. (2011)
Ne	1.898(3)	4.63(4)	PV	Kaiser et al. (1979)
Mg	1.7900(4)	5.375(4)	SR	Bauspiess et al. (1978)
Al	1.8322(9)	3.447(5)	SR	Bauspiess et al. (1976)
Al	1.8322(9)	3.449(5)	SR	Bauspiess et al. (1978)
	1.48(3)	3.42(2)	SR	Tomimitsu et al. (1995)
Si	2.36(2)	4.1479(23)	NS	Lemmel and Wagh (2010)
Si	1.9225(18)	4.1571(28)	CV	Tuppinger et al. (1988)
	2.7(2)	4.15071(22)	NS	Ioffe et al. (1998a, 1998b)
Ar	1.898(3)	2.07(2)	PV	Kaiser et al. (1979)
Ti	1.77632(7)	−3.438(2)	SR	Bauspiess et al. (1978)
V	2.107(2)	−0.408(2)	SR	Rauch et al. (1976)
	1.7900(4)	0.3824(12)	SR	Bauspiess et al. (1978)
Co	1.8558(6)	2.53(5)	SS	Kischko et al. (1982)

(*continued*)

Table 3.1 *(continued)*

	λ (Å)	b_c (fm)	Method[a]	Reference
Cu	1.5748(3)	7.7093(86)	SR	Tomimitsu et al. (2000)
	1.9085(20)	7.66(4)	SS	Bonse and Wroblewski (1985)
	1.77632(7)	7.718(4)	SR	Bauspiess et al. (1978)
^{63}Cu	1.5748(3)	6.477(13)	SR	Tomimitsu et al. (2000)
^{65}Cu	1.5748(3)	10.204(20)	SR	Tomimitsu et al. (2000)
Zn	1.77632(7)	5.680(5)	SR	Bauspiess et al. (1978)
^{69}Ga	1.543(7)	8.053(13)	SR	Tomimitsu et al. (1999)
^{71}Ga	1.543(7)	6.170(11)	SR	Tomimitsu et al. (1999)
Kr	1.898(3)	7.52(6)	PV	Kaiser et al. (1979)
	1.81(6)	7.72(33)	PV	Terburg et al. (1993)
	1.81(6)	8.07(026)	PV	Terburg et al. (1993)
Nb	1.786(2)	7.08(2)	SR	Rauch et al. (1976)
	1.7900(4)	7.054(3)	SR	Bauspiess et al. (1978)
Ag	1.77632(7)	5.932(6)	SR	Bauspiess et al. (1978)
	1.8742(6)	5.922(7)	SS	Bonse and Kischko (1982)
^{107}Ag	1.8742(6)	7.555(11)	SS	Bonse and Kischko (1982)
^{109}Ag	1.8742(6)	4.165(11)	SS	Bonse and Kischko (1982)
Sn	1.7860(2)	6.220(2)	SR	Bauspiess et al. (1976)
	1.7860(4)	6.228(4)	SR	Bauspiess et al. (1978)
Te	1.859(2)	5.6(1)	CV	Rauch and Tuppinger (1985)
	1.48(3)	5.49(2)	SR	Tomimitsu et al. (1995)
	1.9330(3)	5.68(2)	SR	Ioffe and Neov (1997)
Xe	1.898(3)	4.69(4)	PV	Kaiser et al. (1979)
Sm	1.859(2)	0.7(2)	SS	Rauch and Tuppinger (1985)
^{149}Sm	1.557(15)	21.17(18)	SS	Word and Werner (1982)
Eu	1.859(2)	5.3(3)	SS	Rauch and Tuppinger (1985)
Gd	1.859(2)	5.1(4)	SS	Rauch and Tuppinger (1985)
Dy	1.859(2)	16.9(3)	SS	Rauch and Tuppinger (1985)

(continued)

Table 3.1 *(continued)*

	λ (Å)	b_c (fm)	Method[a]	Reference
	1.9178(10)	17.3(3)	SS	Tuppinger et al. (1988)
Ho	1.825(1)	8.01(8)	SS	Boucherle et al. (1985)
W	1.5748(3)	4.755(18)	SR	Tomimitsu et al. (2000)
Pt	1.859(2)	9.60(1)	SS	Rauch and Tuppinger (1985)
	1.859(2)	9.48(11)	NS	Rauch and Tuppinger (1985)
Hg	1.5748(3)	12.595(45)	SR	Tomimitsu et al. (2000)
^{202}Hg	1.5748(3)	11.002(43)	SR	Tomimitsu et al. (2000)
Pb	1.92(7)	9.4017(20)	SS	Ioffe et al. (2000)
^{204}Pb	1.92(7)	10.893(78)	SS	Ioffe et al. (2000)
^{206}Pb	1.92(7)	9.221(69)	SS	Ioffe et al. (2000)
^{207}Pb	1.92(7)	9.286(16)	SS	Ioffe et al. (2000)
^{208}Pb	1.92(7)	9.494(30)	SS	Ioffe et al. (2000)
	2.0105(1)	9.494(29)	SR	Ioffe et al. (1994)
	1.085(8)	9.28(13)	SR	Alexandrov et al. (1989)
	1.92(7)	9.518(2)	NS	Ioffe and Vrana (1997)
Bi	1.8264(5)	8.58(5)	SR	Bauspiess et al. (1976)
	1.7900(4)	8.503(12)	SR	Bauspiess et al. (1978)
	1.9225(24)	8.521(4)	NS	Rauch et al. (1987)
	1.9225(18)	8.508(21)	CV	Tuppinger et al. (1988)
	1.9225(18)	8.5165(62)	SS	Tuppinger et al. (1988)
Th	1.839(6)	10.53(3)	SS	Boeuf et al. (1985)
U	1.8389(6)	8.417(5)	SS	Boeuf et al. (1982)
^{235}U	1.261(1)	10.50(3)	SS	Kaiser et al. (1986)
	1.642(1)	10.47(4)	SS	Arif et al. (1987)

[a] SR, standard rotation; SS, standard sweep; NS, nondispersive sweep; CV, contrast variation; PV, pressure variation.

interference pattern to first order does not depend on the wavelength spread of the beam, and, it remains visible up to very high interference orders. Very high phase sensitivities have been achieved with this method $\Delta\chi/\chi = 2.2 \times 10^{-5}$, which certainly can be increased to 10^{-6} by using longer measuring periods and more advanced sample orientation methods. At extremely high orders the parallel shift of the beam causes a defocusing effect (see Section 4.2.2). From simple

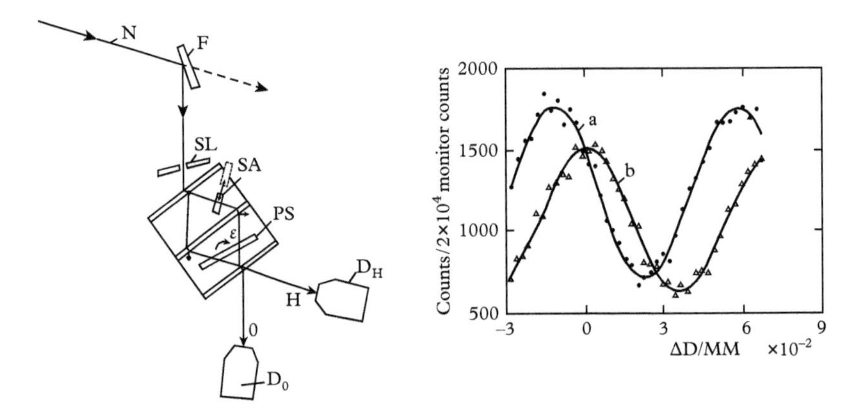

Figure 3.4 *Typical results of the sweep measurement technique when the sample is alternatively inserted and removed from one beam and the interference pattern is measured with a rotating phase flag Bonse and Kischko 1982, with kind permission from Springer Science and Business Media.*

geometrical consideration it follows that a sample of thickness D_0 and with a λ-thickness D_λ causes a defocusing of

$$\Delta t = \frac{D_0 \cdot d_{hkl}}{D_\lambda \sin \theta_B \tan \theta_B}. \tag{3.30}$$

The defocusing phenomena becomes effective when $\Delta t \cong \Lambda_0/10$, where Λ_0 is the Pendellösung length of the Si interferometer crystal (see Section 11.7.1). This corresponds to interference orders up to $m \cong 0.1 \times \Lambda_0 tg\theta_B/d_{hkl} \sim 2 \times 10^4$. Before this defocussing factor becomes effective, several other factors become more dominant. Such factors are the beam attenuation which follows from the imaginary part of the index of refraction (Eqs. 1.25 and 3.19) or more precisely from the optical theorem of general diffraction theory (Sears 1985). This gives the imaginary part of the phase shift as

$$\chi'' = \frac{N\sigma_t}{2} \frac{D_0}{\sin \theta_B}, \tag{3.31}$$

where σ_t is the total attenuation cross-section. Other contrast reduction factors occur due to the roughness of the surfaces of the sample or due to a misalignment of the sample relative to the reflecting planes. The effectiveness of the method is demonstrated by the results shown in Fig. 3.6, which were used to extract the coherent scattering length for Bi with high accuracy ($b_c = 8.508(4)$ fm; Rauch et al. 1987). It should be mentioned that some contrast remains even up to values where Δt equals the width of the Borrmann fan (Section 4.2.2; Petrascheck 1987). The relation of these measurements to questions of coherence properties of the beam is discussed in Chapter 4.

Ioffe and Vrana (1997) tested a method to align the surface of the non-dispersive phase shifter to be precisely parallel to the crystal planes of the interferometer (Fig. 3.6). When a declination ε of the surface and the crystal planes exists, the effective thickness of the phase shifter becomes

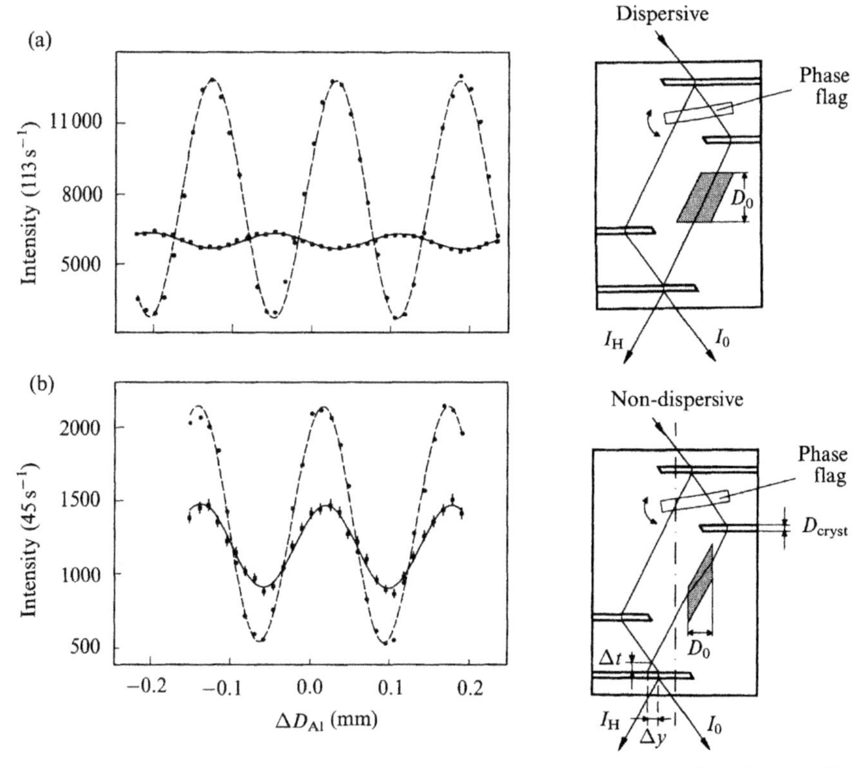

Figure 3.5 *Dispersive and non-dispersive sample arrangement and typical results obtained around the 250th interference order. The dashed lines indicate the interference pattern at low order (Rauch et al. 1987)*

$$D_{\text{eff}} = \frac{D_0}{\sin(\vartheta_B + \varepsilon)\cos\gamma}, \tag{3.32}$$

where γ denotes the vertical (tilt) misalignment. The basic idea for an optimal adjustment is to record the interference patterns as a function of ε when the phase shifter is put into beams I and II alternatively, i.e., for a set of positive and negative values of ε. Expending Eq. (3.32) for small ε-values one gets a phase difference for the sample placed in beam path I and then translated into beam path II

$$\Delta\chi(\varepsilon,\gamma) = \chi_I - \chi_{II} = \frac{2d\,Nb_c\,D_0}{\cos\gamma}\left[2 + \varepsilon^2(1 + 2\cot g^2\vartheta_B)\right]. \tag{3.33}$$

From the minimum of this parabolic function, one gets the scattering length

$$b_c = \frac{(\chi_I - \chi_{II})_{\min}}{4dND_0}. \tag{3.34}$$

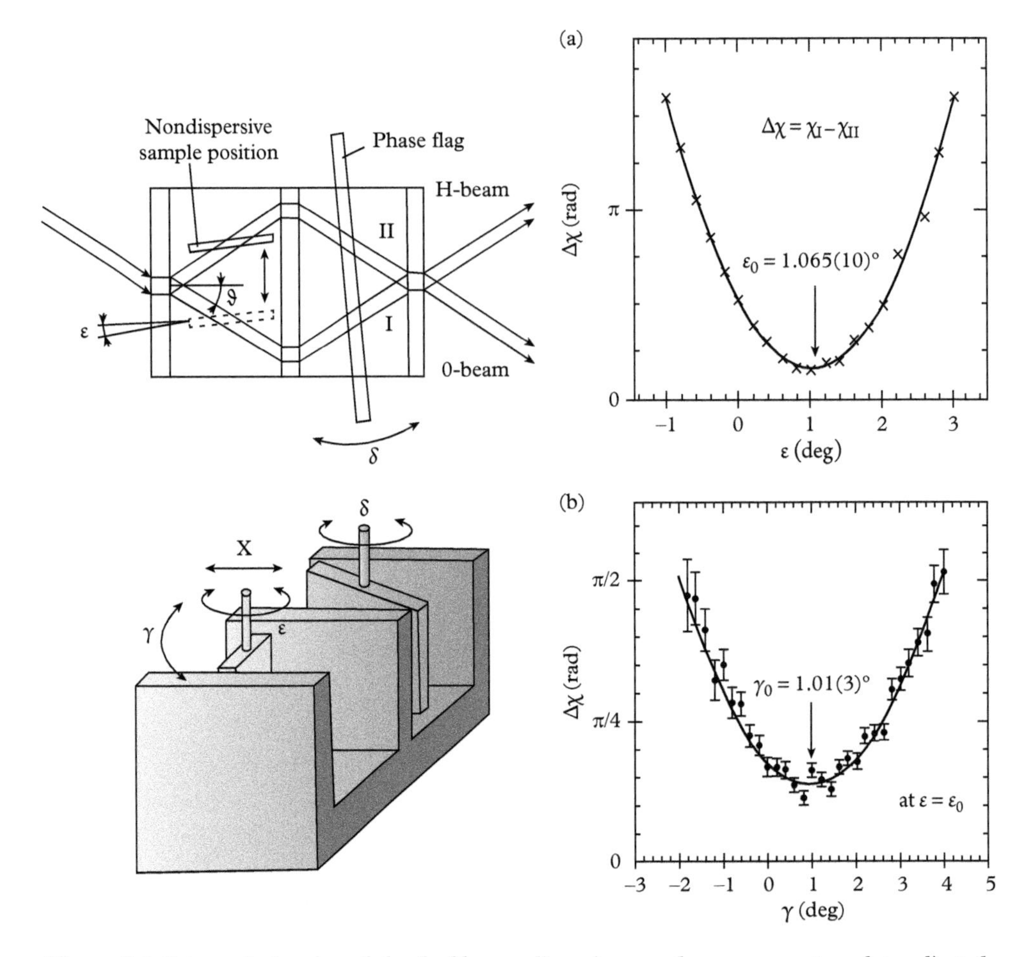

Figure 3.6 *Schematic drawing of the double non-dispersive sample arrangement used to adjust the sample surfaces to be parallel to the reflecting lattice planes (Ioffe and Vrana 1997) and adjustment curves for horizontal and vertical alignment. Reprinted from Ioffe and Vrana 1997, copyright 1997, with permission from Elsevier.*

The formula (3.33) shows that the phase difference $\Delta\chi$ is a symmetric function in ε and γ. This has been verified experimentally (Fig. 3.6). An auxiliary phase shifter (phase flag) is used to determine χ_I and χ_{II}. Accuracies of order $\Delta b_c/b_c \sim 10^{-5}$ have been achieved. A test measurement with a ^{208}Pb sample yielded $b_c = 9.518(2)$ fm, where the accuracy limit was mainly determined by the uncertainty of the sample composition (Ioffe and Vrana 1997). A very high precision measurement of the coherent scattering length of pure silicon used this method and resulted in a value $b_c = 4.15071(22)$ fm (Ioffe et al. 1998a, 1998b).

The same measuring methods as developed for solid samples can be applied to liquid substances when they are put into proper flat containers. The influence of the container itself can easily be subtracted by a separate measurement without the liquid. Related measurements will be discussed in the following sections in connection with a phase contrast method and with measurements on highly absorbing substances.

3.1.2.2 Measurements with Gaseous Samples

In principle, the same measuring methods as described in the previous section for solid and liquid samples can be applied for measurements on gaseous substances. Since pressure and density are related, pressure changes make a pressure variation method feasible. In this case the pressure p (in bars) of the gas inside a flat container is varied. This causes a variation of the particle density according to the thermodynamical equation of state

$$N = N_A \left[\frac{V_M.1.01325}{p} \left(A + Bp + Cp^2 \right) \right]^{-1}, \tag{3.35}$$

where N_A is the Avogadro number, V_M is the molar volume and A, B, C are the temperature-dependent Virial coefficients. The coherent scattering lengths are obtained from the periodicity of the interference pattern according to Eq. (3.26). Interferograms for several gases are shown in Fig. 3.7 (Kaiser et al. 1979) and the extracted values for the coherent scattering lengths are included in Table 3.1. The error bars are mainly due to the purity of the gases and by the precise determination of the particle density according to Eq. (3.35).

Values of the scattering lengths for light elements can be compared with calculated values from nucleon few-body theories and, therefore, they are of special interest. The calculations are based on parameterized advanced nucleon–nucleon potentials and take two- and three-body forces

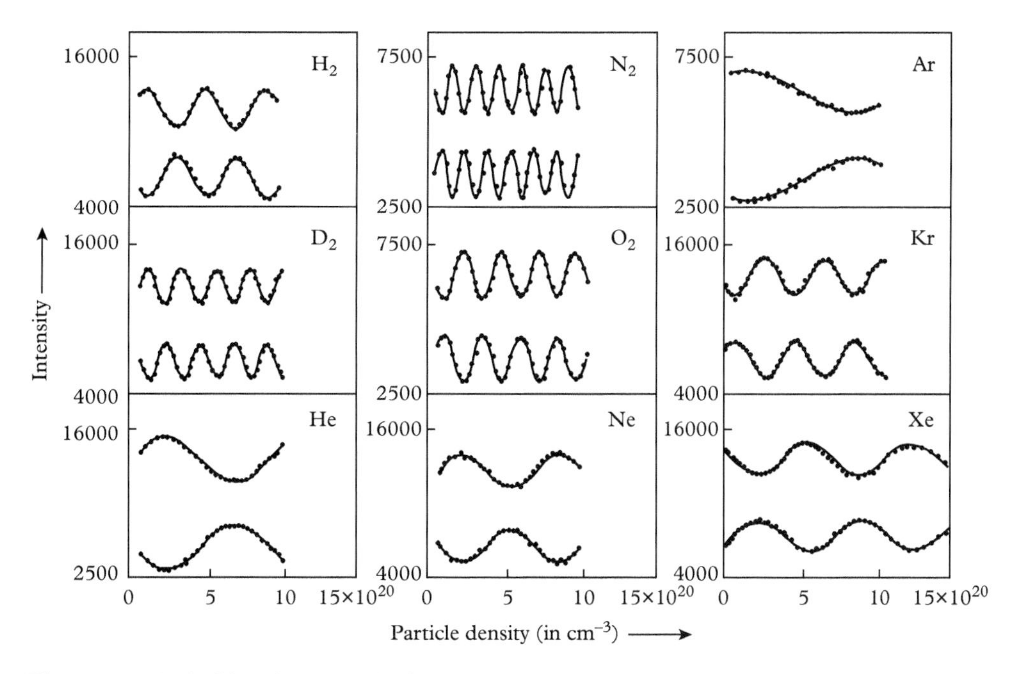

Figure 3.7 *Typical interferograms obtained in the measurement of scattering lengths of various gases (Kaiser et al. 1979)*

into account. The absence of a long-range electromagnetic interaction makes them more reliable than for charged particle interaction, where the inner cutoff radius of the Coulomb part remains uncertain.

The most relevant fundamental neutron–nucleon scattering lengths are for the pairs neutron–neutron, proton–proton, and neutron–proton. Due to the lack of neutron targets the neutron–neutron scattering length b_{nn} is accessible only by indirect measurements (neutron–deuterium, pion–deuterium, etc.) and the proton–proton scattering length b_{pp} is masked by the strong Coulomb contribution. In both cases at low energies only the singlet interaction exists due to the Pauli principle. The related data analysis and necessary approximations cause rather large errors on these quantities. The world averages are (Gardestig 2009)

$$\begin{aligned} b_{nn} &= 37.8(8)\,\text{fm} \\ b_{pp} &= 34.6(8)\,\text{fm}. \end{aligned}$$

Their difference indicates charge symmetry breaking.

The neutron–proton scattering length is accessible by several neutron optical methods and many results have been published in the past. In this case the coherent scattering length is composed of the singlet b^s and triplet b^t scattering lengths in the form

$$b_c = \frac{1}{4}b^s + \frac{3}{4}b^t. \tag{3.36}$$

The world average value is $b_c^{np} = -3.7405(9)\,\text{fm}$. Neutron interferometric measurements contributed to that value (Kaiser et al. 1979, Schoen et al. 2003). The related singlet scattering length must be compared with the b_{nn} and b_{pp} values discussed earlier (e.g., Wiringa et al. 1995).

A very elaborate interferometric method has been used by Schoen et al. (2003) to obtain accurate scattering lengths for light gases. Figure 3.8 shows the arrangement of the gas cells within the interferometer and indicates the sweep method used for these experiments. Many parameters had to be taken into account to extract an accurate value for the coherent neutron–proton scattering length, namely temperature, pressure, cell expansion, virial coefficient, internal phase variations, molecular binding, etc. Finally they got a value of $b_c = -3.7384\ (20)$ fm.

Black et al. (2003) and Schoen et al. (2003) used the same system they had used for the hydrogen measurement (Fig. 3.8) for the neutron–deuterium system. Since the deuteron has the nuclear spin $I = 1$ the coherent scattering length is related to the doublet and quartet scattering length.

$$b_c = \frac{1}{3}b^d + \frac{2}{3}b^q. \tag{3.37}$$

They got a value of $b_c(D) = 6.665(4)$ fm and under the assumption that the quartet scattering length is $b^q = 6.346(7)$ fm they got for the doublet scattering length a value of $b^d = 0.9680(45)$ fm. These values are in very good agreement with theoretical values when three-body forces are taken into account ($b_c^{(\text{theory})} = 6.665$ fm, Kievsky 1997; $b_c^{(\text{theory})} = 6.666$ fm, Chen et al. 1991). Whether this can be used as an unique verification of nuclear three-body forces must be taken with care, but it is one of the strongest indications of the existence of three-body forces in nuclear interaction. The quartet scattering length is rather insensitive to the details of the nucleon–nucleon potential since the three nucleons in this channel exist in a spin-symmetric state and therefore the scattering is completely determined by the long-range part of the nucleon interaction potential. On the other hand the doublet scattering length is rather sensitive to the details of the interaction

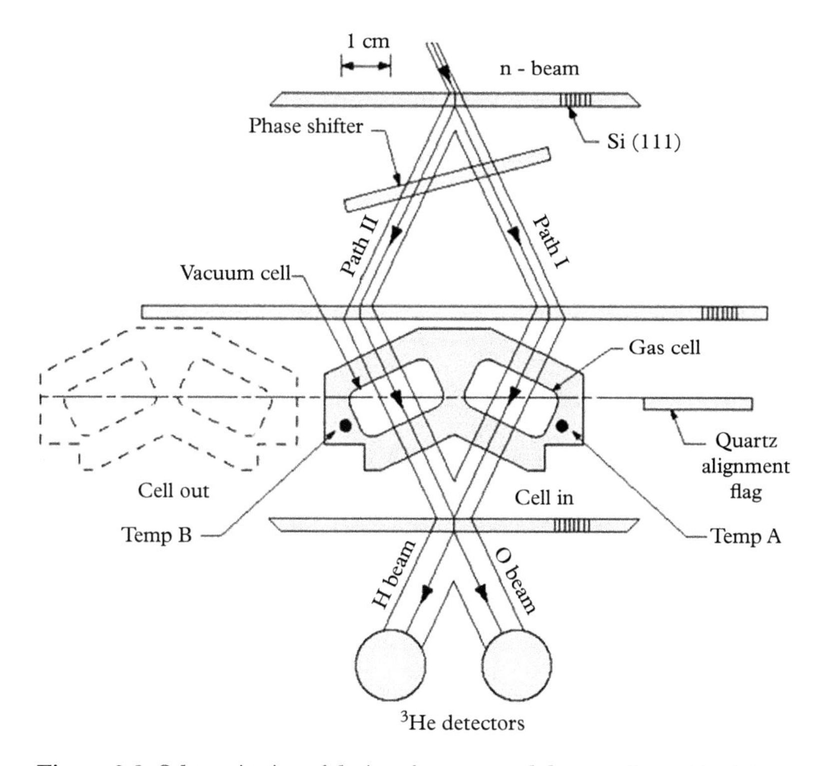

Figure 3.8 *Schematic view of the interferometer and the gas cells used by Schoen et al. (2003) in their measurements of the scattering lengths of H and D*

potential since the Pauli principle does not enter and short range and three-body contributions of the potential become important (Schoen et al. 2003). The measured doublet scattering length b^{d} = 0.9680(45) fm also agrees well with the calculated values Kievsky (1997) and Chen et al. (1991). Huffman et al. (2007) analyzed the data also in their relation to the triton binding energy and found reasonable agreement with the so-called Phillips line.

The scattering length of ^{3}He has been measured with the pressure variation method as well. In this case the index of refraction (Eq. 3.16) must be treated as a complex quantity due to the rather high absorption cross-section of ^{3}He ($\sigma_{\mathrm{a}} = 5327b$ for $\lambda = 1.8$ Å). This absorption process takes place in the singlet state only ($\mathcal{J} = I - ½ = 0$) which facilitates the use of polarized ^{3}He targets as broad band polarization filters (Tasset et al. 1992, Heil et al. 1998). Therefore, the interference pattern is influenced by the macroscopic attenuation cross-section ($\Sigma_t \cong N\sigma_t$) which varies with the gas pressure (see Eqs. 3.20 and 3.35). Using Eq. (3.19) and adding a non-interfering part of the intensity one must use a fitting procedure involving the more complicated formula in Eq. (3.19)

$$I = (C_1 - C_2)\left[C_4 + (1 - C_4)\, e^{-\Sigma_t D}\right] + C_2\left[\left(e^{-\Sigma_t D} + 1\right)/2 + e^{-\Sigma_t D/2}\cos(\chi + C_3)\right], \tag{3.38}$$

instead of Eq. (3.26), where the C_i are quantities characterizing the interferometer. The measured interference pattern is shown in Fig. 3.9 together with the optimal fit curve. A rather precise value

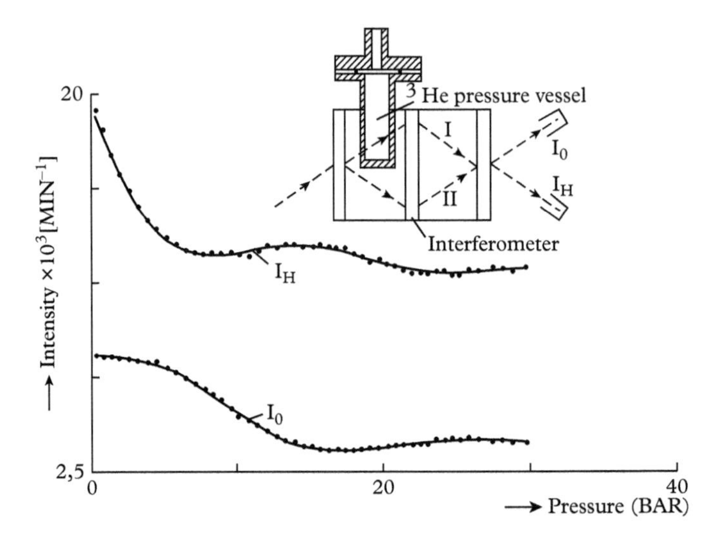

Figure 3.9 *Interference pattern of the highly absorbing ³He gas (Kaiser et al. 1997)*

for the coherent scattering length of $b_c = 5.74(7)$ fm has been obtained from these data for this strongly absorbing nucleus (Kaiser et al. 1977, 1979). In this connection, it has been shown how the relative sign of the scattering length can be determined by observing the intensity variation as a function of varying the phase with an auxiliary phase shifter when the sample is alternately moved in and out of the interferometer.

The neutron-^{3}He scattering length has been re-measured at NIST (Huffman et al. 2004) and at ILL (Ketter et al. 2006). In both cases a sweep method has been used, but the results are somehow different, $b_c = 5.853(7)$ fm and $b_c = 6.000(9)$ fm, respectively.

Substantial efforts have also been made to obtain a reliable value for the coherent scattering length for tritium which is of similar interest to that in the ^{3}He scattering length for the understanding of the nuclear four-body problem. In the tritium case the radiation hazards of the sample made the application of the sweep method necessary because it avoids pressure variations. The phase shift of the tritium gas is a factor of about 100 smaller than that of the tritium vessel. It had to be extracted from two measuring runs with and without tritium which are separated in time by several days due to the required tritium handling procedure. Two interferometric measurements whose error bars, unfortunately, do not overlap have been performed up till now (Hammerschmied et al. 1981, Rauch et al. 1985). Figure 3.10 shows the scheme and the results of the most recent measurement which leads to a value of $b_c = 4.792(27)$ fm. The amount of tritium gas used was only about 65 mg at a pressure of about 20 bars. In combination with results from total cross section measurements the spin-dependent scattering lengths were also determined.

3.1.2.3 Irregularly Shaped Samples—Christiansen Filter Method

Often it is impossible to get properly flat, slab-shaped samples. Therefore, a different measuring method which uses a phase-contrast variation method has been developed (Rauch and Tuppinger 1985). It is similar to the Christiansen filter method which uses the small-angle scattering intensity as a signal for coherent scattering length measurements (Koester and Knopf 1971, Koester 1977).

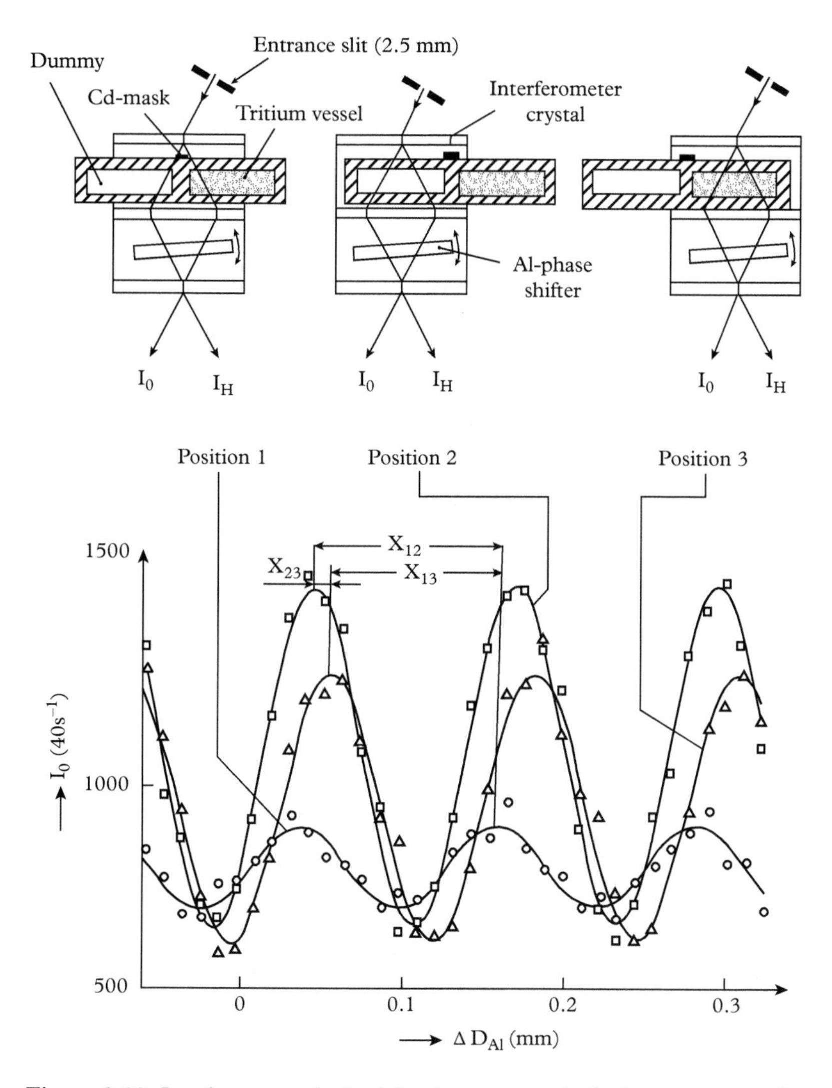

Figure 3.10 *Interferograms obtained by the sweep method when measuring the scattering length of tritium (Rauch et al. 1985c)*

Here the contrast of the interference pattern (*B* in Eq. 3.26) is used as a measuring signal. The contrast peaks when the index of refraction of the irregularly shaped sample (s) matches the index of refraction of the surrounding liquid (L), which occurs when

$$(Nb_c)_s = (Nb_c)_L. \tag{3.39}$$

$(Nb_c)_L$ can be changed in a known manner by varying the composition of the liquid, e.g., various H_20–D_20 mixtures. The typical arrangement and results for measurements with Si powder are shown in Fig. 3.11 (Tuppinger et al. 1988). Accuracies on the order of 0.1% have been achieved. The reduction of the contrast for $(Nb_c)_s \neq (Nb_c)_L$ is partly caused by a non-uniform phase shift and partly by the resulting small-angle scattering effect. A detailed theoretical treatment of these phenomena is still missing. The accuracy which can be achieved depends, in addition to the purity of the materials, on the grain size of the solid sample and on the wetting of the surrounding liquid.

This method has been extended to powdered samples with various grain sizes whose index of refraction is matched by a surrounding liquid with variable scattering length density. Instead of measuring the resulting small-angle scattering intensity, the reduction of the contrast of the interference pattern is measured. This interferometric Christiansen filter method has its highest sensitivity for rather large grain sizes, which is opposite to the sensitivity of the standard small-angle scattering Christiansen filter method. Figure 3.12 shows a typical result obtained for an irregularly shaped Te sample whose dimension was on the order of millimeters (Rauch and Tuppinger 1985).

All interferometric measured coherent scattering lengths are summarized in Table 3.1. The error bars give an idea of the accuracies which can be achieved by the different methods. Complete tables of recommended values of coherent scattering lengths can be found in the literature (Koester et al. 1977, Mughabghab et al. 1981, Mughabghab 1984, Sears 1986, Koester et al. 1991, Rauch and Waschkowski 1998).

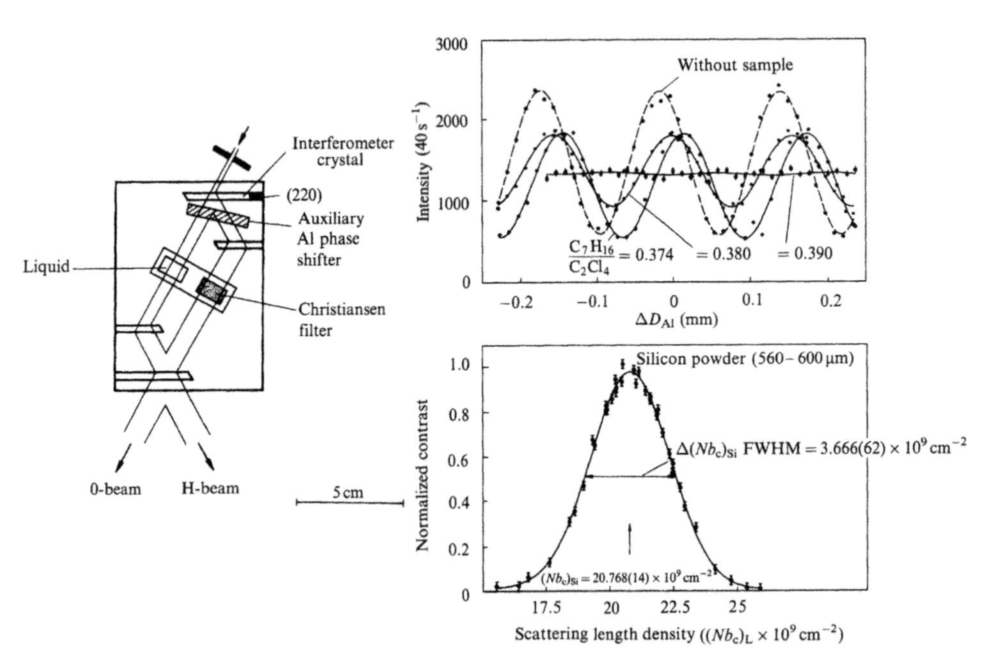

Figure 3.11 *Scheme of the interferometric Christiansen filter method and typical results for Si-powder (Tuppinger et al. 1988)*

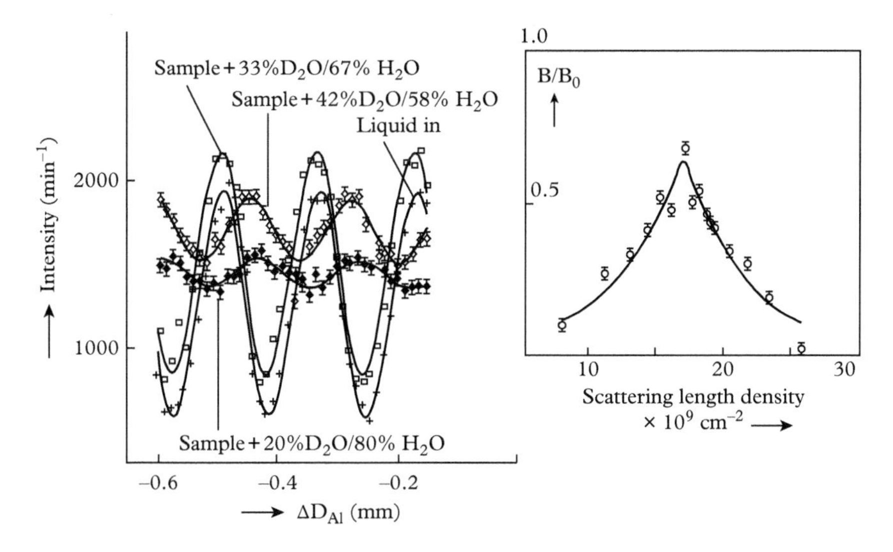

Figure 3.12 *Christiansen filter interferograms for a large grained Te-sample (Rauch and Tuppinger 1985)*

3.1.2.4 Measurements of Spin-Dependent Scattering Lengths

The strong nuclear interaction is spin-dependent, resulting in spin-dependent scattering lengths (Section 3.1). The related formulas are shown in Eqs. (3.12)–(3.15). In many cases such spin-dependent values can be extracted when total scattering cross-sections (Eq. 3.22) or/and incoherent scattering cross-sections are available $\left(\sigma_{\mathrm{inc}} = 4\pi b_{\mathrm{inc}}^2\right)$ in combination with coherent scattering lengths values.

Direct measurements of the spin-dependent scattering lengths require nuclear-oriented targets and polarized neutrons. Singlet and triplet scattering lengths are the basic quantities for a comparison with related nuclear few-body calculations. Such spin rotation experiments within a spin-echo system and with a nuclear-oriented ^{3}He gas target were used to extract spin-dependent neutron–^{3}He scattering lengths (Zimmer et al. 2002). The related neutron spin rotation angle φ of a nuclear-oriented target with polarization f_K of thickness D and particle density N follows from Eq. (3.24) and is given as

$$\varphi = \gamma B^* t = \gamma B^+ D/v = \sqrt{\frac{4I}{I+1}} \lambda N f_K D b_{\mathrm{inc}}. \tag{3.40}$$

This rotation is independent of the magnetic moment of the neutron, indicating that it is caused by the spin-dependent strong nuclear force. The experiments yielded for ^{3}He an incoherent scattering length of $b_{\mathrm{inc}} = -2.365(29)$ fm.

An interferometric method has also been used to measure the singlet and triplet scattering of ^{3}He (Huber et al. 2009a, 2009b). Polarized neutrons and a polarized target have been used (Fig. 3.13). From Eqs. (3.13) and (3.25) one obtains the phase shift $\Delta\phi$ between the interference pattern measured with spin-up and spin-down neutrons respectively, namely

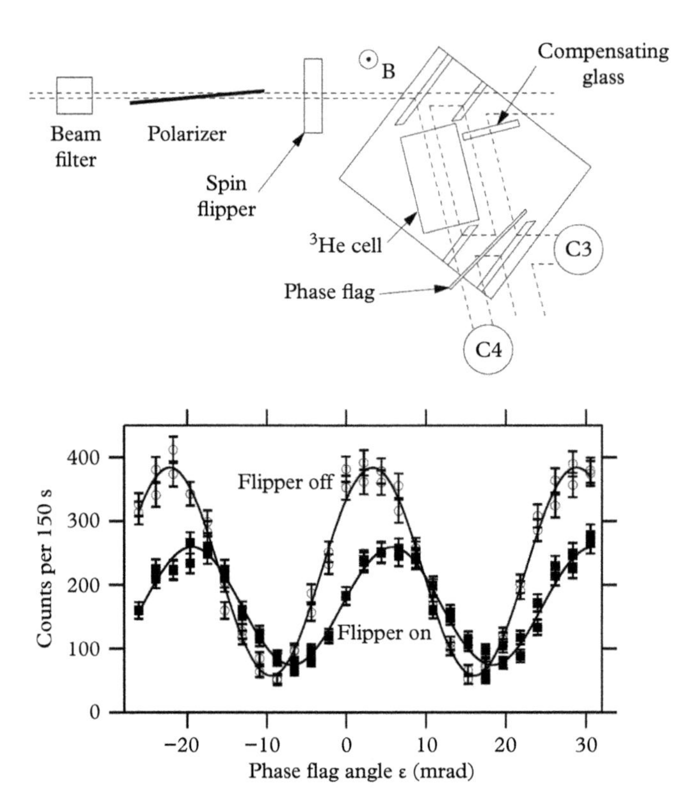

Figure 3.13 *Experimental arrangement (top) and results (bottom) of the $(b_+ - b_-)$ measurement of ^{3}He. Reprinted with permission from Huber et al. 2009, copyright 2009 by the American Physical Society.*

$$\Delta\phi = -\frac{2I+1}{\sqrt{I(I+1)}}\lambda N f_K b_{\mathrm{inc}}; \tag{3.41}$$

the analysis ideas are similar to the spin-rotation angle of the previous experiment. The incoherent scattering length b_{inc} is defined as being the difference between the singlet and triplet scattering lengths (see Eq. 3.12). The final result is $b_{\mathrm{inc}} = -2.429 \pm 0.012 \pm 0.014$ fm, where the statistical and systematical errors are given separately. A more recent experiment by Huber et al. (2014) gives a more accurate value of $b_{\mathrm{inc}} = -2.346 \pm 0.014 \pm 0.017$ fm. In combination with independent measurements of the coherent scattering length the singlet and triplet scattering lengths can be extracted and compared with values calculated from few-body theories. The present situation of this extraction and a comparison is given in Fig. 3.14, which shows that sufficient agreement between experimental and theoretical values and between each of them does not yet exist.

3.2 Electromagnetic Interaction

The magnetic moment of the neutron $\boldsymbol{\mu}$ and the internal electromagnetic structure of the neutron gives rise to additional interactions of the neutron with the atomic magnetic and electric fields arising from the nuclear and electronic charges. The range of this interaction is given by the

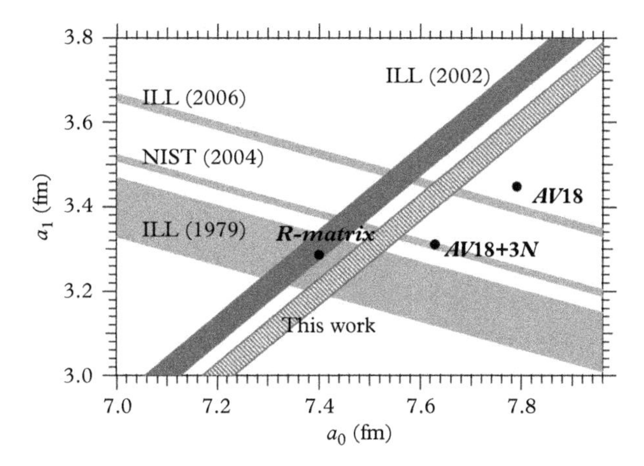

Figure 3.14 *Summary of triplet and singlet free neutron scattering length from experimental and theoretical investigations. The calculated values come from Hofmann and Hale (2003), the experimental ones from Kaiser et al. (1979), Huffman et al. (2004), Ketter (2006). Reprinted with permission from Huber et al. 2009, copyright 2009 by the American Physical Society.*

dimensions of the atoms which are comparable to the wavelength of thermal neutrons. Therefore, the related scattering effects become momentum dependent, which is described by appropriate wave vector-dependent form factors $f(\boldsymbol{Q})$ where $\boldsymbol{Q}$ is the momentum transfer given by the scattering angle $\theta(Q = 2k \cdot \sin\theta/2)$. Various contributions which arise from the magnetic coupling (V_m), from the intrinsic electromagnetic structure of the neutron (V_e), and from the electric polarizability of the neutron (V_p) can be distinguished. Various detailed theoretical analyses can be found in the literature (Sears 1986, Byrne 1994).

3.2.1 Magnetic Terms

The magnetic interaction (V_m) arises from the magnetic dipole moment of the neutron moving in static or slowly varying electric and magnetic fields within and surrounding the atoms. In the non-relativistic, low velocity limit of the Dirac equation this results in three leading terms (Schwinger 1948; Foldy 1951, 1958; Byrne 1994):

$$V_m = -\boldsymbol{\mu} \cdot \boldsymbol{B} - \frac{\hbar}{mc} \boldsymbol{\mu} \cdot (\boldsymbol{E} \times \boldsymbol{k}) - \frac{\hbar\mu}{2mc} \mathrm{div}\boldsymbol{E} = V_Z + V_S + V_F. \tag{3.42}$$

Here V_Z denotes the Zeeman term, V_S the Schwinger spin–orbit interaction, and V_F the Foldy interaction. The Zeeman term, or magnetic dipole interaction, is the largest contribution if the atom or ion carries a permanent magnetic moment. Within the Amperian current loop model for the electromagnetic origin of the magnetic dipole moment of the neutron, this interaction arises from three terms: a change in the total field energy, in the induction forces of the field, and of the particle (Shull et al. 1951; Mezei 1986, 1988). It should be pointed out that the additional factor of $1/2$ that occurs for the electron in the spin–orbit term and also in the Foldy (or Darwin) term to account for Thomas (1926) precession does not appear for the neutron. The Thomas precession occurs for the electron because of its charge and its consequential acceleration eE/m by the electric fields within the atom. This is clearly discussed in the book on electromagnetics by Feynman (1961). The Zeeman interaction arising from the neutron spin and magnetic moment is

$$V_Z = -\mu\boldsymbol{\sigma} \cdot \boldsymbol{B}(\boldsymbol{r}), \tag{3.43}$$

where $\boldsymbol{B}(\boldsymbol{r})$ is the magnetic induction field caused mainly by the unpaired electrons of a magnetic atom and $\boldsymbol{\sigma}$ are the Pauli spin matrices

$$\sigma_x = \begin{pmatrix} 0 & 1 \\ 1 & 0 \end{pmatrix} \quad \sigma_y = \begin{pmatrix} 0 & -\mathrm{i} \\ \mathrm{i} & 0 \end{pmatrix} \quad \sigma_z = \begin{pmatrix} 1 & 0 \\ 0 & -1 \end{pmatrix}. \tag{3.44}$$

The components of $\boldsymbol{\sigma}$ obey the commutation relations

$$[\sigma_x, \sigma_y] = 2\mathrm{i}\sigma_z, \quad [\sigma_y, \sigma_z] = 2\mathrm{i}\sigma_x, \quad [\sigma_z, \sigma_x] = 2\mathrm{i}\sigma_y, \tag{3.45}$$

and the anticommutation relations

$$\sigma_x\sigma_y + \sigma_y\sigma_x = 0, \qquad \sigma_y\sigma_z + \sigma_z\sigma_y = 0, \qquad \sigma_z\sigma_x + \sigma_x\sigma_z = 0. \tag{3.46}$$

Experimental verification of these anticommutation relations has been obtained by interferometric and polarimetric methods as discussed in Section 6.4, Fig. 6.11 (Hasegawa et al. 1997, Wagh et al. 1997, Hasegawa and Badurek 1999). An experimental demonstration of a universally valid error-disturbance uncertainty relation in spin measurements attacts the standard Heisenberg uncertainty relation (Erhart et al. 2012). In a neutron optical experiment the error of a spin-component measurement and the disturbance caused on another spin-component are recorded and fulfill a new error-disturbance relation first formulated by Ozawa (2003, 2005).

These properties of the Pauli spin matrices cause a magnetic field to couple to all components of the spin, which shows that this interaction does not become zero when the neutron magnetic moment is perpendicular to the magnetic field, that is

$$V_z = -\mu\begin{bmatrix} B_z & B_x - \mathrm{i}B_y \\ B_x + \mathrm{i}B_y & -B_z \end{bmatrix}. \tag{3.47}$$

However, when $< \boldsymbol{\mu} >$ is perpendicular to $\boldsymbol{B}$, the expectation (mean) value of the Zeeman energy taken over the spin-up and spin-down parts of the state $< \alpha \mid$ becomes zero ($<\alpha \mid -\mu\boldsymbol{\sigma} \cdot \boldsymbol{B} \mid \alpha> = 0$).

The corresponding magnetic scattering length is obtained from the Fourier transform of the interaction potential

$$p(Q) = \frac{m}{2\pi\hbar^2}\int e^{\mathrm{i}Q\cdot r}\, V(\boldsymbol{r})\, d\boldsymbol{r}, \tag{3.48}$$

which can be written in terms of the atomic magnetic form factor $f_m(\boldsymbol{Q})$ (Halpern and Johnson 1939, Marshall and Lovesey 1971) as

$$p\,(\boldsymbol{Q}) = -\frac{\gamma e^2}{m_e c^2} f_{\mathrm{m}}(\boldsymbol{Q}) < \boldsymbol{s} > (\boldsymbol{s} \cdot \boldsymbol{q}), \tag{3.49}$$

where γ is the magnetic moment of the neutron in units of the nuclear magneton ($\gamma = -1.91304308(58)$; Greene et al. 1979). $< \boldsymbol{s} >$ denotes the effective electronic spin of the magnetic atom, m_e is the electron mass, and $\boldsymbol{q} = \hat{h} - (\hat{h}\cdot\hat{e})\hat{e}$ is the magnetic interaction vector given by the unit vectors $\hat{h}$ and $\hat{e}$ pointing in the direction of the spin $< \boldsymbol{s} >$ and the scattering vector $\boldsymbol{Q}$, respectively. The magnetic form factor $f_m(\boldsymbol{Q})$ is determined by the unpaired electrons of the magnetic atoms or ions and it is the subject of many experimental and theoretical investigations (e.g., Izyumov and Ozerov 1970). This magnetic form factor is normalized to unity at $\boldsymbol{Q} = 0$,, i.e., $f_m(0) = 1$. It can be understood as the Fourier transform of the spin density surrounding the magnetic atom or ion. The factor $r_0 = e^2/m_e c^2 = 2.82$ fm is the classical electron radius. It determines the magnitude of the magnetic scattering length of a magnetic atom. For a spin S = ½ magnetic atom (one Bohr magneton/atom) oriented with its moment perpendicular to the scattering vector $\boldsymbol{Q}$, the magnetic scattering length becomes $p = 2.69$ fm at $\boldsymbol{Q} = 0$, i.e., comparable to nuclear scattering lengths. The index of refraction n is again easily obtained by taking the mean magnetic potential of the neutron within a magnetic material to be $\overline{V} = \pm \mu B$, where B is the magnetic induction field. Using Eq. (3.16) and including the nuclear term one obtains

$$n = \sqrt{1 - \frac{\overline{V} + \overline{V}_m}{E}} \cong 1 - \lambda^2 \frac{N\,(b_c \pm p(0))}{2\pi}, \tag{3.50}$$

where the forward magnetic scattering length $p(0)$ is related to the mean magnetization of the sample and to the total atomic magnetic moment μ_a as

$$p(0) = \mp \frac{\mu m B}{2\pi N \hbar^2} = \mp \frac{2m\mu}{\hbar^2}\,\mu_a. \tag{3.51}$$

We have used $B = 4\pi M = 4\pi N \mu_a$ to relate the magnetization M to the magnetic field B. The related influence of magnetism on the interference pattern is obtained in a way similar to that for a pure nuclear interaction. In the case of unpolarized incident neutrons an intensity and polarization modulation is obtained by an incoherent summation of the effects for spin-up (+) and spin down (–) incident neutrons. Related calculations for interferometry experiments, including the nuclear and magnetic interaction have been made by Eder and Zeilinger (1976). Their results for unpolarized incident neutrons give for the 0-beam interferogram (see Eqs. 5.8 and 5.9)

$$I = \frac{I_0}{2}\left(1 + \cos\chi\,\cos\frac{\alpha}{2}\right), \tag{3.52}$$

which has a phase-dependent polarization

$$\boldsymbol{P} = \frac{< \psi | \boldsymbol{\sigma} | \psi >}{< \psi \,|\, \psi >} = \frac{\sin\chi\,\sin(\alpha/2)}{1 + \cos\chi\,\cos(\alpha/2)}. \tag{3.53}$$

Here χ is the nuclear phase shift ($\chi = -N\lambda b_c D$) and $\alpha/2$ is the magnetic phase shift. The Larmor precession angle $\alpha = \gamma \int \boldsymbol{B}.d\boldsymbol{s} = 2\lambda p(0)ND$. Early measurements using a perfect crystal interferometer and independent components for the nuclear and the magnetic phase shift were performed by Badurek et al. (1976). The results have shown the characteristic beat effects of the intensity and of the polarization as predicted by Eqs. (3.52) and (3.53) (see Fig. 3.15). In the case of a strong intensity oscillation, the polarization modulation becomes weak and vice versa. The magnetic field inside a material was used by Klein and Opat (1976) for the measurement of the 4π-symmetry

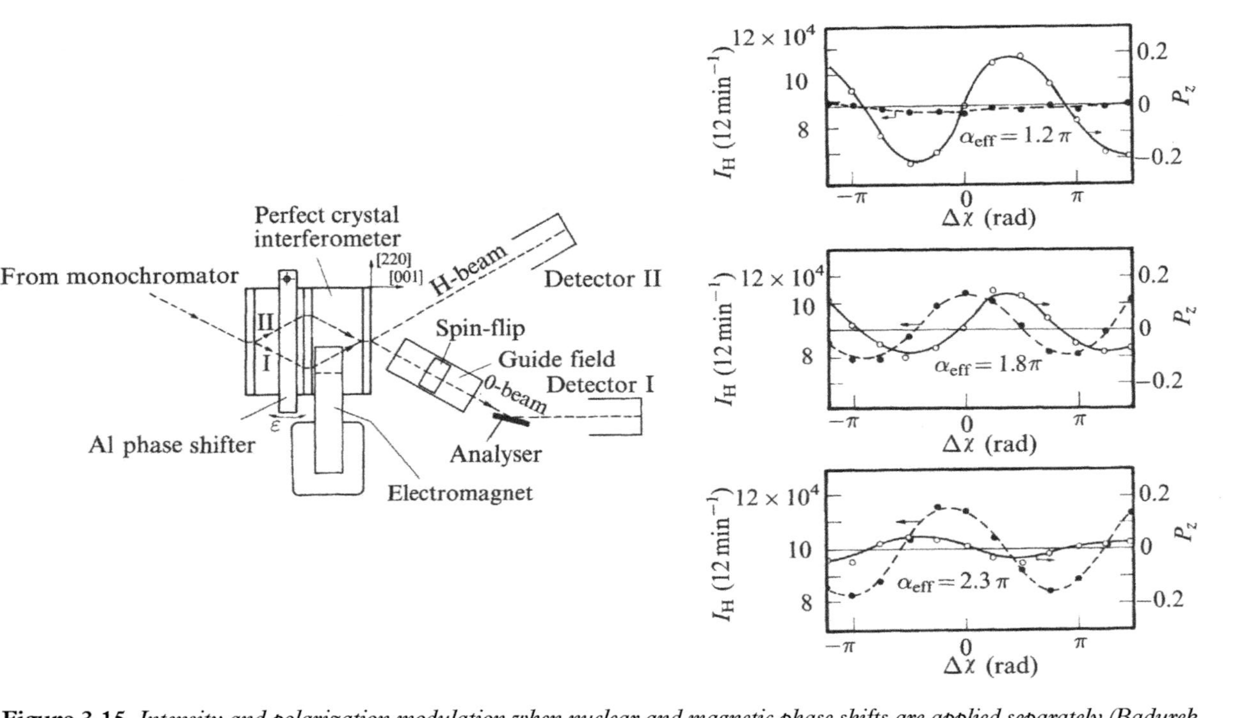

Figure 3.15 *Intensity and polarization modulation when nuclear and magnetic phase shifts are applied separately (Badurek et al. 1976)*

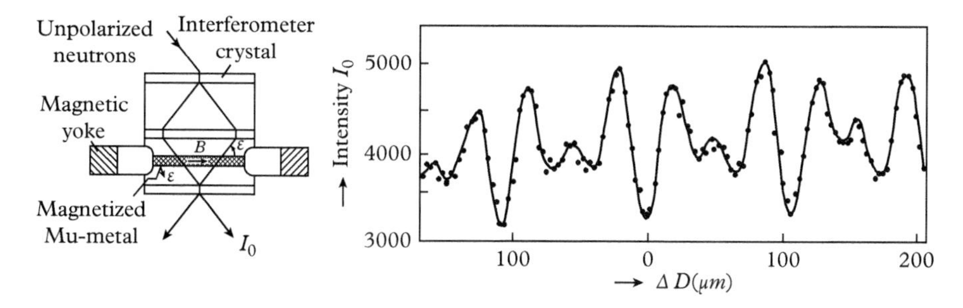

Figure 3.16 *Intensity modulation of a magnetized sample rotated in both beams (Rauch 1980)*

of spinor wave functions (Section 5.1). In both cases the nuclear phase shift was kept constant in both beams ($\chi = 0$) and only the magnetic phase shift was varied.

The Larmor precession angle α (or $p(0)$ or B) can also be obtained by macroscopic measurements of the sample magnetization and, therefore, only little effort has until now been made for such measurements. The result of a related test measurement is shown in Fig. 3.16 (Rauch 1980). A magnetized Mu-metal sheet was rotated within both coherent beams of a perfect crystal interferometer. Corrections for the effects of stray fields outside the sample must be made carefully before precise $p(0)$ or B values can be extracted. An accuracy up to 10^{-3} can be achieved if a closed magnetic yoke with small stray fields is used. This method has also been used for a precise measurement of the 4π-symmetry of spinor wave function (Section 5.1, Rauch et al. 1978a). In this experiment the nuclear phase shift χ was compensated by an opposite rotation of the magnetized sheet in both beams. Various permalloy samples have been investigated in a similar way by Nakatomi et al. (1991, 1996). It should be mentioned that magnetic scattering lengths can also be obtained by magnetic prism deflection, where Schneider and Shull (1971) found for Fe a value of $p(0) = 5.89(6)$ fm, corresponding to a magnetic moment of 2.2 Bohr magnetons per iron atom.

Magnetic inhomogeneities (domains) cause an inhomogeneous phase shift which reduces the contrast of the interference pattern drastically as has been demonstrated in various test measurements (Rauch 1979, 1980). The phase variation is closely related to the neutron depolarization phenomenon frequently used for magnetic domain investigations (Halpern and Holstein 1941, Rekveldt 1973). Before using the reduction of the interference contrast as a relevant measure for domain structure properties, the small-angle scattering effect and the response of the interferometer crystal must be adequately taken into account (see also Section 9.1).

3.2.2 Electrostatic Interactions

The Schwinger and Foldy interactions given in Eq. (3.42) are smaller than the nuclear and the Zeeman interaction and generally are less than a few percent of V_Z. They depend on the electric field produced by the charge density of the nucleus and of the electrons. In many cases the charge distribution can be approximated by a homogeneously charged sphere inside ($r < R$) and by an exponentially decreasing charge distribution for the atomic electrons outside the nucleus; that is

$$\rho(r) = Ze\left[\frac{3}{4\pi R^3}\,\theta(R-r) - \frac{1}{4\pi R_e^{\,2}}\,\frac{1}{r}\,\exp(-r/R_e)\right] \tag{3.54}$$

where R_e is related to the mean square radius of the atomic charge cloud by $R_e^2 = \langle r^2 \rangle_c/6$, and the step function $\theta(R-r) = 1$ for $r < R$, and zero for $r > R$. From the charge distribution the related form factors can be found:

$$f(Q) = \frac{1}{eZ}\frac{4\pi}{Q}\int r\rho(r)\sin(Qr)\,dr. \tag{3.55a}$$

For thermal neutrons ($QR << 1$) the nuclear form factor can be taken to be $f_N(Q) = 1$. For epithermal neutrons it can be expanded as

$$f_N(Q) = 1 - \frac{1}{10}(QR)^2 + \ldots. \tag{3.55b}$$

The atomic form factor can often be approximated by

$$f(Q) = 1 - \left[1 + (QR_e)^2\right]^{-1}. \tag{3.55c}$$

From the charge distribution, the electrostatic potential V_e and the electric field $\boldsymbol{E}$ follow from Poisson's equation

$$\nabla^2 V_e = -4\pi\rho(\boldsymbol{r}), \tag{3.56a}$$

and

$$\boldsymbol{E}(\boldsymbol{r}) = -\text{grad}\, V_e(\boldsymbol{r}). \tag{3.56b}$$

These quantities are needed for the calculation of the Fourier transform of the related interaction potentials (see Eq. 1.21 for the spin–orbit potential). This procedure yields the related scattering lengths as (e.g., Marshall and Lovesey 1971, Squires 1978, Byrne 1994) at a wave vector $\boldsymbol{Q}$ and a scattering angle θ, namely

$$b_s(\boldsymbol{Q}) = -ib_F^0(\boldsymbol{\sigma}\cdot\boldsymbol{n})\, Z\big(1 - f(\boldsymbol{Q})\big)\, ctg\frac{\theta}{2} \tag{3.57}$$

and

$$b_F(\boldsymbol{Q}) = b_F^0 Z(1 - f(\boldsymbol{Q})) \qquad \text{with} \qquad b_F^0 = \gamma e^2/2mc^2. \tag{3.58}$$

$\boldsymbol{n}$ is a unit vector perpendicular to the scattering plane $\boldsymbol{n} = \boldsymbol{k}' \times \boldsymbol{k}/(k^2\sin\theta)$ and $f(\boldsymbol{Q})$ is the total atomic form factor normalized again to unity at $\boldsymbol{Q} = 0$ as $f(0) = 1$. Therefore, there is no contribution in the forward direction $\mathbf{Q} = 0$, which reflects the fact of the electric neutrality of the entire atom. Both the Schwinger and Foldy contributions are proportional to the Foldy scattering length $b_F^0 = \gamma e^2/2mc^2 = -1.4679709(37) \times 10^{-3}$ fm, which is determined by fundamental constants only. Thus, the influence of both terms must be considered for measurements at $\boldsymbol{Q} \neq 0$, which is more important for heavy elements. The Foldy interaction arises due to the "Zitterbewegung," which causes the charge of the particle not to be concentrated in a point but to be spread out over a region of space of a dimension approximately equal to the electron's reduced Compton wavelength $\hbar/m_e c$. For bound electrons in atoms, this spreading corresponds to the classical electron radius $r_e = e^2/m_e c^2$ which permits the definition of an equivalent spherical potential well of depth $\overline{V_F}$ and radius r_e with (Fermi and Marshall 1947, Byrne 1994)

$$\overline{V}_F = -3\mu \frac{c^6 m_e^3}{e^6} = 4072.73 \text{ eV}. \tag{3.59}$$

The Schwinger scattering length b_s is purely imaginary and reflects a right–left asymmetry which is a characteristic feature of the spin–orbit interaction. The sign of the scattering length depends on the direction of the spin and, therefore, polarization effects characteristic of this interaction are expected (Obermair 1967). Related neutron scattering experiments verified the Schwinger contribution (Shull 1963, 1967; Felcher and Peterson 1975). A more detailed analysis including molecular and strong interaction contributions has been given by Gericke et al. (2008). For non-centrosymmetric crystals it causes additional effects in the dynamical diffraction pattern from perfect crystals (e.g., α-quartz), a spin-dependent Pendellösung phase shift (Alexeev et al. 1989), and finite spin rotation effects (Forte and Zeyen 1989). The Schwinger term also influences the scattering of fast (MeV) neutrons at small angles (Benenson et al. 1973, Rimawi and Benenson 1975, Dyumin et al. 1980, Baryshevskii and Zaitseva 1990).

A static electric field also produces a phase shift due to the spin–orbit interaction. It leads to a topological phase shift known as the Aharonov–Casher effect (1984) which was first observed by neutron interferometry (Cimmino et al. 1989) and it is discussed in Section 6.1.

3.2.3 Electrostatic Terms

These terms arise due to the intrinsic electromagnetic structure of the neutron. The non-vanishing magnetic dipole moment indicates an effective charge density distribution $\rho(r)$ inside the neutron, which can be understood in terms of the quark model of hadrons (e.g., Gottfried and Weisskopf 1984). By a multipole expansion of the interaction potential one gets (Foldy 1958, Sears 1986)

$$V_E(r) = (q_0 + \boldsymbol{q}_1 \cdot \boldsymbol{v} + \varepsilon \nabla^2)\ V_e(r), \tag{3.60}$$

which involves three moments of the charge distribution, namely

with $\quad q_0 = \int \rho(\boldsymbol{r}) d\boldsymbol{r}, \quad \boldsymbol{q}_1 = \int \boldsymbol{r}\rho(\boldsymbol{r}) d\boldsymbol{r},$

and $\quad \varepsilon = \frac{1}{6}\int r^2 \rho(\boldsymbol{r}) d\boldsymbol{r} = \frac{1}{6}\ e < r_c^2 > .$

Here $V_e(r)$ is the internal electrostatic potential of the neutron which is assumed to be constant over the dimension of the neutron. Equation (3.60) can be reformulated as

$$V_E = e_n V_e - d_e \boldsymbol{\sigma} \cdot \boldsymbol{E} - 4\pi\varepsilon\rho(r), \tag{3.61}$$

where $\rho(r)$ is the charge distribution of the whole atom. The related scattering lengths follow again from the Fourier transform as

$$b_E = \frac{2mZe}{h^2}\left[\frac{e_n}{Q^2} - i\frac{d_e}{Q^2}\boldsymbol{\sigma} \cdot \boldsymbol{Q} - \varepsilon\right](1 - f(Q)), \tag{3.62}$$

where e_n is a fictitious electric charge of the neutron whose limit is at present $e_n \leq (-0.4 \pm -1.1) \times 10^{-21}$ e (Baumann et al. 1988) and d_e is an assumed static electric dipole moment of the neutron whose limit is at present $d_e = < 2.9 \times 10^{-26}$ e.cm with 90% confidence (see Altarev et al. 1981, Pendlebury et al. 1984, Harris et al. 1999, Dubbers and Schmidt 2011). The limit of their contributions to the scattering lengths are $< 10^{-11}$ and $< 10^{-8}$ fm, respectively, which will be neglected in our further discussion. Nevertheless, it should be mentioned that any contribution due to a fictious electric dipole moment would contribute to the imaginary part of the scattering length only and it becomes zero for the forward direction due to the effective optical potential seen by the neutron:

$$b_{\mathrm{ed}} = -\frac{\mathrm{i}d_e\hat{\sigma}(\bar{h}-\bar{h}')e}{2m\hbar^2 Q^2}(1-f(Q)). \tag{3.63}$$

Such imaginary parts give rise to spin rotation effects which are measurable by neutron interferometry (Forte 1982). Various proposals to observe this effect have been given. A non-vanishing electric dipole moment can in principle be observed by dynamical diffraction effects from non-centrosymmetric perfect crystals (Fedorov et al. 1995, Zeyen et al. 1996). In such crystals interplanar electric fields up to 10^9 V/cm can be anticipated. No direct interferometric measurements along this line exist at the present time.

The remaining third term arises from the internal charge separation within the neutron and can be formulated as

$$b_{\mathrm{E}} = -\frac{me^2}{3\hbar^2} < r_c^2 > Z(1-f(Q)) = -b_e Z(1-f(Q)). \tag{3.64}$$

Here $< r_c^2 >$ is the mean square charge radius of the neutron. These mean square radii characterizing the distribution of the charge and the magnetization are related to the Sachs form factors G_e and G_m which in turn are related to the Dirac (F_1) and Pauli (F_2) form factors (Byrne 1994, Phillips 2007, Smith 2010):

$$\begin{aligned} G_e(Q^2) &= F_1(Q^2) + \left(\frac{\hbar Q}{2mc}\right)^2 F_2(Q^2). \\ G_m(Q^2) &= F_1(Q^2) + F_2(Q^2) \end{aligned} \tag{3.65}$$

For small momentum transfers a dipole fit can be used to relate the form factors to the charge (ρ_e) and the magnetization (ρ_m) distributions of the neutrons. The leading terms in the expansion of the integral are

$$\begin{aligned} G_e(Q^2) &= \int \rho_e e^{\mathrm{i}Qr}\mathrm{d}r = -\frac{1}{6}\langle r_e^2\rangle Q^2 + \ldots. \\ F_1(Q^2) &= \frac{1}{6}\langle r_i^2\rangle Q^2 + \ldots\ldots \\ F_2(Q^2) &= F_2(0)\left(1-\frac{1}{6}\langle r_i^2\rangle Q^2 + \ldots\ldots\right), \end{aligned} \tag{3.66}$$

where $G_e(0) = F_1(0) = 0$ has been used since the neutron has zero charge and $F_2(0) = \mu/\mu_N$ stands for normalization due to the anomalous magnetic moment. From these relations, it follow that the mean square charge radius of the neutron is

$$< r_e^2 > = -6\frac{\mathrm{d}G_e(0)}{\mathrm{d}Q^2} \cong -6\frac{\mathrm{d}F_1(0)}{\mathrm{d}Q^2} + 6\left(\frac{\hbar}{2mc}\right)^2 F_2(0) = \langle r_\mathrm{I}^2\rangle + \langle r_\mathrm{F}^2\rangle. \tag{3.67}$$

The second term arises from the Pauli form factor (Eq. 3.65) due to relativistic corrections associated with the magnetic moment of the neutron. It is known as Foldy contribution (Foldy 1958; Thomas et al. 1981; Byrne 1993a,b; Isgur 1999):

$$6\left(\frac{\hbar}{2mc}\right)^2 F_2(0) = \frac{\hbar^2}{2me^2}\, b_{\mathrm{F}} = \langle r_{\mathrm{F}}^2 \rangle = -0.127\ \mathrm{fm}^2. \tag{3.68}$$

This gives the related scattering lengths by Fourier transformation

$$b_{\mathrm{e}} = b_{\mathrm{i}} + b_{\mathrm{F}} = \frac{me^2}{3\hbar^2}\left(\langle r_{\mathrm{i}}^2 \rangle + \langle r_{\mathrm{F}}^2 \rangle\right). \tag{3.69}$$

Therefore, any deviation of a measured b_{e} value from the Foldy term can be used for an assessment of an intrinsic charge radius of the neutron. In an experiment, the neutron–electron contribution must be separated from the neutron–nuclei interaction which manifests itself in the Q-dependence of the electromagnetic part. Data evaluation of precise epithermal neutron transmission and scattering and cold neutron total reflection experiments on Pb and Bi yield $b_e = -1.33(3) \times 10^{-3}$fm and, therefore, $b_{\mathrm{i}} = 0.16(3) \times 10^{-3}$ fm. This results in an intrinsic charge radius of the neutron of $r_{\mathrm{c}} = 0.118(13)$ fm (Koester et al. 1986, 1995; Kopecky et al. 1997), which is in a fairly good agreement with a Russian estimate based upon measurements on tungsten samples (Alexandrov et al. 1985). This gives a mean square charge radius of $< r_{\mathrm{c}}^2 >= -0.113(3)\ \mathrm{fm}^2$. How the results of different measurement procedures compare with each other is discussed by Leeb and Teichtmeister (1993), Koester et al. (1995), Ioffe et al. (1996), and Kelly (2002). The mean square charge radius $< r_{\mathrm{c}}^2 >$ is related to the slope of the Sachs form factor of the neutron at zero four-momentum transfer ($q^2 = 0$) (Foldy 1958, Frauenfelder and Henley 1974), which can be obtained from form factor measurements on the basis of electron–deuteron scattering (Trubnikov 1981) and, as discussed later, from polarized electron ^{3}He measurements (Meyerhoff et al. 1994). The values from both methods are consistent but caution seems to be advisable because uncertainties in the experiments and in the data analysis are still substantial. Therefore, more precise measurements are needed for an unambiguous identification of the neutron charge radius, which may provide a crucial test for various quark theories. According to simple quark theories the ratio of the neutron to proton mean square charge radius is predicted as $< r_{\mathrm{c}}^2 >_{\mathrm{n}} / < r_{\mathrm{c}}^2 >_{\mathrm{p}} = -0.16$ (Isgur and Karl 1977, 1978), which gives, with the well-established value for the proton $< r_{\mathrm{c}}^2 >_{\mathrm{p}} = 0.862(12)\ \mathrm{fm}^2$ (Simon et al. 1980), a value for the neutron $< r_{\mathrm{c}}^2 >_{\mathrm{n}} = -0.138\ \mathrm{fm}^2$ in fairly good agreement with the experimental value. An extraction of the mean charge radius from electron–deuteron scattering is not very reliable mainly because of difficulties associated with the structure of the deuteron. The negative sign of the mean square charge radius may be attributed to an excess of positive charge near the center of the neutron and a negative charge further out (Smith 2010). The slight difference between the measured and the Foldy term can be used to extract a non-vanishing derivative of the Dirac form factor (Eqs. 3.65) and to define a corresponding Dirac charge radius

$$< r_{\mathrm{c}}^2 >_{\mathrm{D}} = -6\, \mathrm{d}F_1(0) / \mathrm{d}Q^2 = 9.5(2.2) \times 10^{-3}\ \mathrm{fm}^2. \tag{3.70}$$

Although the meaning of this quantity is not completely clear it may reflect relativistic effects of the internal neutron structure and the impossibility of a general separation of the Foldy and the Dirac terms (Aleksandrov 1994, Koester et al. 1995). Therefore, it is justified to identify the mean charge radius directly to the slope of the electric (Sachs) form factor (Eq. 3.67), which gave in a recent transmission measurement covering an energy range between 0.1 and 1000 eV with liquid ^{208}Pb a value $< r_c^2 > = -0.119 \pm 0.004 \pm 0.003$ fm^3 in good agreement with previous values (Kopecky et al. 1994, 1995).

Another possible method uses the fine structure of multi-Laue rocking curves for the extraction of the Q-dependent scattering amplitude. Related experiments for the X-ray case were quite successful (Bonse and Teworte 1980; Teworte and Bonse 1984; Deutsch and Hart 1985; Lemmel 2007, 2013). The question arises whether similar measurements are feasible using neutron interferometry. In any case measurements at $Q \neq 0$ must be included, and from the difference to the scattering length at $Q = 0$ the neutron–electron contribution can perhaps be separated. There exists an interferometric attempt for this extraction by rotating a piece of silicon as a phase shifter within the interferometer through a Bragg reflection (Graeff et al. 1978). In this case the phase shift changes drastically according to the variation of the dispersion surface (Chapter 11), which, in principle, enables the extraction of the scattering amplitude at a finite Q value (Fig. 3.17). If a perfect crystal sample is rotated in the interferometer through a Bragg position, the index of refraction deviates from its normal value $n = 1 - \delta$ (Eq. 3.20) to $n \cong 1 - 2\delta$ at the edges of the Darwin reflection curve at $|y| = 1$ (see Chapter 11). Related calculations have been done by Wietfeldt et al. (2005) and Lemmel (2007). The phase shift as a function of the deviation from the exact Bragg position is shown in Fig. 3.18 for a (220) reflection from a 1-mm-thick silicon crystal. Whether these curves can be used to extract small Q-dependent phase shifts, like electromagnetic or gravitational terms, has yet to be proven experimentally (Rauch 1989a). Related measurements have been reported by Springer et al. (2010b). They achieved qualitative agreement with theoretical predictions, but the sensitivity was not high enough to extract Q-dependent phase shifts. The very high angular sensitivity of about 10^{-6} arcsec may also be used for measurements of the neutron–electron interaction, and perhaps of short-range gravitational forces and accurate measurements of the Coriolis force (Zawisky et al. 2011).

When larger rotations are considered the effect of additional Bragg reflections must be taken into account, which makes the calculations somehow more difficult (see e.g., Lemmel 2013).

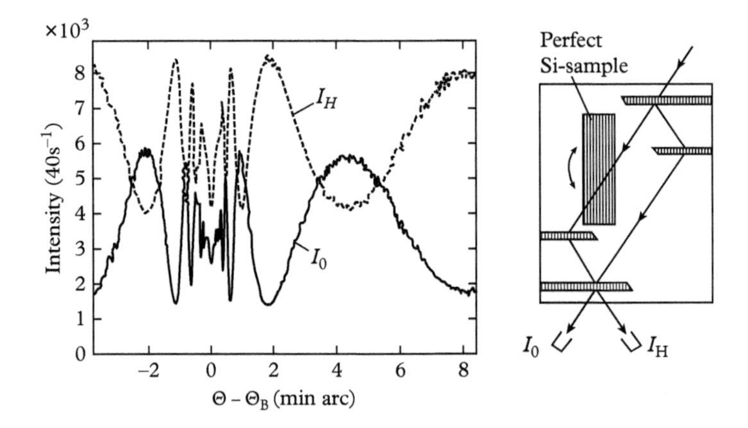

Figure 3.17 *Intensity modulation when a perfect crystal sample is rotated near to Bragg diffraction in one of the coherent beams (Graeff et al. 1978, Rauch 1989a)*

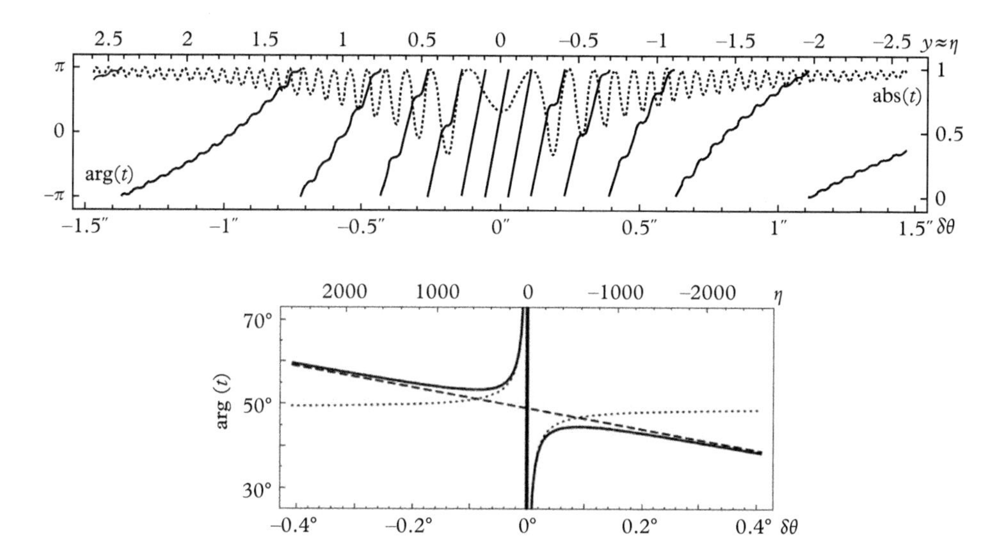

Figure 3.18 *The argument angle of the transmission function near to a Bragg diffraction as a function of the beam deviation from the exact Bragg setting. Reprinted with permission from Lemmel 2007, copyright 2007 by the American Physical Society.*

3.2.4 Electrical Polarizability Term

The internal charge distribution of the neutron is caused by its quark structure and it can therefore be expected that the neutron may be polarized by an electric field $\boldsymbol{E}$. Measureable effects are expected only if the interaction with the electric field near to the atomic nucleus is considered. All laboratory fields give negligible contributions. The induced electric dipole moment is assumed to be proportional to the electric field strength, that is $p_{\text{Induced}} = \alpha \boldsymbol{E}$. The corresponding interaction Hamiltonian for this induced polarization (IP) effect is

$$V_{\text{IP}} = \frac{1}{2} \alpha E^2, \tag{3.71}$$

where α is the electric polarizability of the neutron. The electric field inside and outside the nucleus is caused by the nuclear charge and the electron cloud. Because this term is quadratic in $\boldsymbol{E}$, the related scattering length does not vanish for $Q \to 0$ and it is evident that the Coulomb field near to the nucleus gives the dominant contribution. Thus, one sees that the related scattering length is the Fourier transform of $V_{\text{IP}}(\boldsymbol{r})$ (Thaler 1959, Leeb et al. 1984, Sears 1986, Schmiedmayer et al. 1988)

$$b_p = \frac{2m}{\hbar^2} \int V_{\text{IP}}(r) \frac{\sin Qr}{Qr} r^2 \mathrm{d}r = \frac{Z^2 e^2}{\hbar^2} \frac{m\alpha}{R} \left[\frac{6}{5} - \frac{1}{4} \pi \, QR + \ldots \right], \tag{3.72}$$

where R is the radius of the nucleus. Only the leading term contributes for thermal neutrons.

Experimental attempts in the determination of the electrical polarizability are based on epithermal neutron transmission or scattering experiments. A slight energy (or Q) dependence is expected due to the second term in Eq. (3.72). Aleksandrov (1983), Koester et al. (1986), and Schmiedmayer et al. (1988, 1991) reported a value definitively different from $0\,(\alpha = (1.20 \pm 0.15 \pm 0.20) \times 10^{-3}\ \text{fm}^3)$. A remeasurement with a pure ^{208}Pb target yielded $\alpha = (1.71 \pm 0.24 \pm 0.43) \times 10^{-3}\ \text{fm}^3$ (Riehs et al. 1994). Elastic gamma-deuteron Compton scattering gave a value of $\alpha = (1.13 \pm 0.07 \pm 0.1) \times 10^{-3}$ fm (Griesshammer et al. 2007), which is in rather good agreement with the neutron–lead scattering experiment. It should be mentioned that there are recent measurements and alternative data evaluations which are still in agreement with $\alpha = 0$ (Koester et al. 1995, Aleksejevs et al. 1997). Another method uses the analogy to the proton where from photoabsorption data and the related sum rules, the polarizability term can be extracted (Bernabeau and Ericson 1972). The most precise measurements and calculations support values for α of about $1 \times 10^{-3}\ \text{fm}^3$, which is similar to the related value for the proton and which is in rough agreement with various calculations based on simple quark models (Arnold 1973, Dattoli et al. 1977, Schröder 1980, Weiner and Weise 1985, Schöberl and Leeb 1986). Lattice gauge theory calculations of the polarizability are improving (Christensen et al. 2005). The value $\alpha = 1 \times 10^{-3}$ fm gives a contribution to the measured scattering length of $b_p = -0.0006$ fm for C and $b_p = -0.06$ fm for U. For heavy elements this contribution is larger than the error bars of the measured values. Their experimental separation from the nuclear part is not really solved at present because it behaves in the low-energy limit like the nuclear part.

3.3 Parity Violating Interactions

The beta decay of nuclei and of the neutron itself demonstrate that weak interaction effects exist between strongly interacting particles as well. This causes additional contributions to the coherent forward scattering amplitude which can be written as (e.g., Heckel 1989, Byrne 1999):

$$f_{\text{pnc}} = f_1 \boldsymbol{\sigma} \cdot \hat{k} + f_2 \hat{I} \cdot \hat{k} + f_3 \boldsymbol{\sigma} \cdot (\hat{k} \times \hat{I}). \tag{3.73}$$

The various scattering amplitudes f_1, f_2, and f_3 can be extracted through the reversal of the appropriate vectors for the neutron polarization $\langle\boldsymbol{\sigma}\rangle$, the nuclear polarization $\langle\hat{I}\rangle$, or the direction of the momentum, $\hbar\mathbf{k}$.

The f_1 term gives a helicity dependence of the index of refraction (Michel 1964; Stodolsky 1974, 1982; Krupchitsky 1989). This results in a rotation of the neutron polarization vector about the momentum of the beam

$$\phi_{\text{pnc}} = -2\,N\,\lambda D\,\text{Re}\,f_1. \tag{3.74}$$

A simple Born approximation estimate gives values of $\phi_{\text{pnc}} \sim 4 \times 10^{-6}$ rad/cm for medium heavy elements. The imaginary part can be obtained from the transmission asymmetry for the two helicities

$$A = (2\lambda/\hbar)\,\text{Im}\,f_1/\sigma_t, \tag{3.75}$$

where σ_t denotes the total attenuation cross-section. Neutron spin polarimeters have been used to measure the rotation effect. The largest effects have been observed in nuclei (^{117}Sn, ^{139}La) where a narrow p-wave resonance lies in the proximity of a nearby s-wave resonance which causes a strong s- and p-wave mixing (Forte et al. 1980, Kolomensky et al. 1981). The enhancement factors depend on the resonance parameters of these resonances (Bunakov and Gudkov 1981)

$$f_1 = \sqrt{\frac{\Gamma_n^n(E_s)\,\Gamma_p^n(E_p)}{k_s k_p}}^{3} \cdot \frac{< s\,|H_w|\,p > \hbar\, k}{(E - E_s + i\Gamma_s/2)\,(E - E_p + i\Gamma_p/2)}. \tag{3.76}$$

H_w denotes the weak interaction Hamiltonian between these states. For ^{227}Sn a parity violating rotation of $\phi_{pnc} = -36.7(2.7) \times 10^{-6}$ rad/cm has been measured (Forte et al. 1980), which corresponds to a matrix element for weak interaction on the order of 1 meV. The enhancement factor due to level mixing can become 10^3. The most direct and interesting experiment would be with a hydrogen target. For a para-hydrogen target of 20 cm length one may expect $\phi_{pnc} \sim 2.10^{-7}$ rad, which is below the current experimental sensitivities.

The product of the f_2 and f_3 term leads to parity conserving but time reversal symmetry violating observables. They are expected to be extremely small for thermal neutrons, but enhancement factors near s- and p-wave resonances are expected as well (Stodolsky 1982).

Direct interferometric investigations concerning parity violating phenomena have not been carried out yet. The separation of such small effects from effects arising from paramagnetic impurities, residual magnetic fields, or external influences is difficult to achieve. The search for time symmetry violating terms is difficult mainly due to the requirement of a precise control of the nuclear polarization. Therefore, most efforts in this direction are still focused on the search for an electric dipole moment of the neutron (e.g., Byrne 1994, Dubbers and Schmidt 2011).

3.4 External Influences

Quantum systems cannot be completely isolated from the environment and any optical component of an instrument is influenced by external effects as well. In the case of interaction, the disturbance of the quantum system increases for large momentum transfers and large coherent separations of parts of the wave function. These decoherence phenomena are discussed in Section 4.6. But there are also other more classical influences upon the phase shifts within an interferometer, which will be discussed here. When they are controlled and a proper feedback mechanism is applied these effects can be compensated.

3.4.1 Atmospheric Effects

In most cases the phase shifts are measured in comparison with the phase shift of air (ϕ_a). Therefore, the measured phase shifts χ_m must be corrected for the air displaced by the sample to obtain the phase shift of the sample in relation to vacuum χ_s

$$\chi_s = \chi_m + \phi_a, \tag{3.77}$$

where

$$\phi_a = -\lambda\, D_{eff} \sum_i N_i\, b_{ci}.$$

Thus, one needs to know the composition of the air expelled by the sample. The mol-volume is $V_m = 22.4146$, which gives the number of air molecules in 1 m^3 to $N_A = L/V_m(L = 6.022 \times 10^{23}\ \text{mol}^{-1}$, Loschmidt number). In case of humidity saturation one has $N_{H_2O} = 1.079$ mol H_2O in

1 m^3. When one defines the mean coherent scattering length of an air molecule, $b_{cA} = \sum_i N_i b_{ci}/N =$ 17.07 fm and that of water $b_{cH_2O} = -1.675$ fm, one gets

$$\overline{b_c} = \frac{N_A - N_{H_2O}\varepsilon}{N_A} \cdot b_{cA} + \frac{N_{H_2O}\varepsilon}{N_A} b_{cH_2O} \tag{3.78}$$

$$(\varepsilon = \text{humidity}, 0 \leq \varepsilon \leq 1)$$

Air under normal atmospheric conditions can be considered as an ideal gas

$$pV_m = RT, \tag{3.79}$$

which gives the phase shift of humid air as

$$\phi_a = -\lambda D_{eff} \frac{L.p}{R.T} \overline{b_c}, \tag{3.80}$$

where $R = 8.314\ \text{JK}^{-1}\text{mol}^{-1}$. This gives for air under normal atmospheric conditions (in vol%: N_2, 78.084; O_2, 20.946; CO_2, 0.033; Ar, 0.934) for $T = 25°$C and a pressure of 992 mbar and different humidities (0, 40, and 100%)

$$\begin{aligned} \phi_a^{(0)} &= -0.411 \times D\,[\text{cm}] \times \lambda\,[\text{Å}] \\ \phi_a^{(40)} &= -0.407 \times D\,[\text{cm}] \times \lambda\,[\text{Å}] \\ \phi_a^{(100)} &= -0.4005 \times D\,[\text{cm}] \times \lambda\,[\text{Å}]. \end{aligned} \tag{3.81}$$

The related λ thicknesses are about $D_\lambda \cong 15.2\ \text{cm}/\lambda\,[\text{Å}]$. Figure 3.19 shows typical dependences of this correction function. This air correction is for most materials on the order of 10^{-4} compared to the phase shift of a compact sample but it becomes more important when less dense samples (gases) are measured or when the sample sweep method is used, where phase differences are measured (e.g., Hammerschmied et al. 1981). In the non-dispersive sample position the phase shift due to air becomes wavelength-independent as well. For an interferometer using the (220)-reflection and for normal atmospheric conditions as mentioned earlier one gets (Eq. 3.29)

$$\phi_a^{(40)} = -2 \sum_i N_i\, b_{ci}\, d_{hkl}\, D_0 = -1.551 \times .\, D_0\ [\text{cm}]. \tag{3.82}$$

From the numbers shown, it follows that for precision measurements of scattering lengths the atmospheric conditions must be considered or such measurements must be done in comparison to vacuum.

3.4.2 Temperature Effect

The atoms in matter are never at rest and their actual positions $\boldsymbol{R}_j = \boldsymbol{R}_j^0 + \boldsymbol{u}_j$ are distributed over a thermal cloud with a mean square displacements $< u^2 >_T$, about equilibrium positions $\boldsymbol{R}_j^0$ in solids. These displacements depend on the temperature and can be different for different directions in the crystal. The interaction of neutrons with the nucleus in the thermal cloud causes a $\boldsymbol{Q}$-dependent

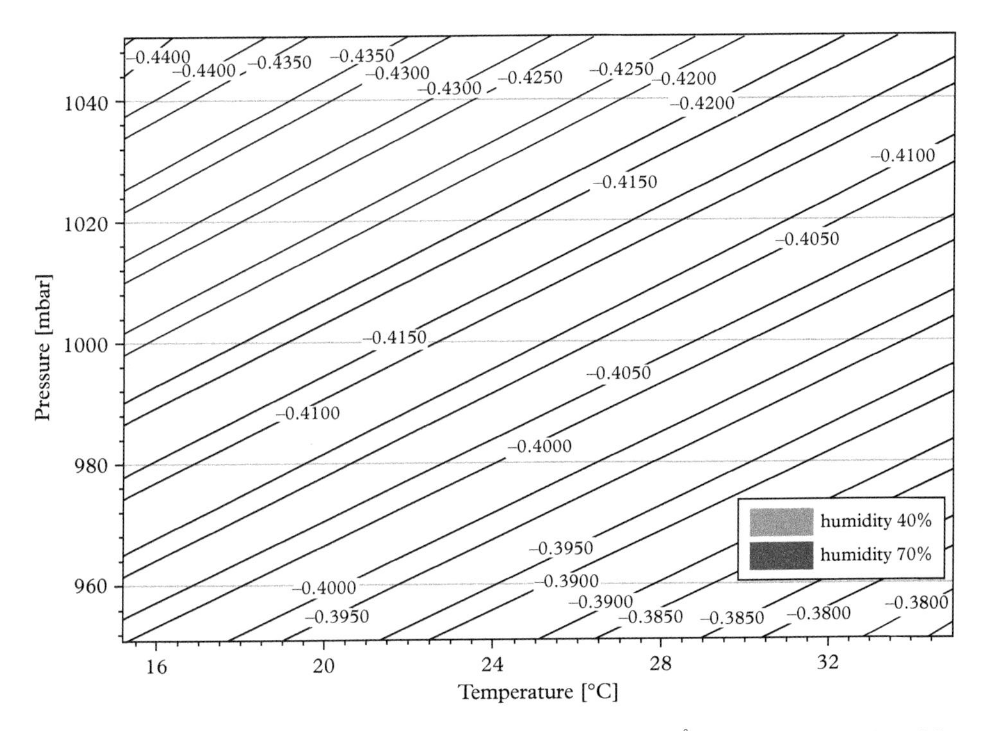

Figure 3.19 *Phase shift (parameters of the curves) of 1 cm air for 1 Å neutrons as a function of the temperature and the pressure for two values of the humidity*

form factor which is given by the well-known Debye–Waller factor requiring a replacement in all of our formulas of

$$b_j \rightarrow b_j e^{-W_j}. \tag{3.83}$$

See Debye (1913), Waller (1923), and Marshall and Lovesey (1971).

W_j is related to the mean square displacement of the atoms along the scattering vector, that is

$$W_j = \frac{1}{2} < (\boldsymbol{Q} \cdot n\boldsymbol{u}_j) >_T, \tag{3.84}$$

where a thermal average over the equilibrium distribution of phonons at temperature, T, is implied by the notation $< >_T$. For a harmonic crystal with a cubic Bravais lattice this factor can be written as

$$W(Q) = \frac{\hbar}{12M} Q^2 \int \frac{g(\omega)}{\omega} \coth(\hbar\omega/2k_B T)\, d\omega, \tag{3.85}$$

where M is the mass of the scattering atom and $g(\omega)$ is the lattice-vibrational frequency spectrum. If this spectrum is approximated by a Debye spectrum ($g(\omega) = \omega^2/\omega_D^3$ for $\omega \leq \omega_D = k_B\theta_D/\hbar$) one obtains for temperatures far above and below the Debye temperature θ_D

$$W(Q) = 3\hbar^2 Q^2/8k_B\theta_D M\,, \quad \text{for} \quad T << \theta_D, \tag{3.86a}$$

and

$$W(Q) = \left(3\hbar^2 Q^2 / 2k_B\theta_D M\right)(T/\theta_D), \tag{3.86b}$$

respectively.

These formulas show that there is still a contribution of the phonons to the scattering length at $T = 0$ due to the zero point fluctuations. There is no contribution to b_j at $Q \to 0$ where neutron interferometric measurements are usually made. This factor must be taken into account carefully if scattering lengths are extracted from measurements at $Q \neq 0$, for example from Bragg diffraction data, and especially if samples with a low Debye temperature are measured. A simple model for the calculation of the Debye–Waller factor for elemental crystals was developed by Sears and Shelley (1991), which gives typical accuracies of 2–3%. A collection of data for cubic crystals and their relation to bulk properties can be found in the literature (Butt et al. 1988, 1993).

The Debye–Waller factor for an anharmonic crystal can be written as

$$W(Q) = \frac{1}{2} <(\boldsymbol{Q}\cdot\boldsymbol{u})^2>_T - \frac{1}{6} <(\boldsymbol{Q}\cdot\boldsymbol{u})^3>_T - \frac{1}{24}\left[<(\boldsymbol{Q}\cdot\boldsymbol{u})^4>_T - 3<(\boldsymbol{Q}\cdot\boldsymbol{u})^2>_T^2\right] + \ldots, \tag{3.87}$$

which again goes to zero for $Q \to 0$. Similar situations exist for liquids and gases, where the effects of thermal motion on the measurement of coherent scattering amplitudes, say by neutron interferometry, do not play a role (Champenois et al. 2008).

There is a difference to the situation of atoms or molecules interacting with gas atoms or molecules where the real part of the scattering amplitude exhibits distinct resonances and therefore a gas does not act as a homogeneous phase shifter; but the atoms also as an individual decoherence object (Chapman et al. 1995, Hornberger et al. 2003, Cronin et al. 2009). In the neutron case there is a collective interaction with many gas atoms and the resulting wave function can be described by the index of refraction formalism. Thus, in the atom–atom case the loss of coherence is more related to decoherence, whereas in the neutron–atom case the collective interaction described by an index of refraction is the major effect. Mixtures of both effects are feasible (Champenois et al. 2008, Davidovic et al. 2010).

3.4.3 Magnetic Field Effects of Paramagnetic Substances

The magnetic scattering lengths of magnetic atoms has been discussed in Section 3.2.1. Here it should be remembered that dia- and paramagnetic atoms cause a magnetic contribution to the forward scattering length only if a magnetic field H is present. The related spin-dependent scattering length is

$$p(0) = \mp \frac{2\mu m \chi_m H}{N\hbar^2}, \tag{3.88}$$

where χ_m is the magnetic susceptibility (e.g., Martin 1967). This gives for a field of $H = 10$ kG a contribution of ∓ 0.00135 fm for bismuth and ± 0.00123 fm for vanadium. For most other para- and diamagnetic materials the contributions are considerably smaller. Using the polarized neutron diffraction technique this small polarization can be used to measure the magnetic form factor, since $p(\boldsymbol{Q}) = p(0)\,\mathbf{f}(Q)$ (Shull and Ferrier 1963, Shull and Wedgwood 1966, Stassis 1970).

A magnetic field produces an additional effect due to its action on the nuclear polarization (see Eq. 3.24). In most situations the condition $\mu_K H/k_B T << 1$ is fulfilled and one gets

$$f_K = \frac{<I>}{I} = \frac{1}{3}\frac{\mu_K H}{k_B T}\frac{(I+1)}{I}, \tag{3.89}$$

which gives for the coherent scattering length according to Eq. (3.24):

$$(b_c)_p = b_c \pm \frac{b_+ - b_-}{3}\frac{\mu_K H}{k_B T}\frac{I+1}{2I+1}. \tag{3.90}$$

The additional term is small at room temperature even for vanadium (0.000062 fm) but it must be considered for measurements at very low temperatures.

3.4.4 Local Field and Holography Effects

It is known from light optics that the decay properties of an atom depend on the surrounding environment of the atom (Dexhage 1974, Goy et al. 1983, Grangier et al. 1983, Hulet et al. 1985, Cook and Milonni 1987). This can be explained by a variation of the available phase space or the mutual interaction between the atom and the radiation field. In the field of neutron optics a similar effect exist if the scattering properties of a nucleus is considered together with the effect of its surrounding. The nucleus sees not only the incident wave but also the waves scattered by all the other nuclei. Therefore local field effects must be included in a complete description. It is also generally known that the familiar expression for the index of refraction (Eq. 3.16) does not fulfill the optical theorem of scattering theory (e.g., Sears 1982a). Several attempts have been made to solve this problem on the basis of the theory of dispersion (Lax 1951, 1952; Sears 1982a) or by using Green's function methods (Ekstein 1953, Edwards 1958, Dietze and Nowak 1981, Nowak 1982, Warner and Gubernatis 1985). Here we follow the formalism of Sears (1985, 1996), who also discussed the consequences for interferometric measurements. According to the general theory of dispersion which includes local field effects the index of refraction must be written as

$$n^2 = 1 - \frac{\lambda^2 N}{\pi} b \cdot c, \tag{3.91}$$

where b also contains an imaginary term, linear in k, besides the usual imaginary part due to absorption ($b_c = b'_c - ib''_c$ with $b''_c = \sigma_a/2\lambda$). That is,

$$b = b_c - ikb_c^2 - \ \ldots, \tag{3.92}$$

and c is the local field correction $c = (1-\mathcal{J})^{-1}$ with

$$\mathcal{J} = Nb_c \int e^{i\mathbf{k}\cdot\mathbf{r}} \frac{e^{ikr}}{r}\,[1-g(r)]\,d\mathbf{r}. \tag{3.93}$$

This equation simplifies for very slow neutrons ($k \to 0$) to

$$\mathcal{J} = 2\pi N b_c \int_0^\infty [1-g(r)]\,dr\,. \tag{3.94}$$

The pair correlation function $g(r)$ can be taken from the literature for various substances (e.g., Marshall and Lovesey 1971, Bacon 1975). The values for $\mathcal{J}$ depend on the interparticle correlation and on the coherent scattering length b_c in the form $\mathcal{J} \cong b_c/r_0$, where r_0 denotes the nearest-neighbor distance in solid and liquid samples and the mean-free path in gaseous samples. Typically $\mathcal{J}$ is on the order of 10^{-4} (for liquid $D_2 : \mathcal{J} = 1.84 \times 10^{-4}$; for Pb : $\mathcal{J} = 0.89 \times 10^{-4}$). In neutron interferometry the real part of the index of refraction determines the intensity modulation. The phase shift can always be written as

$$\chi = -\lambda N b'_{\text{eff}} D. \tag{3.95}$$

Local field effects are contained in the effective scattering length b'_{eff}, which is related to the true scattering length b_c by

$$b'_{\text{eff}} = b_c \left(1 + \mathcal{J}' + \frac{\lambda^2}{4\pi} N b_c\right). \tag{3.96}$$

At present in most cases, the correction factors are below the sensitivity limits of the experiments. No dedicated experiment is known to verify these local field effects whose magnitude has now been reached in the level of advanced interferometric precision measurements.

A special situation exists if a perfect crystal is inserted in the interferometer as a sample. In this case, the wave vectors inside the material change according to the dispersion relation of the dynamical diffraction theory (Fig. 3.17, Chapter 11). At the Bragg position this variation can be written as

$$n \cong 1 - \frac{1}{2\pi} \lambda^2 N b(1 + y) \cdot \left[1 \pm \sqrt{1 - \frac{2}{1 + y}}\right], \tag{3.97}$$

where y is the characteristic parameter for dynamical diffraction which is directly proportional to the deviation from the exact Bragg condition $y \propto (\theta - \theta_B)$. The index of refraction far from the Bragg position ($y \rightarrow \infty$) and that for the edges of the Darwin reflection curve at $|y| = 1$ vary according to (see Figs. 3.17 and 3.18)

$$\frac{n_{y=1} - 1}{n_\infty - 1} \cong 2. \tag{3.98}$$

An imaginary part appears for $|y| \leq 1$, which must be treated in a way similar to an absorption term. The strong variation of the index of refraction in the range of the dynamical reflection curve may be interpreted as an enhanced internal field effect due to the coherence of the scattered waves (Lemmel 2007). It causes a drastic change of the oscillation frequency of the interference pattern as demonstrated in the measurements shown in Fig. 3.17. It might be used in the future to extract Q-dependent scattering lengths where the neutron–electron interaction contributes (Eq. 3.62).

The local field effect can become strong when distinct nuclei are dispersed within the sample. This provides the basis for *neutron holography*. Two experimental methods which either put the radiation source inside and the detector outside the object or vice versa can be utilized. The first method requires strongly incoherent scattering nuclei (hydrogen) and the second method uses strongly absorbing nuclei (cadmium) inside an object composed of coherently scattering nuclei. Under these conditions, the incident wave interferes with the scattered wave and the interference pattern can be detected with an outside detector; or alternatively the scattered waves produce an

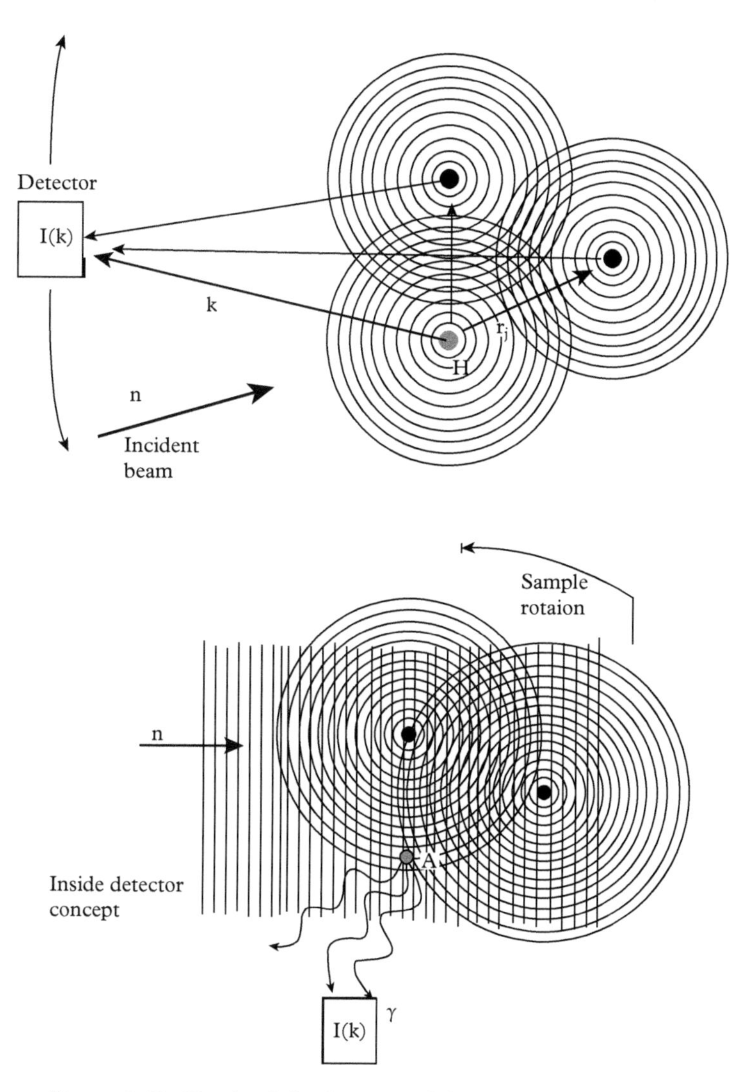

Figure 3.20 *Sketch of the in-source (left) and in-detector (right) methods for neutron holography. Reprinted from Ioffe and Cser et al. 2004, Copyright 2004, with permission from Elsevier.*

interference pattern at the position of the absorbing atoms, which can be detected by the intensity of the resulting gamma decay radiation (Fig. 3.20). By means of these techniques the position of atoms near to the scattering or absorbing atoms can be detected (Sur et al. 2001; Cser et al. 2001, 2002, 2004).

4
Coherence and Decoherence

Coherence phenomena play an important role in any kind of interferometry (Gabor 1956; Glauber 1963; Mandel and Wolf 1965, 1995; Born and Wolf 1975; Walls and Milburn 1994; Mandel and Wolf 1995). In this chapter we summarize some known results, add some new ones obtained with neutrons, and analyze them in terms of general quantum optics, which can be applied to photons and matter waves as well. Indeed, neutrons have many well-known particle properties, but in interference experiments they behave like wave fields which provide the connection to the quantum optical terminology. Coherence appears as a system property (neutron + apparatus) which persists as long as it does not become destroyed by statistical or dissipative effects. These effects are known as decoherence phenomena and are treated in Section 4.6.

In the course of this chapter it will become clear that coherence and interference properties are closely connected to the phase-space formulation of quantum mechanics which has its roots in the classic work of Wigner (1932). The main tool of that formalism is the introduction of phase-space "quasi-probability" distribution functions, from which the Wigner distribution function is the best known one. The connection to quantum optics has been formulated by Glauber (1963, 1965) and summarized by Klauder and Skagerstam (1985) and Mandel and Wolf (1995). Various kinds of phase-space distribution functions can provide useful physical insights providing various practical advantages. A general review including dynamical features of quantum distribution functions demonstrates that advantage (Lee 1995). Here, the neutrons within an optical device will, in many cases, be considered to behave as coherent states, known in quantum optics as states which behave most similarly to classical states (Lamb 1995). They exhibit Gaussian distribution functions, obey a minimum uncertainty relation, and show Poissonian counting statistics. Squeezing phenomena will also be discussed as they demonstrate how non-classical states can be made out of coherent states. The coherence in the neutron case is limited to the related coherence volume of the beam and does not concern the whole beam cross-section as it exists for laser beams, and as it more recently has also been achieved for synchrotron radiation (Abernathy et al. 1998).

One feature of non-classical quantum states is related to the squeezing phenomena. In optics, squeezing is mainly treated between the self-adjoint canonical operators of the field amplitude and the phase of the field. Here we deal with the self-adjoint canonical operators of space and momentum variables which fulfill the Heisenberg commutation relation in the form

$$[x, p] = i\hbar, \tag{4.1}$$

where these quantities can be written in terms of creation (a^+) and annihilation (a^-) operators as

$$x = \sqrt{\frac{\hbar}{2m\omega}}(a^- + a^+), \tag{4.2}$$

Neutron Interferometry. Second Edition. Helmut Rauch and Samuel A. Werner.

and

$$p = \mathrm{i}\sqrt{\frac{m\hbar\omega}{2}}(a^{+} - a^{-}).$$

Here ω represents an arbitrary positive scaling factor often interpreted as the eigen frequency of a harmonic oscillator. The coherent state feature of a free-moving radiation field has also been addressed by Lamb (1995) for many optical configurations including the Mach–Zehnder interferometer setup. In this terminology the spatial and momentum spread of a free particle beam are related to the uncertainties of a harmonic oscillator as

$$(\delta x)^2 = \frac{\hbar}{2m\omega},$$
$$(\delta p)^2 = \frac{m\hbar\omega}{2}. \tag{4.3}$$

These equations can also be obtained in the language of stochastic quantization where a random force is added to the fundamental Langevin equation of a harmonic oscillator (see Chapter 12, Namiki and Kanenaga 1998). These basic uncertainties also cause finite coherence properties associated with all real quantum experiments.

The effect of finite coherence of matter waves was first studied in electron biprism interferometry, where high-order interferences (above 100,000) are feasible (Möllenstedt and Dücker 1954, Gabor 1956, Lenz and Wohland 1984). Coherence is also the basic feature for any electron holography application (Tonomura 1987). Many features are connected to coherence lengths and coherence times which reflect a certain kind of self-consistency of the wave function within a certain spatial volume or time interval. It will be recognized that the coherence features of the various kinds of radiation are quite different, resulting in a variation of optical phenomena. The coherence lengths in neutron optics range from several ångstrom to several micrometers, whereas the coherence lengths in optical interferometry are many orders of magnitude larger. Those in atom interferometry are usually considerably smaller (several ångstroms) due to the limited monochromaticity of atom beams (e.g., Ekstrom et al. 1992, Miniatura et al. 1992, Riehl et al. 1992, Chapman et al. 1995, Berman 1996, Scully and Zubairy 1997). Early comprehensive treatments of coherence property measurements for the neutron case have been given by Petraschek (1987) and Rauch et al. (1996).

4.1 Basic Relations

4.1.1 Mach–Zehnder Interferometer in Second Quantization

We begin here by analyzing the Mach–Zehnder interferometer in a formal way (Zeilinger 1981b, Dunningham and Vedral 2011). A neutron wave function describing a neutron initially in beam path $|1>$ becomes behind a 50:50 beam splitter (Fig. 4.1)

$$\frac{1}{\sqrt{2}}\left(|1\rangle_2|0\rangle_3 + \mathrm{i}|0\rangle_2|1\rangle_3\right). \tag{4.4}$$

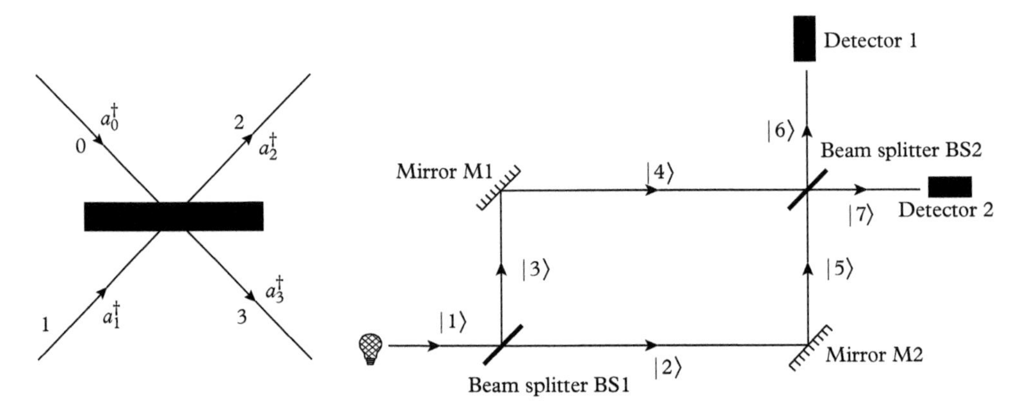

Figure 4.1 *A beam splitter and a Mach–Zehnder interferometer. Reprinted with permission from Dunningham and Vedral 2011, © Imperial College Press.*

The factor "i" is due to the $\pi/2$ phase change in the case of reflection. This can be reformulated when we write the initial state as

$$|0\rangle_0|1\rangle_1 = a_1^+|0\rangle|0\rangle \tag{4.5}$$

and Eq. (4.4) can be written in terms of creation and annihilation operators

$$\frac{1}{\sqrt{2}}\left(|1\rangle_2|0\rangle_3 + \mathrm{i}|0\rangle_2|1\rangle_3\right) = \frac{1}{\sqrt{2}}(a_2^+ + \mathrm{i}a_3^+)|0\rangle|0\rangle, \tag{4.6}$$

which gives

$$a_1^+\,|0\rangle|0\rangle \rightarrow \frac{1}{\sqrt{2}}(a_2^+ + a_3^+)\,|0\rangle|0\rangle \tag{4.7}$$

or

$$a_1^+ \rightarrow \frac{1}{\sqrt{2}}(a_2^+ + \mathrm{i}a_3^+). \tag{4.8}$$

Here we are using the rotation of a forward arrow (→) to mean "evolves into"; generally, the arrow can be replaced with an equal sign without loss of meaning. Similarly when we start with a neutron in beam |0>

$$a_0^+ \rightarrow \frac{1}{\sqrt{2}}(\mathrm{i}a_2^+ + a_3^+). \tag{4.9}$$

The mirror crystals in the middle of the interferometer change the operators, such that

$$\begin{aligned} a_2^+ &\rightarrow \mathrm{i}a_5^+ = -a_5^+ \\ a_3^+ &\rightarrow \mathrm{i}a_4^+ = -a_4^+ \end{aligned} \tag{4.10}$$

and the second beam splitter makes

$$a_4^+ \to \frac{1}{\sqrt{2}}(ia_6^+ + a_7^+)$$
$$a_5^+ \to \frac{1}{\sqrt{2}}(a_6^+ + \mathrm{i}a_7^+). \tag{4.11}$$

This gives

$$\frac{1}{2}\left[\mathrm{i}(a_6^+ + ia_7^+) - (\mathrm{i}a_6^+ + a_7^+)\right] = -a_7^+. \tag{4.12}$$

In this case we always find the neutron in outgoing beam |7>, i.e., in the forward direction.

When a phase shifter is introduced into beam |4> we get the operator describing the outgoing beams:

$$\frac{1}{2}\left[\mathrm{i}(a_6^+ + \mathrm{i}a_7^+) - e^{\mathrm{i}\chi}(\mathrm{i}a_6^+ + a_7^+)\right] = \sin\left(\frac{\chi}{2}\right)a_6^+ - \cos\left(\frac{\chi}{2}\right)a_7^+. \tag{4.13}$$

From this operator one immediately gets the well-known intensity relations of the two outgoing beams:

$$P_6 = \sin^2\left(\frac{\chi}{2}\right) = \frac{1}{2}(1 - \cos\chi)$$
$$P_7 = \cos^2\left(\frac{\chi}{2}\right) = \frac{1}{2}(1 + \cos\chi), \tag{4.14}$$

where χ is the phase shift introduced by the phase shifter. This result is, of course, identical to our initial discussion of the Mach–Zehnder interferometer in Chapter 2 (Eq. 2.2). An interesting matrix formalism to describe the action of beam splitters has been introduced by Zeilinger (1981b). Input and output channels are connected by Pauli spin matrices which connect interferometry to quantum q-bits used in quantum communication.

4.1.2 Coherence Function

The concept of coherence follows from the description of field properties by wave functions as they are used routinely in quantum physics and quantum optics. Here we focus our attention on first-order coherence phenomena of Schrödinger quantum fields which are described by the Schrödinger equation (Eq. 1.2)

$$\mathcal{H}\psi(\mathbf{r},t) = \mathrm{i}\hbar\frac{\partial\psi(\mathbf{r},t)}{\partial t}. \tag{4.15}$$

The propagation of waves in free space from a source to a detector is described by a wave packet (Eq. 1.27)

$$\psi(\mathbf{r},t) = \int a(\mathbf{k})e^{\mathrm{i}(\mathbf{k}\cdot\mathbf{r}-\omega_k t)}\,\mathrm{d}^3\mathbf{k}. \tag{4.16}$$

For neutrons such a packet is shown in Fig. 1.1. The amplitude factor $a(\boldsymbol{k})$ stems from creation a_k^+ and annihilation a_k^- operators which create or annihilate a mode with the corresponding $\boldsymbol{k}$-vector. The kth mode of the field is mechanically analogous to a one-dimensional simple harmonic oscillator. Thus, k labels a mode, i.e., a degree of freedom equivalent to the wave vector of the neutrons (Lamb 1995). The quantization steps of the coherent field between the source and the detector (a distance L apart) are extremely narrow ($\Delta k \sim 2\pi/L$) and, therefore, the integral form of the wave function can be used (e.g., Walls and Milburn 1994). Such a wave packet describes a multimode coherent state, which can be seen as a quasi-classical state (e.g., Deutsch 1991).

The first-order, two-point-two-time autocorrelation function relating the physical situation at (r, t) and (r', t') is given by (Glauber 1963)

$$G^{(1)}(\mathbf{r}, t; \mathbf{r}'t') = \mathrm{Tr}\{\hat{\rho}\psi^*(\mathbf{r}, t) \cdot \psi(\mathbf{r}', t')\}, \tag{4.17}$$

where the density operator can be written as

$$\hat{\rho} = \int p(r, t)\psi^*(r, t) \cdot \psi(r, t)\mathrm{d}^3 r\mathrm{d}t. \tag{4.18}$$

The function $p(\mathbf{r}, t)$ describes the classical probability of finding the quantum system at $\mathbf{r}$ and t; that is, $p(r, t)$ describes the beam profile (see Fig. 2.9) and a possible time structure of the beam and shows that the coherence features are limited to the coherence features of the wave packet. The whole beam is an incoherent superposition of coherent beams described by the wave function. The density operator develops according to the von Neumann (1931) equation (quantum Liouville equation, Eq. 1.51)

$$\mathrm{i}\hbar\frac{\partial\hat{\rho}}{\partial t} = [\mathcal{H}, \hat{\rho}], \tag{4.19}$$

which gives a constant phase space density when only conservative forces are acting on the system. For a constant p distribution one obtains $\rho \propto \int |a(k)|^2\mathrm{d}^3k$, where $|a(\mathbf{k})|^2 = g(k)$ is the density of states in $\boldsymbol{k}$-space. The auto-correlation function $G^{(1)}$ has the general features

$$G^{(1)}\,(r, t;\; r, t) \geq 0, \tag{4.20}$$

which represents the intensity and

$$G^{(1)}(\mathbf{r}, t; \mathbf{r}, t) \cdot G^{(1)}(\mathbf{r}', t'; \mathbf{r}', t') \geq \left|G^{(1)}(\mathbf{r}, t; \mathbf{r}', t')\right|^2. \tag{4.21}$$

These self-correlation functions can be measured by several interferometric methods where parts of the wave function can be spatially or temporally shifted compared to a reference beam. For neutron matter waves, this can be accomplished by various interferometers where the wave function behind the interferometer is composed of a linear superposition of the wave functions originating from beam paths I and II (Eq. 2.1). In the case of an empty interferometer, these two contributions to the wave function in the forward direction (0-beam) behind the interferometer are equal in amplitude and phase. This follows from symmetry considerations because they are transmitted–reflected–reflected (trr) and reflected–reflected–transmitted (rrt), respectively (Eq. 2.1). Thus, in the 0-beam

$$\psi_0 = \psi_0{}^{\mathrm{I}} + \psi_0{}^{\mathrm{II}}\,. \tag{4.22}$$

The related intensity at any point $\boldsymbol{R}$ is composed of waves arising from beam paths I and II, which experience interactions along the optical paths described by the vectors $\boldsymbol{r}$ and $\boldsymbol{r}'$. In the case of time-dependent interactions along the beam path they exhibit a different time-of-flight (t and t') as well, thus

$$\begin{aligned} I_0 &= \mathrm{Tr}\,\{\rho\psi_0{}^*(\mathbf{R},t)\psi_0(\mathbf{R},t)\} \\ &= G^{(1)}(\boldsymbol{r},t;\,\boldsymbol{r},t) + G^{(1)}(\boldsymbol{r}',t';\,\boldsymbol{r}',t') + 2\mathrm{Re}\; G^{(1)}(\boldsymbol{r},t;\,\boldsymbol{r}',t'). \end{aligned} \tag{4.23}$$

If we write the self-correlation function $G^{(1)}$ for $(\mathbf{r},t)\neq(\mathbf{r}',t')$ as a complex function

$$G^{(1)}(\boldsymbol{r},t;\,\boldsymbol{r}',t') = \left|G^{(1)}(\boldsymbol{r},t;\,\boldsymbol{r}',t')\right| e^{i\chi\,(\boldsymbol{r},t;\,\boldsymbol{r}',t')}, \tag{4.24}$$

we then see that, in terms of the phase χ, the intensity is

$$I = G^{(1)}(\mathbf{r},t;\,\boldsymbol{r},t) + G^{(1)}(\mathbf{r}',t';\,\mathbf{r}',t') + 2\left|G^{(1)}(\mathbf{r},t;\mathbf{r}',t')\right|\cdot\cos\chi\,(\mathbf{r},t;\mathbf{r}',t'). \tag{4.25}$$

One should note that $G^{(1)}(\mathbf{r},t;\mathbf{r},t)$ and $G^{(1)}(\mathbf{r}',t';\mathbf{r}',t')$ denote the intensities originating from beam paths I and II, respectively (Eq. 4.17). That is, all space-time points on path I are given by $\boldsymbol{r},t$ and those along path II by $\boldsymbol{r}',t'$. We have suppressed the subscript "0" on the intensity I because a similar relation holds for the H-beam as well.

The fringe visibility (contrast) of the interference pattern is related to the normalized correlation (coherence) function $\Gamma^{(1)}(\boldsymbol{r},t;\boldsymbol{r}',t')$, that is

$$\Gamma^{(1)}(\mathbf{r},t;\mathbf{r}',t') \equiv \frac{G^{(1)}(\mathbf{r},t;\mathbf{r}',t')}{\left[G^{(1)}(\mathbf{r},t;\mathbf{r},t)\;\cdot\;G^{(1)}(\mathbf{r}',t';\mathbf{r}',t')\right]^{1/2}}\,. \tag{4.26}$$

By combining Eqs. (4.16), (4.17), and (4.25), the complex degree of mutual coherence can be written as

$$\Gamma^{(1)}\,(\boldsymbol{r},t;\boldsymbol{r}',t') \propto \int |\,a(\mathbf{k})|^2\, e^{i[(\boldsymbol{r}-\boldsymbol{r}')\cdot\mathbf{k}-(t-t')\,\omega_k]}\,\mathrm{d}^3\mathbf{k}\,. \tag{4.27}$$

This can be simplified by using the spatial and temporal translation invariances ($\mathbf{r}-\mathbf{r}'=\boldsymbol{\Delta}, t-t'=\tau$) and the free-space dispersion relation $\omega_k=\hbar k^2/2m+mc^2/\hbar$ (Eq. 1.9), such that

$$\Gamma^{(1)}(\boldsymbol{\Delta},\tau) = \int \rho(\mathbf{k})e^{i(\mathbf{k}\cdot\boldsymbol{\Delta}-\omega_k\tau)}\mathrm{d}^3\mathbf{k}, \tag{4.28}$$

which is a special form of the van Cittert (1934)–Zernike (1938) theorem used in light and X-ray optic for a quantitative description of coherence properties (e.g., Born and Wolf 1980, Paganin 2006). Note, in this context, that partially coherent radiation may be produced by an incoherent source, by the act of free-space propagation. For stationary neutron beams and stationary interactions ($\omega=0$) one gets the more familiar form

$$\Gamma^{(1)}(\boldsymbol{\Delta},0) = \Gamma(\boldsymbol{\Delta}) \mathrel{\hat{=}} \int g(\boldsymbol{k})\; e^{i\mathbf{k}\boldsymbol{\Delta}}\mathrm{d}\mathbf{k}, \tag{4.29}$$

where $g(k)$ denotes the momentum distribution of the beam ($g(k) = |a(k)|^2$). We point out that the analogy to the van Hove formalism of neutron scattering from condensed matter is striking (van Hove 1954, Marshall and Lovesey 1971, Squires 1978). We can now write the interference pattern in the form (Eq. 4.25)

$$I(\mathbf{\Delta}, \tau) = I_0\left(1 + \left|\Gamma^{(1)}(\mathbf{\Delta}, \tau)\right| \cos(\mathbf{k} \cdot \mathbf{\Delta} - \omega_k \tau)\right). \tag{4.30}$$

In case that the intensities from the two beam paths are different, the intensity after superposition can be written more generally as

$$I(\mathbf{\Delta}, \tau) = I_1 + I_2 + 2\sqrt{I_1 I_2}\,\left|\Gamma^{(1)}(\mathbf{\Delta}, \tau)\right| \cos(\mathbf{k} \cdot \mathbf{\Delta} - \omega_k \tau). \tag{4.31}$$

The visibility of the interference pattern becomes

$$V = \frac{I_{\text{Max}} - I_{\text{Min}}}{I_{\text{Max}} + I_{\text{Min}}} = \frac{2\sqrt{I_1 I_2}}{I_1 + I_2}\left|\Gamma^{(1)}(\mathbf{\Delta}, \tau)\right|. \tag{4.32}$$

For a completely coherent field, $\left|\Gamma^{(1)}(\mathbf{\Delta}, \tau)\right|$ equals unity, whereas this function may become zero for any $\mathbf{\Delta} \neq 0$ and $\tau \neq 0$ for a completely incoherent field. Any real experimental arrangement provides partially coherent fields where the coherence functions tend toward zero for $\mathbf{\Delta} \to \infty$ and $\tau \to \infty$. The coherence lengths $\mathbf{\Delta}_c$ and the coherence time τ_c are usually defined when the coherence function has decayed to a value $1/e$. It should be mentioned that a damped oscillatory behavior occurs in certain cases where the more general definition for the coherence length should be used (e.g., Perina 1973, Mandel and Wolf 1995):

$$\Delta_c{}^2 = \frac{\displaystyle\int \Delta^2 \left|\Gamma^{(1)}(\Delta)\right|^2 \mathrm{d}\Delta}{\displaystyle\int \left|\Gamma^{(1)}(\Delta)\right|^2 \mathrm{d}\Delta}. \tag{4.33}$$

Equations (4.27) and (4.28) show that the space- and time-dependent correlation functions are the Fourier transforms of the related momentum (wave vector) and energy (frequency) distribution of the beam (density of states). This will provide the basis for the Fourier spectroscopy method using coherent beams (Section 10.15).

For a cross-spectrally pure field, i.e., when the temporal structure of the beam varies slowly over the measurement interval, such that it behaves quasistatically or even is time independent, the corresponding time (τ) variations of $\Gamma^{(1)}(\mathbf{\Delta}, \tau)$ can be separated from the spatial ($\mathbf{\Delta}$) correlations. In this case the mutual coherence function $\Gamma^{(1)}$ may be written as a product:

$$\Gamma(\mathbf{\Delta}, \tau) = \Gamma(\mathbf{\Delta}, 0)\Gamma(0, \tau) = \Gamma(\mathbf{\Delta})\Gamma(\tau). \tag{4.34}$$

For Gaussian momentum distributions having widths δk_i in each of the three orthogonal directions ($i = x, y, z$) one obtains a Gaussian coherence function in the form

$$\Gamma(\mathbf{\Delta}) = \prod_{i=x,y,z} e^{-\left[(\Delta_i \delta k_i)^2/2\right]}, \tag{4.35}$$

and $\Gamma(0,\tau)$ is unity for all times. This is the case for continuous wave (cw) experiments. In this case the coherence lengths Δ_i^c are directly related to δk_i by the Heisenberg uncertainty relation

$$\Delta_i^c \, \delta k_i = \frac{1}{2}. \tag{4.36}$$

Since we are only concerned with first-order coherence here, we have suppressed the superscript (1) on the coherence function in Eq. (4.35).

The coherence function is an average (trace in Eq. 4.17) over different beam paths and/or over varying environment conditions $\varepsilon(\eta)$ along both beam paths. The effect of the environment on coherence will be discussed in detail in Section 4.6. One should keep in mind that any measured coherence property is an average over the beam cross-section and the measuring time (Section 2.1, Fig. 2.9). Here we will give some of the fundamental ideals and concepts of the decoherence theory. If we include a hypothetical wave function of the environment ε depending on the coordinates η we get for the overlapping wave function (Stern et al. 1990)

$$\psi = \psi^I(\boldsymbol{r}) \times \varepsilon^{\mathrm{I}}(\eta) + \psi^{\mathrm{II}}(\boldsymbol{r}) \times \varepsilon^{\mathrm{II}}(\eta) \tag{4.37}$$

and an interference term

$$2\,\mathrm{Re}\left[\int \mathrm{d}\boldsymbol{r}\, \psi^{\mathrm{I}}(\boldsymbol{r})\, \psi^{\mathrm{II}}(\boldsymbol{r}) \int \mathrm{d}\eta\, \varepsilon^{\mathrm{I}}(\eta)\, \varepsilon^{\mathrm{II}}(\eta)\right]. \tag{4.38}$$

Thus, the interference term reduces to 0 when the environment's state coupling to one beam path becomes orthogonal to the one coupling to the other beam path. The disappearance of the interference term introduces at the same time irreversibility, because the time-reversed Schrödinger equation cannot reproduce the same beam path probabilities (Gottfried 1966).

In case that only real phase shifts occur, the integral of Eq. (4.38) takes the form equivalent to an average phase factor

$$< e^{\mathrm{i}\chi} > = \int \mathrm{d}\chi\, P(\chi)\, e^{\mathrm{i}\chi}. \tag{4.39}$$

Since $e^{\mathrm{i}\chi}$ is periodic in χ, $< e^{\mathrm{i}\chi} >$ tends to 0 if the distribution function $P(\chi)$ is slowly varying over a region larger than 2π. Introducing the average phase shift $< \chi >$ and the mean squared fluctuation of phase $< \delta\chi^2 >$ according to the standard definitions

$$< \chi > \equiv \int \chi P(\chi) \mathrm{d}\chi, \tag{4.40}$$

and

$$< \delta\chi^2 > \equiv \int (\chi - < \chi >)^2 P(\chi) \mathrm{d}\chi, \tag{4.41}$$

one can rewrite the average phase factor as

$$< e^{\mathrm{i}\chi} > = e^{\mathrm{i}<\chi>} \int \mathrm{d}\chi\, P(\chi)\; e^{\mathrm{i}(\chi - <\chi>)} \cong e^{\mathrm{i}<\chi> - <\delta\chi^2>/2}. \tag{4.42}$$

The last formula is strictly correct for narrow distribution functions only. One notices that $<\chi> = (1/\hbar) \oint V(\boldsymbol{r}) \mathrm{d}t = -(1/\hbar) \oint \delta p . \mathrm{d}s$, which represents the average over the potential energy along the beam paths (compare Eq. 1.38). A very simple case is a phase-shifting slab with thickness variations δD. One then gets $<\chi> = Nb_c\lambda_0 D_0$ and in terms of the λ-thickness of the sample $D_\lambda = 2\pi / Nb_c\lambda_0$ a reduction factor $\exp[-(2\pi\delta D/D_\lambda)^2/2]$ of the interference contrast. Wavelength (momentum) or atom density variations can be treated in a similar fashion.

For a diffusion-like behavior of the environment where the distribution function becomes a normal distribution which decays to 0 with a characteristic time τ_{int} much shorter than the duration of the experiments, one also gets the result given by Eq. (4.42), which agrees with a previous model developed for the coupling of a quantum system to a heat bath by Feynman and Vernon (1963) and by Caldeira and Leggett (1983). In many cases one can use a quasistatic approximation where the phase fluctuation becomes $<\delta\chi^2> \cong V^2\tau_{int}{}^2/\hbar^2$. A rather comprehensive overview of decoherence and dephasing effects has been given by Guilini et al. (1996) and this topic is treated in more detail in Section 4.6.

4.2 Coherence Measurements

In neutron interferometry a spatial shift of the wave packets between the two coherent beams can be provided by a phase-shifting slab which changes the optical path length according to its index of refraction n and its thickness D_0. The boundary condition of quantum mechanics requiring continuity of the wave function allows only the normal ($\hat{s}$) component of the momentum to change at the slab surface, resulting in a spatial shift of the wave packet (e.g., Born and Wolf 1957, Sears 1989, Lemmel and Wagh 2010):

This results in a phase shift:

$$\chi = D(K_\perp - k_\perp) = Dk\left(\sqrt{n^2 - \sin^2\varphi_0} - \cos\varphi_0\right) = \boldsymbol{K}\cdot\boldsymbol{\Delta}, \tag{4.43}$$

where $\boldsymbol{\Delta}$ is the spatial shift perpendicular to the slab surface. It is given by

$$\boldsymbol{\Delta} = -sD\left(\frac{k_\perp}{K_\perp} - 1\right) = -sD\left(\frac{\cos\varphi_0}{\sqrt{n^2 - \sin^2\varphi_0}} - 1\right). \tag{4.44}$$

Here n denotes the index of refraction $n = K/k$, $k_\perp = k\cos\phi_0$ and $K_\perp = K\cos\phi = nk\cos\phi$. Within a first-order expansion of n around unity, one obtains

$$\chi \approx -Dk\frac{1-n}{\cos\varphi_0} = -Nb_c\lambda D_{eff}. \tag{4.45}$$

The relation of the components of the wave vectors $\boldsymbol{k}$ and $\boldsymbol{K}$ to the spatial shift $\boldsymbol{\Delta}$ and the effective thickness is shown in Fig. 4.2. Here N and b_c are the atom density and the coherent scattering length of the phase shifter material, and $D_{eff} = D_0/(\hat{k}\cdot\hat{s})$ is the neutron path length inside the material slab. Absorption (σ_a), incoherent scattering (σ_{incoh}), small-angle scattering (σ_{SAS}) effects, fluctuations of the thicknesses and of the density of the phase shifter, and imperfections of the interferometer crystal itself cause the modulation of the interference pattern to be incomplete. That is, $|\Gamma(0,0)|$ is always less than unity. Thus, the observed interference pattern has the general form similar to Eq. (2.5)

$$I(\boldsymbol{\Delta}) = I_0\,[A + B\cos(\boldsymbol{\Delta}\cdot\boldsymbol{K} + \varphi_0)], \tag{4.46}$$

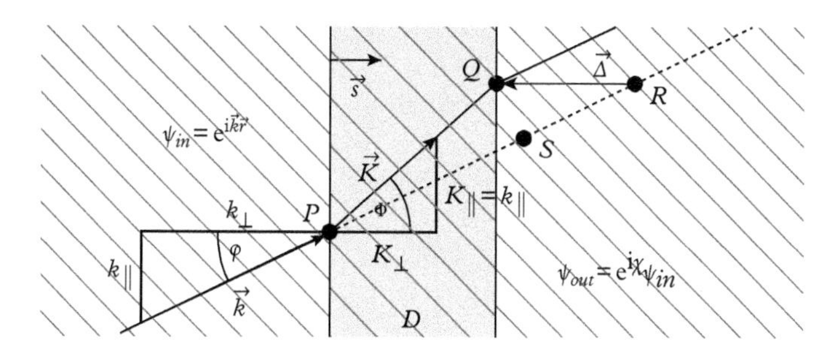

Figure 4.2 *Angular, momentum, and phase shift relations in case of planar phase shifters. Reprinted with permission from Lemmel and Wagh 2010, copyright 2010 by the American Physical Society.*

where I_0A correspond to $I_1 + I_2$ in Eq. (4.31) and I_0B to $2\sqrt{I_1I_2}\,|\Gamma(\mathbf{\Delta})|$. The relative intensities arising from both beam paths depend on the beam attenuation cross-section ($\sigma_t = \sigma_a+\sigma_{incoh}+\sigma_{SAS}$). For beam path II we must replace the intensity I_2 by

$$I_2'(D_{eff}) = I_2(0)\exp[-N\sigma_t D_{eff}]\,. \tag{4.47}$$

Variances of the thickness (δD) and density (δN) of the phase-shifting material cause a variance of phase ($<\chi>^2 - <\chi^2>$) which gives rise to a damping factor (see Eq. 4.42). These influences on the coherence factor take the form

$$\begin{aligned} |\Gamma(\mathbf{\Delta})| &= \Gamma_t(\delta D)\Gamma_d(\delta N)|\Gamma(\mathbf{\Delta})|_0 \\ &= \exp\left\{-\left[\left(\frac{\delta D}{D_0}\right)^2 + \left(\frac{\delta N}{N_0}\right)^2\right](\Delta_0 k_0)^2/2\right\}|\Gamma(\mathbf{\Delta})|_0, \end{aligned} \tag{4.48}$$

where the subscript "0" indicates the value of the coherence factor when $(\delta D) = 0$ and $(\delta N) = 0$. These additional factors do not depend on the overall thickness of the phase shifter but only on the quality of it. δD scales with the number n of surfaces, i.e., $\delta D = \sqrt{n}\,\delta D_0$. In addition the interferometer crystal also cannot be absolutely perfect in its geometry and its internal structures, which results in $|\Gamma(0)|_0 \neq 1$. These considerations justify the definition of a normalized degree of coherence

$$\gamma(\mathbf{\Delta},\tau) = \frac{|\Gamma(\mathbf{\Delta},\tau)|}{|\Gamma(0,0)|}. \tag{4.49}$$

All parameters entering Eq. (4.46) can be measured separately and they are sensitive to differing disturbing environmental effects in different ways. Especially φ_0, the internal phase, is very sensitive as shown in Fig. 2.10. When no phase shifter is inserted (other than a rather perfect phase flag), one measures $|\Gamma(0)|$ and φ_0; and the average intensity I_0A. The internal (empty interferometer) phase φ_0 can be caused by internal strains or geometry factors of the interferometer crystal or/and due to fields acting on the neutron sub-beam (gravitation, magnetic, etc.).

The coherence volume v_c of the beam is given by the product of coherence lengths $v_c = \Delta_x\Delta_y\Delta_z$. When it is transformed into the k-space it defines the resolution function in the spectrometric use of neutrons (e.g., Felber et al. 1998). The coherence properties of a beam discussed

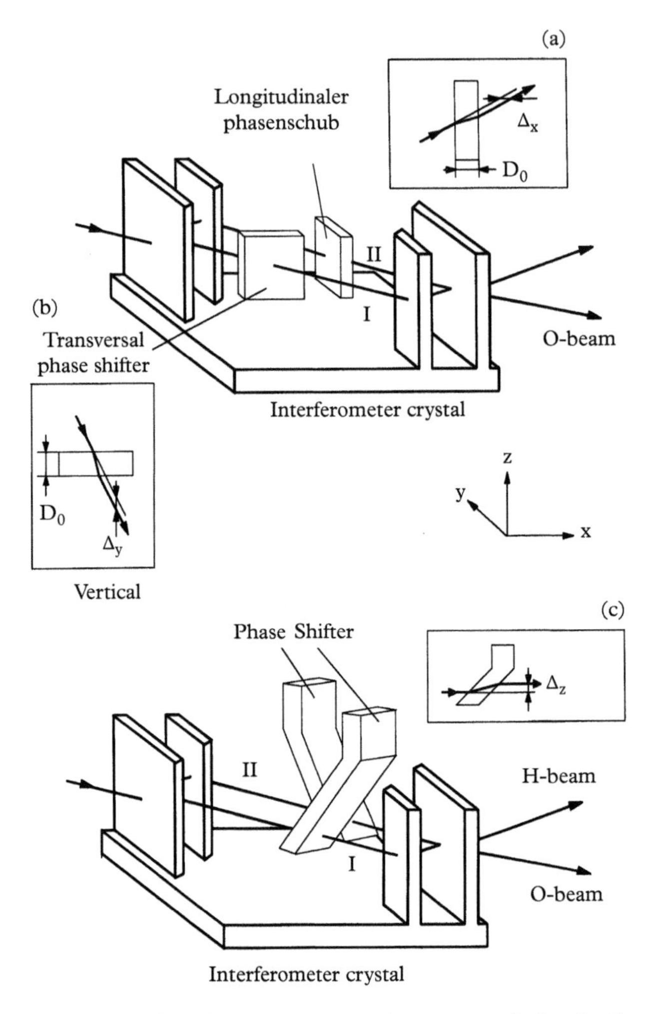

Figure 4.3 *Sample arrangement used to measure the longitudinal (a), transverse (b), and vertical (c) coherence lengths*

here are also of essential interest for the interpretation of any scattering or diffraction experiment because they determine the region of interaction of the neutron within a sample (Bernhoeft et al. 1998). Various measurement geometries are indicated in Fig. 4.3.

4.2.1 Longitudinal Coherence, *x*-Direction

In order to measure the coherence length in the *x*-direction (see Fig. 4.3a), the surface of the phase shifter is perpendicular to the reflecting lattice planes of the Si-crystal interferometer. The phase shifter displaces the wave packet in a direction where the perfect crystal does not influence the original momentum distribution function. The related spatial shift of the wave packets becomes (Eq. 4.44)

$$\Delta_x = -Nb_c\lambda^2 D_0/2\pi, \tag{4.50}$$

and the corresponding phase shift is

$$\chi = \boldsymbol{\Delta} \cdot \boldsymbol{K} = -Nb_c\lambda D_0/\cos\Theta_B, \tag{4.51}$$

where Θ_B denotes the Bragg angle for the interferometer crystal corresponding to wavelength λ. Related observations have been carried out since the first neutron interferometry experiments. They showed a more or less continuous reduction of the fringe visibility at high order (Rauch 1979, Kaiser et al. 1983) which results from the nearly "Gaussian-shaped" momentum distribution in the incident beam (Fig. 4.4). The calculated values shown in the figure are obtained using Eqs. (4.35) and (4.48). The related coherence function is obtained when plotting the amplitudes of the interference pattern as a function of the spatial phase shift Δ_x.

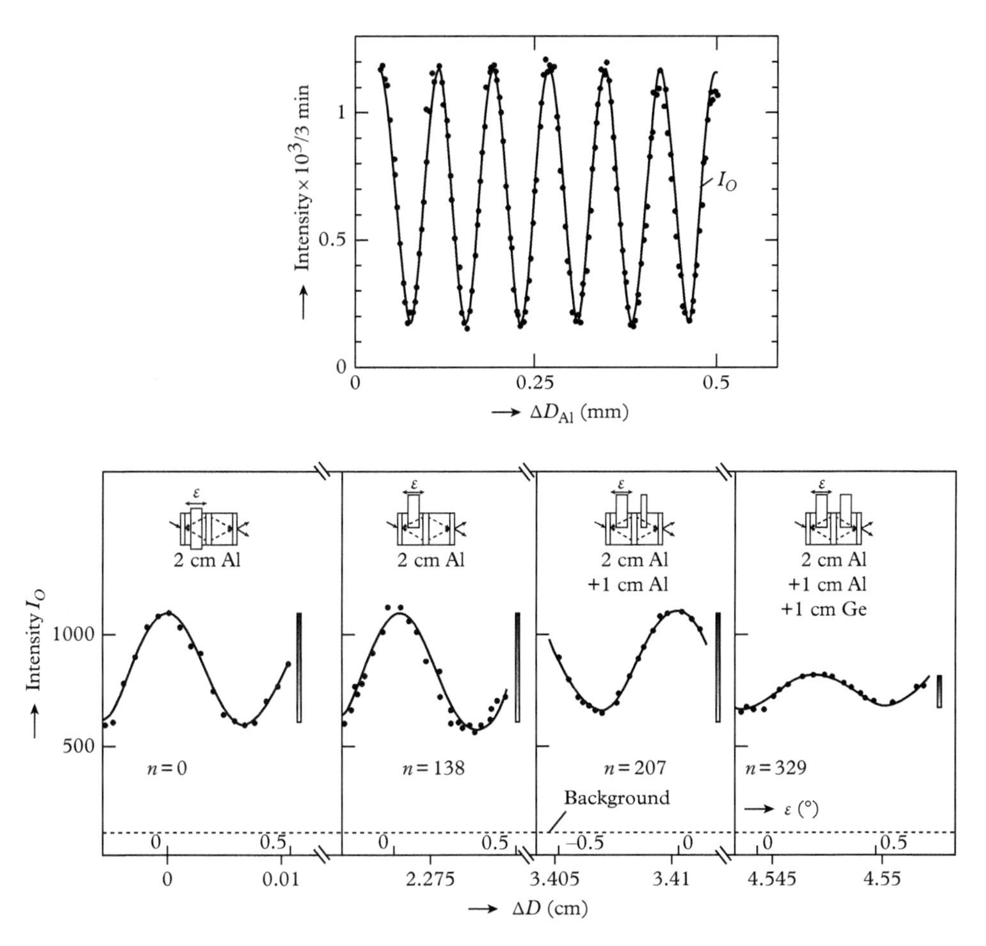

Figure 4.4 *High contrast (above) and high-order interferences (below) in the case of longitudinal phase shifts (Rauch 1979)*

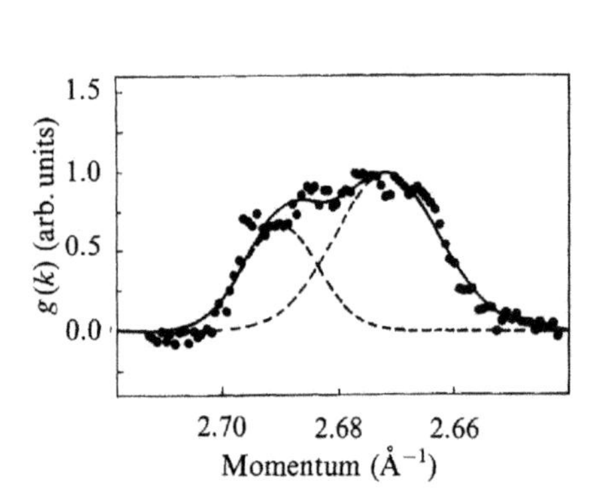

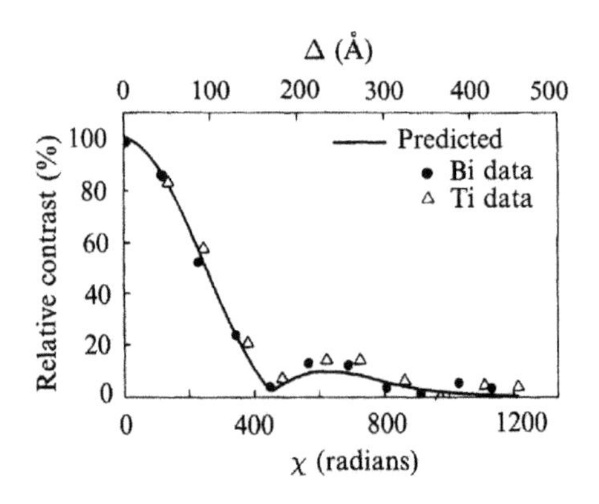

Figure 4.5 *Measured momentum distribution (left) and measured coherence function for the longitudinal direction (Clothier et al. 1991)*

Figure 4.5 shows such results of an experiment carried out at the MURR reactor where the momentum (wavelength) distribution had a non-Gaussian shape as it was determined separately by rocking an additional silicon analyzer crystal in the dispersive position through the I_0 beam (Clothier et al. 1991, Kaiser et al. 1992). The full lines are the mutual Fourier transforms as they are expected from Eq. (4.29).

The interference pattern can even be partly recovered after superposition at the third plate, when a four- or five-plate interferometer is used (Rauch 1984a, Heinrich et al. 1988a; see Section 4.5.6). The interference pattern can also be restored if the beam monochromaticity is increased behind the interferometer by a time-of-flight or a crystal diffraction system (see Section 4.5.4).

Macroscopic inhomogeneities of the sample can be described by a spatially dependent index of refraction. This also causes a reduction of the contrast and of the degree of coherence of the beams since coherence is defined as an average over the whole beam cross-section (Eq. 4.18, Fig. 2.9). If defined structured objects (such as periodic media) are used, some parts of the beam become labeled—i.e., carrying beam path information—due to diffraction effects into distinct orders. Those neutrons are usually lost for contributing to the interference because they are within a preparatory stage for a possible path identification and detection (see Section 4.3.2). Looking more closely into this phenomenon by using a position-sensitive detector device one can recover parts of the contrast and still show the intrinsic coherence properties. A somewhat different situation exists for statistically distributed inhomogeneities of a sample. These cause rather random phase shifts which yield the well-known small-angle scattering phenomenon. In this case the coherence cannot be restored due to the statistical nature of the scattering within the sample (Eqs. 4.42 and 4.48). These conclusions are in agreement with a theoretical analysis of Morikawa and Otake (1990), who showed that a loss of contrast can, but need not be, an indication of a loss of coherence. The measurement of the loss of contrast over the whole beam cross-sections can be used to obtain information about the inhomogeneous structure of the sample. Related measurements have been made for porous graphite (Bauspiess et al. 1974), for precipitates in Al–Mg–Zn alloys (Rauch 1979), in metal-hydrogen systems (Rauch et al. 1978b), and for ferromagnetic domain structures (Rauch 1980). Within a zero-order approximation these effects can be described by a statistical fluctuation of the scattering-length density ($\Delta N b_c$) or by the neutron depolarization formalism (e.g., Rekveldt 1973). Besides this smearing effect of the phase, sample inhomogeneities can cause a remarkable beam deflection due to small-angle scattering, an effect which must be included in the calculation of the related wave functions. The measured effect depends in this case on the acceptance angle of the successive perfect crystals which makes a quantitative analysis difficult. This effect provides the basis for solid state physics applications of neutron interferometry and will be discussed in more detail in Section 9.1.

4.2.2 Transverse Coherence, *y*-Direction

In this case, the surface of the phase shifter is parallel to the reflecting lattice planes where the momentum distribution becomes strongly influenced due to the dynamical diffraction effects within the perfect crystal (Rauch and Petrascheck 1978; Chapter 10). The resulting momentum

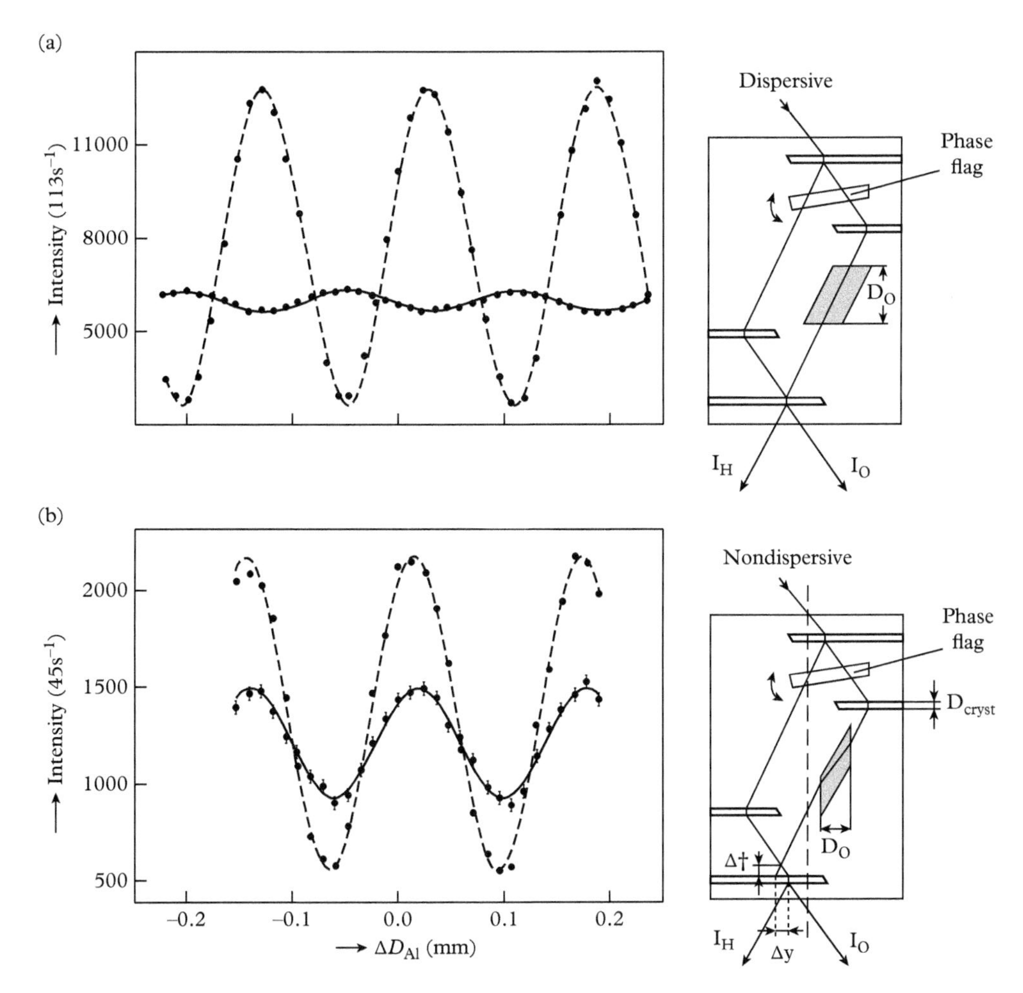

Figure 4.6 *Comparison of the high-order interference pattern for phase shifts in the longitudinal (dispersive) and transverse (non-dispersive) directions (Rauch et al. 1987)*

distribution becomes rather narrow ($\Delta k_y/k_0 \cong 10^{-5}$) exhibiting an oscillatory structure. The related spatial shift of the wave packet becomes (Eq. 4.44, Figs. 4.3b and 4.6)

$$\Delta_y = -Nb_c\lambda^2 D_0/2\pi, \tag{4.52}$$

and the phase shift is

$$\chi = -2Nb_c D_0 d_{hkl}, \tag{4.53}$$

where the Bragg relation $\lambda = 2d_{hkl}\sin\Theta_B$ has been used with the Si interferometer lattice plane spacing d_{hkl}. Thus we see that the phase shift is independent of the neutron wavelength, which means an accurate measurement of the wavelength is not necessary. This phase shift behaves non-dispersively up to rather high interference orders and, therefore, the visibility of the interference fringes is correspondingly enhanced compared to the case of a longitudinal phase shifter. Related experiments have been performed at the ILL reactor and have verified this behavior (Rauch et al. 1987); see Fig. 4.6. The contrast is shown around the 250th interference order when the path length of the neutron beam inside the phase shifter is 33.8 mm and an auxiliary thin Al-phase flag is rotated inside the interferometer. The reduction of the contrast in the dispersive (longitudinal) x-direction is mainly caused by the influence of the finite momentum spread on the coherence function $|\Gamma(\mathbf{\Delta})|$; the effects of beam attenuation and phase shift fluctuations are smaller and are comparable to their values in the dispersive case (Eq. 4.48). The reduction of the constrast in the non-dispersive (transverse) y-direction case is caused nearly exclusively by beam attenuation and phase shift fluctuation effects, but only very little by the coherence function. The coherence function for the non-dispersive position has been calculated within the framework of spherical

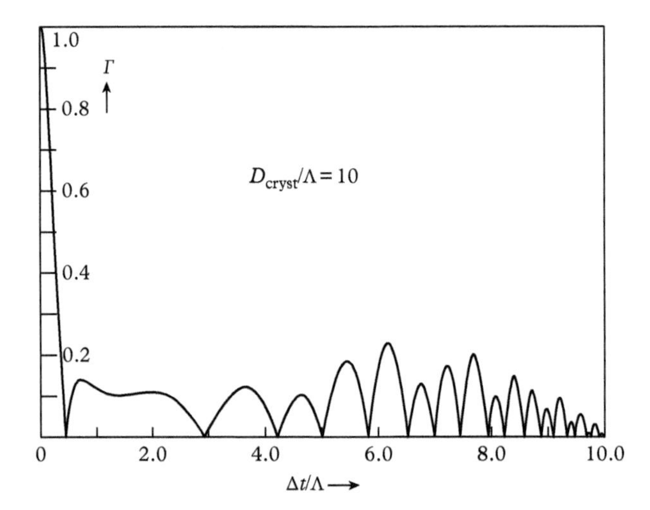

Figure 4.7 *Calculated coherence function in the transverse direction Bonse and de Kaat 1971, with kind permission from Springer Science and Business Media. Reprinted from Petrascheck 1988, copyright 1988, with permission from Elsevier.*

diffraction theory where the contrast of a defocused interferometer is evaluated (Bonse and de Kaat 1971, Bauspiess et al. 1976, Petrascheck and Folk 1976). An interpretation in terms of a mutual coherence function has been given by Holy (1980) and Petrascheck (1987, 1988). There is a non-vanishing contrast over the whole width of the Borrmann fan as shown in Fig. 4.7. The plotted results are for a thickness of the perfect crystal D_{cryst} which is ten times the characteristic length $\Lambda = (2d\,b_c N)^{-1}$, which amounts to about $10\,\mu\text{m}$ for most silicon reflections. The coherence length must be extracted by using Eq. (4.33) but a distinct influence of the coherence function appears only upon the $\sim 10^4$ interference order, i.e., at thicknesses where all other damping factors usually dominate. In many cases variations of the thickness of the interferometer crystal plates smear out the high-order modulations of the coherence function, and the related coherence function can be written as

$$|\Gamma(\Delta_y)| = \left|1 + \frac{\Delta t}{\Lambda} - \frac{5}{9}\left(\frac{\Delta t}{\Lambda}\right)^2\right| \exp(-\Delta t/\Lambda), \tag{4.54}$$

with $\Delta_y = 2\Delta t/\tan\Theta_B$, where Δt denotes the defocusing distance. Direct measurements applying a variable defocusing onto one beam path become feasible (Fig. 2.4). Transverse coherence lengths up to $175\,\mu\text{m}$ have been reported by Wagh et al. (2011) by using Fankuchen-cut crystals to produce a narrow and nearly plane wave of neutrons. An interesting non-defocusing and non-dispersive sample position will be discussed in Section 4.2.5.

4.2.3 Vertical Coherence, *z*-Direction

In order to measure the vertical coherence length, a shift Δ_z of the trajectories perpendicular to the plane of the interferometer is achieved by a phase-shifting slab whose surface is tilted with respect to the horizontal plane by an angle φ, as shown in Fig. 4.3c. This small spatial shift, due to refraction in the tilted slab is given by the laws of geometrical optics as

$$\Delta_z = -\frac{\lambda^2 N b_c}{2\pi} D_0 \tan\varphi, \tag{4.55}$$

and the corresponding total phase shift becomes

$$\chi = -N b_c \lambda D_0/\cos\varphi. \tag{4.56}$$

Such a phase shifter produces phase shifts in other directions too, which must be balanced by a proper phase shifter put into the reference beam and which compensates for beam attenuation, as well.

Experiments using phase shifters with different thicknesses and tilt angles were performed at the MURR reactor (Rauch et al. 1996; Fig. 4.8). At this interferometer setup, a twin focusing monochromator made up of pyrolytic graphite (PG) crystals was used which produced a double humped momentum distribution in the vertical direction as it was measured by scanning a horizontal slit (1 mm) through the intensity distribution behind a static slit (1 mm) placed at the interferometer table (Fig. 4.8, top). These measurements were performed at different beam heights and averaged subsequently. The contrast was extracted from interferograms obtained by rotating an auxiliary phase shifter around a vertical axis with the various tilted phase shifters and compensator phase shifter slabs inserted into the two beams of the interferometer. This contrast (fringe visibility) directly yields the coherence function as it is plotted in Fig. 4.8. The full lines in this figure correspond to an optimal fit to the data and they are related to each other by their mutual Fourier transformations, which verifies Eq. (4.29).

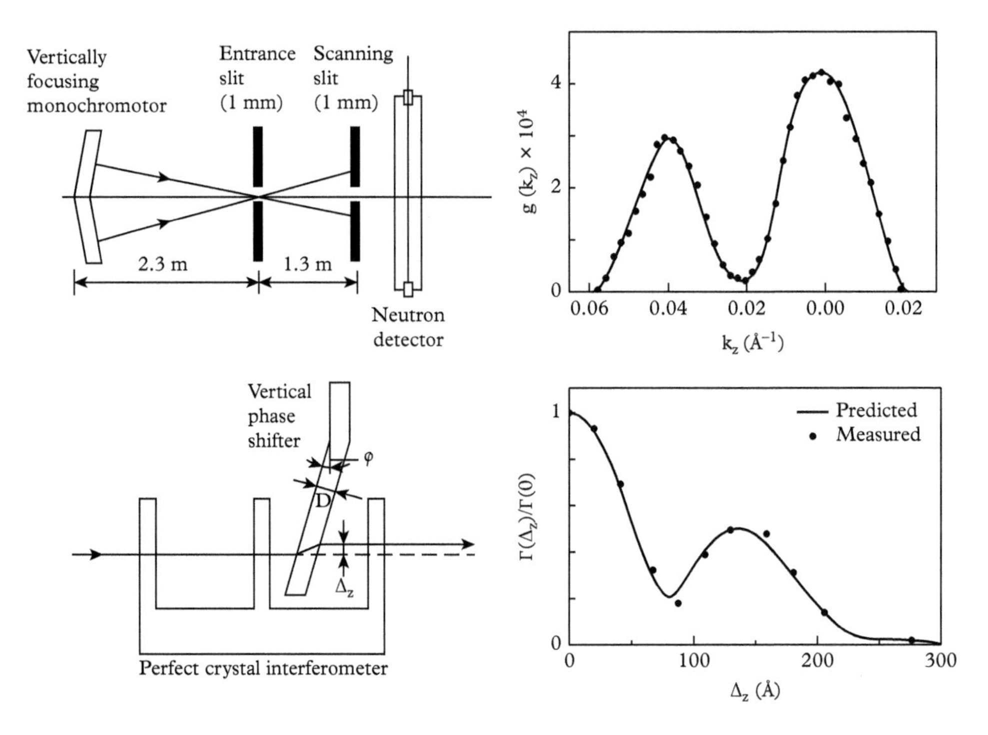

Figure 4.8 *Measured momentum distribution and measured coherence function in the vertical direction when a double-peak momentum distribution exists due to a twin vertically focusing monochromator (Rauch et al. 1996)*

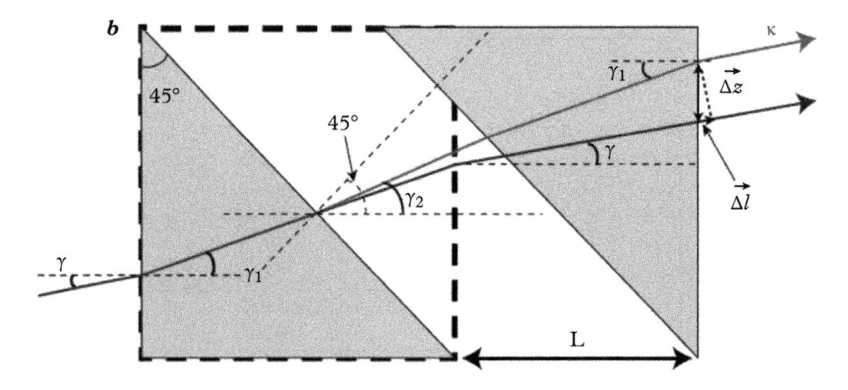

Figure 4.9 *Vertical phase shifter where the vertical shift can be controlled by the distance* L *of the two wedges. Reprinted with permission from Pushin et al. 2008, copyright 2008 by the American Physical Society.*

Double prism systems have been used by Pushin et al. (2008) to measure the vertical coherence function. In one beam the prisms are separated and in the other they are not, which balances the horizontal phase shift. The principle of this phase shift is shown in Fig. 4.9. The vertical beam shift can be calculated by means of geometrical optics. When the beam divergences are small ($k_z/k << 1$) one gets

$$\Delta z \approx L(1-n), \tag{4.57}$$

and the related phase shift as $k_z \Delta z$. The coherence functions for different vertically focusing monochromators have been measured and good agreement between the Fourier transform of the related momentum distributions and the coherence functions has been shown. Analogies to the situation addressed in the next section are evident.

4.2.4 Phase-Echo and Spin-Echo Experiments

Phase echo is a similar technique to spin echo, which is routinely used in neutron spectroscopy (Mezei 1972, 1980; Badurek et al. 1980a). A large phase shift ($\Delta > \Delta_c$) can be applied in one sub-beam of the interferometer, which can be compensated by a negative phase shift acting in the same sub-beam or by the same phase shift applied to the other beam path (Kaiser et al. 1983, Clothier et al. 1991). According to Eq. (1.38), the phase shift is additive and the coherence function depends only on the net phase shift. Thus, the interference pattern can be restored as it is shown schematically and in form of an experimental example in Fig. 4.10. Here, the bismuth slab presents a positive optical potential, and the titanium presents a negative optical potential to

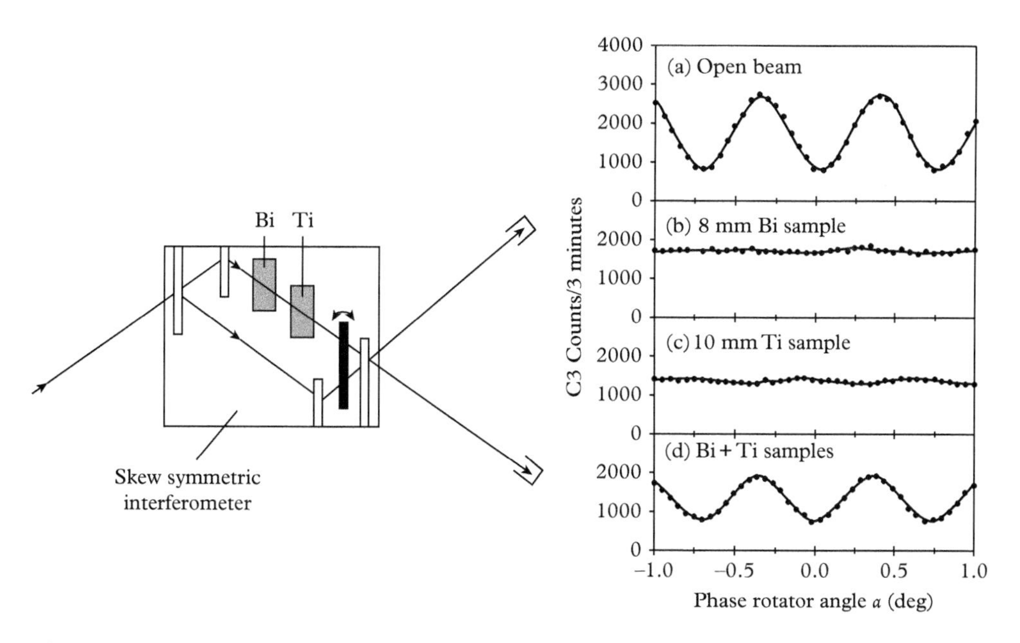

Figure 4.10 *Phase-echo experiment using phase-shifting slabs with positive (Bi) and negative (Ti) coherent scattering lengths (Clothier et al. 1991)*

the neutron de Broglie wave. Thus, the wave packet is slowed down in traversing the Bi slab, and speeded up in traversing the Ti slab. This gives the phase-echo condition

$$\chi_1 \cong \chi_2. \tag{4.58}$$

The phase-echo method can alternatively also be applied behind the interferometer loop when multiplate interferometers are used (Heinrich et al. 1988). A peculiar situation exists when the spin echo and the phase condition are fulfilled simultaneously (see Section 2.4):

$$\begin{aligned} B_1 L_1 &\cong B_2 L_2 \\ \chi_1 &\cong \chi_2 . \end{aligned} \tag{4.59}$$

This situation has been tested by a Jamin configuration as shown in Fig. 4.11 (Ebisawa et al. 1998a). It should be mentioned that in this case longitudinal and transverse shifts of the packets are compensated by the echo conditions. Coherence revival experiments are also known from electron interferometry where Wien filters are used to shift the wave packets without a resultant deflecting force due to the matched action of an electric and a magnetic field (Möllenstedt and Wohland 1980, Nicklaus and Hasselbach 1993, Hasselbach 1995).

In Section 4.5 we will discuss the coherence recurrence situation in much more detail. Here we interpret the situation in perspective. Information first appearing in spatially shifted wave packets becomes transferred into a momentum modulation, and then can be retrieved again in ordinary space modulation effects. However, it becomes intrinsically more difficult to restore the original contrast the wider the separation of the wave packets in ordinary space. These results show that a vanishing contrast does not intrinsically indicate a loss of coherence, but it is associated with

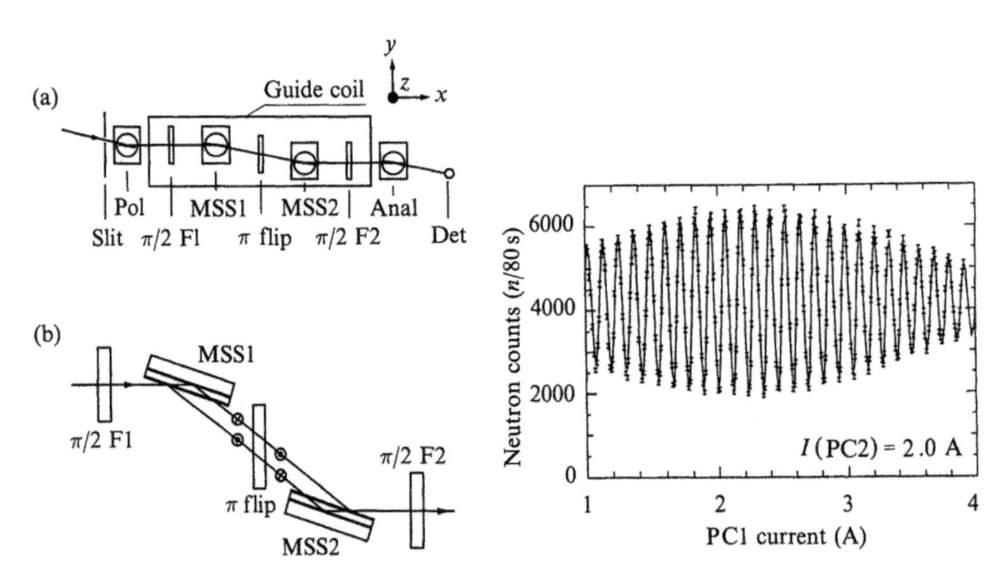

Figure 4.11 *(Left) Arrangement of a phase-spin-echo interferometer using two identical multilayer spin splitters mounted within precession fields 1 and 2. (Right) Measured spin-echo profile. Reprinted with permission from Ebisawa et al. 1998b, copyright 1988 by the American Physical Society.*

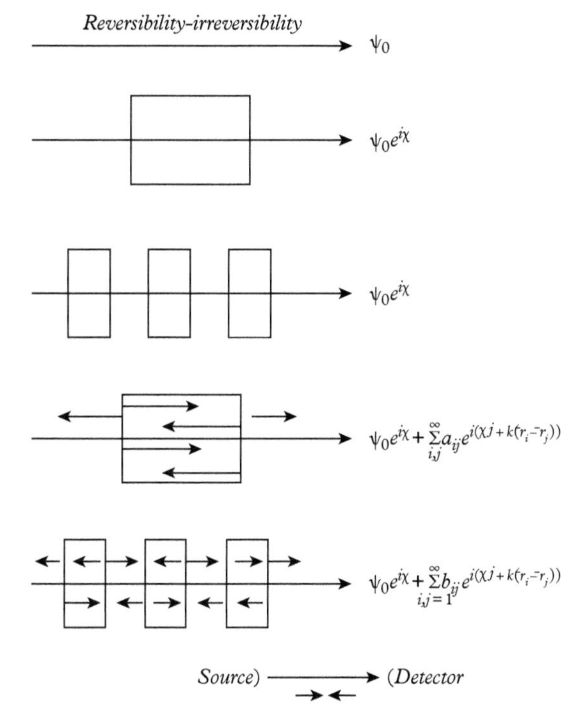

Figure 4.12 *Approximate and complete wave functions when various shapes of phase shifters are used*

at least a slight reduction in fringe visibility because unavoidable quantum losses appear. When one takes into account the physical situation very carefully, one notices that each arrangement produces different wave functions and that the additive forward part is the leading, but not the only term (Fig. 4.12). There are always multiple reflections and traversals of the wave function. This indicates also that a complete retrieval of the wave function becomes impossible. This conclusion also follows from the beam attenuation and phase variation factors of Eq. (4.48). Each phase shifter represents a potential barrier which produces unavoidable and non-revivable separations of the whole wave function. A lossless transmission is only possible for an infinitely small wavelength band near to the resonance, but this then implies zero intensity (see, e.g., Piron 1990). The situation of polarized neutrons crossing various magnetic potentials has been treated by Barut et al. (1987). Its consequences for the feasibility or non-feasibility of observing an ideal quantum Zeno effect has been discussed by Nakazato et al. (1995).

In a more complete and accurate measurement, more and more parts of the complete wave functions, which contain more and more of the detailed history the quantum system has experienced between the source and the detector, become visible. This indicates a basic irreversibility process caused not by parasitic effects like absorption or incoherent scattering processes but by the appearance of an infinite number of additional terms in the wave function. This indicates that the original state cannot be restored completely by any means. This conclusion was also drawn by Englert et al. (1988) in connection with the splitting and recombination of a spin-½ system in a Stern–Gerlach magnet (see also Schwinger et al. 1988). When a quantum system evolves, so that it returns at least approximately to its initial state, it acquires a memory effect expressed in terms of additional terms in the wave function, which can be a loss in intensity, the appearance of a geometric phase, etc. (Section 6.3; Anandan 1992). This subject is also closely connected to the decoherent histories approach to quantum mechanics (Caldeira and Leggett 1985; Buzek et al. 1995a,b) and the Weyrl entropy approach (Weyrl 1978). In addition, unavoidable fluctuations (even zero-point fluctuations) cause an irreversibility effect which becomes more significant for widely separated Schrödinger cat-like states (Paz et al. 1993, Rauch and Suda 1995). It has been shown that decoherence also occurs at absolute zero temperature (Sinha 1997). In this case, a harmonic oscillator which is coupled to an environment of harmonic oscillators at zero temperature where a temperature-independent power-law loss of quantum coherence was obtained has been considered. The off-diagonal density matrix elements die out with time as

$$\overline{\rho}(x, x', t) \sim t^{-\alpha}, \tag{4.60}$$

where α depends on the mass, the dissipative coefficient (γ; see Eq. 4.149), and the square of the spatial separation $\alpha = (2/\pi\hbar)m\gamma(x-x')^2$. A more detailed discussion of the decoherence process is given in Section 4.6. All these effects can be described by an increasing entropy inherently associated with any kind of interaction (Lorentz 1927). This also supports the idea that irreversibility is a fundamental property of nature and reversibility is only an approximation, a conclusion stated by several authors (e.g., Haag 1990, Prigogine 1991, Cini and Serva 1992, Blanchard and Jadczyk 1993, Venugopalan and Ghosh 1995, Englert 2013). At the same time changes in entropy are closely connected to the quantum measurement problem (Chapter 12; Giulini et al. 1996, Namiki et al. 1997). The appearance of entropy associated with decoherencing effects reflects the presence of an arrow in time in quantum theory, i.e., a fundamental irreversibility in the formalism of the theory itself. This irreversibility comes into play only through initial and boundary conditions in our universe.

This shows that irreversibility and, therefore, the measurement process starts with the first interaction that the quantum system experiences in the experimental setup. The assignment of a source and a detector region define the direction of increasing entropy. This may be summarized in the statement "All quasi-classical phenomena, even those representing reversible mechanics, are based on de facto irreversible decoherence" (Zeh 2001).

4.2.5 Non-dispersive and Non-defocusing Phase Shifters

In the previous sections we dealt with phase shifters placed into only one gap of the interferometer. There are other possibilities as well which have distinct advantages as pointed out by Lemmel and Wagh (2010). Figure 4.13 shows the standard and advanced phase-shifter arrangements. The different arrangements cause different phase shifts and different spatial shifts of the wave packets. The spatial shift $\boldsymbol{\Delta}$ of the neutron wave packet when passing through a homogeneous slab is given by Eq. (4.44) with a corresponding phase shift χ. When more than one phase shifter is placed in the interferometer, say one of the dual phase-shifter geometries in Fig. 4.13, various interesting trajectories which are non-dispersive and also focusing are possible. The total phase accumulation along each beam path is always given by the line integral of the canonical momentum along the *unperturbed* beam paths (see Section 1.2)

$$\Phi_{\mathrm{I\ or\ II}} = \int \boldsymbol{k}(\boldsymbol{s}) \cdot \mathrm{d}\boldsymbol{s} \tag{4.61}$$

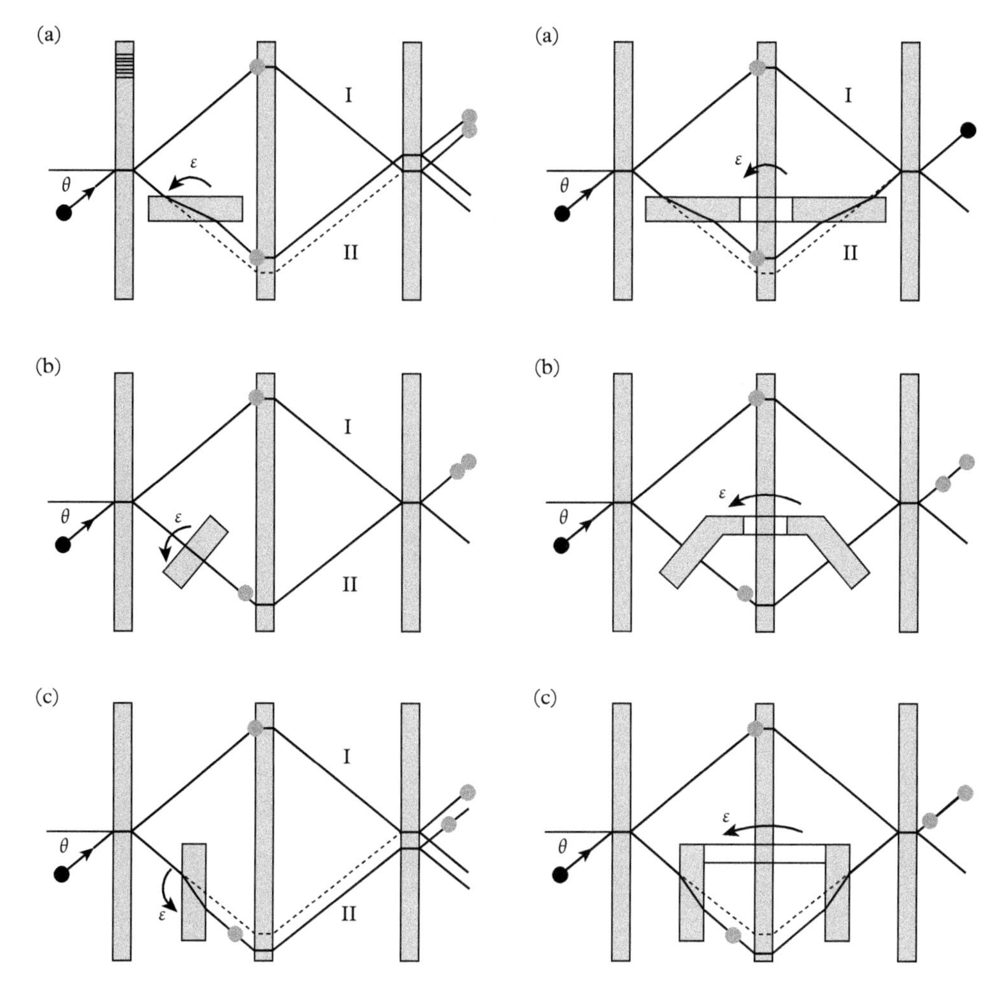

Figure 4.13 *Phase shifter configurations within the interferometer. The bullets along the beam paths indicate the packets and their spatial position at various times. All (a) are non-dispersive. Right (a) is non-dispersive and non-defocusing. Reprinted with permission from Lemmel and Wagh 2010, copyright 2010 by the American Physical Society.*

and the phase difference is (see Eq. 4.43)

$$\Delta\Phi = \Phi_{II} - \Phi_I = D(K_\perp - k_\perp) = \boldsymbol{K} \cdot \boldsymbol{\Delta}. \tag{4.62}$$

In the dual non-dispersive phase-shifter configuration (Fig. 4.13a, right) the equal but opposite spatial shifts $\boldsymbol{\Delta}$ in the two slabs cancel out and yield a focusing geometry. However, since $K_\perp$ also reverses sign in the second gap and $\boldsymbol{K} \cdot \boldsymbol{\Delta}$ has the same sign in both gaps, the phase shifts add up. This results in the remarkable situation that there is no spatial shift of the wave packets upon recombination, but there is a strong non-dispersive phase shift χ. It can be shown that the dispersive action of the dual non-dispersive phase shifter is strongly enhanced by orders of magnitude compared to a single non-dispersive phase shifter (Fig. 4.13a, left). There is also no defocusing. This allows for phase measurements up to very high order and therefore measurements with extreme sensitivity.

Thus, in the geometry of the dual phase shifter in Fig. 4.13a (right) the wave packets traversing path I and path II do not experience a net relative spatial shift at the point of recombination. That is, they wind up right on top of each other. Nevertheless, there is a net large phase shift of -2χ. In some ways, this is reminiscent of the observation of Aharonov–Bohm effects by various techniques including neutron interferometry (Werner and Klein 2010). Because of refraction in the phase-shifting slabs, there is a net path length shortening. One might think that this will cause an additional phase shift. In fact, it does. However, this phase shift is exactly canceled by a change in the phase of the de Broglie waves in traversing the middle silicon crystal plate. One is left with a phase shift of -2χ, coming only from the phase changes due to the optical potential of the slabs. That is, one should integrate the canonical momentum along the *unperturbed* beam paths. This subtlety of calculating the phase shift has been discussed recently by Lemmel (2014). Since the conclusions reached here are not at all obvious, or for that matter agreed upon by all people involved in this field of research, a serious experimental test of these ideas and predictions would appear to be necessary.

4.2.6 Neutronic Coherence Features

The results of the dedicated coherence measurements show that spatial coherence is a basic three-dimensional phenomena, and that related coherence functions can be obtained from the contrast of the interference pattern when variously oriented phase-shifting slabs are inserted into the interferometer. Properly shaped magnetic fields could be used as well. The coherence function in a certain direction is the Fourier transform of the related momentum distribution in that direction. Thus, it is determined by the collimation and monochromatization defining the beam. In this respect, the coherence function represents beam properties rather than single particle properties. Nevertheless, within quantum mechanics, the related wave function (Eq. 4.16) can also be attributed to a single neutron, but this quantum system should also always be considered as part of a certain beam configuration. That is, the wave function contains at the same time properties of the quantum system and of the apparatus as well, which should be seen as their mutual entanglement and as a basic feature of quantum mechanics (see Chapter 12). As mentioned at the beginning of this chapter the coherence areas $(\Delta_x \Delta_y)$ as given by the wave functions are much smaller than the beam cross-section and, therefore, one performs measurements successively with many coherent beams simultaneously.

The coherence features extracted from neutron experiments are rather analogous to those of light, X-ray, or electron coherence experiments where the coherence properties are also attributed to the beam properties (monochromaticity, collimation) even though there is only self-interference involved (Keller 1961, Mandel and Wolf 1965, Lenz 1972, Francon 1979, Möllenstedt and Wohland 1980, Tonomura 1987, Ishikawa 1988, Haroche and Raimond 2006). Various theoretical attempts to introduce Glauber's definition of quantum coherence into neutron beam experiments show the analogy between phenomena observed with different radiations (Sears 1979, Klein et al. 1983, Ledinegg and Schachinger 1983, Byrne 1994).

The coherence experiments with neutrons described above have been performed with perfect crystal interferometers. For representative values for the collimation and monochromatization, the measured coherence length in the longitudinal direction is about 100 Å and the vertical direction about 50 Å. In the transverse, i.e., in perpendicular direction to the reflecting lattice planes, the coherence length is about 50,000 Å, due to the action of the perfect crystal (Chapter 11; Bonse and de Kaat 1971, Petrascheck 1988). These values define the phase space volume and from the measured intensity the related phase space density (a dimensionless quantity) of about 10^{-14} neutrons can be extracted, which corresponds to the expected phase space density immediately outside of a thermal moderator of a standard neutron source. The size of the coherent packet describes, in a certain sense, the volume which the neutron "sees" when it interacts with its environment. This phenomenon has also been elucidated in an experiment where the wave packet was sent through an absorbing lattice which was oriented in various ways in relation to the three axes of the packet (Summhammer et al. 1987; see Section 4.3.2). The size of the coherence volume also influences the scattering properties of a sample when their coherence volume (e.g., grain size) become comparable. High-resolution X-ray investigations touch this limit more easily than neutron investigations but forthcoming improvements of the experimental conditions will make the inclusion of coherence properties of the beam essential (Bernhoeft et al. 1998, Gaehler et al. 1998, Sinha et al. 1998).

The question may arise as to whether the coherence vanishes when the absolute value of the coherence function becomes zero at large phase shifts. The phase echo and certain post-selection experiments for neutrons and electrons (Kaiser et al. 1983, 1992; Clothier et al. 1991; Nicklaus and Hasselbach 1993; Jacobson et al. 1994) have shown that this is not the case and that interference fringes and coherence phenomena can be revived when a proper position, momentum, or time selection is applied to the beam, even subsequent to the superposition of the two coherent beams in the last crystal of the neutron interferometer. In the case of large spatial separations of the interfering packets ($\Delta >> \Delta^c$), when $|\Gamma(\Delta)|$ becomes zero and the interference fringes disappear, the coherence phenomena manifest themself in momentum space by an intrinsic modulation of the momentum distribution (Rauch 1993, Jacobson et al. 1994). This provides the basis for phase-echo experiments described earlier in Section 4.2.4 and for multiplate interferometry (Section 4.5.6). All these phenomena indicate that coherence cannot be destroyed by any Hamiltonian interaction, but only by stochastic and dissipative effects which are related to a transition from a quantum to a classical world (Chapter 12 and Section 4.6; Walls and Milburn 1985, Schleich et al. 1991, Rauch and Suda 1995). Such effects become more significant the larger the spatial separation of the (potentially) interfering wave packets, which thereby provides a natural limit on how far so-called coherent Schrödinger cat-like states can be spatially separated. The interaction of the neutron quantum system with the environment, the phase shifter or the detector must not be of purely statistical nature, but can cause a quantum entanglement between the system and the experimental setup. In this case the random-average model of coherence loss provides a proper description of the loss of coherence (Wooters and Zurek 1979, Stern et al. 1990, Tan and Walls 1993). In addition, environment-induced decoherence becomes also more and more accepted in the modern measurement theories (Namiki and Pascazio 1992; Zurek 1993, 1998a) and part of the epistemological interpretation of quantum mechanics (Chapter 12).

Only the spatial coherence phenomena have been treated so far, but it should be mentioned that temporal coherence properties can also be elucidated by neutron interferometry. In this case, energy is exchanged differently in the two beams. This can be achieved by applying a Zeeman energy exchange between the neutron and a resonator coil (Section 5.3; Badurek et al. 1986) or multiphoton exchange in an oscillating field (Section 4.6.2; Summhammer et al. 1995,

Sulyok et al. 2012). The related coherence function $|\Gamma(\tau)|$ depends on the coherence time of the beam $\tau_c = \Delta_c/v$, which is on the order of nanoseconds in usual cases. It can also be shown that this quantity is connected to the beam monochromaticity δE by the uncertainty relation $\tau_c \delta E \cong \hbar$ and that the temporal coherence function becomes observable in the contrast of an interference pattern when the energy transfer (ΔE) to the beams becomes comparable to the energy width (δE) of the beam ($|\Gamma(\tau)| \cong \exp[-(\Delta E\tau/\hbar)^2/2]$). Diffraction of neutrons from fast vibrating mirrors causes a diffraction-in-time phenomenon (Moshinski 1952). In principle, this can give the basis for the measurement of the time-dependent coherence function (Felber et al. 1996).

4.3 Partial Beam Path Detection

4.3.1 Stochastic and Deterministic Beam Attenuation

Here, we consider the influence of the imaginary part of phase shift (Eq. 3.18) which gives rise to beam attenuation effects and can be attributed to a partial beam path detection. These measurements are closely related to Einstein's version of the double-slit experiment where one retains a surprisingly strong interference pattern by not insisting on a 100% reliable beam path detection (e.g., Jammer 1974). The interference contrast will become reduced because the intensities arising from both beam paths become different and because the coherence function depends on how the absorption takes place (Eq. 4.31). Stochastic absorption in this sense means that every neutron has the chance to be absorbed and thereby detected, whereas deterministic absorption means that only neutrons in a certain time or space interval become absorbed.

The absorption of neutrons is in any case a measurement of the particle's location because a compound nucleus is formed with an excitation energy of about 7 MeV from which the decay products (mainly gamma rays) can be detected quite easily. Although the total number of absorbed neutrons can be the same for either a stochastic or a deterministic situation, the effect on the contrast of the interference pattern can be quite different.

Stochastic absorption exists when an absorber sheet is inserted in one of the beams of the interferometer which, in addition to causing a phase shift, also causes a beam attenuation due to the imaginary part of the index of refraction (Eq. 1.24). The related experiments were performed at low interference order where the reduction of the contrast due to the influence of the coherence function can be omitted. The change of the wave function must be calculated with a complex index of refraction (Eqs. 1.25 and 3.18)

$$\psi' = \psi\, e^{i(\chi' + i\chi'')} = \psi \sqrt{a}\, e^{i\chi'}, \tag{4.63}$$

where $\chi'' = \sigma_r ND/2$ and $a = I/I_0 = \exp(-\sigma_r ND)$ denotes the beam attenuation. After superposition with the undisturbed beam one gets for the 0-beam intensity (compare Eq. 3.19)

$$I_0 = \left|\psi_0{}^{\mathrm{I}} + \psi_0{}^{\mathrm{II}}\right|^2 \propto \left|\psi_0{}^{\mathrm{I}}\right|^2 \left((a+1) + 2\sqrt{a}\cos\chi'\right). \tag{4.64}$$

All other factors influencing the contrast as they are discussed in Section 4.2 also exist in this case, but they do not vary during such measurements. That is, the analysis of data is done according to Eq. (4.46) where it is shown how the various effects can be isolated. Due to beam attenuation, the amplitude of the interference fringes $2\sqrt{a}$ is proportional to the square root of the probability that the particle has passed the absorber region and, therefore, the beam modulation is higher than the intensity of the weaker of the two interfering beams. If "absorber-detectors" are inserted in both beams of the interferometer one gets

$$I_0 \cong \left|\psi_0{}^{\mathrm{I}}\right|^2 \left[(a+b) + 2\sqrt{ab}\cos\chi\right], \tag{4.65}$$

which shows an unchanged visibility $V = 2\sqrt{ab}/(a+b) = 1$ exists for the case of equal beam attenuation $a = b$, although the transparency of the whole system decreases as given by $T = (a+b)/2$. Scattering lengths measurements with highly absorbing phase shifters actually use this effect (Rauch and Tuppinger 1985).

A deterministic absorber—for instance a slowly rotating periodic chopper—leaves the beam unaffected, or absorbs it with an efficiency near to 100%. The on/off ratio of the chopper determines the transmission probability $a = t_{\mathrm{open}}/(t_{\mathrm{open}} + t_{\mathrm{closed}})$. The situation of deterministic absorption, which is equivalent to a deterministic beam path detection, can also be achieved with unequal slits introduced into the coherent beams. For a homogeneous beam a is determined in this case by the ratio of the area of the smaller slit to that of the larger slit. In the quantum mechanical sense this means an either/or measurement because the beam path is determined, or not determined, for a certain time or space interval. Thus, in this case the intensity behind the interferometer is the weighted sum of the open and closed conditions of the chopper (or the relative areas of slit assemblies) and is given by

$$I \propto \left[(1-a)\left|\psi_0{}^{\mathrm{II}}\right|^2 + a\left|\psi_0{}^{\mathrm{I}} + \psi_0{}^{\mathrm{II}}\right|^2\right] \propto \left|\psi_0{}^{\mathrm{I}}\right|^2 \left[(a+1) + 2a\cos\chi'\right], \tag{4.66}$$

where the amplitude of the interference fringes $2a$ now depends linearly on the mean probability of the neutron to pass the absorber region. The average intensity arising from the deterministic and stochastic situations can be the same but the contrast and, therefore, the visibility (Eq. 4.32) can be quite different, especially for very small values of a. The different outcomes are related to the different amount of information one can deduce in these two different situations. According to quantum mechanics in the case of the stochastic absorber one can only say that the wave function of the particle has spread with different amplitudes over both beam paths, whereas in the case of the deterministic time-dependent absorber the wave function has spread out over both beam paths with the same amplitude *or* it was only in one beam path.

The related experiments performed with absorbing phase shifters and a slowly rotating chopper disk yielded the expected results (Rauch and Summhammer 1984b). These experiments are also related to delayed choice experiments of the Wheeler type, which are discussed in Section 4.5.5. They are also closely related to "which way" information in the quantum interference of unstable particles, for example, the decaying neutron (Krause et al. 2014). The deterministic situation approaches the stochastic one when the particular history gets lost due to a very fast or very narrow slit chopper, which approaches the quantum limit to be discussed later. In the lower part of Fig. 4.14 the normalized amplitudes of the interference pattern are shown (Rauch and Summhammer 1984b, Summhammer et al. 1987).

The difference between the stochastic ($\sqrt{a}$) and the deterministic (a) behavior of the fringe visibility becomes especially obvious at very small transmission probabilities—a regime which is more difficult to access (Rauch 1989, Rauch et al. 1990). The measured values are also shown in the lower part of Fig. 4.14. In the limit $a \to 0$ an interference pattern has an oscillation amplitude orders of magnitudes larger than the residual beam intensity coming from the beam path with the absorber. Namiki and Pascazio (1990) showed that before reaching this limit additional effects must be considered. In this respect, the amplitude reduction must be calculated, including

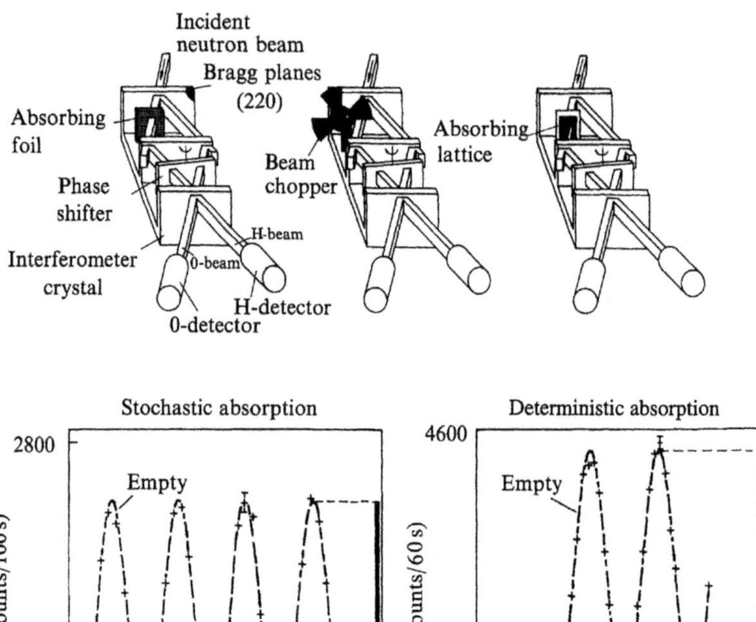

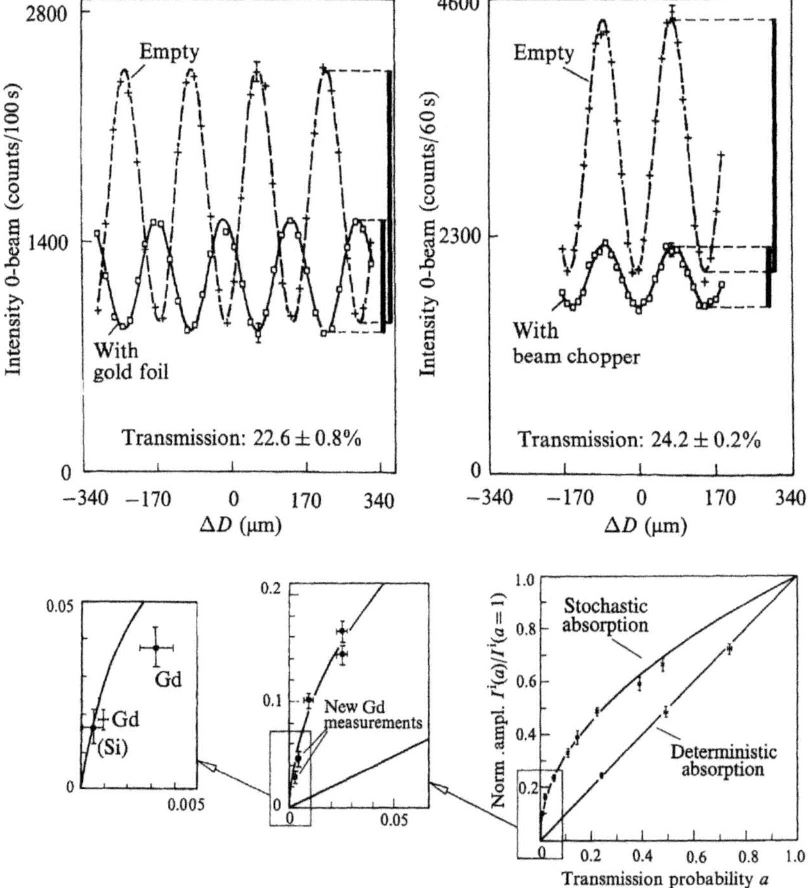

Figure 4.14 *Various beam attenuation methods. (Above) Absorbing material, slowly rotating chopper, and absorption lattice; (middle) observed interference pattern for a stochastic and a deterministic absorber and a compilation of the measured contrast as a function of the transmission probability (below; Summhammer et al. 1987)*

variations due to thickness or density variations of the absorber plate ($\chi'' \rightarrow \chi''_0 + \delta\chi''$). Averaging the exponential factor $\exp(-\chi''_0 + \delta\chi'')$ over a Gaussian distribution with a standard width $\delta\chi''$ yields an amplitude for the stochastic case, which depends upon the square root of the transition probability, namely

$$\sqrt{a} = \sqrt{a_0}\, e^{(\delta\chi'')^2/2}. \tag{4.67}$$

For the deterministic case, the absorption factor ($\exp[-2(\chi''_0 + \delta\chi'')]$) must be averaged, giving

$$a = a_0\, e^{2(\delta\chi'')^2}, \tag{4.68}$$

since $\sqrt{a} < \sqrt{a_0}$, the measured values lie below the $\sqrt{a}$ curve (see Fig. 4.14). Similar experiments with laser light have been reported by Awaya and Tomita (1997). Hasegawa and Kikuta (1994) investigated this effect with synchrotron radiation with strong gas absorbers put into an X-ray interferometer. They observed the $\sqrt{a}$ behavior down to a transmission probability of $2.6(9) \times 10^{-4}$. Hafner and Summhammer (1997) used such an arrangement to demonstrate "interaction-free" measurements with neutrons. A strong absorber put into one of the beam paths produces a measurable output in an interferometer channel (0- or H-beam) even though this channel has been adjusted to zero intensity in the interferometer mode. In this sense "interaction-free" means that a negative result of a quantum measurement apparently modifies the wave function of the non-detected object. Therefore, it is perhaps a triviality to state that some kind of interaction plays an important role which is a kind of entanglement between the quantum probe and the object (Geszti 1998, Karlsson et al. 1998). This can, in some sense, be considered as a modification of the boundary conditions. Interferometry is a proper tool for contrast enhancement by means of producing dark field images of an object. This leads to the concept of "weak measurements" (Aharonov and Vaidman 1990, Vaidman 2014) where one can assign definite values to observables when they are correlated (entangled) in a contextual sense (Chapter 7; Tollkansen 2007). Along these lines quantum Zeno tomography can be established where only a few neutrons are needed to achieve high resolution images (Section 10.18; Facchi et al. 2002).

The above absorber measurements have also been analyzed by de Muynck and Martens (1990) and Tang et al. (2013) in terms of joint non-ideal measurements of the interference and path observables. Using the formalism of positive-operator-valued measures, they reproduced the results described earlier. They also identified a region of unaccessible joint non-ideal measurements which is based on the quantum limit of an inaccuracy relation, which is a kind of a generalized Heisenberg uncertainty relation. It is shown that in stochastic absorption the absorption process is fundamentally quantum mechanical, hence wiping out phase relations less effectively. Deterministic absorption represents the consequences of an either/or nature in the experiment. An analysis based on Barut's (1988) compatible statistical interpretation of quantum mechanics which favors de Broglie's trajectories through the interferometer is given by Bozic and Maric (1991). A computer simulation based on many-Hilbert-spaces theory also elucidates the situation (Namiki 1988, Murayama 1990a, de Raedt et al. 2012). Their views show how models which reveal how the interference pattern builds up event by event can be developed, and then how individual deviations cause the reduction of the contrast.

The situation of a partially absorbing chopper wheel or of a partial absorbing sheet inserted partially into one of the sub-beams of the interferometer is described by Hasegawa and Kikuta (1991). The results show an intermediate behavior where the interference pattern becomes less pronounced at an intermediate position than in the case of a complete insertion of the partially absorbing sheet. X-ray experiments also verified that prediction. Related experiments with photons show a similar behavior independent of whether the mean number of photons inside the interferometer was smaller (0.4) or considerably larger (400) than unity (Awaya and Tomito 1997). This is because the interference phenomena in the classical wave picture are formally the same as the case of a single-particle interference state (Loudon 1983).

In this discussion we have used the following semi-classical definitions for stochastic and deterministic absorption processes:

- *An absorber is stochastic* when the experimenter has no means, not even in principle, to predict whether the neutron will be absorbed at any given point in the absorber region at any given instant of time.
- *An absorber is deterministic* when, in principle, it is known with certainty what will happen at any point in the absorber region at any instant of time.

These are idealizations defining a homogeneous gray absorber as stochastic and one consisting of black-and-white sections as deterministic. The transmitted fraction of the beam, a, was derived as 1 minus the absorbed fraction. While formally this is true, it needs closer inspection in the quantum limit, as was done by Kaloyerou et al. (1992) and Hussain et al. (1992). When speaking of the quantum limit we mean either the chopper period becomes smaller than the coherence time τ_c or the spatial period of an absorbing lattice becomes shorter than the coherence length Δ_c. Thus, a reasonably consistent definition of transmission through an absorber leading to a common picture of stochastic and deterministic absorbers in the quantum limit will be needed (Rauch and Summhammer 1993). The new attempt must provide a smooth transition from very wide slits to very narrow ones (atomic distances) and from very slow rotating choppers to very fast ones (quantum choppers). The results discussed here have also been analyzed by de Raedt et al. (2012) on the basis of a pure event-based particle model, where they could show good agreement between calculated and measured values without using any wave equation.

The difference between deterministic and stochastic behavior can be enhanced by using multiplate interferometers. Similar calculations, as shown in Section 4.5.6, give for a double-loop interferometer with an absorber ($T_d = \exp(-\sigma_a ND)$) in beam "d" and a phase shifter ($\chi_f = \Delta_f k_0$) in beam "f" for the deterministic case the visibility is

$$V_{\det 2\Delta_f} = \frac{4T_d \cos(\Delta_f k_0/2)}{4\cos^2(\Delta_f k_0/2) + T_d} \tag{4.69}$$

and for the stochastic case the visibility is

$$V_{\mathrm{sto}2\Delta_f} = \frac{4\sqrt{T_d}\cos(\Delta_f k_0/2)}{4\cos^2(\Delta_f k_0/2) + T_d} \tag{4.70}$$

as shown in Fig. 4.15 (Suda et al. 2004). One notices that high visibilities can be achieved even for small transmission probabilities through the absorber. When $T_d = 4\cos^2(\chi_f/2)$ is chosen the visibility approaches 1, which indicates a kind of homodyne situation for the measurement of weak signals (Freyberger et al. 1995, Leonhardt 1997). It can also be considered as a kind of coherence enhancement due to absorption.

4.3.2 Quantum Limit of Stochastic and Deterministic Absorption

In order to complete the discussion of the previous section, we propose to take into account the finite coherence volume associated with a real beam (Section 4.2.6). Plane wave pictures are only a

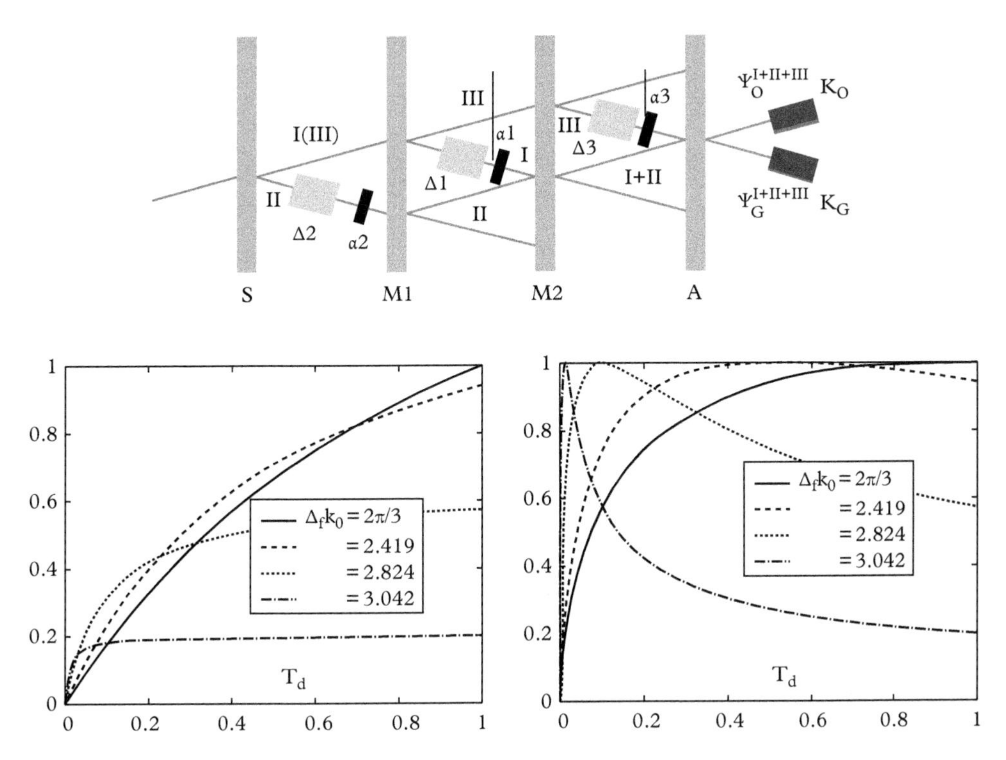

Figure 4.15 *Double-loop interferometer and visibilities for deterministic (right) and stochastic (left) absorption (Suda et al. 2004)*

limiting case that may result in misleading conclusions. Wave-packet formulations must be used to account for the finite momentum spreads defining the original momentum space $(\Delta k_x, \Delta k_y, \Delta k_z)$. For the sake of simplicity we focus on the forward, 0-beam, behind the Mach–Zehnder-type interferometer used in the experiments. It is a superposition of two beams occupying the same phase space volume with equal density within the appropriate coherence volume.

Attenuation is achieved either by neutron–nuclear absorption or by coherent or incoherent scattering. An important distinction is whether scattering occurs within the initial coherence volume defined by the momentum space volume $(\Delta k_x, \Delta k_y, \Delta k_z)$ or into other differential elements of the momentum space. Here, we shall consider only the latter case as attenuation of the initial beam, because in the former case a momentum filter cannot eliminate momentum components that were changed by the attenuating object. Therefore, the *non-removal probability* of neutrons from the initial phase space volume must be considered. This is in contrast to the previous more semi-classical definition where all neutrons re-emerging from the absorbing region were considered transmitted, and it implies, for instance, that the attenuation of an absorbing lattice of given transmission depends on the lattice period and on the coherence volume of the beam and not only on the open-to-closed ratio. Experimentally, the new definition implies that the detector must only be sensitive to the phase space volume of the original beam. The purpose is to exclude from the detector all experimentally accessible path information carried by the beam right behind the interferometer. This information can, of course, be picked up by other detectors. Applying these criteria to the case of an absorbing lattice, the critical quantity is the momentum transfer due to diffraction: $k_y(n) = k_y(0) + 2\pi n/s$, where s is the spatial periodicity of the lattice. Our criterion is whether $k_y(n) - k_y(0)$ is larger or smaller than the momentum width Δk_y. That is, the deterministic limit applies if

$$k_y(n) - k_y(0) = 2\pi n/s << \Delta k_y \text{ for } n > 0, \tag{4.71}$$

and the stochastic limit applies if

$$k_y(n) - k_y(0) = 2\pi n/s >> \Delta k_y. \tag{4.72}$$

The wave function behind a lattice can be calculated and contains separated diffraction peaks if the incident momentum width is smaller than the separation of the first diffraction peaks.

The intensity modulation as a function of the phase shift χ is thus solely due to the superposition of the non-scattered component $k_y(0)$ of the two sub-beams of the interferometer upon recombination. Because of the finite width of the momentum distribution Δk_y, different momentum states must actually be understood as such distributions, each centered around a different $k_y(n)$ (see Fig. 4.16). In the case of an absorbing grating with a large spacing s in comparison to

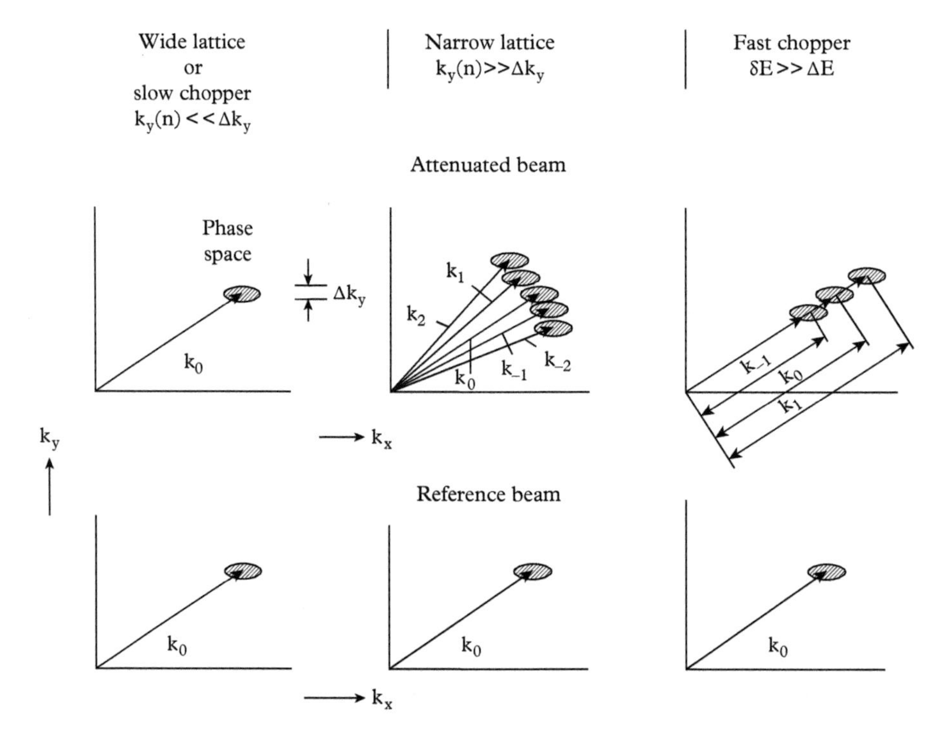

Figure 4.16 *Comparison of phase spaces when various absorption lattices or choppers are inserted into one beam of the interferometer with an undisturbed reference beam*

the coherence length in that direction (Eq. 4.54), the momentum distributions arising from both beam paths overlap and in a hypothetical momentum measurement it is not possible to distinguish between scattered and non-scattered components of the beam. The scattered components cannot be considered "labelled." Therefore, only the truly absorbed part of the wave function is removed from the original phase space volume. This leads to the deterministic behavior of the interference pattern (Figs. 4.14 and 4.17). In the opposite (quantum) limit of a very small period of the absorbing grating (Eq. 4.72), the distributions around the different $k_y(n)$ are all separate and can be resolved by a hypothetical momentum measurement. In phase space this means that at the detector only the momentum distribution around $k_y(0)$ still overlaps with the momentum distribution of the reference beam, resulting in the stochastic behavior of interference (Fig. 4.17). An interesting situation occurs in the transition region. Here, the momentum distributions around the different $k_y(n)$ partially overlap, but are already partly distinguishable from the original one. Accordingly, the interference pattern lies between stochastic and deterministic behavior. This situation is related to unsharp wave–particle measurement (Wooters and Zurek 1979, Mittelstaedt et al. 1987). This behavior has been observed by rotating an absorbing lattice from the horizontal (nearly deterministic) to the vertical (nearly stochastic) position (Fig. 4.17). In these lattice experiments (Summhammer et al. 1987, Rauch and Summhammer 1992) the intermediate case between deterministic and stochastic beam attenuation has been approached, because the coherence length in the horizontal plane is rather large, as discussed in Section 4.2.2 ($\cong 10\,\mu\text{m}$), whereas the coherence lengths for the other directions are considerably smaller (~ 100 Å; Petrascheck 1987; Rauch et al. 1987, 1996). With vertical slits the momentum change due to diffraction becomes comparable to the momentum width of the beam. Equivalently, the effective slit width ($s = 50\,\mu\text{m}$) becomes comparable to the coherence length ($\Delta_y \cong 10\,\mu\text{m}$) in that direction. This causes an increased beam attenuation in the $n = 0$ phase space and an increased visibility of the interference pattern, thus moving toward the stochastic limit.

From Fig. 4.16 it is also apparent that other phase space situations can be envisioned by manipulating the reference beam properly. For example, if the same lattice is also introduced in the reference beam, neutron states are shifted out of the phase space volume in the same manner

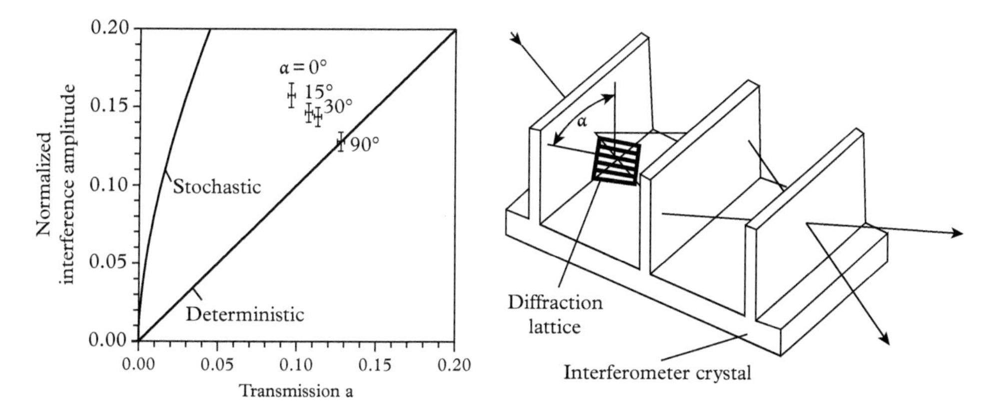

Figure 4.17 *Contrast as a function of the angle of an absorbing lattice in relation to the vertical and horizontal coherence lengths of the neutron beam. The results indicate a transition from a deterministic to a stochastic situation (Rauch and Summhammer 1992)*

in both beams, thus recovering full interference with a concomitant loss of any path information contained in the scattered components of the wave function. Similarly an inhomogeneous phase shifter with a large fluctuation of the phase shift χ causing small-angle scattering represents a higher phase space removal probability than a homogeneous one. Then, compensation with another inhomogeneous phase shifter in the other beam is not possible because of the random mixture of different phase space elements. Thus, dephasing occurs as described in theories on the measurement process (Section 4.6, Namiki and Pascazio 1991b, Zurek 1991). Computer simulations concerning fluctuations of the number and of the distance between scattering centers and/or the strength of the individual scattering center interaction with the neutron wave function also demonstrate a reduction of the fringe visibility (Murayama 1990b). Although these results cannot be taken literally due to the rather limited number of scattering centers which can be handled numerically, they do show the increasing loss of contrast with increasing fluctuations.

A fast chopper may be considered in analogy to a narrow absorbing lattice. Such a chopper produces bursts Δt shorter than the coherence time τ_c (quantum chopping), which is given by the related coherence length Δ_c and the neutron velocity

$$\tau_c = \Delta_c/v\,. \tag{4.73}$$

In cases where $\Delta t \leq \tau_c$, "diffraction in time" occurs and the lengths of the k vectors change due to energy transfer, thereby broadening the spectral width of the beam (Fig. 4.16). Diffraction in time can be treated in a manner similar to diffraction in real space as discussed earlier (Moshinsky 1952, Gähler and Golub 1984, Nosov and Frank 1991). The first diffraction-in-time effects for neutrons were seen in the form of sidebands from reflecting surfaces vibrating at a frequency of 700 kHz (Felber et al. 1996), and also with fast mechanical choppers (Hils et al. 1998). These methods could produce inelastic scattering effects as indicated in Fig. 4.16. In these cases, time-dependent transmission operators must be used to calculate the wave functions behind the chopper unit (Hussain et al. 1992, Imoto 1996).

All interferometric experiments performed so far with mechanical choppers have dealt with the "deterministic" case where $\Delta t >> \Delta t_c$ (Heinrich et al. 1989, Rauch et al. 1992). From the experimental point of view it should be mentioned that the quantum limit can be much more easily reached by fast vibrating surfaces as mentioned earlier, or by means of a high-frequency resonance spin flipper where the phase of the neutron wave packet can be varied by rapidly changing the resonance magnetic field (phase chopping), i.e., $\Delta t_{HF} = 1/\nu_{HF} = h/2\mu B_0 < \tau_c$ (Section 2.4; Alefeld et al. 1981, Badurek et al. 1983). This condition is equivalent to the constraint that the energy shift of the beam is larger than the energy width ΔE of the incident beam ($\delta E_{HF} = 2\mu B_0 > \Delta E$). This also explains why in the proposed "Einweg" (there is a beam path) experiment of Rauch and Vigier (1990) the coherence properties become washed out in the case $\Delta E_{HF} > \Delta E$ (see Section 5.3). This phenomenon can also be elucidated by the temporal coherence of the beam (as briefly discussed in Section 4.1.2).

In conclusion, the present description of the neutron interferometer partial absorber experiments provides a smooth transition from deterministic behavior in the classical limit (characterized by scattering within the original phase space volume), to stochastic behavior in the quantum limit. The key physical ingredient is that the pure transmission probability is replaced by a phase space non-removal probability. The complementarity principle between interference and beam path detection is thereby preserved.

In light optics two-slit experiments in the time domain have been reported (Sillitto and Wykes 1972). In this case electro-optical shutters were antiphase-modulated such that an opening occurs more than once during the coherence time of a low intensity laser beam. Thus, it is to be expected that interference effects are likewise possible in a neutron interferometer equipped with an appropriate shutter mechanism keeping one beam path closed when the other one is open and vice versa (Brown et al. 1992). In the plane wave treatment where the coherence time of the incoming wave is always larger than the open–close period of the chopper ($t_c = 2\pi/\omega_c$) a significant beam modulation is anticipated. That is, an interferogram will take the general form

$$I \propto \left(1 + 2\sum_{-\propto}^{\propto} c_n^2 (-1)^n \cos(\chi(k_n))\right), \tag{4.74}$$

with

$$k_n{}^2 = 2m(\omega_0 - n\omega_c)/\hbar,$$

$$c_n = \frac{\sin(n\pi/2)}{n\pi},$$

which indicates the necessity of a phase shifter in a dispersive position (Section 3.1.2). The rather short longitudinal packet (coherence) length of any practical neutron beam makes the realization of such a double-slit diffraction in time rather difficult.

4.3.3 Unsharp Wave–Particle Behavior

All the coherence and beam attenuation measurements discussed in Sections 4.2 and 4.3 can be related to the concept of unsharp wave–particle measurements or equivalently to an unsharp wave–particle preparation. According to this view, complementary variables can at least be measured simultaneously in an approximate sense (Wooters and Zurek 1979, Bartell 1980, Busch 1987, Mittelstaedt et al. 1987, Tang et al. 2013). Thus, for example, a surprisingly strong interference pattern can be retained although the beam path has been observed to a high degree by putting an absorber detector into one of the interferometer beams (Fig. 4.14). In some other cases a preparatory stage is achieved which can be brought to a real measurement with an associated collapse of the wavefield afterwards, but which can also be brought back together coherently (at least approximately) to interfere by various phase-echo methods. A preparatory stage can be achieved by coherent diffraction or phase-shift effects which produce states without overlap with the state of the reference beam of the interferometer. One has different information about the wave- and particle-like behavior in the various coherence experiments discussed above and, therefore, a formulation within the framework of the Shannon information-theoretic entropy approach seems likely. Some of the related formulations proposed in the literature will be discussed (Zeilinger 1986a,b; Kraus 1987; Greenberger and Yasin 1988; Rauch 1989b; Englert 1996, 1999).

Here, a general formulation based only on wave function properties is given. For this more general discussion the interference pattern which delivers information about wave and particle properties can be written as (Eq. 4.46)

$$I = <\left|\psi^{\mathrm{I}} + \psi^{\mathrm{II}}\right|^2> = T\left[1 + V\cos(\Delta\chi)\right], \tag{4.75}$$

where T is the generalized transparency of the system which is given by various attenuation and labeling effects, e.g., by $T = (a+b)/2$, and V is the generalized visibility (Eq. 4.49). In the case of statistical absorbers as $V_s = 2\sqrt{ab}/(a+b)$ and for deterministic absorption $V_d = 2ab/(a+b)$ (see Eqs. 4.65 and 4.66). The visibility is also reduced at high order due to the finite coherence length of the beams. This effect can be taken into account with $V_c = \exp[-\chi^2(\Delta k/k)^2/2]$ (Eq. 4.35). The transparency and visibility can be influenced by various other deterministic and stochastic effects.

A more general view of the wave–particle dualism appearing in interference experiments can be given by means of a Poincaré sphere (or circle) representation (Mittelstaedt 1989). In this case, the accessible area is projected onto a hemisphere of the Poincaré sphere. In this case, the wave properties are described by the square of the fringe visibility and the path distinguishability in terms of the diagonal terms of the density operator P_D (Wooters and Zurek 1979; Greenberger and Yasin 1988; Mittelstedt 1989; Mandel 1991; Jaeger et al. 1995; Englert 1996; Ghose 1999, 2009, Badurek et al. 2000, Tang et al. 2013). This connection is known as the Greenberger–Englert relation (duality relation)

$$P_D^2 + V^2 \leq 1. \tag{4.76}$$

Purely coherent (wave-like) states lie at the equator while the pure particle character appears at the poles (Fig. 4.18). Phase differences can be defined in any equatorial plane but not at the poles. Mixed states lie inside the sphere which provides also a formal representation of the background of any interference experiment. The Poincaré representation will also be used in Section 6.8.4 to describe absorption as a generalized phase phenomenon. A discussion of interferometric complementarities including two-particle interferences has been given by Jaeger et al. (1995). Their conclusion is rather similar to that of Mittelstaedt (1989). Englert (1996) formulated Eq. (4.76) as an inequality which describes interferometric duality. In this terminology P denotes the predictability or distinguishability of the paths through the interferometer. Whether Eq. (4.76) can be related to the uncertainty principle or to the complementary feature only is still under discussion (Busch and Gallawy 2006). Dürr and Rempe (2000) showed that explicitly for some suitably chosen observables. The observation of an interference pattern and the acquisition of which-way information are mutually exclusive, which defines duality in a very general

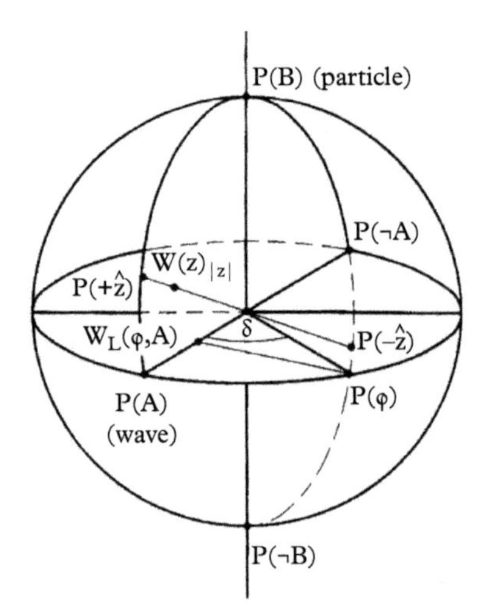

Figure 4.18 *Presentations of the particle-like and the wave-like behavior of a quantum system inside the interferometer*

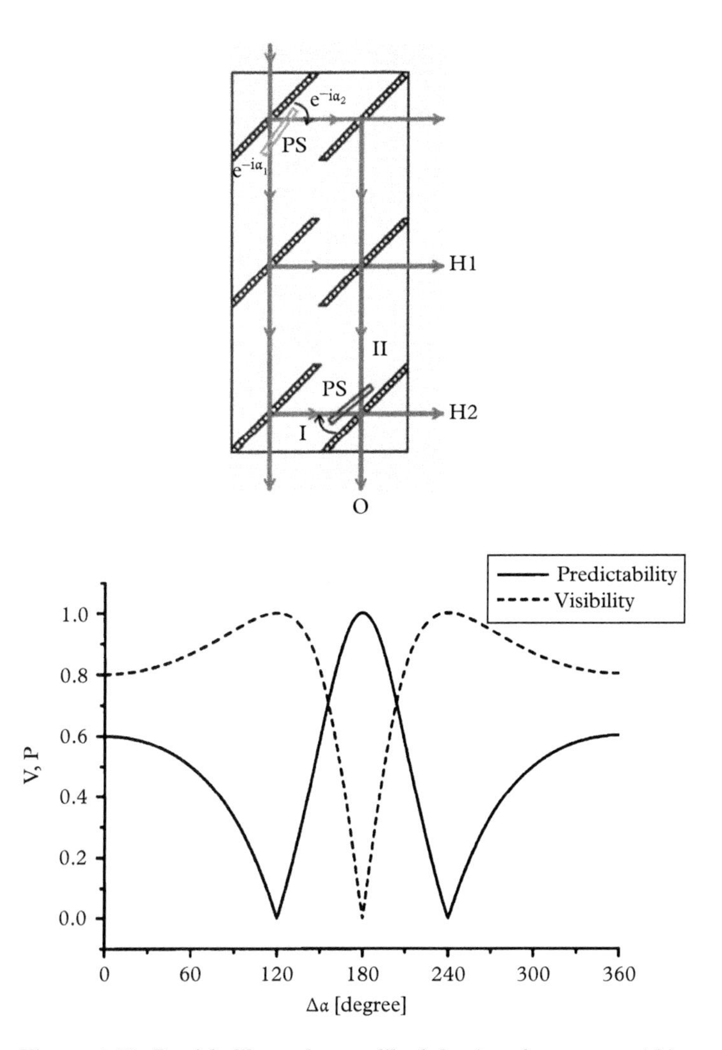

Figure 4.19 *Particle-like and wave-like behavior of neutrons within a double-loop interferometer (left) and a comparison between calculated and measured visibility and predictability (distinguishability) values (right). Reprinted with permission from Zawisky et al. 2002, copyright 2002 by the American Physical Society.*

sense. Experiments performed with atom interferometers support this view (Dürr et al 1998). Dedicated neutron interferometry measurements have been performed with a double-loop interferometer where visibility and distinguishability can be controlled by means of phase shifters in loop I and loop II, respectively (Fig. 4.19). Good agreement between theory and experiment has been achieved (Zawisky et al. 2002, Zawisky 2004).

The measure for the distinguishability as needed for Eq. (4.76) can also be obtained by an analog of Pauli's spin operators (Englert 1996), where the two possible paths through the interferometer are labeled as +1 (up) and −1 (down) states, respectively. The predictability can be related to the likelihood L for each and every event that the particle took the most likely path ($P_D = 2L - 1$). In this respect a close connection between complementarity, the uncertainty relation, and the superposition principle has been established (Björk et al. 1999). For orthonormal states of the two paths through the interferometer we can define L from the probabilities w_+ and w_-, for the two events ($L = (1 + |w_+ - w_-|)/2; w_+ + w_- = 1$). Holladay (1998) introduced a "which-value-interference" form of wave–particle complementarity which is claimed to be rooted only on the formal structure of quantum theory. A connection of the wave–particle behavior with the quantum erasure features has been shown by Englert and Bergou (2000) and for the neutron Interferometric case by Badurek et al. (2000).

The smooth transition between wave and particle properties is closely related in any measurement theory. It always involves the question of the wave function collapse and whether a classical observer is essential as in the Copenhagen interpretation of quantum mechanics; or, if it is possible that a pure state can evolve to a mixed state without resorting to classical concepts (Machida and Namiki 1980, Guilini et al. 1996). In the latter case, a dephasing process is postulated which causes increasing decoherence to reach the limit of a complete measurement (Namiki and Pascazio 1991, Kono et al. 1996). In connection with stochastically and deterministically driven beam splitters of Mach–Zehnder light interferometers, DeMartini et al. (1992) gave the division between particle- and wave-like aspects in terms of the Shannon information entropy which can be adapted to the description given earlier. It has been shown that a stochastic beam splitter results in a decrease of information entropy related to the particle trajectories and a deterministic beam splitter causes an increase in information entropy.

Afshar et al. (2007) performed related double-slit experiments with photons and used a grid placed at the minimum of the interference pattern while measuring which-way information and concluded a persisting interference pattern remains from the small reduction of the overall intensity. Thus, visibility has been determined with a minimum wave function perturbation and afterwards which-way information has been obtained by a destructive measuring process. They report a violation of the Greenberger–Englert relation (Eq. 4.76) by a factor of 1.35. In this case the measurement of P_D and V happened at different places and different times. Consequently, one may argue that the particle and the wave are not present simultaneously. Nevertheless, the authors claim that they are present simultaneously when the photons pass through the pinholes. Whether the Englert–Greenberger–Yasin inequality is applicable in this case is still not settled.

An additional effect arises when the neutron is taken as an unstable particle, where the decay process acts like the statistical absorption discussed earlier. In this case some "which path" information can be obtained and interference can be preserved. Such effects are based on the fact that the time spent by the neutron in both beams may be different due to gravity or due to a phase shifter. The expected magnitude of the effect is on the order of the ratio of the coherence time ($\tau_c = \Delta/v$, Eq. 4.73) divided by the decay time (τ_d; Table 1.1), which is extremely small and not yet accessible (Bonder et al. 2013, Krause et al. 2014).

4.4 Counting Statistics

4.4.1 General Relations

In any neutron interferometer experiment the primary source is a thermal source represented by the moderator of a reactor or a spallation neutron source. Neutrons are produced in fission or spallation processes having a broad energy distribution with a mean energy of about 2 MeV. These neutrons are successively slowed down by collisions with the moderator atoms where they lose on average an energy per collision of

$$\Delta E = \frac{E}{2} \frac{4A}{(A+1)^2}. \tag{4.77}$$

Here A is the mass ratio of moderator atom/neutron. The moderation in hydrogenous materials requires on the average about 20 collisions for a slowing down to thermal energies ($E_{th} = k_B T = 25$ meV) and a time of about 10 μs. The neutrons reach thermal equilibrium with the moderator by multiple up and down scattering events, finally reaching a Maxwellian energy distribution given by

$$\phi(E) = \phi_{th} \frac{E}{(k_B T)^2} e^{-E/k_B T} \left[\text{neutrons/m}^2\text{/s/eV}\right]. \tag{4.78}$$

The number of collisions in the thermal regime is determined by the ratio of the mean lifetime of neutrons within the moderator ($\ell_0 = (\sigma_a N v_{th})^{-1}$), which is influenced by the leakage and to the mean-free path length for collisions ($\ell_s = (\sigma_s N v_{th})^{-1}$). For light water the number of collisions of thermal neutrons is about 500. The total thermal flux can reach values $\phi_{th} \cong 10^{15}\text{cm}^{-2}\text{s}^{-1}$. This is many orders of magnitudes smaller than that of most light or electron sources.

The number of counts N registered in a detector in a certain time interval T obeys a Poissonian distribution (Glauber 1963, Walls and Milburn 1994)

$$P(N) = \frac{\overline{N}^N}{N!} e^{-\overline{N}}, \tag{4.79}$$

whose variance $(\Delta N)^2 = \overline{N^2} - \overline{N}^2 = \overline{N}$, where the mean counts is $\overline{N} = \overline{I} \cdot \tau$, for an intensity I and a time τ. This Poissonian distribution is also a signature for a coherent state behavior. The time sequence of detection of such neutrons is completely random in contrast to other kinds of radiations. It should be mentioned that sub- and super-Poissonian distributions can be achieved by dead-time effects of the detector or other feedback mechanisms of the system and by various quantum superposition effects, as well. The conditional probability of detecting a neutron within a time interval t when another one was detected at $t = 0$ is given by Mandel (1963) and Glauber (1968):

$$W(t) = \overline{I}\, e^{-\overline{I}\, t}. \tag{4.80}$$

This is also the expected time-interval distribution function representing the pair-correlation function of the beam.

A coherent state $|\phi>$ is defined as an eigenstate of the annihilation operator $\hat{a}|\phi> = \alpha|\phi>$ and it may be expanded in terms of number (Fock) states $|n>$ as (Glauber 1963, Suderashan 1963)

$$|\phi> = e^{-|\phi|^2/2} \sum_n \frac{\phi^n}{(n!)^{1/2}} |n>, \tag{4.81}$$

where $|\phi|^2 = \overline{N}$ denotes the mean particle number, interpreted in a time-averaged sense. The number states are formally created by successive applications of the creation operator on the vacuum state (Glauber 1963, Walls and Milburn 1994):

$$|n> = \frac{(a^+)^n}{(n!)^{1/2}} |0>. \tag{4.82}$$

A coherent state can be regarded as a state with the most classical behavior, exhibiting minimum Gaussian distributions for the conjugate variables and it exhibits Poissonian particle statistics (e.g., Zurek et al. 1993). Lamb (1995) has also addressed the equivalence of a radiation field in free space with a system of simple quantum-mechanical oscillators representing a coherent state. The superposition of coherent states can produce non-classical states exhibiting squeezing phenomena. These squeezed states have been extensively discussed in quantum optics (see also Section 4.5.2). Properties of coherent states of free particles are built from a continuous spectrum of states and have additional features similar to harmonic oscillator states (Lamb 1995, Spiridonov 1995). The analogy between coherent states and free, but coherently coupled particle motion inside an interferometer needs further justification (see also remarks at the beginning of this chapter). By weakening the harmonic potential which is generally used to define coherent states, the characteristic level structure disappears and reaches the limiting case of a momentum distribution function which characterizes a freely moving particle beam. The non-spreading of the coherence lengths which are determined by the momentum distributions can be seen as an analog to the non-spreading wave-packet phenomena in coherent atomic states. In addition, in a beam experiment a time average must be taken instead of an ensemble average. These averages are equivalent in any ergodic interpretation of quantum mechanics, which has been tested explicitly with atoms (Huesmann et al. 1999). Coherent states exist for boson and fermion fields, as well. However, since the bosonic and fermionic algebras commute, their coherent states can only be constructed separately (Klauder 1960, Zhang et al. 1990).

4.4.2 Analysis of the Neutron Counting Statistics in Interferometry

Consider the case where neutrons are registered repetitively many times for a certain fixed time interval; or better yet for a certain number of incident beam monitor counts, which compensates for long-term variations of the reactor power. The related histograms for a mean total particle number $\overline{N} \cong 2$ and $\overline{N} \cong 50$ measured at the maximum and minimum of the interference pattern of the 0-beam are shown in Fig. 4.20 (Rauch et al. 1990). The agreement with the predicted Poissonian distribution is excellent (Eq. 4.79).

Similar measurements have been made where the total number of counts in the main detector was fixed instead of the monitor counts. In this case, the Poissonian distribution changes to a binominal one due to the constraint of a maximal count number (N_{max}) (e.g., Walls 1966)

$$P_{\mathrm{B}}(N) = \frac{N_{\mathrm{max}}!}{N!(N_{\mathrm{max}} - N)!} f^{N}(1 - f)^{N_{\mathrm{max}}-N}, \tag{4.83}$$

where f denotes the average fraction of counts of the detector in the forward (0-beam) direction where the measurements were made. The agreement between predicted and measured distributions is again very good.

The measured Poissonian distributions are a fingerprint of coherent states. Non-classical states can be constructed by quantum superposition of a finite number of coherent states (Walls 1983, Yurke and Stoler 1986, Loudon and Knight 1987, Schleich et al. 1991). Superposition of such states can produce squeezed states, which are non-classical states with fluctuations of one quantum conjugate quantity below the coherent state value. This manifests in several squeezing phenomena as they will be discussed in Section 4.5.2 (see Fig. 4.31). The related distribution functions can be narrowed or widened or can take an oscillatory structure depending on the mean particle number and the squeezing factor r, which describes the change of the variance of the squeezed

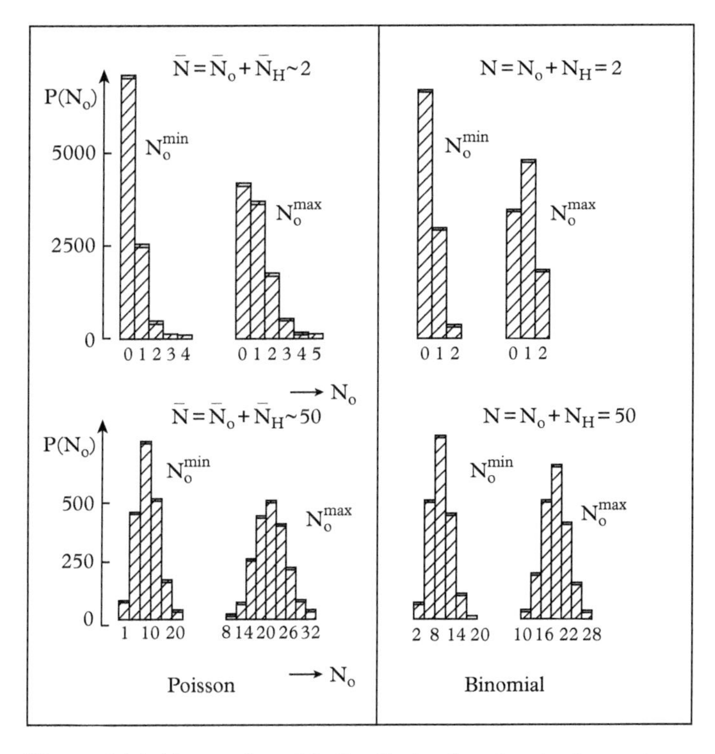

Figure 4.20 *Measured particle distribution functions in the maximum and minimum of the interference pattern for different mean particle numbers. Poissonian distributions appear when a constant measuring time is chosen (left) and binomial distribution appears when the number of counted neutrons is kept constant (right; Rauch et al. 1990)*

state compared to the coherent state (Yuen 1976; Buzek and Knight 1991; Dodonov et al. 1994, 1995). In optimal cases, the number fluctuations can be suppressed (s) by squeezing as

$$(\Delta N)_s^2 = \overline{N}\, e^{-2r}. \tag{4.84}$$

In this case, the distribution function is more complicated:

$$P(n) = (n!\mu)^{-1}\left(\frac{\nu}{2\mu}\right)^n \left|\left(H_n\left(\frac{\beta}{\sqrt{2\mu\nu}}\right)\right)\right|^2 \exp\left[-\beta^2 + \frac{\nu}{2\mu}\,\beta^2 + \frac{\nu^*}{2\mu}\,\beta{*}^2\right], \tag{4.85}$$

where

$$\nu = \sinh r\, e^{2\,i\phi}\ , \mu = \cosh r,$$
$$\beta = \mu\,\alpha + \nu\,\alpha{*},$$

and H_n are the Hermitian polynominals. Even and odd coherent states appear, being characterized by having zero odd or even particle states, respectively (Buzek and Knight 1995b). The expected

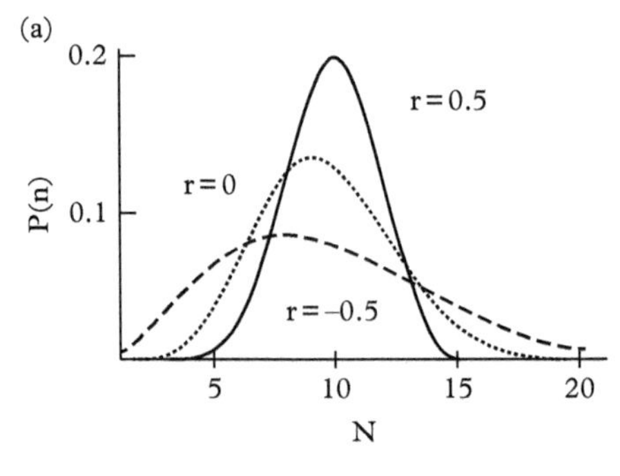

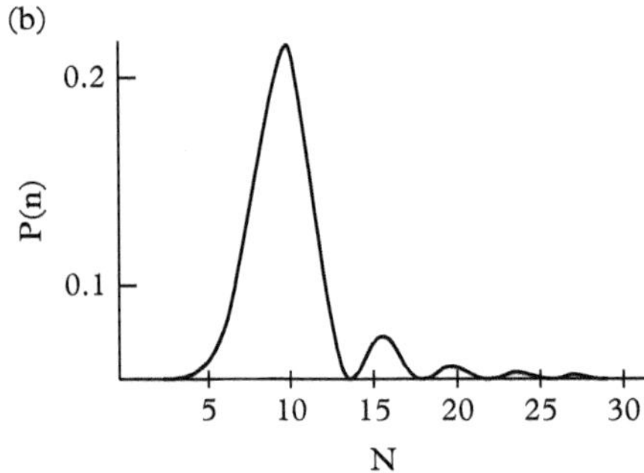

Figure 4.21 *Calculated particle distribution functions in case of squeezing. Walls and Milburn 1994, with kind permission from Springer Science and Business Media.*

change of the distribution function (Fig. 4.21) is quite analogous to the observed modulation of the momentum distribution in the case of Schrödinger cat-like states appearing at high-order interferences (Section 4.5.2). All these phenomena show characteristic features of non-classical states which arise from interference in phase space (Schleich and Wheeler 1987). The complementary situation concerning the phase and number distribution of angular momentum states has been elucidated by Agarwal and Singh (1996).

4.4.3 Particle Number–Phase Uncertainty Relation

The problem of quantum phase is as old as quantum mechanics itself. Its measurement requires a continuous spectrum of registered particles. The accuracy of the phase determination is related to the number of registered particles. The most commonly used particle number–phase uncertainty relations are given by Carruthers and Nieto (1968).

Here we deal first with a classical statistical analysis of measured interferometry data. We will focus on the limit of rather small particle numbers where the particle number–phase uncertainty relation plays an important role. In a realistic experiment, the intensity behind the interferometer is measured at M different positions j of the phase flag for a certain period of time or a number of monitor counts. Interferometry theory predicts that

$$N_j = \overline{N}\left[1 + V\cos(\phi_j)\right], \tag{4.86}$$

as given by Eq. (4.46). Standard statistical theories yield an estimate of ϕ_j with an uncertainty $\Delta\phi_j$. If the mean counting rate $\overline{N}$ and the fringe visibility $V(V = \langle|\Gamma(\Delta)|/|\Gamma(0)|\rangle)$ are known from other experiments, then

$$(\Delta\phi_j)^2 = \frac{N_j}{\overline{N}^2(V^2-1) + 2\overline{N}N_j - N_j^2}, \tag{4.87}$$

which follows directly from writing $(\delta N_j)^2 = \left(\frac{\partial N_j}{\partial \phi_j}\right)^2 (\delta\phi_j)^2 = N_j$. From the M different phase flag settings one obtains an estimate for the total phase uncertainty $\Delta\phi$ by χ^2 optimization, namely

$$(\Delta\phi)^2 = \left[\sum_{j=1}^{M} (\Delta\phi_j)^{-2}\right]^{-1}. \tag{4.88}$$

Approximating the sum by an integral over a definite number of interference fringes one obtains a particle number–phase uncertainty relation, namely

$$(\Delta\phi)^2(\Delta N)^2 = \frac{1}{1-\sqrt{1-V^2}}, \tag{4.89}$$

which approaches unity as the visibility $V \to 1$. Here ΔN denotes the standard deviation of the total counting rate, N, registered at this detector in the 0-beam. The related distribution obeys Poissonian statistics, so that $\Delta N = \sqrt{N}$, which is a basic feature of the thermal source emission process.

If one includes the counting rates observed by the other detector (the H-beam), or if one uses the constraint $N_j^0 + N_j^H = N^B$ (binomial scan), one obtains

$$(\Delta\phi)^2 N_t = \left[\frac{fV^2}{2\pi}\int_0^{2\pi} \frac{\sin^2\alpha \, d\alpha}{1-f+V(1-2f)\cos\alpha - fV^2\cos^2\alpha}\right]^{-1}, \tag{4.90}$$

where f is the average fraction of the counts the first detector (0-beam) receives and $N_t = MN^B$ is the total number of neutrons counted. $(\Delta\phi)N_t$ approaches $\left[1+\sqrt{1-2f}\right]^{-1}$ as the visibility $V \to 1$; and furthermore it approaches unity if $f = ½$. Comparison of Eqs. (4.89) and (4.90) shows that using the counts of both detectors decreases the uncertainty of the phase measurement by a factor of $(N/N_t)^{1/2}$. A coupling of the interference counting rate to the monitor counting rate increases the uncertainty product up to 1% (for a monitor efficiency of 10^{-4}). Similar results have also been obtained by Yurke (1986) for an ideal Mach–Zehnder interferometer.

If the visibility is considerably smaller than 1 the term $\left(1-\sqrt{1-V^2}\right)^{-1}$ in Eq. (4.89) can be approximated by $2/V^2$, which yields for monitor scans (Poissonian fluctuations)

$$(\Delta\phi)^2 N = \frac{2}{V^2} \tag{4.91}$$

and for binomial scans

$$(\Delta\phi)^2 N_t = \frac{2\overline{N}^H}{\overline{N}^0 V^2}. \tag{4.92}$$

The difference between this result and the exact formula (Eq. 4.88) is of order 5% for $V = 0.5$.

Related experiments have been performed at the interferometer setup at the 250-kW TRIGA reactor in Vienna, which certainly is a proper place for the investigation of low counting-rate phenomena (Rauch et al. 1990). Interference patterns were taken for different fixed times (actually fixed monitor counts). Figure 4.22 shows some representative results. These measurements were

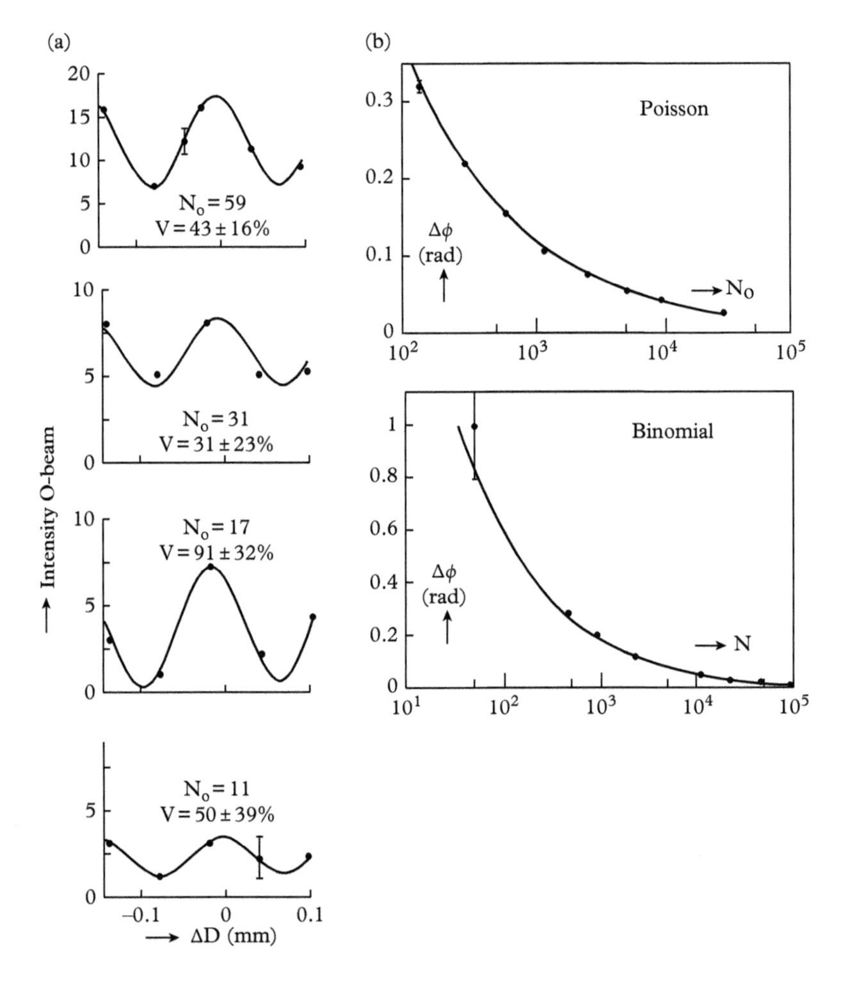

Figure 4.22 *Low particle interference pattern with least squares fit curves (left) and measured particle number–phase uncertainties (right) in case of a constant time (Poisson) and a constant count (binomial) measuring method (Rauch et al. 1990)*

repeated manifold and the data points were fit by means of a least squares fit procedure according to Eq. (4.86). From the repetitive measurements, a set of interferograms yielded values for ΔN and $\Delta\phi$. The results obtained are compared with the expected behavior in Fig. 4.21. A least squares fit procedure which must be completed by a maximum likelihood prescription has been used, because at rather low particle numbers systematic errors may occur. The agreement with the predictions of Eqs. (4.89) and (4.80) is again fairly good. At the same time these experiments demonstrate the build-up of the interference pattern from single-neutron events as has been demonstrated in a similar fashion for electron interference patterns (Lichte 1988, Tonomura et al. 1989).

The formulation of the particle number–phase uncertainty relation in this chapter is based on standard probability theory and must be completed in the case of a more rigorous treatment including quantum mechanical effects (Carruthers and Nieto 1968, Nieto 1977, Pegg and Barnett 1988, Lee 1995). The fermion character of neutrons does not show up directly because the low intensities involved in any kind of neutron interference experiment make particle–particle interaction negligible. In this case Fermi–Dirac statistics approach Bose–Einstein statistics because successive neutrons are generally separated in time, space, and momentum (Ledinegg and Schachinger 1983). Such systems are generally described by coherent states whose associated number states are Poissonian distributed (Eq. 4.78) and whose phase difference–particle number uncertainty relation can be written as (Gerhard et al. 1973, Nieto 1977)

$$(\Delta\phi)^2 = 1 - e^{-N} - N^2 e^{-4N}\left[\sum_{n=0}^{\infty} \frac{N^n}{n!\sqrt{n+1}}\right]^4, \tag{4.93}$$

which reduces for $N >> 1$ to the statistical limit discussed earlier (Eqs. 4.75 and 4.78) but which deviates from that value for $N \to 0$ due to projection states onto the vacuum state. This behavior has also been verified in the analysis of low counting rate photon experiments (Gerhard et al. 1973, Nieto 1977). The construction of coherent Fermion states is described by Zhang et al. (1990).

The classical limit (Eq. 4.91, $\Delta\phi \geq (\Delta N)^{-1} = (N)^{-1/2}$) corresponds to the so-called shot-noise limit or the coherent state limit and comes from the Heisenberg uncertainty relation (Dirac 1927). On the other hand, quantum mechanic does not set any restriction to the fluctuations ΔN, other than it cannot exceed the mean number of particles, which sets an upper limit

$$\Delta\phi \geq \frac{1}{<N>}. \tag{4.94}$$

This Heisenberg limit follows also from general arguments of the complementarity principle, but it requires non-classical quantum states as discussed in detail by Yurke et al. (1986) and Ou (1997). The Heisenberg limit (Eq. 4.92) can also be approached by using multipath interferometer arrangements (Zernike 1950, d'Ariano and Paris 1997). In this case classical (coherent) radiation can be used and the phase sensitivity increases by the number of available paths M

$$\phi = \frac{1}{\sqrt{NM}}. \tag{4.95}$$

This technique can be used in case of multiplate interferometers and it should be useful when measuring small phase shifts. There are no fundamental limits on the accuracy when measuring small phase shifts but there are technical problems of realizing multipath interferometers.

A comparison between light (laser) and matter wave phase sensitivity is given by Scully and Dowling (1993), which results in the statement that matter wave interferometers may be superior

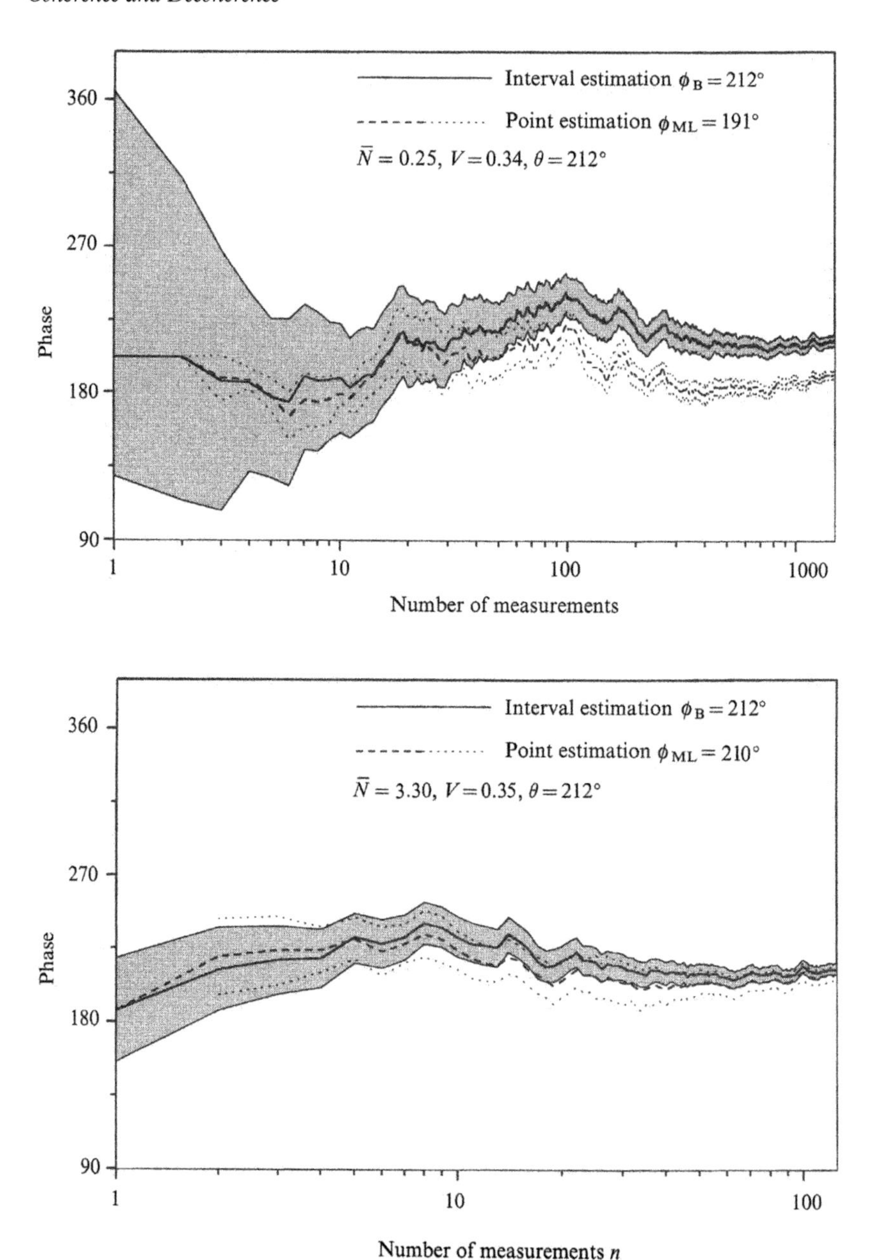

Figure 4.23 *Comparison of the Bayes (interval) and maximum likelihood (point) phase estimation in the case of various mean particle numbers as a function of the number of measurements made (Zawisky et al. 1998)*

by a factor of 10^4 when they are used as gyroscopes to detect rotation effects (Sagnac effect—see Section 8.2). A more detailed analysis of interferometric phase measurements must take into account the phase distribution functions, which can be obtained by a multiple measurement of interferograms (Beck et al. 1993, Smithey et al. 1993, Hradil et al. 1996, Zawisky et al. 1998, Theo et al. 2011). The influence of the phase distribution function becomes stringent for rather small counting rates where root-mean-square estimations can fail. Particularly, the difference between measurements with and without accumulation of counted data may appear crucial. Figure 4.23 shows the difference of the phase estimation based on a Bayesian analysis (interval estimation) and maximum likelihood (point) estimation for a set of interference experiments performed with rather low mean counting rates, $\overline{N} = 0.25$ and $\overline{N} = 3.03$, respectively (Zawisky et al. 1998). The visibility of the actual experiment has been included as well. Such investigations require a much more complicated measuring and data-handling procedure but they are capable of detecting any systematic failure of an experiment.

The extraction of the quantum phase from restricted data sets is a long-standing problem. The operational approach of Noh et al. (1992, 1993) is adapted to a multiport homodyne detection method. This method can be used for a Mach–Zehnder interferometer situation, when an additional 0- and $\pi/2$-phase shifter is applied. Slight improvements to this operational approach are feasible when non-Gaussian statistics and phase-sensitive signal and noise exist (Rehacek et al. 1999, Teo et al. 2012). In all interferometer experiments phase distribution functions play an important role, which indicates that in all pragmatic interpretations of quantum mechanics the results of measurements depend irreducibly on both the state preparation and the measurement apparatus.

4.4.4 Intensity Correlation Experiments

In this case one must distinguish between correlations on a time scale much larger than the coherence time τ_c, and also comparable with the coherence time (Hanbury-Brown and Twiss effect). First we deal with $t >> \tau_c = \Delta_c/v$. See Eq. (4.71).

Higher order correlations in quantum optics are generalizations of the two-point-two-time correlation function (Eq. 4.17; Glauber 1963)

$$\begin{aligned} &G^{(n)}(\boldsymbol{r}_1, t_1; \boldsymbol{r}_2, t_2; \ldots\ldots; \boldsymbol{r}_n, t_n; \boldsymbol{r}'_1 t'_1; \ldots\ldots \boldsymbol{r}'_n t'_n) \\ &\quad = Tr\left[\rho\psi^*(\boldsymbol{r}_1, t_1)\ldots\ldots\psi^*(\boldsymbol{r}_n, t_n) \bullet \psi(\boldsymbol{r}'_1, t')\ldots\ldots{}_1\ldots\ldots\psi(\boldsymbol{r}'_n t'_n)\right]. \end{aligned} \tag{4.96}$$

Interest in higher correlations ($n = 2$) is mostly connected to the intensity or particle number correlation, which is given by

$$G^{(2)}_{1,2}(\boldsymbol{R}, t) = Tr\left[\rho\, I_1(0,0))I_2(\boldsymbol{R}, t)\right], \tag{4.97}$$

where $\boldsymbol{R} = \;<(r_1 - r') - (r_2 - r_2')>$ and $t = \;<(t_1 - t_1') - (t_2 - t_2')>$. These times are much larger than the coherence times τ_c defined by the related coherence length (Section 4.1, Eqs. 4.33 and 4.73). These coherence lengths and times are related to the self-correlation function of the neutron (see Section 4.1). For large times ($t >> \tau_c$) the pair correlation function can be measured, which gives the conditioned probability that a neutron arrives in a detector of a time interval t later than another has arrived at $\tau = 0$. For a Poissonian beam this function is given by Eq. (4.80) as (Mandel 1963, Glauber 1968)

$$G^{(2)}(0, t) = \langle I_1(0,0)\; I_2(0,t)\rangle = \overline{I}\, e^{-\overline{I}t}, \tag{4.98}$$

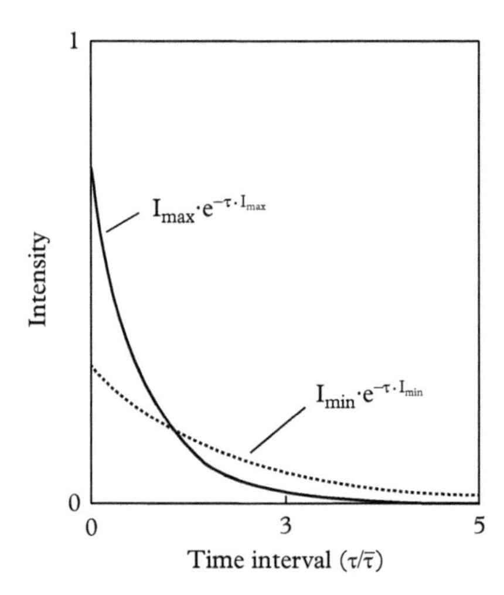

Figure 4.24 *Time pair correlation function for Poissonian beams with different count rates*

which exhibits an exponential behavior depending on the mean intensity $\bar{I}$. Thus, in an interference experiment it depends on whether this correlation function is measured at the maximum, minimum, or any other position of the interference pattern (Fig. 4.24). This provides the basis for post-selection experiments in time, which are described in Section 4.5.4 (Zawisky et al. 1994). The counting statistics for pairs of neutrons arriving within a time interval ($0 \leq t \leq \bar{\tau} = 1/\bar{I}$) and ($1.5\bar{\tau} \leq t \leq = \alpha$) are shown in Fig. 4.25. It is visible by eye that these distributions are more regular than the Poissonian distribution. This indicates sub-Poissonian and non-classical features. Related experiments have been performed at the 250-kW TRIGA reactor in Vienna where the arrival time of each neutron was registered, permitting the identification of time correlation in the course of data analysis. Related experiments will be discussed in connection with post-selection experiments (Section 4.5.4; Fig. 4.36).

For times smaller than the coherence time ($t < \tau_c$) self-correlation effects like the Hanbury-Brown and Twiss (1956) effect are expected; but they are difficult to approach due to the smallness of τ_c (see Eq. (4.7.3) and Section 10.9). In this case coincidence measurements of split beams are performed with high resolution detectors. The coincidence rate can be written as (Mandel and Wolf 1995, Klein and Furtak 1996, Fox 2006)

$$\overline{C} = \int_{-T/2}^{+T/2} P_2(r_1, t; r_2, t + \tau)\mathrm{d}\tau, \tag{4.99}$$

where P_2 denotes the joint detection probability at position r_1 and r_2. It depends on the efficiencies of the detectors η_i, their sizes S_i, and the intensities at these positions ($I(r_1, t_1)$ and $I(r_2, t_2)$). Here

$$P_2(r_1, t_1; r_2, t_2) = \eta_1\eta_2 S_1 S_2 \langle I(r_1, t_1)I(r_2, t_2)\rangle = \eta_1\eta_2 S_1 S_2 \langle I(r_1)\rangle \langle I(r_2)\rangle \Delta t \Delta\tau \left[1 - |\gamma(r_1, r_2, \tau)|^2\right]. \tag{4.100}$$

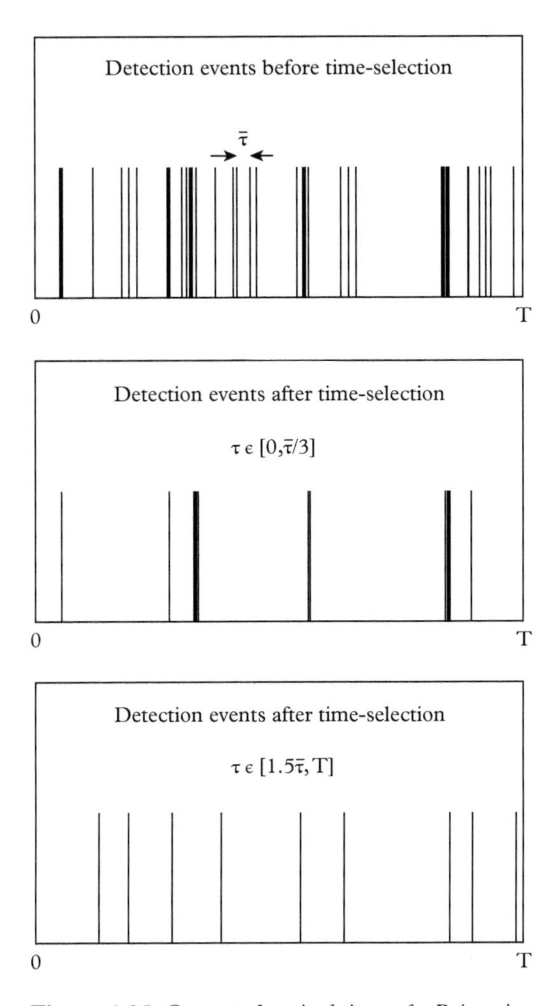

Figure 4.25 *Computed arrival times of a Poissonian distributed beam (top) and pairs whose arrival times lie within a short interval (middle) and within a late interval (bottom; Zawisky et al. 1994)*

When a stationary and a cross-spectral pure situation for fermions is considered one gets

$$\gamma(r_1, r_2, \tau) = \frac{\langle \Psi^*(r_1, 0)\,\Psi(r_2, \tau)\rangle}{\sqrt{\langle I(r_1)\rangle\,\langle I(r_2\rangle}} = \gamma(r_1, r_2, 0)\gamma(\tau) \tag{4.101}$$

and with Eq. (4.98)

$$\overline{C} = \eta_1\eta_2 S_1 S_2 T\left[1 - \frac{\tau_c}{T}|\gamma(r_1, r_2, 0|^2\right], \tag{4.102}$$

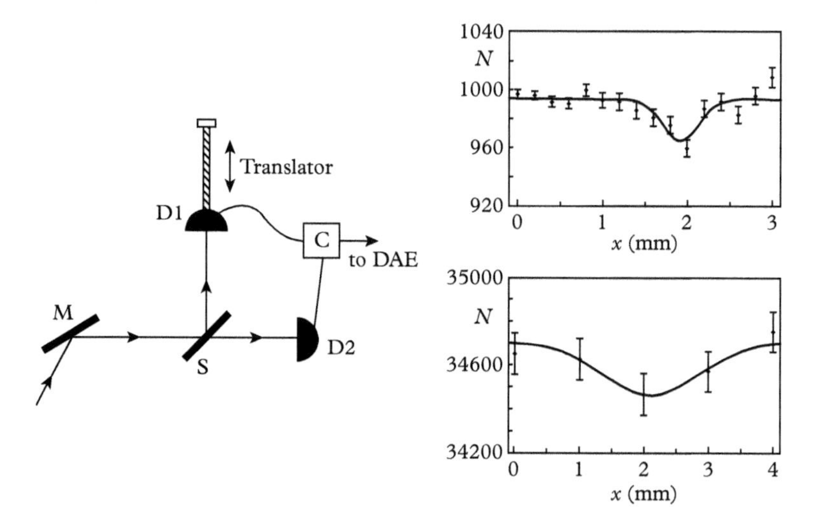

Figure 4.26 *Experimental setup and results of a Hanbury-Brown and Twiss experiment demonstrating the fermion anti-bunching effect. Reprinted with permission from Iannuzzi et al. 2006, copyright 2006 by the American Physical Society.*

where it has been assumed that $\tau_c \leq T/2$, which gives

$$\int_{-T/2}^{T/2} |\gamma(\tau)|^2 d\tau \approx \int_{-\infty}^{\infty} |\gamma(\tau)|^2 \mathrm{d}\tau = \tau_c. \tag{4.103}$$

An experiment related to this cross-correlation of detection in time at two positions has been performed by Iannuzzi et al. (2006) at a neutron back-scattering instrument and with a mosaic graphite single crystal as beam splitter at the Institute Laue-Langevin in Grenoble. The coherence time was about $\tau_c \approx 16$ ns, the detection time resolution was about $T \approx 1.1\mu$s. A small dip has been observed when the flight paths to both detectors were equal (Fig. 4.26). It may happen that the small and perfect mosaic blocks act as coherent 50:50 beam splitters and as a microscopic source, resulting in a macroscopic coherence area at the position of the detectors. A similar experiment has also been done by the same authors when they used a high-resolution, position-sensitive detector. They again found a dip structure between neighboring pixels (Iannuzzi et al. 2011). The observed effect is small and certainly needs more attention for verification in the future. Some discussion about these measurements can by found in the literature (Varro 2008, Yuasa et a. 2008).

4.5 Post-selection Measurements

Various post-selection measurements in neutron interferometry have shown that interference fringes can be restored even in cases when the overall beam does not exhibit any interference fringes due to spatial shifts larger than the coherence lengths of the interfering beams. This

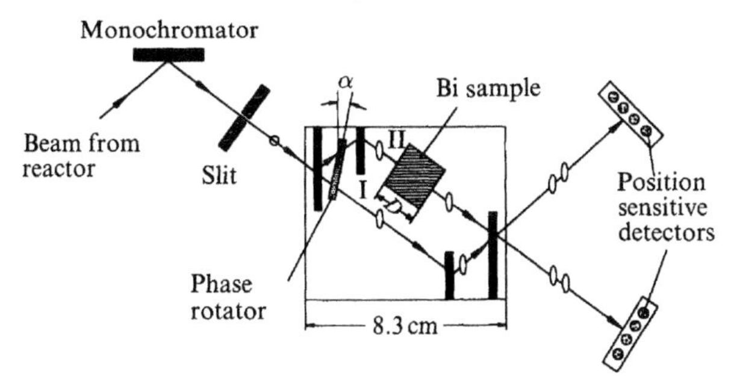

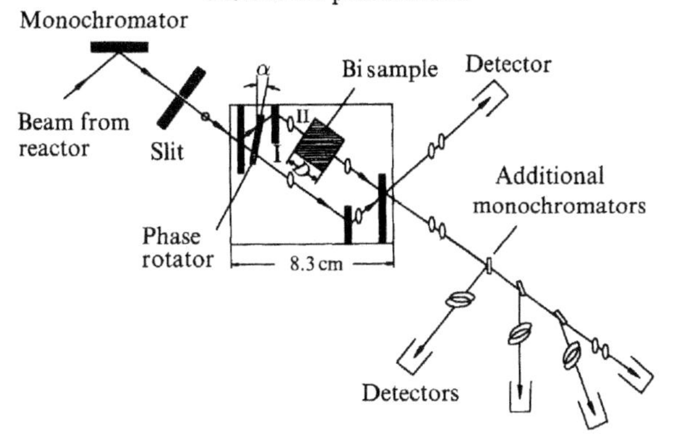

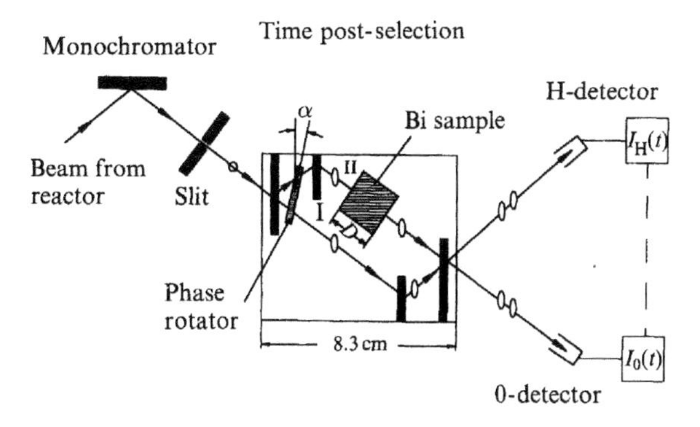

Figure 4.27 *Schematic diagrams for various post-selection experiments which permit the extraction of additional information behind the interferometer after interference and recombination*

indicates that the simple picture which predicts interference only when wave packets spatially overlap is not quite true. Thus, the degree of coherence as defined in Eq. (4.49) can be essentially zero but coherence can actually persist. A rather misleading language has evolved in this field. Interference actually occurs no matter how large the optical path difference between two sub-beams or two wave packets may be. From classical optics it has been known for many years that the coherence properties manifest themselves in a spatial intensity variation for spatial shifts smaller than the coherence length and in a spectral intensity variation for large phase shifts (Mandel 1962, 1965; Heineger et al. 1983; James and Wolf 1991; Zou et al. 1992; Agarwal and James 1993). A detailed experimental study of the spectral modification in an optical Mach–Zehnder interferometer has been reported by Rao and Kumar (1994). This phenomenon becomes more apparent for less monochromatic beams and can cause overall spectral shifts (Faktis and Morris 1988, Wolf 1989) and even squeezing phenomena (Jansky and Vinogradov 1990, Schleich et al. 1991). The related phenomena for matter waves have been discussed by Rauch (1993a, 1995b) and investigated experimentally by Jacobson et al. (1994). Various methods of postselection are indicated in Fig. 4.27. We will see that in all cases of post-selection measurements, more information can be extracted from the experimentally observed events than is usually done.

4.5.1 Post-selection in Ordinary Space

The intensity, the visibility, and the intrinsic phase (see Eq. 4.46) depend on the position within the beam cross-section, as shown in Fig. 2.9. The overall parameters of these quantities are mean values obtained from the local distribution. For the operation of a good interferometer an essentially flat distributions is desirable, which can be achieved by a well-balanced interferometer operated under stable environmental conditions.

In any interference experiment, the beam leaving the interferometer is adjusted to obtain maximum contrast (i.e., maximum fringe visibility, V). However, the contrast varies across the beam cross-section and across the Borrmann fan due to the fundamental dependence of the crystal reflectivities on the deviation of the individual momentum k from the momentum k_B fulfilling the exact Bragg condition; and also due to unavoidable imperfections of the crystal and different sensitivities against various strains, vibrations, and stray fields (Fig. 2.9). The intensity profile behind the interferometer can be calculated in detail on the basis of plane wave or spherical wave theory of dynamical diffraction from perfect crystals. These calculations are summarized in Chapter 11, where also an overview of the profiles inside and behind the interferometer is shown. Figure 4.28 shows the intensity profile measured by scanning a narrow slit across the Borrmann fan and the measured contrast at different positions (Bauspiess et al. 1978). One notices that the contrast and the internal phases (see Eq. 4.46) are spatially dependent, each part contributing differently to the overall contrast (in this case 42%). Now, modern position-sensitive detectors can be used and the contrast per pixel V_i can be measured. This provides more information than the overall contrast ($\Sigma V_i \cdot I_i/I > V_0$). Future applications of this feature of a general neutron interferometry experiment is expected to lead to phase contrast microscopy of density fluctuations in material media. When testing new interferometers and optimizing them for a high contrast, it is obligatory to scan the contrast of the exit beam across its cross-section. More detailed measurements dealing with spatial post-selection are discussed in Chapter 2, Fig. 2.9.

4.5.2 Post-selection in Momentum Space

In the course of several neutron interferometer experiments it has been established that smoothed out interference properties at high interference order can be restored even behind the

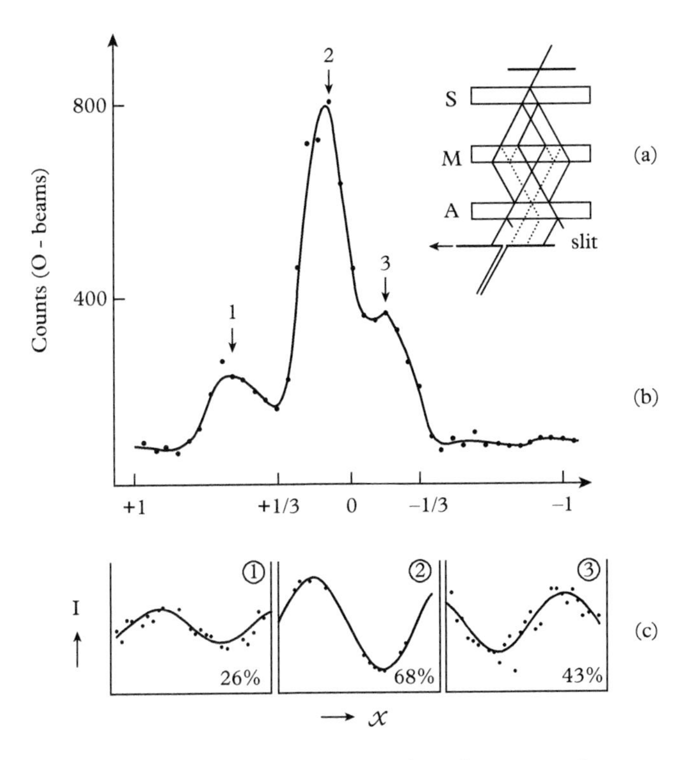

Figure 4.28 *Post-selection experiment in ordinary space by measuring the intensity profile and the related interference pattern (Bauspiess et al. 1978)*

interferometer when a proper spectral filtering is applied (Werner et al. 1991, Kaiser et al. 1992, Rauch et al. 1992, Jacobson et al. 1994). The experimental arrangement with a schematic indication of the wave packets at different parts of the interference experiment is shown in Fig. 4.27 (middle). An additional post-monochromatization (analyzing filter) is applied behind the interferometer by means of a single crystal brought into the Bragg reflecting position, thereby defining a narrower momentum band than the pre-monochromator.

Using Eqs. (4.16) and (4.31) in a time-independent form, the momentum-dependent intensity is

$$I_0(\boldsymbol{\Delta}, \boldsymbol{k}) = \left|\psi_0{}^{\mathrm{I}}(\boldsymbol{r}, \boldsymbol{k}) + \psi_0{}^{\mathrm{II}}(\boldsymbol{r} + \boldsymbol{\Delta}, \boldsymbol{k})\right|^2 \propto \left|a(\boldsymbol{k})\right|^2 (1 + \cos(\boldsymbol{\Delta}(\boldsymbol{k})\cdot\boldsymbol{K})), \tag{4.104}$$

whereas the overall beam intensity is given by an integration over k

$$I_0(\boldsymbol{\Delta}_0) \propto 1 + |\Gamma(\boldsymbol{\Delta}_0)| \cos \boldsymbol{\Delta}_0 \cdot \boldsymbol{K}_0. \tag{4.105}$$

Here $K = nk$ and $\boldsymbol{\Delta}_0$ is the spatial shift for the central $\boldsymbol{k}_0$-component of the packet due to a phase shifter (Eq. 4.44, Fig. 4.2). This formula also shows that the overall interference fringes disappear for spatial shifts larger than the coherence lengths $[\Delta_i \geq \Delta_c{}^i = 1/(2\delta k_i)]$ (see Eq. 4.35).

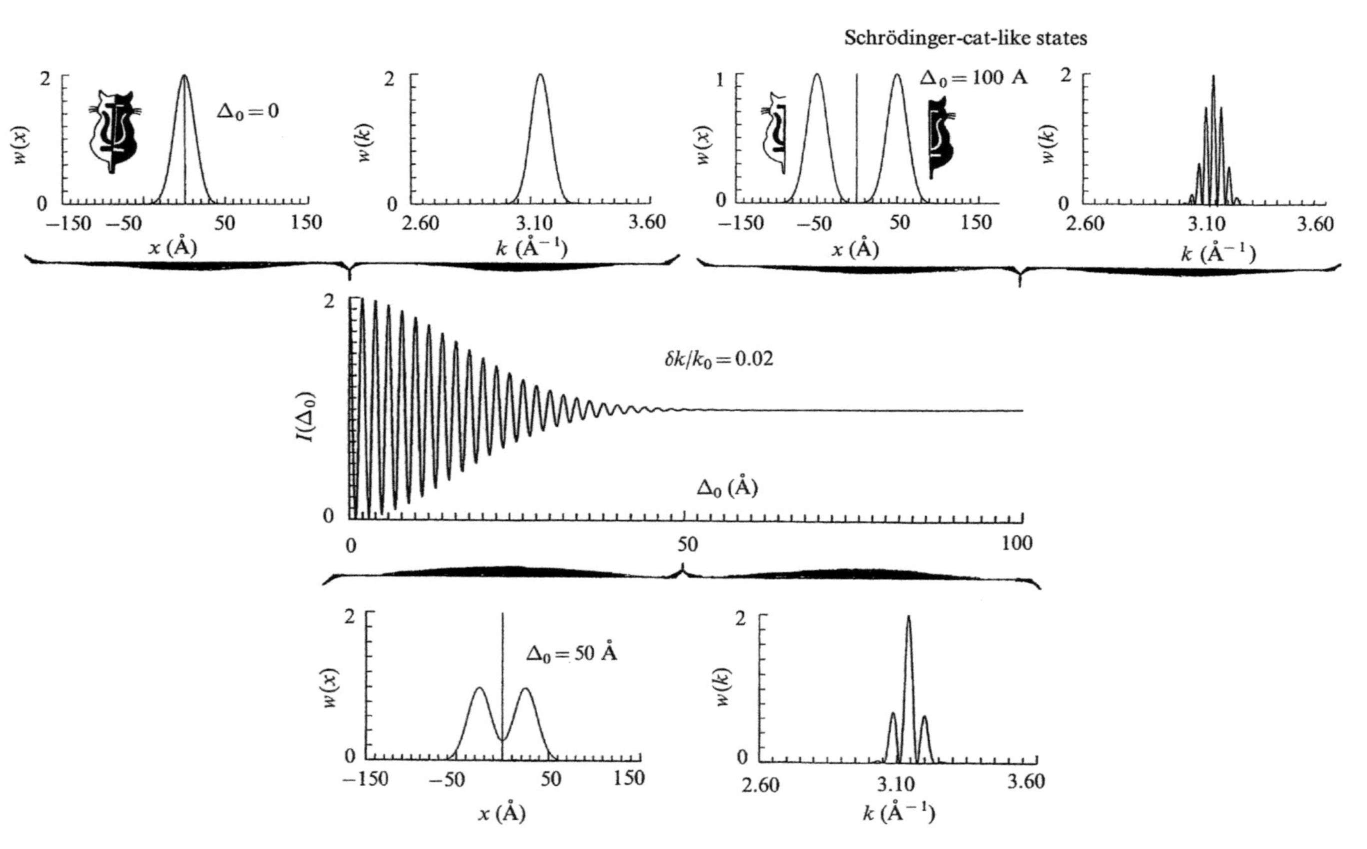

Figure 4.29 *Calculated interference pattern (middle) and calculated wave packets in ordinary and momentum space at different phase shifts. The appearance of Schrödinger cat-like states at high interference orders is indicated (Rauch 1993a)*

The coherence properties along the interferometer axis (x) were investigated in a dedicated experiment (Jacobson et al. 1994). In this case, the (transverse) components of the momentum vectors (and coherence length) do not change due to the Bragg diffraction. According to basic quantum mechanical laws, the related momentum distribution for the 0-beam follows from Eq. (4.102) and can be rewritten for Gaussian packets in the form

$$I_0(k) = \exp[-(k-k_0)^2/2\delta k^2]\left\{1 + \cos\left(\chi_0 \frac{k_0}{k}\right)\right\}. \tag{4.106}$$

Here the mean phase shift is introduced ($\chi_0 = k_0\Delta_0 = -N b_c \lambda_0 D_{eff}$). The surprising feature is that the 0-beam intensity, $I_0(k)$, becomes oscillatory in momentum space for large phase shifts where the interference fringes described by Eq. (4.106) disappear (Rauch 1993a; Fig. 4.29). This indicates that interference in phase space must be considered (Schleich et al. 1978, 1988) rather than the simple wave function overlap criterion described by the coherence function (Eqs. 1.29 and 4.29). The second beam, the H-beam, behind the interferometer shows the complementary modulation $I_H = I_{total} - I_0$, as it must in order to conserve neutron intensity.

The amplitude function of the packets arising from beam paths I and II determines the spatial shape of the packets behind the interferometer

$$\begin{aligned} I_0(x) = \left|\psi(x) + \psi(x+\Delta)\right|^2 &= \exp[-x^2/2\delta x^2] + \exp[-(x+\Delta_0)^2/2\delta x^2] \\ &+2\exp[-x^2/4\delta x^2]\,\exp[-(x+\Delta)^2/4\delta x^2]\cdot\cos\chi_0. \end{aligned} \tag{4.107}$$

This result separates for large phase shifts into two peaks. $I_0(k)$ and $I_0(\Delta)$ are Fourier transform pairs (e.g., Levy-Leblond 1990); see Fig. 4.28. For Gaussian packets, having a spatial width, δx, corresponding to the coherence length Δ_c, the minimum uncertainty relation $\delta x \delta k = ½$ is fulfilled. For an appropriately large displacement ($\Delta >> \Delta_c$), the related state can be interpreted as a superposition state of two macroscopically distinguishable states; that is, a stationary Schrödinger cat-like state (Leggett 1984, Yurke et al. 1990, Schleich et al. 1991). These states—widely separated in ordinary space, but oscillating in momentum space—seem to be notoriously fragile and sensitive to dephasing effects (Zurek 1981, 1991, 1998a; Walls and Milburn 1985; Glauber 1986; Namiki and Pascazio 1991). The detailed structure of these wave packets in ordinary and momentum space is seen in more detail in Fig. 4.30.

Measurements of the wavelength spectrum behind the interferometer were made with an additional silicon post-monochromator crystal with a rather narrow mosaic spread which reflects in the parallel position relative to the pre-monochromator a very narrow band of neutrons ($\delta k'/k_0 \approx 0.0003$). Scanning this analyzer crystal, placed in the 0-beam (see Fig. 4.27, middle), through the Bragg position gives the wavelength distribution. Scanning the phase shifter (at a given setting of the analyzer crystal) gives the enhanced visibility of the interference pattern. The related results are shown in Fig. 4.31 for various phase shifts (Jacobson et al. 1994). This feature shows that an interference pattern can be restored even behind the interferometer by means of a proper post-selection procedure. In this case the overall beam does not show interference fringes anymore and the wave packets originating from the two different beam paths have no substantial overlap. These results clearly demonstrate that the predicted spectral modulation (Eq. 4.106) appears when the interference fringes of the overall beam essentially disappear. The modulation is somewhat smeared out due to averaging processes across the beam due to various unavoidable imperfections, existing in any experimental arrangement. Nevertheless, the predicted spectral modulation is clearly observed. The contrast of the empty interferometer was 60% in this case.

Each peak in the momentum distribution shown in Fig. 4.31 corresponds to a different number of 2π phase shifts experienced by the neutrons of that wavelength band during its passage

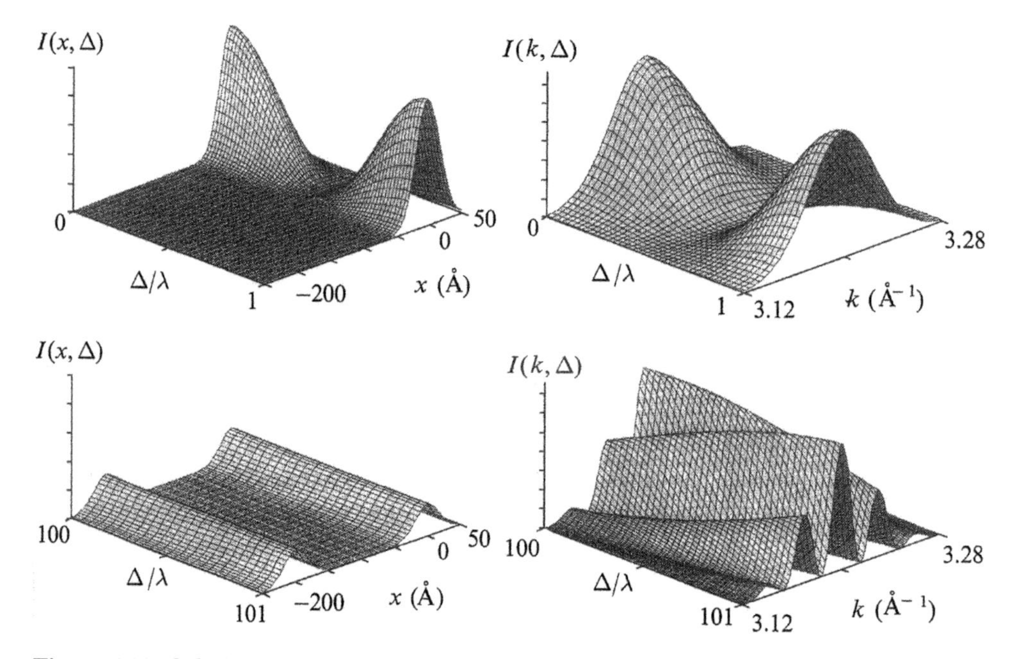

Figure 4.30 *Calculated wave packets in ordinary (left) and momentum (right) appropriate for low (above) and high (below) interference orders*

through the interferometer. In terms of quantum optics this means that different parts of the wave function obey a different number of 2π phase shifts. In this sense, the peaks represent new quantum entities with distinguishable properties from the other parts of the spectrum. This kind of labeling shows that constructive interference is restricted to a certain wavelength band. This is a situation similar to that where new states have been created due to lattice diffraction inside the interferometer (Section 4.3.2, Fig. 4.16).

The new quantum states created behind the interferometer can be analyzed with regard to their uncertainty properties. Analogies between a coherent state behavior and a free but coherently coupled particle motion inside the interferometer have been addressed (Rauch et al. 1990). In such cases, the dynamical conjugate variables x and k minimize the uncertainty product with identical uncertainties $(\Delta x)^2 = (\Delta k)^2 = ½$ (in dimensionless units). Using $I_0(k)$ and $I_0(x)$ (Eqs. 4.106 and 4.107) as distribution functions we obtain for Gaussian packets (for $\delta k/k_0 << 1$)

$$\begin{aligned} < (\Delta x)^2 > &= < x^2 > - < x >^2 \\ &= (\delta x)^2 \left[1 + \frac{(\Delta_0/2\delta x)^2}{1 + e^{-(\Delta_0/2\delta x)^2/2} \cos(\Delta_0 k_0)} \right], \end{aligned} \tag{4.108}$$

and

$$\begin{aligned} < (\Delta k)^2 > &= < k^2 > - < k >^2 \\ &= (\delta k)^2 \left\{ 1 - \left(\frac{\Delta_0}{2\delta x} \right)^2 \frac{e^{-(\Delta_0/2\delta x)^2/2} \cos(\Delta_0 k_0) + e^{-(\Delta_0/2\delta x)^2}}{\left[1 + e^{-(\Delta_0/2\delta x)^2/2} \cos(\Delta_0 k_0) \right]^2} \right\}. \end{aligned} \tag{4.109}$$

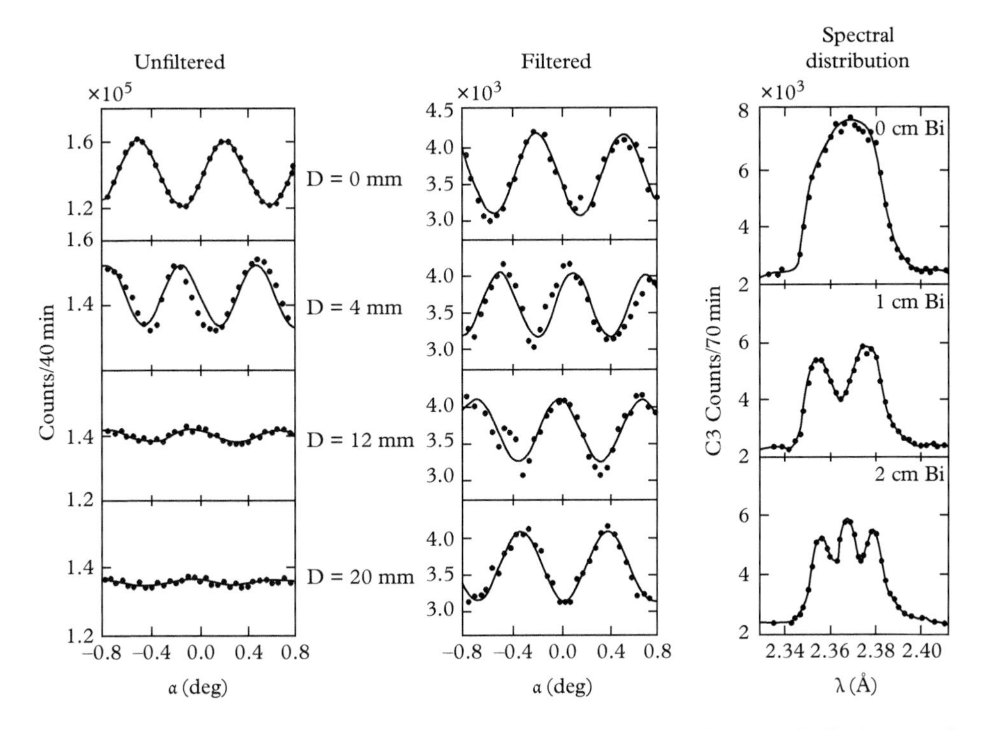

Figure 4.31 *Measured interference pattern of the overall (left) and filtered (middle) beam and measured momentum distribution of the overall beam at different interference orders (from above to below; Jacobson et al. 1994). The phase shifter were bismuth slabs of various thickness,* D

These relations are shown in Fig. 4.32. This indicates that $<(\Delta k)^2>$ values below the coherent state value, δk^2, can be achieved. In quantum optics terminology this can be interpreted as state squeezing (Walls 1983, Braunstein and McLachlan 1987, Loudon and Knight 1987, Jansky and Vinogradov 1990, Buzek and Knight 1991, Schleich et al. 1991). One emphasizes here that a single coherent state does not exhibit squeezing, but a state created by superposition of two coherent states can exhibit a considerable amount of squeezing as displayed in Fig. 4.32. Thus, highly non-classical states are made by the power of the quantum mechanical superposition principle. The degree of squeezing can be further enhanced by multiplate interferometry (Rauch 1995b, Suda 1995, Suda and Rauch 1996). Such situations have been treated for the optical case by Jansky and Vinogradov (1990), Adams et al. (1991), Buzek and Knight (1991), and Szabo et al. (1996). Properly formed squeezed input states can be used in a Mach–Zehnder interferometer to produce entangled states (Paris 1999). It should be mentioned that the general uncertainty relation $\Delta x \Delta k \geq 1/2$ remains valid for squeezed states as well.

A very similar situation exists in neutron spin-echo systems (Section 2.4). In this case the spin-up and spin-down states having slightly different momenta due to the longitudinal Zeeman splitting interfere with each other; this causes the well-known Larmor rotation and a modulation of the momentum distribution similar to that in split beam interferometry (Eq. 4.106). Figure 4.33 shows a dedicated arrangement for such a measurement and typical results (Badurek et al. 2000). In this setup a magnetic field with a strength of 3.5 mT and of length 0.57 m was used.

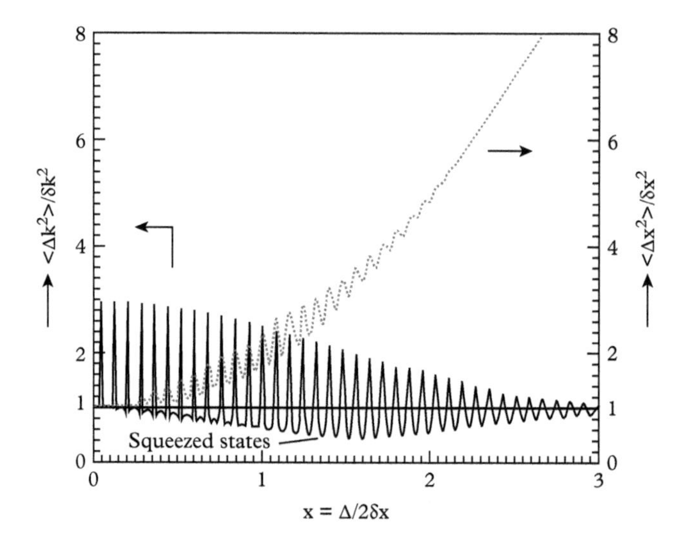

Figure 4.32 *Calculated widths of the spatial and momentum distribution functions indicating squeezing in the momentum domain Rauch 1995b, with kind permission from Springer Science and Business Media.*

The measurement was done by a time-of-flight analysis. A rather wide neutron spectrum ($\delta\lambda/\lambda_0$ = 0.27) was used. Intuitively one may be forced to believe that individual neutrons with different velocities produce such a pattern, but a rigorous quantum optics formulation associates this behavior with the quantum state itself. In existing spin-echo instruments the magnetic fields and the length of the fields are much larger and, therefore, the number of Larmor rotations is much higher, which results in a much finer structured momentum distribution, and it is not easy to resolve by ordinary means.

4.5.3 The Wigner Function

The persistent coherent coupling of the states in phase space even in the case of spatially, well-separated, packets can also be visualized by means of the Wigner quasi-distribution function defined as (Wigner 1932, Walls and Milburn 1994, Buzek and Knight 1995b, Schleich 2001, Suda 2005)

$$W_s(k,x) = \frac{1}{2\pi}\int_{-\infty}^{+\infty} e^{ikx'}\psi_s^*\left(x+\frac{x'}{2}\right)\psi_s\left(x-\frac{x'}{2}\right)\mathrm{d}x'. \tag{4.110}$$

For the beam behind the interferometer the superposition state from both beam paths must be considered, namely (Eqs. 2.1 and 4.104)

$$\psi_s(x) = \psi(x) + \psi(x+\Delta). \tag{4.111}$$

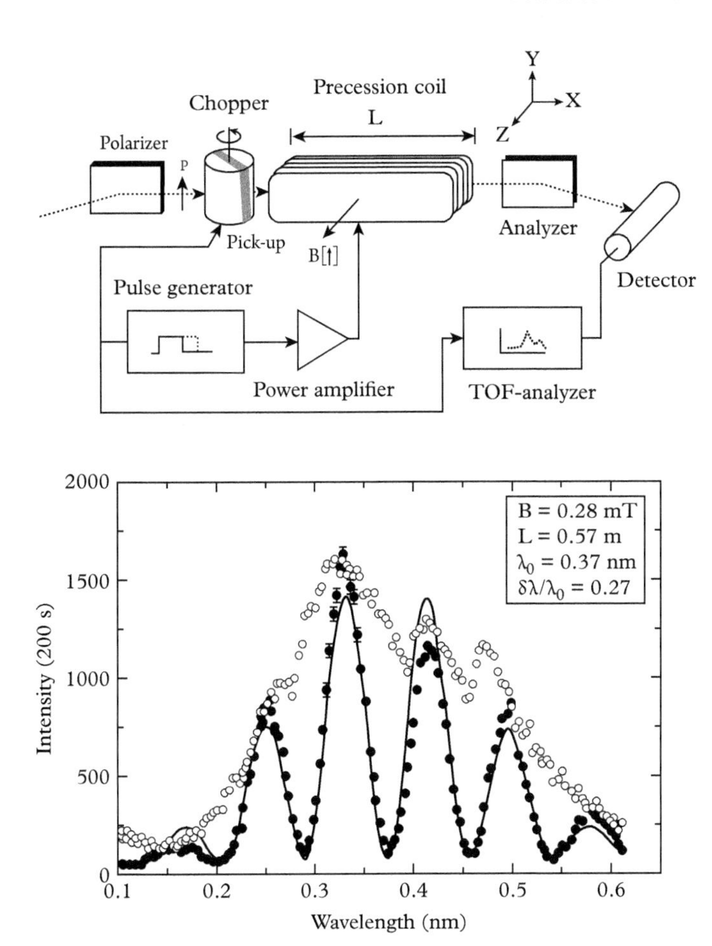

Figure 4.33 *Experimental arrangement and typical results of a measurement of the momentum distribution of a definite spin state in a neutron spin-echo arrangement. Open circles: primary neutron spectrum; full circles: modulated spectrum after spin precession (Badurek et al. 2000)*

After several steps the calculation of the Wigner function gives the following result (written for one dimension only):

$$
\begin{aligned}
W_s(x,k,\Delta) &= W(x,k) + W(x+\Delta,k) + 2W\left(x+\frac{\Delta}{2},k\right)\cos(\Delta\cdot k) \\
&\propto \exp\left[-(k-k_0)^2/2\delta k^2\right]\left\{\exp(-x^2/2\delta x^2)\right. \\
&\quad + \exp[-(x+\Delta)^2/2\delta x] + 2\exp\left[-\left(x+\frac{\Delta}{2}\right)^2/2\delta x\right] \\
&\quad \left.\times\cos(\Delta\cdot k)\right\}.
\end{aligned}
\tag{4.112}
$$

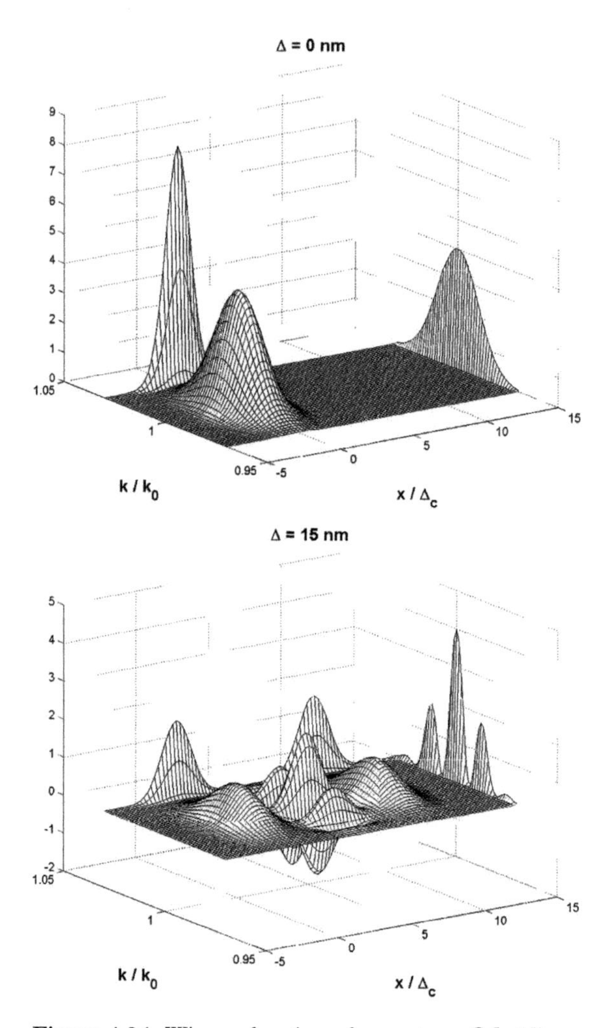

Figure 4.34 *Wigner functions for neutron Schrödinger cat-like states at zero-order (left) and at high-order interference (right) and their projections onto the x- and k-coordinate, respectively (Rauch and Suda 1995)*

Several algebraic steps are necessary to derive this result. Thus, the Wigner function of the coherent Schrödinger cat-like states becomes the sum of the Wigner function of the two spatially shifted wave packets and a cross term oscillating more rapidly in the case of increasing phase shifts. Typical examples are shown in Fig. 4.34 for zero phase shift and for path shifts comparable with the coherence length of the beam. More details about the features of such Wigner functions and other state representations for the neutron case can be found elsewhere (Rauch 1995b, Suda 1995). Schrödinger cat-like neutron states exist not only in split beam interference experiments but in spin-echo systems as well. Both situations can be described properly by the Wigner function formalism (Rauch and Suda 1998). It is important to note that integration over the momentum variable k gives the spatial density distribution (Eq. 4.107) and integration over the spatial variable gives the momentum density distribution (Eq. 4.106), that is:

$$\begin{aligned} \int W_s(x,k)\mathrm{d}k &= |\psi(x)|^2 \\ \int W_s(x,k)\mathrm{d}x &= |\psi(k)|^2 \end{aligned} \quad . \tag{4.113}$$

This opens the possibility of quantum state tomography because both quantities $|\psi(x)|^2$ and $|\psi(k)|^2$ can be measured. This permits an interferometric measurement of the Wigner function which is equivalent to the knowledge of the state wave function (Vogel and Risken 1989, Freyberger et al. 1997). This can be achieved because a quantum state description represents a set of potentialities that are revealed by certain appropriate experimental conditions. A single measurement performed on a quantum system reveals a certain aspect of its state, but it will not uncover this state completely. However, when we know how to determine the whole set of potentialities, the quantum state can be reconstructed. This is the way to measure the state of a quantum system by interferometric methods (Iaconis and Walmosley 1996). Thus, the basic assumptions of quantum theory enters into tomographic and endoscopic methods of quantum state reconstructions that measurements on an infinite ensemble or an infinite number of measurements on a single particle must be done to uncover the quantum state of such systems. Wigner functions are quasi-probability distributions symmetrized in x- and k-space. Other visualizations are the Husimi (Q) and the Glauber–Sudarskan (P) representation (Hillery et al. 1984, Lee 1995). When one explicitly takes into account the action of the measuring device, one ends up with operational probability density distributions which are closely related to the von Neumann, Shannon, and Weyrl entropies (Wodkiewicz 1984, 1988; Buzek et al. 1995a, 1995b; Schleich 2001).

The same procedure as applied to split beam interferometry can be used for neutron spin-echo systems (Section 2.4). In that case the spin-up and spin-down components of a state performing Larmor rotation in an external magnetic field form a superposition state whose Wigner function can be calculated using Eqs. (2.30), (2.31), and (4.110). That is

$$\psi = \int a_+(k)e^{i(k+\Delta k)x}\mathrm{d}k\,|z> + \int a_-(k)e^{i(k-\Delta k)x}\,\mathrm{d}k\,|-z>, \tag{4.114}$$

where Δk denotes the related Zeeman momentum shift ($\Delta k = \mu mB/\hbar^2 k$, Eq. 2.30). Figure 4.35 shows the behavior of the Wigner function without and with field fluctuations for the case of unrealistically high magnetic fields or very slow neutrons; $\Delta k/k_0 = 0.25$ (Rauch 1995b, Rauch and Suda 1998, Suda 2005). m_i denotes the number of Larmor rotations of the mean momentum band around k_0 ($m_i = \Delta k.x/\pi$). In existing spin-echo setups $\Delta k/k_0$ is much smaller but the number of Larmor rotations is much larger ($m \cong 10^5$). Therefore, the wiggle structure is much finer and the separation of the Schrödinger cat-like states much larger. It can reach values of about 0.15 μm which must be compared to the coherence length of the beam, which is about 4 nm (Rauch et al. 1999).

As already mentioned, such Schrödinger cat-like states are highly sensitive to any kind of fluctuation and to dissipation (Zurek 1981, 1991, 2003; Walls and Milburn 1985; Glauber 1986). Therefore, dedicated calculations have been made for the neutron case where the Wigner functions (Eq. 4.112) are averaged over Gaussian distributions for density and/or thickness fluctuations ($\delta N, \delta D$) of the phase shifter which causes fluctuations of the spatial delay ($\delta\Delta$). These calculations have shown clearly how sensitive the wiggle structure becomes at high order to such

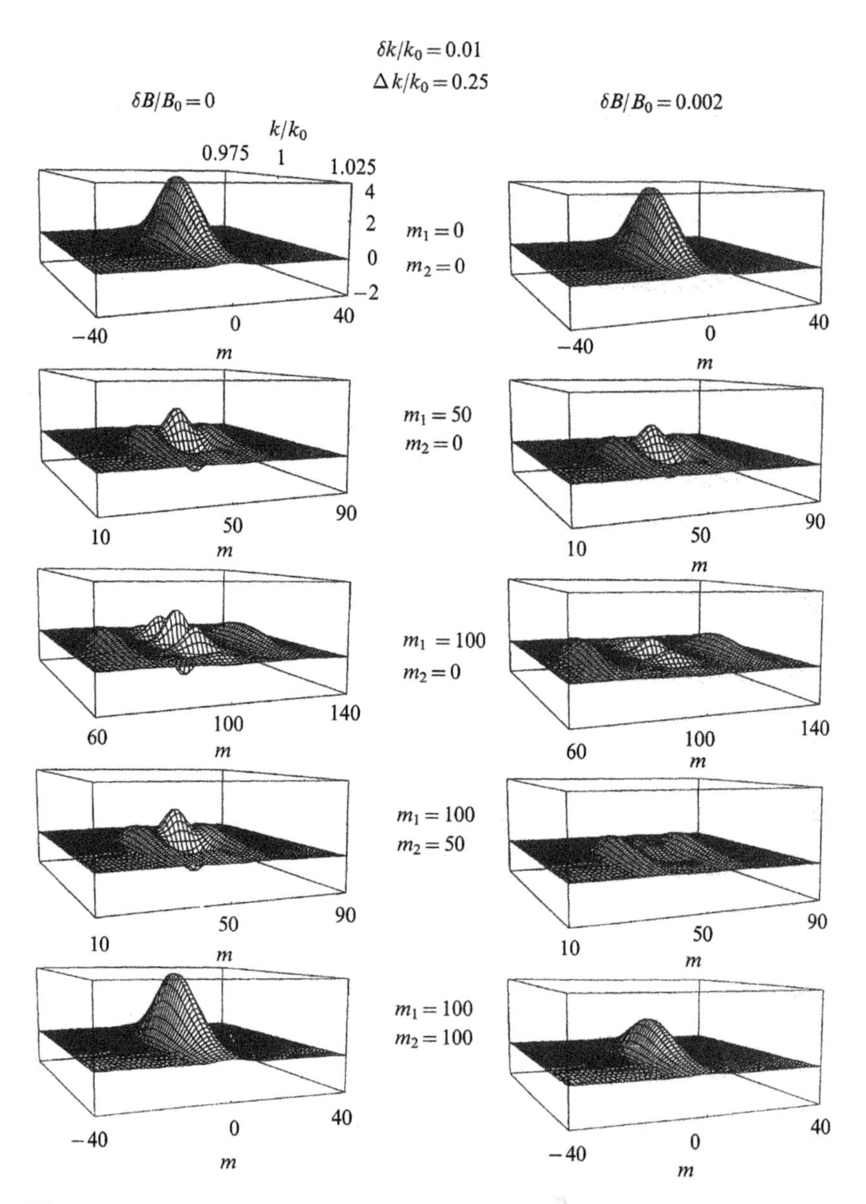

Figure 4.35 *Wigner function without (left) and with (right) field fluctuations within a spin-echo instrument (Rauch and Suda 1998)*

fluctuations (Rauch and Suda 1995). This is shown in Fig. 4.35 (right-hand side). It demonstrates again that the coherent separation of a Schrödinger cat-like state becomes progressively more difficult with increasing spatial separation which will result in an upper limit for the separation when zero-point fluctuations are considered. The vanishing of the oscillating structure of the Wigner function indicates the transition from a superposed coherent state to a mixed state. This is equivalent to the disappearance of the off-diagonal terms of the density matrix (Eq. 4.18). This disappearance can also be understood in terms of the Weyrl (1978) entropy of superposition states, where the decay rate of quantum coherence is proportional to the "distance" in phase space between the coherent components of a superposition state. In this respect, the quantum measurement process is described within the framework of quantum mechanics as a dephasing process (Machida and Namiki 1980; Zurek 1981, 1991, 1998a; Stern et al. 1990; Guilini et al. 1996; Kono et al. 1996; Scully and Zubairy 1997). A linear spatial dependence of the states in phase space means a quadratic dependence of the dephasing parameter in real space. Decoherence phenomena of Schrödinger cat-like states in Stern–Gerlach-type experiments show also a quadratic dependence from the spatial separation of the cat states (Venugopalan 1997). Decoherence phenomena even appear at absolute zero temperature (Sinha 1997). A more detailed discussion of these decoherence phenomena is given in Section 4.6.

4.5.4 Post-selection in the Time Domain

Instead of measuring the interference pattern by scanning the phase shifter, one can also measure the time-dependent intensity correlation function defined in Eq. (4.97)

$$G^{(2)}(\mathbf{\Delta}, t) = \langle I(0,0)\, I(\mathbf{\Delta}, t) \rangle, \tag{4.115}$$

giving the probability of registering a neutron at time t, if there was another one registered at $t = 0$. The probability of detecting a neutron at a time τ after another neutron has arrived is given for a stationary beam from a thermal (Poissonian) source as (Glauber 1968; Section 4.4.4; Eq. 4.98)

$$W(t) = I(\Delta)\, \exp[-t \cdot I(\Delta)]. \tag{4.116}$$

This probability function exhibits an intensity-dependent "decay time" $\tau(\Delta) = [I(\Delta)]^{-1}$ as shown in Fig. 4.36 for the case when the overall interference pattern is tuned to its maximum or minimum values (in this special case, the overall contrast was 41.8%).

One notes that the contrast for neutron pairs arriving within short time intervals τ is higher than the overall contrast. For larger time separations, the contrast vanishes and appears with an opposite sign, reaching values of 100% for widely separated pairs in time. This behavior has been verified experimentally with the arrangement shown in Fig. 4.27 (lower panel) at a low-flux TRIGA reactor (Zawisky et al. 1994). These results show that remarkably higher phase sensitivities can be achieved by using this new measuring technique. We note here that the change of sign of the contrast where it becomes zero in Fig. 4.36 is accompanied by a reversal of 180° in phase. Thus phase jumps appear to be a very general phenomenon (see Section 6.3 and Bhandari 1997).

These results demonstrate that considerably more information can be deduced, even from a Poissonian beam, if the individual arrival times of the neutrons are also registered to define the pair correlation function inherent to the quantum system.

The measured quantity is the registered number of pair events within various time intervals $\tau_1 \le t \le \tau_2$:

$$I(\chi, \tau_1, \tau_2) = I(\chi)\left[e^{-\tau_1 \bar{I}(\chi)} - e^{-\tau_2 \bar{I}(\chi)}\right]. \tag{4.117}$$

This follows from Eq. (4.114) for a given phase shift χ The overall interference pattern and the interference pattern of the correlated pairs are shown in Fig. 4.36, together with the contrast for

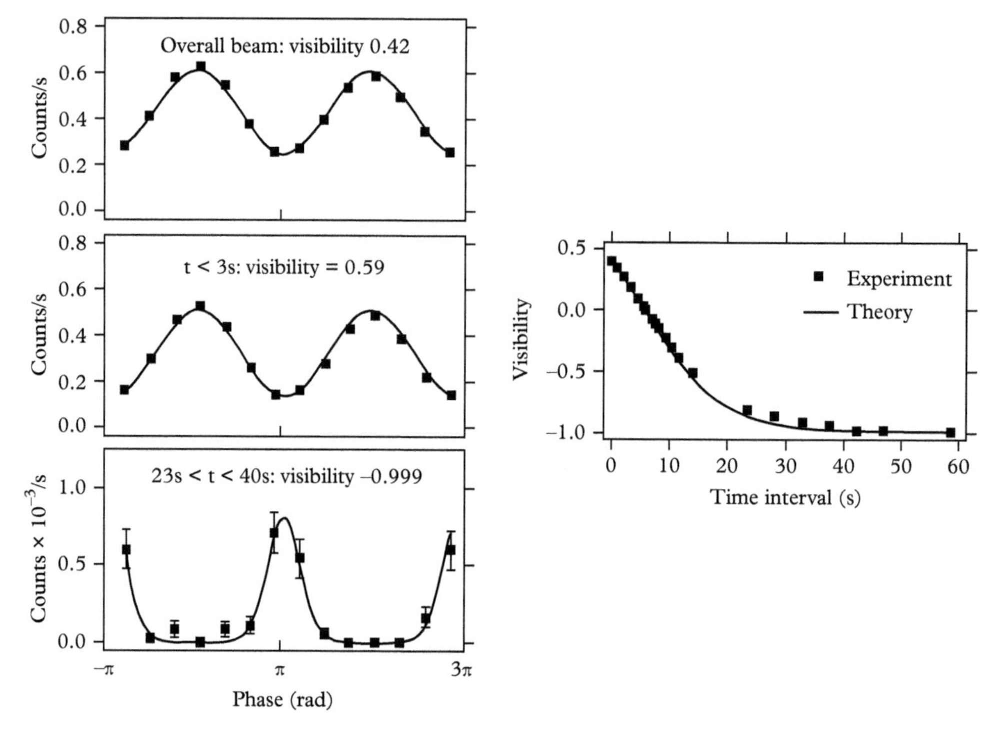

Figure 4.36 *Interference pattern of the overall beam (above) and of neutron pairs arriving within certain time intervals (middle and below). Comparison of the calculated and measured contrast as a function of the time delay of arrival intervals of neutron pairs (right; Zawisky et al. 1994)*

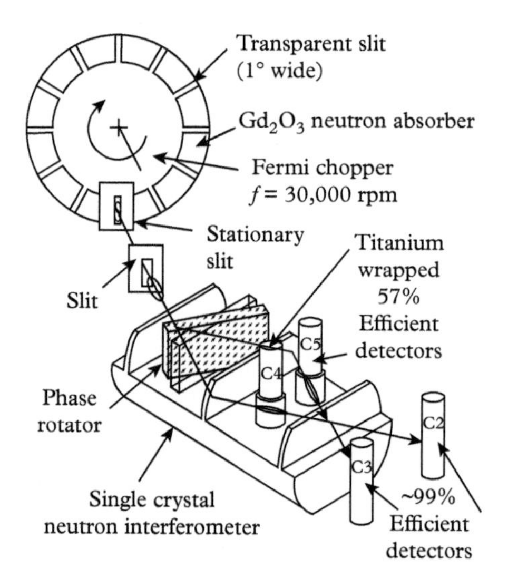

Figure 4.37 *Experimental arrangement for time correlation measurements using a chopped incident beam (Jacobson et al. 1996)*

selected intervals $\tau_1 = 0 \leq t \leq \tau_2$ and $\tau_1 \leq t \leq T > \bar{\tau} = 1/\bar{I}$. A statistical analysis of the data shows that in the range $0.5\ \tau_2 \leq t \leq 8\ \tau_2$ the phase sensitivity is increased compared to a non-time-resolved experiment. This is predicted for cases where the contrast of the overall beam is less than 100%.

These experiments have been repeated with pulsed beams as well (Jacobson et al. 1996). In this case pair correlations have been measured for the beams behind the interferometer and with semi-transparent detectors placed inside the interferometer as shown in Fig. 4.37. In all cases the pair correlation function resulting from Poissonian statistics has been used to verify the data.

4.5.5 Time-of-Flight Post-selection

All experiments described in the previous sections of this chapter, except the last one, were performed with stationary neutron beams where the wave functions inside the interferometer follow from the solution of the time-independent Schrödinger equation. Regardless of whether plane waves or wave packets are used for the description, the wave functions of the two separated beam paths remain connected at the beam splitter and in the region of beam superposition. Therefore, one could argue that some information can be exchanged via these mesh points. In the case of pulsed beams with burst lengths shorter than the dimension of the interferometer completely unconnected wave packets exist inside the interferometer (Fig. 4.38). Such a system also gives the basis for delayed choice experiments where the decision whether to observe the interference or the beam path can be made after the wave packet has passed the beam splitter (Wheeler 1978). More discussion about delayed choice experiments is given in Section 10.5.

Here we will describe experiments which demonstrate how the interference pattern changes in time-resolved measurements where the momentum distribution within certain time slices can be made much narrower than the original one and where frame overlap effects of faster and slower neutrons from neighboring pulses can be observed. The main components of the experimental setup are shown in Fig. 4.38.

The motion of a wave packet in free space is described by the time-dependent Schrödinger equations (1.2) and (4.15). In the case of minimum uncertainty packets with Gaussian widths δx and δk in real and momentum space existing at $t = 0$ ($\delta x\,(0)\,\delta k = ½$) one expects the quantum mechanical spreading of the packet as a function of time to follow the formula (e.g., Messiah 1965)

$$[\delta x(t)]^2 = [\delta x(0)]^2 + \left[\frac{(\hbar/2m)t}{\delta x(0)}\right]^2. \tag{4.118}$$

This minimum uncertainty wave packet is difficult to achieve because $\delta x(0)$ must approach the coherence length $\Delta_x^c \sim (2\delta k)^{-1}$ as discussed previously (Section 4.5.4). This also means that pulse lengths on the order of the coherence time $\tau_c = \Delta_c/v$ must be produced, which requires chopper opening times on the order of nanoseconds. In this case, diffraction in time would play an important role, which is quite analogous to the well-known single-slit diffraction phenomena in ordinary space (Moshinsky 1952, Gaehler and Golub 1984, Nosov and Frank 1991, Brukner and Zeilinger 1997). Thus, Fraunhofer- and Fresnel-like phenomena, which change the energy of the beam accordingly, are expected to occur. For a triangular slit opening for a time Δt one expects in the Fraunhofer limit for an incident plane wave with an energy E_0 and energy spectrum

$$\rho(E) = |A(E)|^2 \propto \left[\frac{\sin((E - E_0)\Delta t/\hbar)}{(E - E_0)/\hbar}\right]^2. \tag{4.119}$$

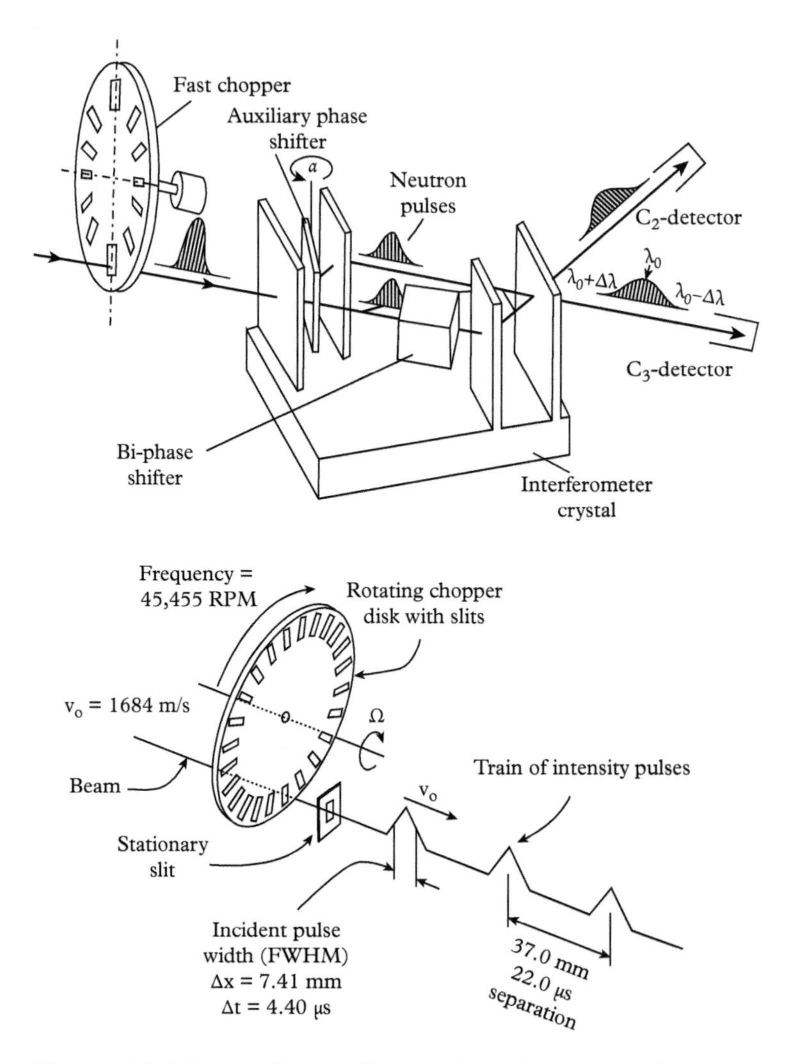

Figure 4.38 *Schematic diagram of the experimental arrangement for interference experiments with a pulsed incident beams (Rauch et al. 1992)*

One sees here that, in this limit, the energy spectrum represents the square of the Fourier transform of the time signal of how the aperture was opened. In the more general case of matter wave diffraction in space and in time both phenomena are correlated to each other. As mentioned in Chapter 1, there is a marked difference to the situation with electromagnetic radiation. The main reason for this can be seen in the different dispersion relations (Eq. 1.6) and because all wave components of a light pulse propagate with the same velocity; that is no spatial spreading (Brukner and Zeilinger 1997).

Various "Gedanken" experiments with very fast choppers causing diffraction-in-time effects (Eq. 4.119) due to the cutting-off part of the coherent wave packet ($\Delta t^c \leq \Delta_c/v$) were discussed at the end of Section 4.3.2. The diffraction-in-time effect was first observed with fast vibrating surfaces (Felber et al. 1996). This quantum limit is rather difficult to achieve by means of mechanical choppers due to the smallness of the coherence time values Δt^c in neutron interferometry. However, Hils et al. (1998) achieved a pulse width of 33 ns and observed the energy transfer given by Eq. (4.119). However, a combination of these fast pulses with interferometry seems to be rather difficult to achieve. It might be easier to approach this regime by high-frequency spin-flipping devices, which then causes a kind of quantum phase chopping (see Sections 5.3 and 5.4). The diffraction-in-time effect has also been investigated by the diffraction of atoms from pulsed evanescent laser waves (Steane et al. 1995, Szriftgiser et al.1996). These results can be described in the Fraunhofer limit, which gives Eq. (4.119).

For long opening times ($\Delta t >> \Delta t^c$) diffraction effects occur only at the edges of the pulse and can be neglected in most cases. Under these conditions, the phenomena can be described by classical distribution functions. Their spreading occurs in a manner similar to the quantum case. For Gaussian pulses, with a velocity spread δv, one gets

$$[\delta x(t)]^2 = [\delta x(0)]^2 + [\delta v t]^2. \tag{4.120}$$

One notices the similarity of this equation to Eq. (4.118) if one uses the de Broglie relation and $\Delta_x{}^c = (2\delta k)^{-1} = \lambda^2/4\pi\delta\lambda$. In terms of the temporal pulse length one obtains

$$[\Delta t]^2 = \left[\Delta t_p\right]^2 + \left[\frac{\delta\lambda}{\lambda} t_0\right]^2, \tag{4.121}$$

where Δt_p is the opening time of the chopper and $t_0 = L/v_0$ is the average time-of-flight of neutrons between the chopper and the detector at distance L. This describes the standard broadening of a neutron burst in any time-of-flight spectrometer (e.g., Willis and Carlile 2009). In fact, Eq. (4.120), which describes the broadening of the intensity pulse is identical to the expression which describes the broadening of a quantum mechanical wave packet (Eq. 4.118), except for the fact that in Eq. (4.120), $\delta x(0) = v\Delta t_p$ does not fulfill the minimum uncertainty relation.

The mean wavelength of neutrons being measured at time t at a detector with an effective thickness Δd becomes in the classical limit ($\delta x(0) >> \Delta_x{}^c$):

$$< \lambda(t) >= \lambda_0 \left[1 + t_0(t - t_0)(\delta\lambda/\lambda_0)^2/(\Delta t')^2\right], \tag{4.122}$$

where $(\Delta t')^2 = (\Delta t)^2 + (\Delta d/v_0)^2$. Because the phase shift of neutrons traversing a material slab is proportional to the de Broglie wavelength (Eqs. 3.20 and 4.51), one gets a similar equation for the phase shift $< \chi(t) > \propto < \lambda(t) >$ of those neutrons detected at time t, namely

$$< \chi(t) > = \chi_0 \left[1 + t_0(t - t_0)(\delta\lambda/\lambda_0)^2/(\Delta t')^2\right]. \tag{4.123}$$

At the detector position at a distance L from the chopper, at a time t, there still exists in a certain time channel a Gaussian wavelength distribution with a restricted spectral width $\delta\lambda'$ given by

$$\left(\frac{\delta\lambda'}{\lambda_0}\right)^2 = \left[\left(\frac{\lambda_0}{\delta\lambda}\right)^2 + \left(\frac{t_0}{\Delta t'}\right)^2\right]^{-1}. \tag{4.124}$$

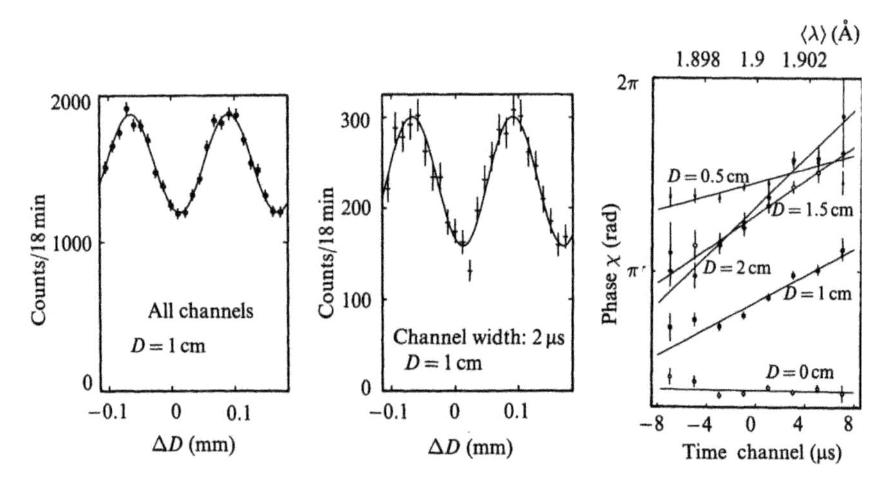

Figure 4.39 *Interference pattern of the overall beam (left) and of the intensity within a narrow time-of-flight channel (middle). The increasing phase shift for later time channels where the slower neutrons arrive is visible (right; Rauch et al. 1992)*

This restricted spectral width $\delta\lambda'$ results in a slower diminution of the interference contrast for measurements with narrow time channels than for the whole beam. This is because the related coherence function becomes wider (Eq. 4.35) and the expected 0-beam modulation is

$$I \propto \left[1 + e^{-\chi_0{}^2(\delta\lambda'/\lambda_0)^2/2} \cos\chi_0\right]. \tag{4.125}$$

If one prefers to write these results in terms of wave vector, clearly $\delta\lambda'/\lambda_0 = \delta k'/k_0$.

Measurements using a setup at the high-flux reactor at Grenoble have shown that the contrast within narrow time channels can exceed the contrast of the overall beam and have shown that the mean phase shift varies as predicted by Eq. (4.123) (Heinrich et al. 1989; Fig. 4.39). These measurements have been continued and completed at the MURR reactor in Columbia, Missouri (Rauch et al. 1992).

At larger distances downstream the edges of each pulse begin to overlap with the preceding and following ones. Gaussian-shaped neutron pulses produced within a time interval T, corresponding to wavelength $\lambda_1 = \lambda_0 - \delta\lambda$ and $\lambda_2 = \lambda_0 + \delta\lambda$, overlap at a distance $L^c = hT/2m\delta\lambda$ downstream from the chopper. The spectral width at the relevant time interval is given by the spectral widths around λ_1 and λ_2. The measured intensity in the related time slice around $t = t_0{}^c(1 - \delta\lambda/\lambda_0)$ is given as

$$\begin{aligned} I(\lambda_1, \lambda_2) &= I(\lambda_1) + I(\lambda_2) \\ &\propto 2 + <|\Gamma(\Delta_1)|> \cos\chi_1 + <|\Gamma(\Delta_2)|> \cos\chi_2. \end{aligned} \tag{4.126}$$

In the case when the nuclear phase shifters are arranged perpendicular to the reflecting planes (dispersive geometry, Fig. 4.3a) and for rather narrow Gaussian wavelength distributions around λ_1 and λ_2 one can reformulate this equation in the form

$$I(\lambda_1, \lambda_2) \propto \left\{1 + \exp\left[-\chi_0{}^2(\delta\lambda/\lambda_0)^2/2\right] \cdot \cos(Nb_c D\Delta\lambda) \cdot \cos(Nb_c D\bar{\lambda})\right\}. \tag{4.127}$$

We have used Eq. (4.35) and

$$<|\Gamma(\Delta_1)|>\approx<|\Gamma(\Delta_2)|> \exp\left[-(\Delta\delta k)^2/2\right]. \tag{4.128}$$

The mean wavelength $\bar{\lambda}$, the mean phase shift χ_0, and the difference wavelength $\Delta\lambda$ in this equation are defined by

$$\begin{aligned} &\bar{\lambda} \equiv (\lambda_1 + \lambda_2)/2\,, \Delta\lambda \equiv (\lambda_1 - \lambda_2)/2\,, \\ &\chi_0 \equiv -Nb_c D\lambda_0 \end{aligned} \tag{4.129}$$

Thus, the total intensity exhibits a series of "beats," modified by a decaying exponential. The contrast of the intensity pattern as a function of the thickness D of the phase shifter is given by

$$C(D) = \frac{I_{\max} - I_{\min}}{I_{\max} + I_{\min}} = \left|\cos(Nb_c D\Delta\lambda)\right| \exp\left[-\chi_0{}^2(\delta\lambda/\lambda_0)^2/2\right]. \tag{4.130}$$

The exponential term is due to the wavelength spread and the cosine term represents a contrast oscillation caused by the overlap of two pulses; i.e., the contrast is expected to vary according to a damped cosine function going through a series of minima and maxima.

Measurements in the overlap region at larger distances and with a less monochromatic beam have been made at the MURR reactor (Clothier 1991, Rauch et al. 1992). The variation of the interference contrast as a function of the phase shift (sample thickness) for the overall pulse and for time slices with and without pulse overlap is shown in Fig. 4.40. The contrast modulation in the case of pulse overlap time slices is clearly visible. We note again that where the contrast goes to 0 in this data, there is a 180° phase jump.

The neutron pulses produced by a fast mechanical chopper were significantly shorter than the dimensions of the perfect crystal interferometer, thus assuring that there is no permanent overlap of wave functions or of plane wave components of wave packets. This experiment again demonstrates very clearly the single-particle interference phenomena in neutron interference experiments because the mean occupation number in a single neutron pulse is much smaller than unity (it is on the order of 10^{-4} in this case). The neutron intensity pulses used in this experiment are considerably longer than the coherence length of the beams and, therefore, diffraction-in-time effects are negligible. Several aspects of neutron wave mechanics have been explored and have been shown to be in agreement with the theoretical predictions. In particular, the spreading of the neutron pulses as well as the spectral narrowing in distinct time slices downstream from the interferometer has been observed.

The excellent agreement between the predicted contrast curves and the measured data, particularly in Fig. 4.40, shows that formulas (4.125) and (4.130) correctly describe the propagation of neutron pulses, the separation of the wavelength components, and the various corrections necessary to extract the experimental results. As in other spectral filtering experiments (Clothier et al. 1991, Kaiser et al. 1991, Werner et al. 1991) these time-resolved experiments show a way of extracting interference fringes and their contrast from a beam which, on the whole, exhibits no contrast. Thus, interference and coherence phenomena can be completely hidden due to general averaging effects but they can be recovered even behind the interferometer if a proper post-selection measuring procedure is utilized.

The observed beam modulation occurring in the overlap region of successive pulses is another example how intrinsic coherence phenomena become manifest, in this case when time resolution

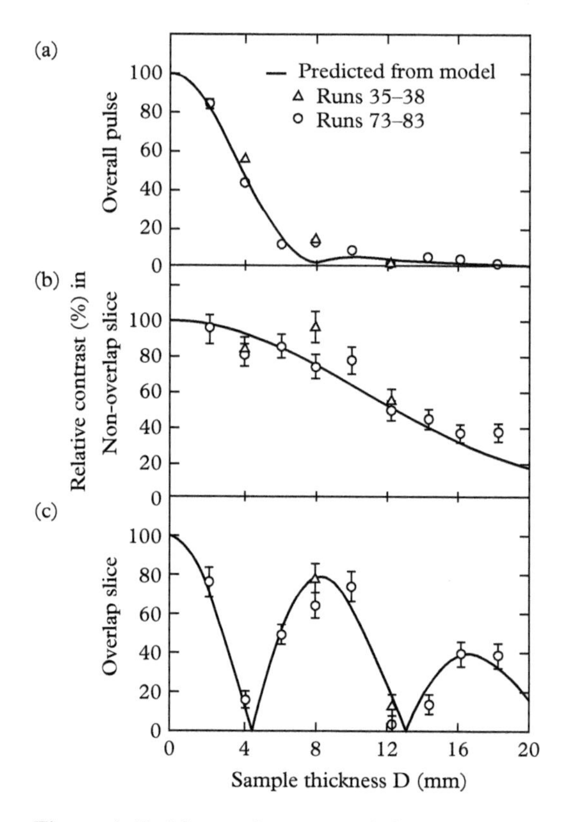

Figure 4.40 *Measured contrast (coherence) of the overall (above) and within time-of-flight intervals without (middle) and with pulse overlap from different neutron bursts (below; Rauch et al. 1992)*

is applied. Even when the average intensity remains constant the contrast shows a marked beating effect (aside from attenuation effects). This behavior is characteristic of the incoherent beam superposition of two different wavelength bands. Here, it becomes obvious that the contrast of an interference pattern cannot always be identified with the coherence function as it is defined in Eqs. (4.29) and (4.49). In this case the time-dependence of the classical probability $p(r, t)$ must be taken into account (see Eq. 4.18).

4.5.6 Interferometric Post-selection—Multiplate Interferometers

We now discuss the case where an additional interferometer loop is used to revive coherence properties that appear hidden behind the first interferometer loop. Such systems engender some interest because there exist coupled interferometer loops which are partly fed with coherent beams, instead of incoherent beams as in the case of the standard interferometers. The wave functions and the intensities can be calculated by extending the methods applied to the triple-plate interferometer to a multiplate system (Heinrich et al. 1988, Suda et al. 2004). There are different possibilities for "focused" four-plate interferometers as shown in Figs. 4.42 and 2.3. The intensity modulation behind the equidistant arrangement (Fig. 4.41) can be shown to be

$$\begin{aligned} I_{B0}(\chi_A, \chi_B) &= K_2 \{ 3 + 2 \ [\cos(\chi_A + \chi_B) + \cos\chi_A + \cos\chi_B]\} \\ I_{BG}(\chi_A, \chi_B) &= K_1 + 2K_3 - 2K_2 \ [\cos(\chi_A + \chi_B) + \cos\chi_B] + 2K_3 \cos\chi_A, \end{aligned} \tag{4.131}$$

where χ_A and χ_B denote the net phase shifts of the interferometer loops A and B respectively. The $\chi_A = \alpha_1 + \beta_2 - \beta_3 - \alpha_2$, etc. and the K_n quantities are given as $K_1 = 417\pi/2048$, $K_2 = 79\pi/2048$, and $K_3 = 65\pi/2048$. Standard plane wave dynamical diffraction theory was used to calculate these parameters (Chapter 11). Similar formulas apply for I_{CO} and I_{CG}. Only the beam I_{B0} shows a maximum theoretical contrast of 100%, whereas the contrast for all other beams remain below this value. Typical interference patterns are shown in Fig. 4.41 where two phase shifters were rotated in opposite senses to cause $\chi_A = \chi_B$ (Heinrich et al. 1988). All the results are in good agreement with theoretical predictions. In the case of absorbing phase shifters one gets, e.g.,

$$I_{B0}(\chi_A, \chi_B) = K_2 F\left[A^2 + B^2 + 1 + 2AB\cos(\chi_A + \chi_B) + 2A\cos\chi_A + 2B\cos\chi_B\right], \tag{4.132}$$

with $A = e^{-\chi_A''}$, $B = e^{-\chi_B''}$, and $F = \exp\left[-2(\alpha_1'' + \beta_2'' + \gamma_3'')\right]$. The double-primed quantities denote the imaginary parts of the phase shifts which are given by the related reaction (absorption) cross-section; for example $\alpha'' = \sigma_r N d/2$ (Eq. 3.19). Beam attenuation is equivalent to a partial beam path detection and, therefore, its influence is different according to where it occurs (see Section 4.3.1). The absorption effect in coupled interference loops can also be interpreted as a topological phase effect, which can be described by a Poincaré sphere representation (see Section 6.8).

Higher order interference maxima become damped due to the wavelength spread of the beam (Eqs. 4.35 and 4.48)

$$\begin{aligned} I_{B0}(\chi_A, \chi_B) = K_2 \Big\{ 3 + 2 \Big[& e^{-(\Delta\lambda/\lambda_0)^2 (\chi_A + \chi_B)^2/2} \cos(\chi_A + \chi_B) \\ & + e^{-(\Delta\lambda/\lambda_0)^2 \chi_A{}^2/2} \cos\chi_A + e^{-(\Delta\lambda/\lambda_0)^2 \chi_B{}^2/2} \cos\chi_B \Big] \Big\}, \end{aligned} \tag{4.133}$$

which shows that a distinct phase echo situation exists for $\chi_A = -\chi_B$, This permits a determination of χ_A even in the case where there is no contrast behind the first interferometer loop A $[(\Delta\lambda/\lambda_0)\chi_A >> 1$ or $\Delta >> \Delta^c]$. These formulas also show that a phase shifter, absorbing or not, placed in a beam with zero intensity has no influence on the interference pattern (e.g., γ if $\chi_A = \pi$), which demonstrates that waves with zero intensity cannot engender physical effects. In laser physics the same conclusion has been drawn (Mückenheim et al. 1988), which seems to show the non-reality of the idea of empty waves with a velocity faster than light as they have been postulated by de Broglie and Einstein (see, e.g., Loschak 1984). A related proposal for an interferometer experiment was made by Croca et al. (1988) to test this idea for neutrons. A second beam from an independent source should be fed into an interferometer loop and the residual interference pattern should be measured in a coincidence mode. It can be shown that due to the incoherent admixture of this beam and the smallness of the coherence volume no additional information about the role of empty waves can be drawn from this proposed experiment. A related experiment with photons verified this conclusion (Wang et al. 1991).

Four-plate interferometer systems can be designed which create two completely modulated beams (I_{A0} and I_{C0}; Bonse and Graeff 1977, Suda et al. 2004). In certain cases the total interfering intensity can be higher than in the standard triple-Laue case interferometer. Multiplate

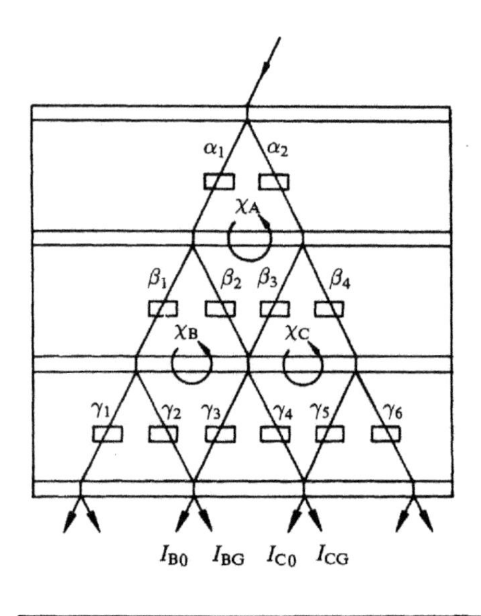

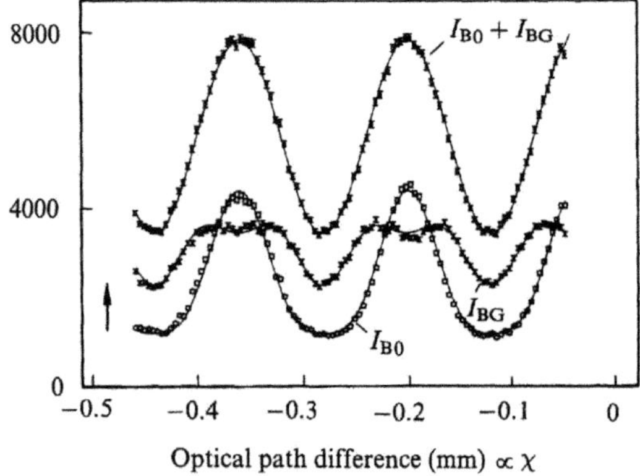

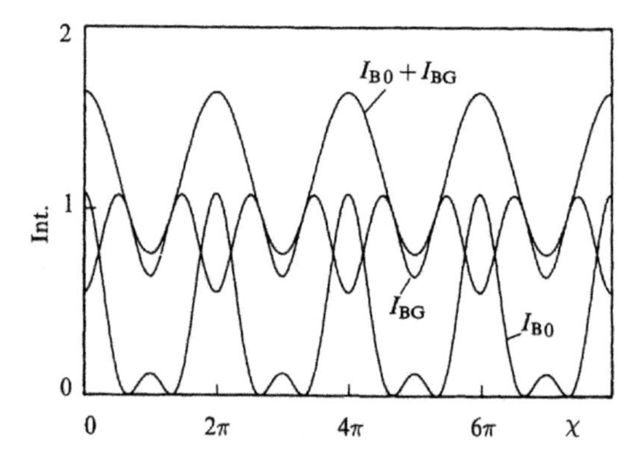

Figure 4.41 *Schematic of a four plate interferometer and a comparison of measured and calculated interference pattern (Heinrich et al. 1988)*

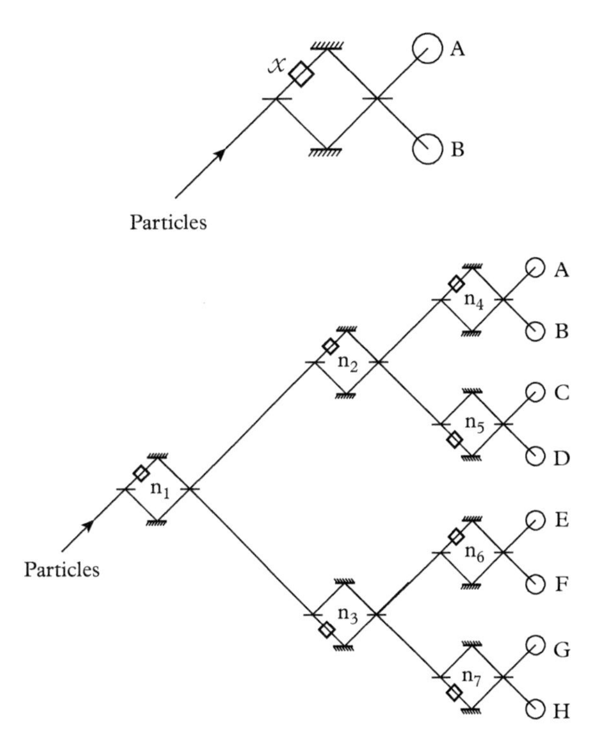

Figure 4.42 *An ideal Mach–Zehnder interferometer where the phase shifts can be varied distinctively and a series of Mach–Zehnder interferometers for parallel computing Reprinted with permission from Summhammer 1997, copyright 2007 by the American Physical Society.*

interferometers have demonstrated again the linear superposition principle of quantum mechanics and have stimulated discussion about coherent beam mixing and unsharp particle or wave property determination (Wooters and Zurek 1979; Mittelstedt et al. 1987; Greenberger and Yasin 1988; Englert 1996, 1999). The formulas for five- and more plate interferometer systems become lengthy and therefore, they are not given here. Such interferometers provide multiple coherent exit ports of coherent radiation. A complete revival of the split beams seems to be impossible when unavoidable loss factors are taken into account.

Mach–Zehnder interferometers can also be used for factoring and Fourier transformation as has been pointed out by Summhammer (1997). Clauser and Dowling (1996) showed that the number of peaks behind a N-slit arrangement identifies the factors of this integer N. In an analog sense a dispersive Mach–Zehnder interferometer can factorize the number N given as the ratio of the coherence length divided by the neutron wavelength (Fig.4.41). The phase shifts in the interferometer are $\chi_j = 2\pi kN/n$ when they are increased in discrete steps $2\pi/n(k = 1, 2, 3\ldots)$. So, at the kth observation the probability of registering the particle at detector A is

$$p(k) = \frac{1}{2}\left[1 + \cos\left(\frac{2\pi kN}{n}\right)\right]. \tag{4.134}$$

Only when N is a multiple of n we get $p(k) = 1$. When we perform n observations with $1 \leq k \leq n$ we do create an intensity at detector A of

$$I_n = \sum_{k=1}^{n} p(k) = \frac{n}{2} + \frac{1}{2} \sum_{k=1}^{n} \cos\left(\frac{2\pi kN}{n}\right). \tag{4.135}$$

This gives $I_n = n$ when n is a factor of N, whereas $I_n \sim n/2$ when n is not a factor of N.

It is also possible to implement a kind of parallel computation with a multiple interferometer setup (Fig. 4.42). Here the numbers $n_1, n_2, n_3, \ldots n_7$ can almost simultaneously be checked for being factors of N. Thus, single-particle but multi-interference arrangements can have useful applications in quantum computation.

4.6 Decoherence and Dephasing

Dephasing and decoherence effects are essential for an understanding of how a classical world arises out from the quantum features of nature. According to the modern view of this topic, decoherence means the irreversible formation of quantum entanglements of a system with its environment. Nevertheless, there is a connection of this topic with measurement theory and the question of the collapses of the wave function.

An experimenter is fighting continuously against dephasing and decoherencing effects since a high contrast of an interference pattern signals a high degree of coherence. This is essential for most experiments. Nevertheless, decoherence is an important topic in the understanding of quantum physics. Dephasing must be distinguished from decoherence although both terms are used in the literature synonymously. In an experiment dephasing and decoherence cause a reduction of the contrast of the interference pattern. In the quantum formalism a disappearance of the off-diagonal terms of the density operator and a smearing of the wiggle structure of the related Wigner function occur (see Fig. 4.35). In the first case a retrieval of the contrast can be achieved by several post-selection methods, whereas in the case of decoherence no reconstruction seems to be possible. The border between both processes determines the border between the classical and the quantum world and is of fundamental interest for understanding quantum physics and especially for understanding the quantum measurement process (e.g., Zeh 1970; Machida and Namiki 1980; Zurek 1981, 2003; Joos and Zeh 1985; Guilini et al. 1996; Scully and Zubairy 1997; Zeh 2001).

Decoherence arises from the entanglement of the quantum object with the environment (Zeh 1970). It does not solve the measurement problem but it tells us why certain objects appear classical when they are observed. The effect derives from standard quantum theory due to the inclusion of the coupling to the environment which causes an entanglement between the quantum system (with states ϕ) and the environment (with its initial state Θ_0). This coupling can be written by means of a von Neumann (1932)-type "measurement" equation

$$\left(\sum_i c_i \,|\phi_i\rangle\right) \Theta_0 = \sum_{n,m} c_{n,m} \,|\phi_n\rangle |\Theta_m\rangle \rightarrow \sum_i c_i \,|\phi_i\rangle |\Theta_i\rangle, \tag{4.136}$$

where $|\Theta_i\rangle$ denote the *pointer states*, telling us the state that the system has been found in. The last step indicates a diagonalization according to the Schmidt eigenbasis $|\Theta_i\rangle$ (Schmidt 1907).

The fact is that there is no totally closed system, except for the whole universe. Decoherence is thus a non-local and an intrinsic feature of all physical processes. Equation (4.136) follows from a unitary evolution where the resulting entanglement becomes practically irreversible when many degrees of freedom are involved. One notices that dephasing and decoherence mean a distortion of the environment rather than a distortion of the quantum system. Coupling and entanglement of the quantum system to an environment cause a random variation of the phase shift and, therefore, a loss of contrast of the interference pattern. Whereas in the first case the coupling is known or can be known in principle, in the second case the coupling remains unknown due to random processes during the interaction. The dynamics of the quantum state in the case of dephasing is unitary, yet stochastic. A profoundly quantum cause for decoherence is entanglement between the quantum system and the environment. The distinction between both cases is related to the status of experimental technique and it is often claimed that a complete decoherence is impossible in principle. One might intuitively expect that a neutron ensemble suffers a greater loss of coherence by interacting with an increasingly disordered environment, which should provoke more randomization of the quantum phase. This is not always true and in some cases the neutron wave function is more robust, even if it has interacted with a disordered medium (Facchi et al. 2001). The observed robustness of the geometric phase within a fluctuating environment may be an example of this stability phenomenon (see Chapter 12; Filipp et al. 2009).

Decoherence must also have been present during the evolution of the universe, where quantum gravity effects may have played an essential role (Joos 1986, Penrose 1986, Kiefer 2000).

4.6.1 Basic Relations

The simplest situation of dephasing effects is static variation of the phase shift χ due to variation of the thickness of the phase shifter, the spread of neutron wavelengths used, or imperfections of the reflecting crystals. In this case one gets for Gaussian distributions of variations (see Eq. 4.42)

$$\left\langle e^{\mathrm{i}\chi}\right\rangle = \int P(\chi)e^{\mathrm{i}\chi}\mathrm{d}\chi \cong e^{\mathrm{i}\langle\chi\rangle-\left\langle\delta\chi^2\right\rangle/2}. \tag{4.137}$$

This behavior has also been discussed in connection with the Wigner presentation of quantum states in Section 4.5.3, Fig. 4.35. This phenomenon can also be described by the disappearance of the off-diagonal terms in the density matrix (Eq. 4.18).

In a more general view, it must be stated that no quantum-mechanical system is totally isolated. Any system interacts with its environment even in cases with a zero-temperature reservoir. In this case the density operator fulfills a zero-temperature master equation of the Born–Markov type (Walls and Milburn 1985, Buzek and Knight 1995b, Sinha 1997)

$$\frac{\partial\hat{\rho}}{\partial t} = \frac{\gamma}{2}(2a^-\hat{\rho}a^+ - a^+a^-\hat{\rho} - \hat{\rho}a^+a^-), \tag{4.138}$$

where γ describes the coupling to the vacuum states. The results show that the off-diagonal terms of the density matrix, which are related to the interference terms of the Wigner function, become rapidly dephased at a rate governed by the energy separation of the coherent states.

Another important model for the description of decoherence phenomena starts with a kind of quantum Boltzmann equation where a damping term due to random walk processes is added to the von Neumann equation (Eq. 1.51; Joos and Zeh 1985)

$$\mathrm{i}\hbar\frac{\partial\rho}{\partial t} = [\mathcal{H},\hat{\rho}] - \mathrm{i}\Lambda[x,[x,\hat{\rho}]]. \tag{4.139}$$

This gives a damping factor due to the non-unitary dynamics of the system

$$\frac{\partial \hat{\rho}(x, x', t)}{\partial t} = -\Lambda (x - x')^2 \hat{\rho}, \tag{4.140}$$

and, therefore,

$$\hat{\rho}(x, x', t) = \hat{\rho}(x, x', 0) \exp[-\Lambda t (x - x')^2]. \tag{4.141}$$

When decoherence is caused by various scattering processes the damping constant Λ becomes

$$\Lambda = \frac{k^2 N v \sigma_{\text{eff}}}{V}, \tag{4.142}$$

where Nv/V is the incoming flux and σ_{eff} the total cross-section. These results can be adapted by the Wigner formalism, which gives a diffusion process in momentum space

$$\frac{\partial W(x, p, t)}{\partial t} = \Lambda \frac{\partial^2 W}{\partial p^2}, \tag{4.143}$$

which gives

$$W(x, p, t) = \frac{1}{\sqrt{4\pi \Lambda t}} \int \mathrm{d}p' \; W(x, p', 0) \; \exp\left[-\frac{(p - p')^2}{4\Lambda t} \right]. \tag{4.144}$$

This also shows the equivalence of the density and Wigner formalism. Other models of decohering effects are described in the book of Guilini et al. (1996) and of Namiki et al. (1997). This yields to a measure of robustness of a quantum state which is defined by its linear entropy

$$S_{\text{lin}} = Tr(\hat{\rho} - \hat{\rho}^2), \tag{4.145}$$

which gives

$$\frac{\mathrm{d}S_{\text{lin}}}{\mathrm{d}t} = 2\Lambda \, Tr(\hat{\rho}^2 x^2 - \hat{\rho} x \hat{\rho} x). \tag{4.146}$$

When the system was initially in a pure state the entropy increases proportional to the spatial separation of the state

$$\frac{\mathrm{d}S_{\text{lin}}}{\mathrm{d}t} = 2\Lambda (< x^2 > - < x >^2), \tag{4.147}$$

which may also be applied to the motion of a free particle where one obtains (Joos 1996)

$$S_{\text{lin}} = 1 - \left[\frac{3(\delta x(0))^2}{4(\Lambda m)^2 (\delta x(0))^2 \tau^4 + 2\Lambda m \tau^3 + 24\Lambda m (\delta x(0))^4 \tau + 3(\delta x(0))^2} \right]^{1/2}, \tag{4.148}$$

where $\tau = t/m$. This gives a valley of robustness at $\delta x(0) = (\tau/2\sqrt{3})^{1/2}$.

When a complete positivity, probability conservation, and entropy increase are postulated, the dynamics of the neutron beam in interferometry experiments can be described by so-called quantum dynamical semigroups (e.g., Spohn 1980). In this case fundamental limits for dissipative terms can be extracted from the contrast features of an interference pattern. In a related analysis Benatti and Floreanini (1999) extracted from an existing interference pattern a magnitude of the dissipative term in Eq. (4.141) as $(\Lambda = 0.71 \pm 0.21) \times 10^{-12}$ eV (see Section 10.13). More complicated models are on the market under the topic "quantum Brownian motion" (e.g., Caldeira and Leggett 1983).

As mentioned earlier, decoherence is closely connected to the quantum measurement problem. The interaction between the quantum system and the environment produces, besides rapidly decohering states, very robust states ($|\phi_t\rangle$), also called preferred, pointer, or einselected states. In the case of a damped harmonic oscillator (basic frequencyΩ, damping γ) they obey a nonlinear evolution equation

$$\frac{\delta\,|\phi_i\rangle}{\delta t} = -i\Omega a^{+} a\,|\phi_i\rangle - \frac{\gamma}{2}\left(a^{+} a - \langle\phi_i|\,a^{+} a\,|\phi_i\rangle\right)|\phi_i\rangle\,, \tag{4.149}$$

whereas the total product states of the system and environment ($|\phi_i\rangle\,|\Theta_i\rangle$) are still solutions of the linear Schrödinger equation. Zurek (1993, 1998b) showed that such *einselected* pointer states are rather robust in their interaction with the environment, but they are very fragile against entanglement between them, which yields to nearly classical states. Einselection means a kind of environment-induced superselection of states, which singles out robust states in the case of open quantum systems. In the interaction between the quantum system and the environment most of the entangled states decohere very rapidly, but these einselected pointer states become very robust. Such pointer states of an apparatus communicate intensively with the environment, indicating a repeated measurement process that stabilizes the quantum system in the spirit of the Zeno phenomenon (Misra and Sudarshan 1977, Joos 2006).

4.6.2 Dephasing and Decoherence Experiments

Static and time-dependent variations within an interferometer can cause dephasing and decoherence. In many cases these effects can be compensated by a proper adjustment or/and proper post-selection procedures (Section 4.5).

Static dephasing in neutron interferometry experiments can be achieved by means of inhomogeneous phase shifters with a spatially varying index of refraction or by magnetic domain structures. In all these cases the interaction Hamiltonian along the beam paths are known or can be measured separately, at least in principle. Spatial post-selection can provide a sectional interference pattern. Inhomogeneous phase shifters produce additional small-angle scattering effects which open an additional access to the features of these inhomogeneities. Some applications of these effects for condensed matter research are discussed in Section 9.1.

Time-dependent fluctuations of the interaction potential give rise to additional energy exchanges between the neutron and the potential. In the case of neutrons, time-dependent potentials can be realized most easily with magnetic fields. Single-mode, multimode, and noise fields can be applied. When the Larmor resonance frequency is chosen, a complete spin-reversal and a single-photon exchange takes place as demonstrated by Alefeld et al. (1981a) and Weinfurter et al. (1988) and in the off-resonance case multiphoton exchange occurs as measured by Summhammer et al. (1995). In these cases a quantum state is transferred to another one and coherence is preserved, as discussed in Sections 5.3–5.5.

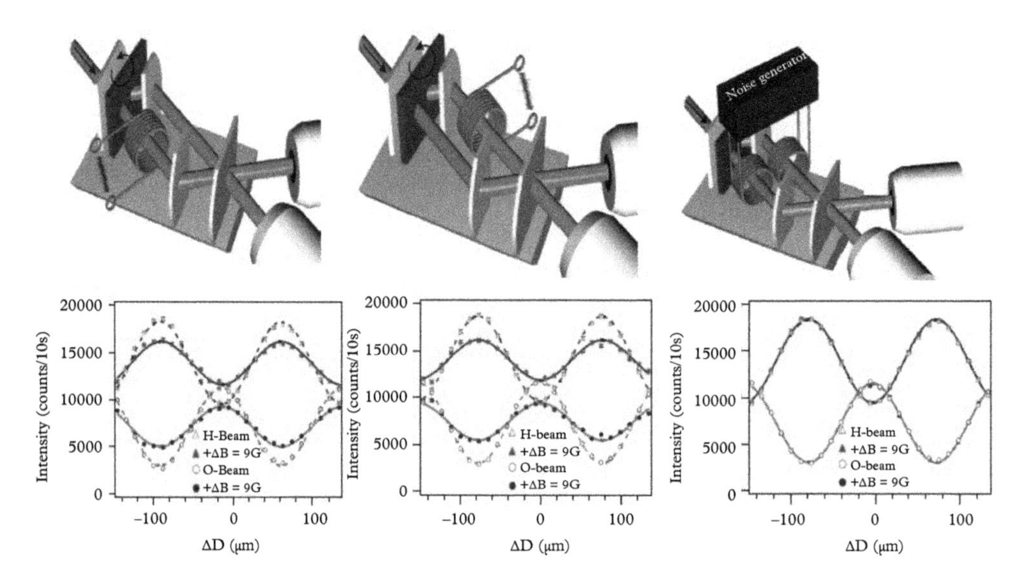

Figure 4.43 *Interference pattern with (full symbols) and without (open symbols) magnetic noise fields applied to the beams separately and synchronously (Baron and Rauch 2011)*

To model decoherence effects different kinds of noise fields have been applied to the beam paths and the interference contrast has been measured (Baron 2005, Sulyok et al. 2010, Baron and Rauch 2011). The loss of contrast depends on the strength and the band width of the noise signal. When the same noise fields are applied to both beam paths the contrast is recovered, as shown in Fig. 4.43. The average of the Gaussian noise field strength was 9 G and the frequency range between 0 and 20 kHz. These results demonstrate that coherence is preserved even in cases where the phases across the beam become mixed statistically.

At high order the contrast vanishes due to the wavelength spread (Eq. 4.105) and a modulation of the momentum distribution appears (Fig. 4.31; Jacobson et al. 1994). When a noise field is applied, the momentum distribution becomes smeared out but the situation remains similar to the case discussed for low-order interferences (Sulyok et al. 2010, Baron and Rauch 2011).

The dephasing behavior for random phase shifts corresponding to related separations Δ_x of Schödinger cat-like states can be written as (see Eq. 4.104):

$$I_0(\Delta_x k_x) = \left|\psi_0^{\mathrm{I}}\right|^2 \left|e^{i\varphi(B)} + e^{ik_x\Delta_x}\right|^2. \tag{4.150}$$

When this equation is averaged over a Gaussian distribution (width ΔB) of magnetic phases $\phi(B) = \mu Bl/\hbar v$ one gets in a quasi-static approximation (Stern et al. 1990, Sulyok et al. 2010):

$$\bar{I}_0(\Delta_x, k_x) = I_0(\Delta_x, k_x)\left[1 + e^{-(\mu l^2/\hbar v)(\Delta B)^2/2}\cos(\Delta_x k_x)\right]. \tag{4.151}$$

This behavior describes a reduction of the contrast at low-order interferences and a broadening of the momentum modulation peaks at high order (Fig. 4.44; Sulyok et al. 2010). The smearing of the momentum distribution has been measured with equipment similar to that shown in Fig. 4.27 (middle). A more detailed description of the connection between noise fields and decoherence

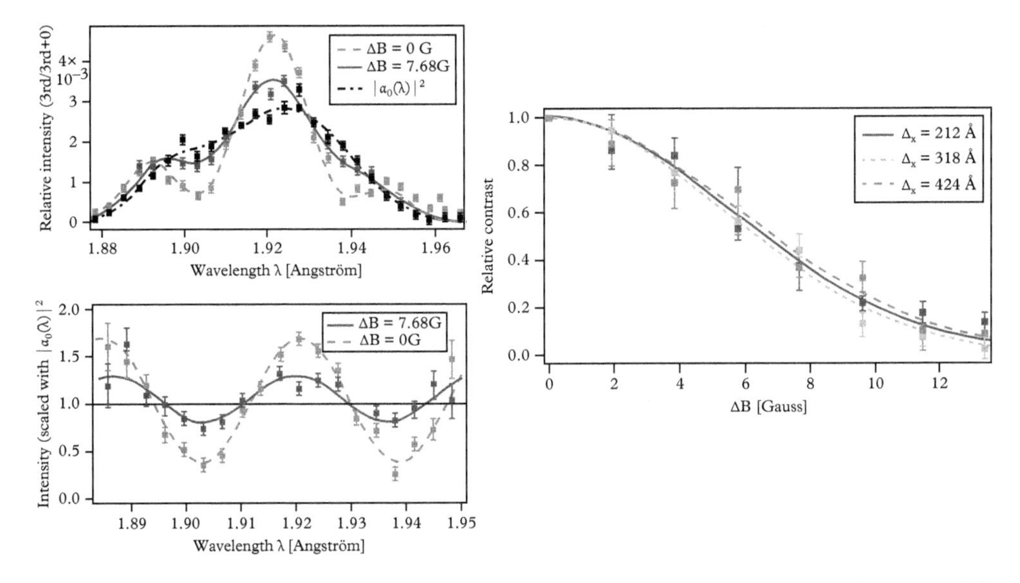

Figure 4.44 *Smearing of the momentum (wavelength) distribution of Schrödinger cat-like states (left) and the dephasing factor for different separations Δ_x as a function of the strength of the magnetic noise field (right) for a noise spectrum between 0 and 20 kHz (Sulyok et al. 2010)*

(dephasing) effects in the framework of Lindblad master equations (Eq. 4.139) has been given by Bertlmann et al. (2006).

In the quasi-static approximation the energy exchange between the neutron and the resonator system is not taken into account. This is justified since the energy change is orders of magnitude smaller than the separation of the modulation peaks. Remarkably, the dephasing effect is stronger for low-frequency bands than for high-frequency bands.

Within experimental errors the loss of contrast is independent of the spatial separation of the Schrödinger cat-like states. This indicates that within the quasi-static approximation the dephasing effect dominates a possible (irreducible) decoherence effect (Zurek 1991). Although many experiments can be described by a dephasing factor, it should be mentioned that in time-dependent fields a photon exchange is always associated with a phase shift. When a time-dependent field in the z-direction acts on the system in a region $x = 0$ to $x = L$ one must solve the time-dependent Schrödinger equation (Eq. 1.2). In analogy to a single mode a rectangular-shaped multimode field is assumed (Haarvig and Reifenstein 1982). See Fig. 4.45. The time dependence is described by

$$\boldsymbol{B}(\boldsymbol{r}, t) = [B_0 + B(t).(\Theta(x) - \Theta(x - L))]\,\hat{z} \tag{4.152}$$

$$B(t) = \sum_{i=1}^{N} B_i \cos(\omega_i t + \varphi_i). \tag{4.153}$$

Tacking into account that the kinetic energy of the neutrons ($\sim$20 meV) is much larger than the maximal potential barrier ($\mu B_i \sim 0.5$ neV) one obtains after some analytical efforts the wave field behind the field region

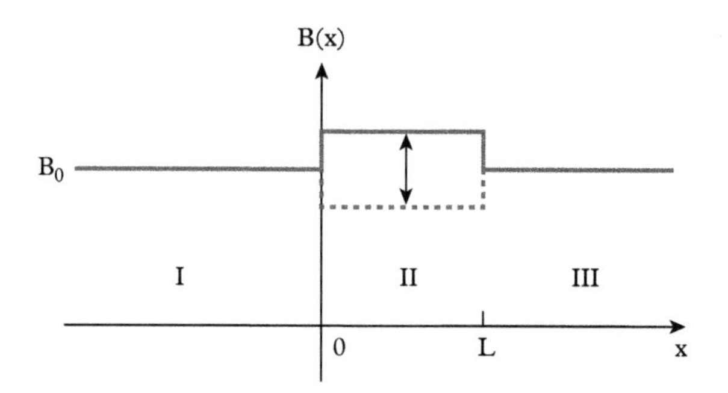

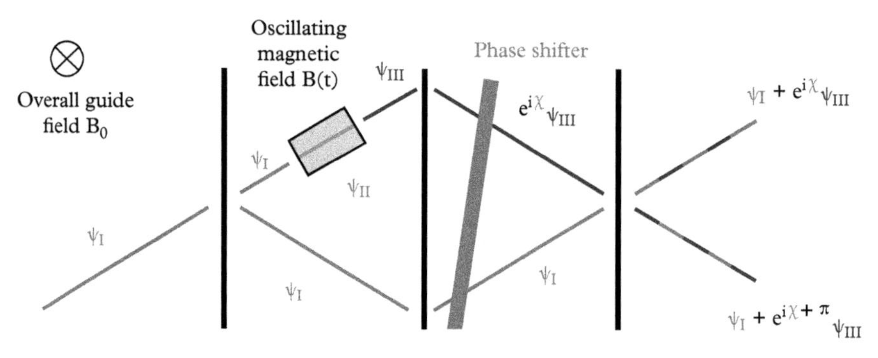

Figure 4.45 *Oscillating magnetic field in the region x = 0 to x = L (left) and sketch of the experimental setup (right)*

$$\Psi_{\mathrm{III}}(x,t)=\sum_{n}\mathcal{J}_{n_1}(\beta_1)\ldots\ldots\mathcal{J}_{n_N}(\beta_N)e^{-n\eta}e^{ik_nx}e^{-i\omega_nt} \tag{4.154}$$

with

$$\omega_n=\omega_0+\boldsymbol{n\omega} \qquad \boldsymbol{k}_n^2=k_0^2-\frac{2m}{\hbar^2}\mu B_0+\frac{2m}{\hbar}\boldsymbol{n\omega}$$

$$\eta_i=\varphi_i+\frac{\omega_iT+\pi}{2} \qquad \beta_i=2\alpha_i\sin\frac{\omega_iT}{2} \qquad \alpha_i=\frac{\mu B_i}{\hbar\omega_i} \qquad T=\frac{L}{v_0}$$

$$\boldsymbol{n}=(n_1\ldots\ldots\ldots n_N), \qquad \boldsymbol{\varphi}=(\varphi_1\ldots\ldots\ldots\varphi_N), \qquad \boldsymbol{\omega}=(\omega_1\ldots\ldots\ldots\omega_N), \qquad \boldsymbol{\eta}=(\eta_1\ldots\ldots\ldots\eta_N),$$

and where $\mathcal{J}_{n_i}(\beta_i)$denote the Bessel functions of order n_i determining the transition amplitudes. From this one gets the interference pattern as

$$I_0(x.t)=\frac{1}{2}\left|\Psi_I(x,t)+e^{i\chi}\Psi_{\mathrm{III}}(x.t)\right|^2=1+\mathrm{Re}\left\{e^{i\chi}\sum_n\mathcal{J}_{n_1}\ldots\ldots\mathcal{J}_{n_N}.e^{in(\xi_i+\omega t)}\right\}$$
$$\text{with } \xi_i=\eta_i-\frac{\omega_ix}{v_0}. \tag{4.155}$$

When the fundamental frequency of all frequencies is ω_f, the interference pattern can be expressed in a Fourier series

$$I_0(x,t)=\sum_{m=-\infty}^{m=\infty}c_m(x)e^{im\omega_ft}, \tag{4.156}$$

where a comparison with Eq. (4.153) gives for unpolarized neutrons where even m-terms remain:

$$c_m = \delta_{m0} + \sum_{\substack{n; n\omega = m\omega_f \\ \sum_i n_i \text{even}}} \mathcal{J}_{n_1}(\beta_1) \ldots\ldots \mathcal{J}_{n_N}(\beta_N) e^{in\xi} \cos\chi. \tag{4.157}$$

This shows that the Fourier coefficient belonging to the frequency $m\omega_f$ contains the same product of Bessel functions as the transition amplitudes for an energy exchange $m\hbar\omega_f$. The argument of

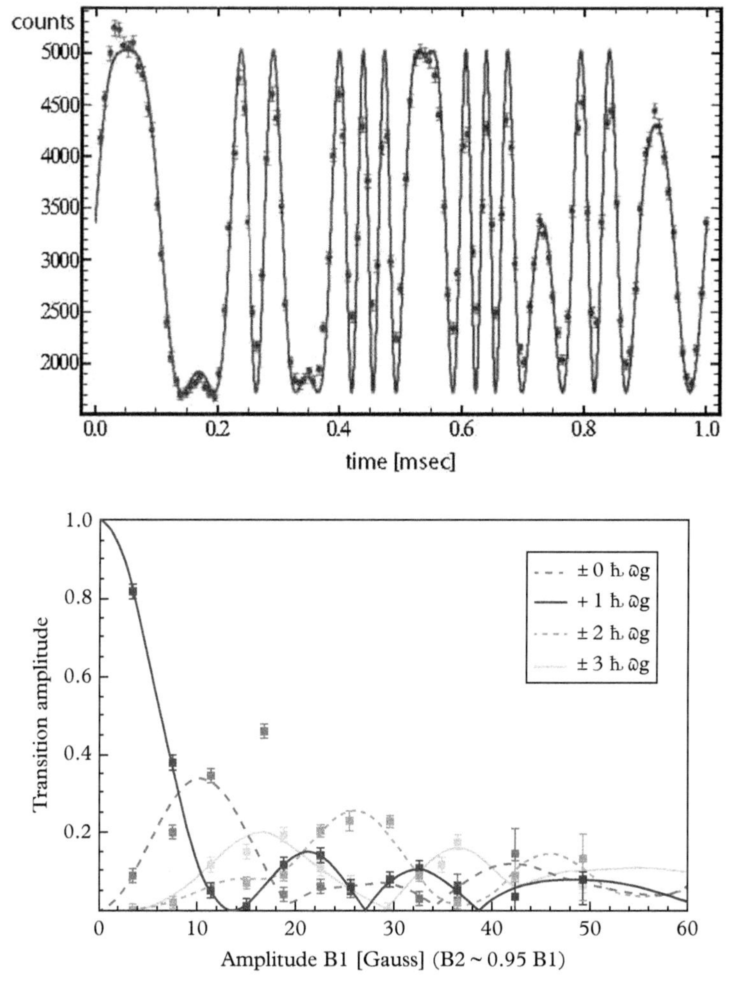

Figure 4.46 *Characteristic results for a 2-mode field with frequencies of 2 and 3 kHz for an amplitude of 30 G. Extracted photon exchange amplitudes in comparison with the calculated ones (right; Sulyok et al. 2012)*

the Bessel functions also contains a $\sin(\omega t/2)$ term defining a "resonance" condition. If the time-of-flight $T = L/v$ through the field region fulfills $\omega_i T = l \times 2\pi$ $(l = 1, 2, 3 \ldots)$ no resulting energy exchange occurs.

Related experiments have been done with a time-resolved analysis of the interference pattern, where the periodicity $\omega_f = 2\pi f_f (f_f = 1\text{ kHz})$ of the interference pattern has been used (Fig. 5.7; Sulyok et al. 2012). The related energy transfers lie in the range of 1 peV and energy gain and energy loss processes are equal. The calculated results show good agreement with the measured values (Fig. 4.46).

To approach the noise limit one must average the results for multimode fields according to a high number of modes ω_i and random phases φ_i. One gets

$$\langle I_0 \rangle = \frac{1}{M} \sum_{i=1}^{M} I_0(t_i) \approx \frac{1}{T_m} \int_0^{T_m} I_0(t_i) dt, \tag{4.158}$$

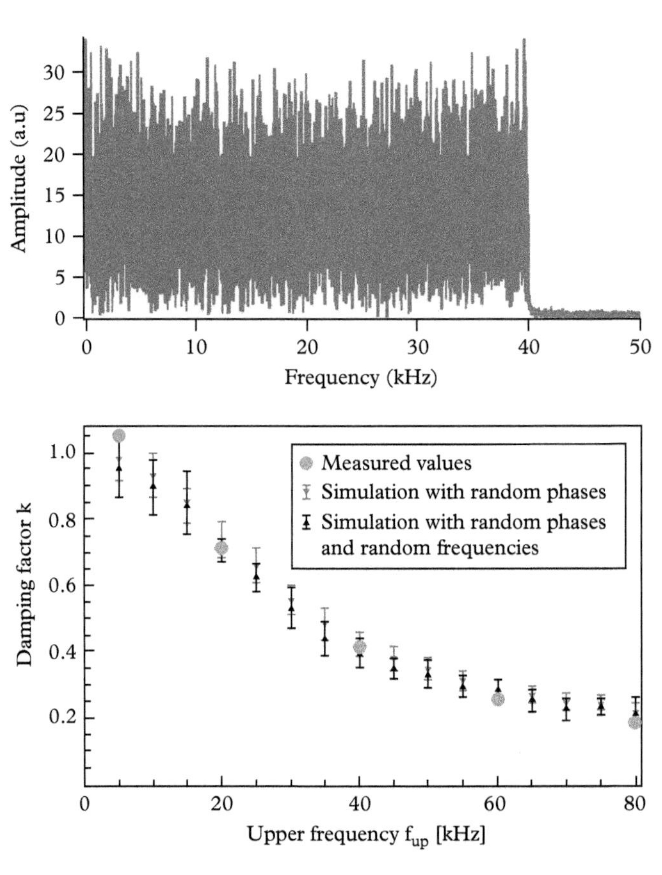

Figure 4.47 *Frequency spectrum ($0 < f < 40$ kHz) and loss of contrast calculated for a multimode noise field, displayed for various frequency bands compared to measured values (Sulyok et al. 2012)*

where T_{m} denotes the measurement time. This corresponds to a contrast of the interference pattern

$$C = \frac{1}{T_{\mathrm{m}}} \int_0^{T_{\mathrm{m}}} \cos\left(\sum_{i=1}^{N} \beta_i \sin(\omega_i t + \xi_i) \mathrm{d}t \right), \tag{4.159}$$

which must be calculated numerically. Figure 4.46 shows a comparison between calculated and measured values, showing again good agreement. It shows that such a high mode field can mimic a real noise field and can be used to model decoherence processes. In all cases multiphoton exchange occurs but a simultaneous measurement of the field and a correlation experiment between the field and neutron counts can restore the interference pattern. Thus, the effect is closer to dephasing than to decoherence.

An additional comment must be made concerning the interaction of neutrons with gas atoms or molecules because several experiments with atoms and molecules use the statistical behavior as a source of decoherence. There is a marked difference between decoherence effects of atoms and molecules and that of neutrons interacting with gas atoms or molecules. In the first case, individual scattering processes due to the much better localization occur, whereas in the second case a collective interaction with many gas atoms occurs, which is described by an index of refraction (Chapman et al.1995, Cronin et al. 2009). Thus, the wave function of atoms or molecules interacting individually with gas atoms remains unknown, whereas in the case of a collective interaction the wave function can be calculated by means of a phase shift resulting from the index of refraction formalism. The first case relates more to decoherence and the second one to dephasing, whereby mixtures of both processes are feasible (Wiseman et al. 1997, Hornberger et al. 2003, Vacchini 2005, Champain et al. 2009). From the discussion of these experiments one must conclude that a general distinction between dephasing and decoherence remains uncertain and a question of the standard of the experiment.

5

Spinor Symmetry and Spin Superposition

In this chapter some basic features of quantum physics are described and their experimental verification will be demonstrated. The 4π-symmetry of spinor wave functions demonstrates a common feature of all spin-½ particles, and spin superposition shows this basic quantum effect on a macroscopic level.

5.1 Spinor Symmetry

The spinor calculus can be regarded as applying a deeper level of structure of space-time than that described by the standard quantum mechanics. It makes transparent some of the subtle properties of space-time phenomena. The essential reason is that the basic spin space is two-dimensional (but complex) rather than the four-dimensional (real) space-time structure. Spinor properties have profound links with the complex numbers that appear in quantum mechanics (e.g., Penrose and Rindler 1984, Shapere and Wilczek 1989).

The magnetic moment $\boldsymbol{\mu}$ of the neutron couples as a magnetic dipole to the magnetic induction field $\boldsymbol{B}$ according to the Hamiltonian (Eq. 1.19)

$$\mathcal{H} = -\boldsymbol{\mu} \cdot \boldsymbol{B} = -\mu\boldsymbol{\sigma} \cdot \boldsymbol{B}. \tag{5.1}$$

Therefore, the wave function within a magnetic field propagates in time as

$$\psi(t) = e^{-i\mathcal{H}t/\hbar}\,\psi(0) = e^{-i\boldsymbol{\mu}\cdot\mathbf{B}t/\hbar}\,\psi(0) = e^{-i\boldsymbol{\sigma}\cdot\boldsymbol{\alpha}/2}\,\psi(0) = \psi(\alpha), \tag{5.2}$$

where α is numerically equal to the Larmor precession angle (Eq. 2.35)

$$\alpha = \frac{2\mu}{\hbar}\int B\,\mathrm{d}t \cong \frac{2\mu}{\hbar v}\int B\,\mathrm{d}s. \tag{5.3}$$

v is the neutron's velocity and ψ in Eq. (5.2) represents the two components of the spinor wave function. The operator $\boldsymbol{\sigma} = \sigma_x\hat{x} + \sigma_y\hat{y} + \sigma_z\hat{z}$, where $\sigma_x, \sigma_y, \sigma_z$ are the Pauli spin matrices (Eqs. 3.44–3.46), and $\int \mathrm{d}s$ denotes the path integral along the neutron's trajectory. Electrodynamically the magnetic dipole moment is identical to an infinitesimal Amperian current loop (Mezei 1986) and not to a permanent magnet with a combination of a north and a south pole separated by a distance δx. It follows immediately that the wave function (Eq. 5.2) displays a characteristic 4π-symmetry. That is,

$$\begin{aligned}\psi(2\pi) &= -\psi(0) \\ \psi(4\pi) &= \psi(0),\end{aligned} \tag{5.4}$$

indicating a characteristic −1 phase factor in the case of a 2π rotation. All expectation values show 2π-symmetry $|\Psi(2\pi)|^2 = |\Psi(0)|^2$. The behavior of the spinor wave function and of its expectation value (polarization) is visualized in Fig. 5.1. This figure shows that the 4π-symmetry of spinor wave functions appears for polarized and unpolarized neutrons as well. Another basic feature of the Pauli spin matrices is that their coupling to the magnetic field remains finite for any direction of the spin polarization (Eq. 3.43) and there is a momentum–spin entanglement which we already saw in Fig. 2.31 and the discussion surrounding this figure.

Aharonov and Susskind (1967) and, independently, Bernstein (1967) predicted the observability of this 4π-symmetry for interferometric experiments and Eder and Zeilinger (1976) gave the theoretical framework for the neutron interferometric realization. Indeed this 4π-symmetry was known from the beginning of quantum mechanics (Dirac 1930), but it was mostly treated as a non-observable property of spinor quantum mechanics. Using Eq. (5.2) the intensity behind the interferometer becomes

$$I_0 = |\psi_0(0) + \psi_0(\alpha)|^2 = 2\left|\psi_0^{\mathrm{I}}\right|^2\left(1 + \cos\frac{\alpha}{2}\right), \tag{5.5}$$

which appears for polarized and unpolarized neutrons and which indicates again the self-interference properties involved in this type of experiments. Experimental verification was achieved in 1975 (Rauch et al. 1975, Werner et al. 1975). The magnetic field was varied and the path integral $\int B\,ds$ was measured by Hall probes. The observed interference pattern (Fig. 5.2) showed the expected 4π-symmetry as it has been also shown in Chapter 1 (Fig. 1.3). Further verification has been provided by a wave-front division interferometer (Klein and Opat 1976), by a molecular beam system (Klempt 1976), and by an NMR system (Stoll et al. 1978, Mehring et al. 1984), and has been repeated with the perfect crystal interferometer using the well-defined magnetic fields within Mu-metal sheets (Rauch et al. 1978a). In the latter experiment the Mu-metal sheets were magnetized in opposite directions and rotated in a sense that the nuclear phase shift

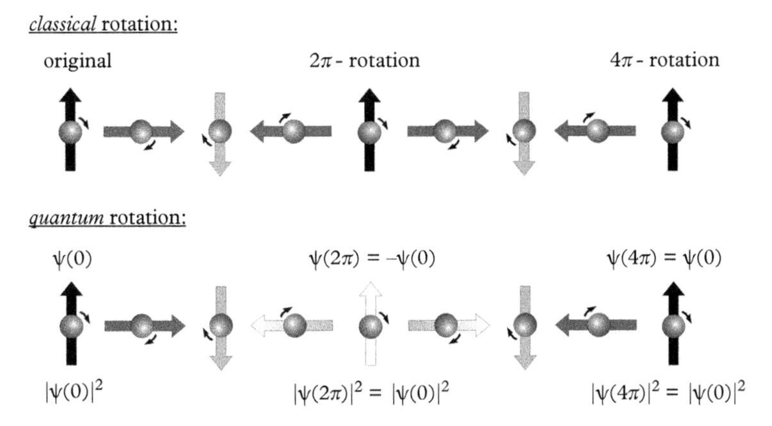

Figure 5.1 *Visualization of the spinor rotation of an unpolarized beam with the spin-up component (above) and the spin-down component (below)*

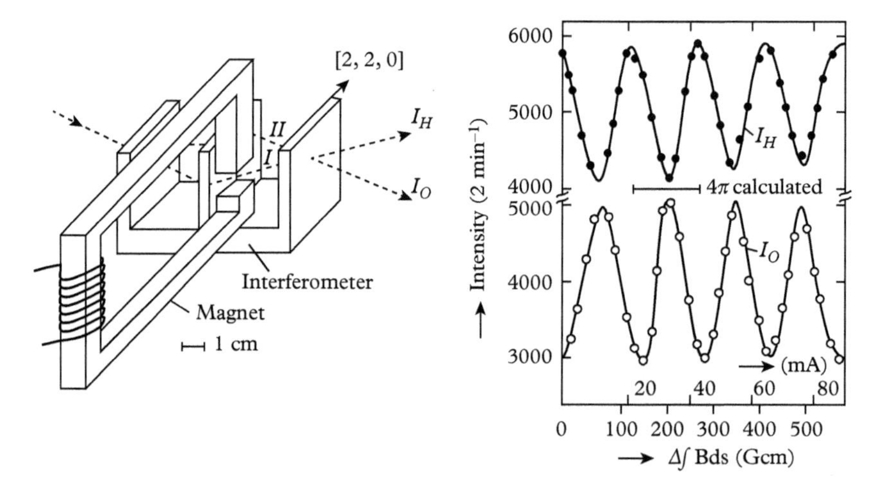

Figure 5.2 *Sketch of the experimental arrangement and first results of the 4π-spinor symmetry experiment (Rauch et al. 1975)*

χ remained 0 (Eq. 3.52). The most precise periodicity factor extracted from these measurements and corrected for up-to-date values of the physical constants is

$$\alpha_0 = (715.87 \pm 3.8)\ \text{deg}, \tag{5.6}$$

which is (within 3σ) in agreement with the predicted 720°, i.e., the 4π-symmetry. The related setup is shown in Fig. 5.3. The accuracy is limited by possible stray fields and the magnetization value of the Mu-metal sheet.

It should be mentioned that Eq. (5.5) can also be obtained by using a different index of refraction for both spin states describing the unpolarized incident beam (Eq. 3.52). In this case the axis of quantization is chosen parallel to the magnetic field and at the entrance in the magnetic field a slight change of the kinetic energy occurs due to the longitudinal Zeeman splitting. This results from energy conservation and has been observed experimentally as well (Zeilinger and Shull 1979, Alefeld et al. 1981a, Weinfurter et al. 1988, Otake et al. 1996). Therefore, the velocity of the two sub-beams (±) within the region of space containing the magnetic field B is slightly different, which accounts for the $\cong$ sign in Eq. (5.3) and gives a correction factor on the order of 10^{-5}, i.e., the ratio of the Zeeman energy to the kinetic energy of the neutrons (Bernstein 1979, Bernstein and Zeilinger 1980, Home et al. 2013). Nevertheless, it should be mentioned that the second step made in Eq. (5.3) is meaningful for rather monochromatic neutrons only ($\delta v/v_0 << 1$). When the axis of quantization is chosen perpendicular to the magnetic field the rotational effect of the states becomes more visible (e.g., Barut and Bozic 1990). In both cases the 4π-symmetry effect is caused by the action of a Hamiltonian ($\mathcal{H} = -\boldsymbol{\mu}\cdot\boldsymbol{B}$). When the axis of quantization is chosen parallel to the field it is related more closely to a dynamical phase and when it is chosen perpendicular to the field it is more closely connected to a topological phase, as discussed in Section 6.3. Even here it should be mentioned that the neutrons not only remember the dynamical phase but also the topological one, which is related to the axis around which the rotation has taken place. Both views are complementary to each other and also depend on the choice of the axis of quantization (Rauch et al.

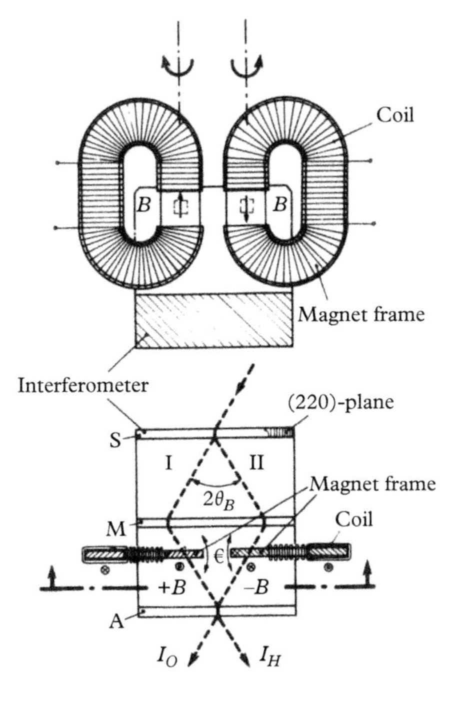

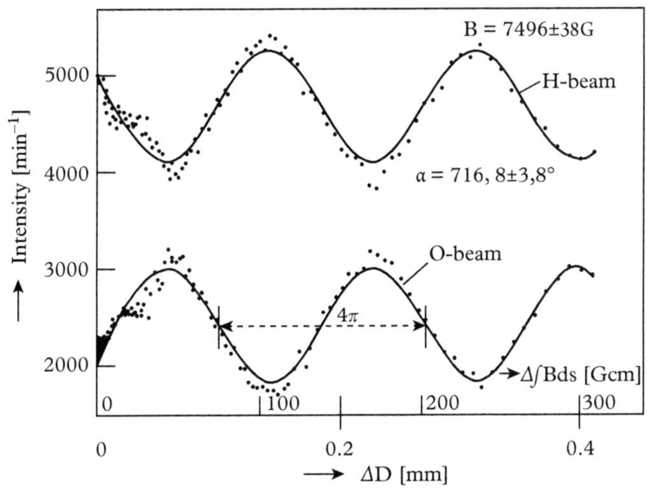

Figure 5.3 *Setup (above) and results (below) of a precise 4π-symmetry experiment (Rauch et al. 1978a)*

1975, Mezei 1988). In the course of such topological phase measurements, the dynamical phase has been investigated again by several authors (Allman et al. 1997, Wagh et al. 1997, Werner 2012).

The 4π-symmetry effect measurements have been widely discussed (Byrne 1978, Mezei 1979, Silverman 1980, Bernstein and Phillips 1981, Zeilinger 1981a, Jordan 1983, Anandan and Stodolsky 1987, Bhandari 1993, Wagh and Rakhecha 1996, Dubbers and Stöckmann 2013, Klepp et al. 2014). The isomorphism of all two-level system, as has been addressed by Feynman et al. (1957), has given a large variety of experimental verifications, including spinor interferometry with polarized light beams (Bhandari 1991, 1993). The 4π-symmetry of the spinor wave function has been measured for unpolarized and polarized beams, as well. The combination of all these experiments demonstrates the intrinsic features of such a quantum system in general and the self-interference property of the neutrons within the interferometer (Badurek et al. 1988).

These experiments also demonstrate the equivalence of spin rotation (spin echo) and interference measurements because spin rotation around a magnetic guide field is caused by the space-superposed spin-up and spin-down wave functions and, therefore, it is an interference experiment itself (Section 2.4; Mezei 1972, 1988; Baryshevskii et al. 1991). Therefore, several symmetry and topological phenomena can be seen in spin-rotation experiments, for example the Berry phase (Sections 6.3 and 6.4). The generalization of this rotational phase factor involves the inclusion of an additional geometrical or topological phase factor, which is geometry dependent, and it matters how the Hamiltonian (Eq. 5.1) acts along the beam paths (Berry 1984). Aerts and Reignier (1991) discussed the 4π-symmetry neutron interferometry experiments in terms of the non-locality aspects of quantum mechanics, and they emphasized the amazing character of the "de-localization" effect for single neutrons. What seems to be so amazing is that the spin angular momentum is a quantized quantity $\hbar/2$ but nevertheless the two wave packets within the interferometer remain connected to the transit of a single neutron and precess separately in each arm of the interferometer. That is, the precession causes the phase shift of the interference pattern.

There is a discussion in the literature concerning the analogy between symmetry effects of fermionic and bosonic systems (Byrne 1978, Zeilinger 1981a, Bernstein 1985). The reason for this analogy may be found in the fact that in certain cases a similar two-dimensional subspace exists for bosonic systems (e.g., the helicity ± 1 for left and right circular polarized light); while it exists intrinsically for fermions (Bhandari 1997). Bernstein (1985) has shown that magnetic precession of neutrons represents indeed a rotation due to the well-established proportionality of the Hamiltonian to the magnetic moment and hence to the spin angular momentum where the time derivative of the spin, $d\mathbf{s}/dt$, is the torque acting on the neutron.

The phenomenon can also be seen as a tunneling effect of spin-½ particles through a magnetic field where the phase space coupling of space and spin variables becomes obvious. Only in the low-field limit ($2\mu B << E$) can the discussion be based equivalently on a rotation phenomenon of the spin variables or on the index of refraction formalism acting on the longitudinal k-vector of the beam. For higher magnetic fields or lower energies of the neutrons the coupling of the space and spin variables must be taken into account according to the non-relativistic Pauli equation. That is, the spinor part of the wave function is entangled with its spatial part. The related reflection and transmission effects at the field boundaries cause the outgoing wave to be composed of a sum of waves which pass through the barriers once or after several internal reflections (Barut et al. 1987, Frank 1989, Home et al. 2013). The acceleration (deceleration) of the spin-up (spin-down) neutron wave packet upon entering a magnetic field region has been described in Section 2.4 and has been measured by Alefeld et al. (1981a) and by Weinfurter et al. (1988) and is discussed in a recent paper by Cappelletti (2012).

Nuclear phase shifts χ and magnetic spinor rotations α can be applied simultaneously to the coherent beams such that the wave function becomes (Eqs. 2.23, 3.52, and 5.2)

$$\psi(\chi,\alpha) = e^{i\chi} e^{-i\sigma\cdot\alpha/2} \psi(0,0). \tag{5.7}$$

After some algebra, one obtains for the intensity and the polarization of the beam in forward direction behind the interferometer (Eder and Zeilinger 1976)

$$I = |\psi(0,0) + \psi(\chi,\alpha)|^2 = \frac{I_0}{2}\left[1 + \cos\chi\cos\frac{\alpha}{2} + \hat{B}\cdot\boldsymbol{P}\sin\chi\sin\frac{\alpha}{2}\right]. \tag{5.8}$$

The polarization $\boldsymbol{P}'$ of this outgoing beam behind the interferometer becomes

$$\begin{aligned} \boldsymbol{P}' &= \frac{[\psi^+(0,0) + \psi^+(\chi,\alpha)]\boldsymbol{\sigma}[\psi(0,0) + \psi(\chi,\alpha)]}{I'} \\ &= \frac{I_0}{2I'}\left\{\left[\cos^2\frac{\alpha}{2} + \cos\frac{\alpha}{2}\cos\chi\right]\boldsymbol{P} + \left[\hat{B}\cdot P\sin^2\frac{\alpha}{2} + \sin\chi\sin\frac{\alpha}{2}\right]\hat{\mathrm{B}}\right. \\ &\quad \left. + \left[\cos\chi\sin\frac{\alpha}{2} + \cos\frac{\alpha}{2}\sin\frac{\alpha}{2}\right](\hat{B}\times \boldsymbol{P})\right\}, \end{aligned} \tag{5.9}$$

where $\boldsymbol{P}$ denotes the polarization of the incident beam. This formula shows characteristic beat effects in the outgoing intensity and also in the polarization even for unpolarized incident neutrons ($\boldsymbol{P} \equiv 0$). These beat effects have been found experimentally to be in agreement with these predictions (Badurek et al. 1976; Fig. 3.16). The influence of slightly different rotation angles with and without material in the beam has also been discussed by Baryshevskii et al. (1991). This method has been used by Nakatani et al. (1992) to extract information about the magnetic domain structure of a Fe-3% Si crystal (Section 9.2). In later work Nakatani et al. (1996) used the combination of nuclear and magnetic phase shifters to control the neutron polarization behind the interferometer. Using a spin analyzing crystal behind the interferometer they could identify neutron polarizations in all directions.

5.2 Spin Superposition

Any superposition state which is created by the linear superposition of coherent orthogonal states exhibits new quantum features which are intrinsically different from the beams before overlap. This is most clearly demonstrated by the superposition of spin-up and spin-down states as shown in Fig. 5.4, where a new pure state is created instead of a classical mixture. In this case, polarized incident neutrons are used, where the polarization-dependent parts of Eqs. (5.8) and (5.9) describe the spin-superposition phenomenon. Special attention has been drawn to the case where

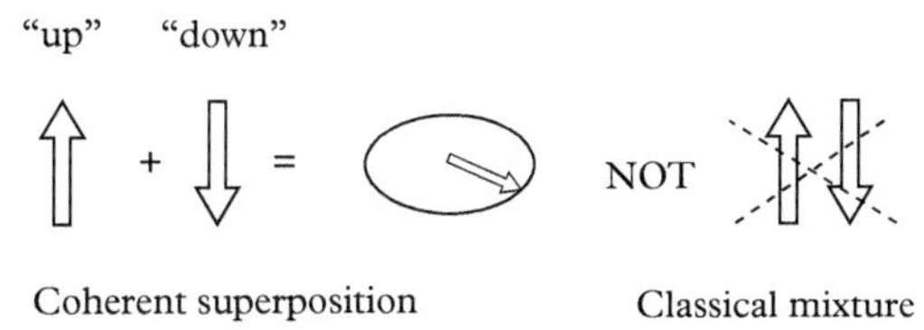

“up” “down”

+ = NOT

Coherent superposition Classical mixture

Figure 5.4 *Coherent superposition of quantum states (left) and of classical systems (right)*

oppositely oriented coherent sub-beams exit prior to superposition at the third interferometer plate. In this case, the spin state in one coherent beam path is inverted, which can be achieved by a static DC flipper (Mezei 1972) or by a resonance flipper (Alvarez and Bloch 1940). Both are standard devices in polarized neutron physics (Krupchitsky 1987, Williams 1988). First, we discuss the situation when a static spin flipper turns the spin around the y-axis (Fig. 5.5), which rotates a beam initially polarized in the $\langle z|$-direction into the $\langle -z|$-direction (Mezei 1972, Zeilinger 1979),

$$\psi(\chi, \pi) = e^{i\chi} e^{-i\sigma_y \pi/2} |z> = e^{i\chi} |-z> \tag{5.10}$$

and the final polarization after superposition becomes

$$\mathbf{P}' = \begin{pmatrix} \cos \chi \\ \sin \chi \\ 0 \end{pmatrix}, \tag{5.11}$$

which lies in the (x, y)-plane, i.e., perpendicular to the two polarization states of the two coherent sub-beams before superposition. Therefore, a related experiment can demonstrate the quantum-mechanical spin-superposition law on a macroscopic scale. Wigner (1963) brought attention to this subject in his famous article on the theory of measurement. It is quite often called the "Wigner phenomenon." The general question, whether a complete reconstruction of a pure state is possible, has been discussed by Schwinger et al. (1988) and by Scully et al. (1989). These papers are known as the "Humpty-Dumpty" papers, connected to the nursery rhyme "Humpty-Dumpty fell off the wall, can Humty-Dumpty be put back together at all?" The general answer is that one can come close enough to recovering spin coherence in a pure state and only realistic field distributions and the finite dimensions of the neutron wave packet cause some marginal incoherence effects.

The experimental arrangement for producing and analyzing the unique spin states corresponding to the Wigner phenomenon was achieved in 1982 using the setup shown in Fig. 5.6 (Summhammer et al. 1982, 1983). A polarized incident beam is produced by magnetic prism deflection, which in combination with a non-dispersive perfect crystal monochromator–interferometer arrangement produces a double-humped rocking curve due to prism deflection, corresponding to the + and – polarization states of the neutrons (Just et al. 1973, Badurek et al. 1979). The spin inversion in one of the coherent beam is achieved by a properly shaped DC coil in combination with a weak guide field in the z-direction. The condition for a complete spin inversion requires $2\mu B\ell/\hbar v = \pi$, where ℓ is the length and B is the effective field strength around which the neutron spin precesses (see Section 2.4, Eq. 2.28). The main part of the polarization analyzer system behind the interferometer is the $\pi/2$ spin turn coil which rotates the y-polarization component into the analyzer (z) direction. The experimental results are in complete agreement with the theoretical prediction and they indicate that every neutron has information about the physical situation in both beam paths, because a pure initial state in the z-direction is transferred into a pure final state having a definite polarization direction within the (x, y)-plane. The coil behind the interferometer was used to demonstrate the equivalence of a nuclear (scalar) phase shift within the interferometer and an additional Larmor precession applied to the final polarization vector.

A discussion regarding the non-locality of the quantum state involved in these kinds of experiments can be found in an article by Dewdney et al. (1988). They used a non-local quantum potential to demonstrate how an initially pure spin state can be transported through a two-path interferometer, thereby forming a new pure spin state perpendicular to the initial spin state. These experiments are also connected with complementarity and "which-way" information and

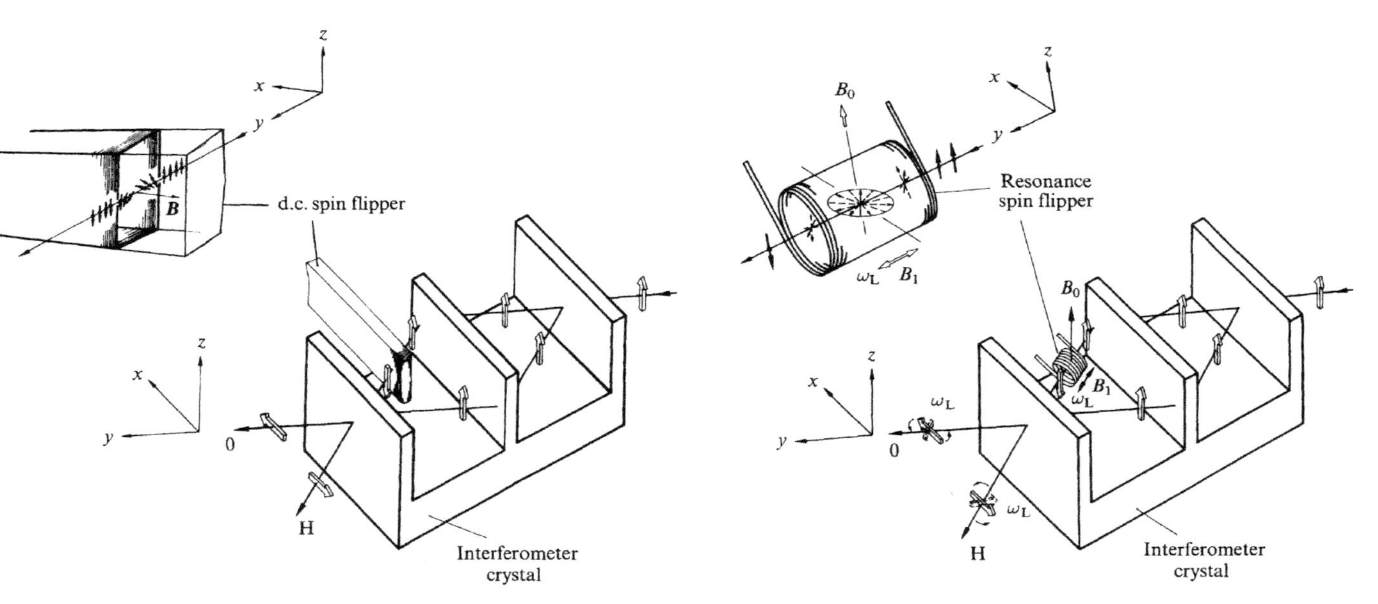

Figure 5.5 *Sketch of the spin-superposition experiments using a static (left) and a time-dependent flipper (right)*

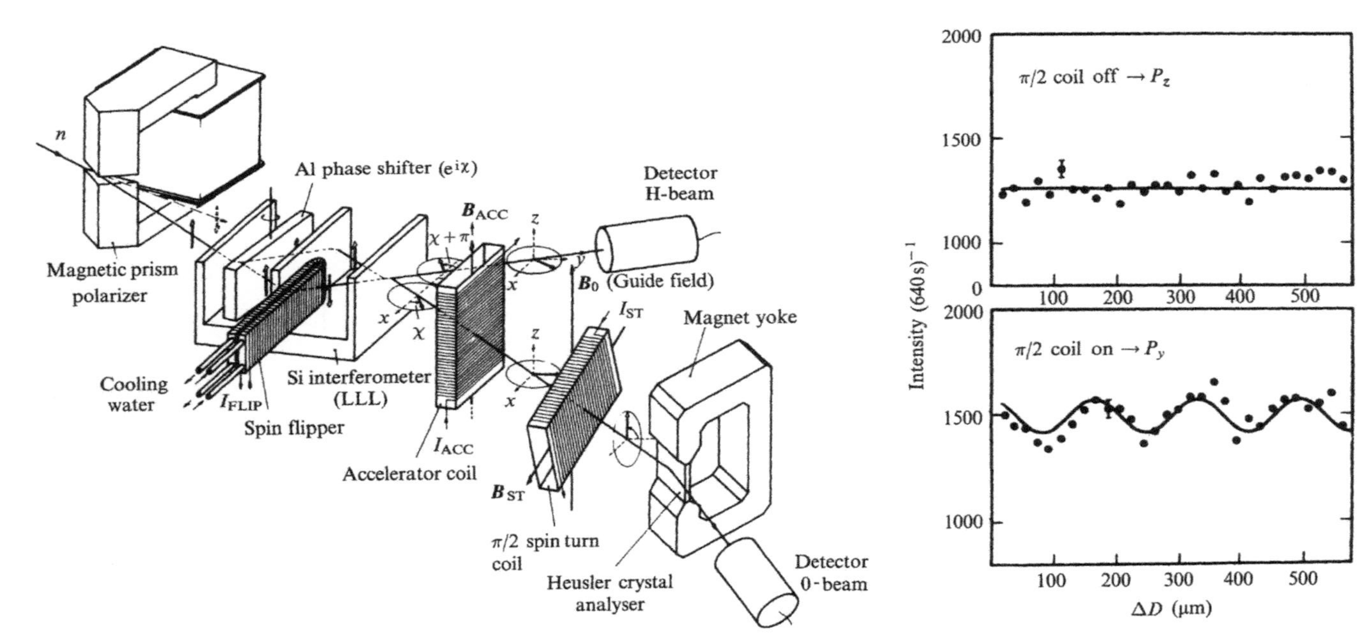

Figure 5.6 *Details of the experimental arrangement of the static spin-superposition experiment (left) and typical results when the polarization was measured in the z-direction (right above) and when the polarization was measured in the y-direction due to activating the π/2-spin turn coil (right below) (Summhammer et al. 1983)*

is seen as a major epistemological breakthrough (Wagh 1999). It can also be seen as a precursor experiment to the Cheshire cat experiment performed much later (Denkmayr et al. 2014; Section 7.2).

5.3 Time-Dependent Spinor Superposition

When a resonance spin flipper is used instead of a static flipper the physical situation changes in an essential and interesting way (Fig. 5.5). Here a time-dependent interaction occurs, necessitating the use of the time-dependent Schrödinger equation to describe the result (Eq. 1.2). In this case, the spin reversal is accompanied by an exchange of a radio-frequency (rf) photon between the neutron and the resonator system. The transferred energy is equal to the difference of the Zeeman energy levels $\hbar\omega_r = 2|\mu|B_0$ of the neutrons within the guide field of strength B_0 as predicted by Drabkin and Zhitnikov (1960), and verified experimentally by Alefeld et al. (1981b) and Weinfurter et al. (1988). The behavior of the polarization vector within such a resonance flipper field has been treated by Kendrick et al. (1970) and that of the wave function by Krüger (1980), Zhang et al. (1994), and Utsuro and Ignatovich (2010). For a complete spin reversal the frequency of the rotating field must match the resonance conditions $\hbar\omega_r = 2|\mu|B_0$ and its amplitude B_1 must fulfill the relation $2|\mu|B_1\ell/\hbar v = \pi$ (Eq. 2.43). For experimental reasons—as in the case of Fig. 5.5—oscillating fields are used instead of rotating ones. Such fields can be seen as two counter-rotating fields. Therefore, the amplitude B_1 must be doubled and a Bloch–Siegert (1940) shift must be considered, which changes the resonance condition slightly, especially in the case of weak guide fields where $B_0 \sim B_1$. Due to the energy exchange, the potential energy, and thereby the total energy change, however, the kinetic energy remains constant as long as the neutron stays in the guide field. Therefore, the wave function behind the flipper at resonance is described as

$$\psi(\chi, \omega_r) = e^{i\chi} e^{-i(\omega-\omega_r)t} |-z>. \tag{5.12}$$

This gives, after superposition with the undisturbed reference wave function of the other beam path, a final polarization vector in the (*x*,i)-plane, but now it rotates with the Larmor frequency ω_r synchronous with the flipper field

$$\mathbf{P}' = \begin{pmatrix} \cos(\chi - \omega_r t) \\ \sin(\chi - \omega_r t) \\ 0 \end{pmatrix}. \tag{5.13}$$

This time-dependent rotation can be detected by a stroboscopic registration of the neutrons synchronized with the phase of the flipper (Fig. 5.7). The results of this experiment show that coherence can be preserved even when an energy exchange occurs (Badurek et al. 1983a, 1983b).

The fact that the spin flip is associated with an energy transfer between the neutron and the resonator system was first verified by Alefeld et al. (1981b) with a high-resolution backscattering instrument and in more detail by Weinfurter et al. (1988) with a perfect crystal diffraction camera (Fig. 5.8).

With regard to the discussions of the double-slit experiment (e.g., Feynman et al. 1965) one might argue that, in addition to the interference pattern, the beam path can be detected by observing the added or missing photon of the resonance circuit or by measuring the change of the kinetic energy of the neutron behind the guide field. But, from the particle number–phase uncertainty relation applied in its most simple form $\Delta N \Delta\phi \geq ½$ (Section 4.4.3; Carruthers

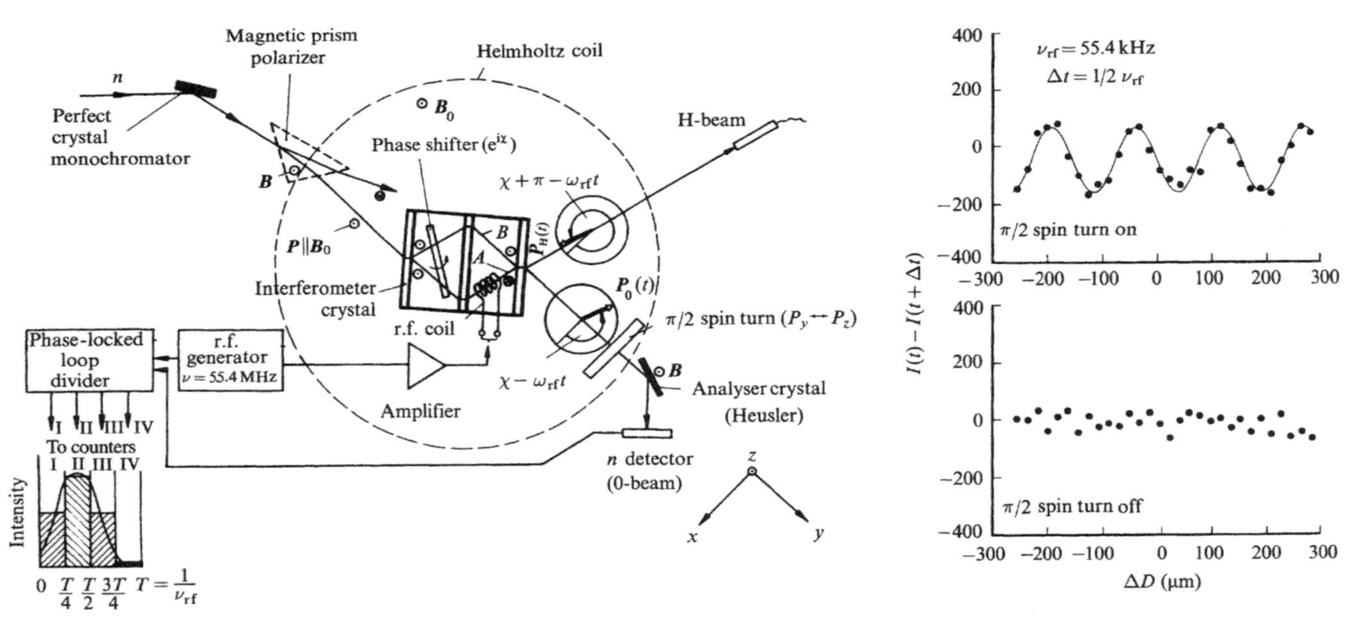

Figure 5.7 *Details of the experimental arrangement for the time-dependent spin-superposition experiment indicating the stroboscopic measurement procedure (left) and typical results when the z-component (right, below) or the y-component of the polarization was measured (right, above) (Badurek et al. 1983b)*

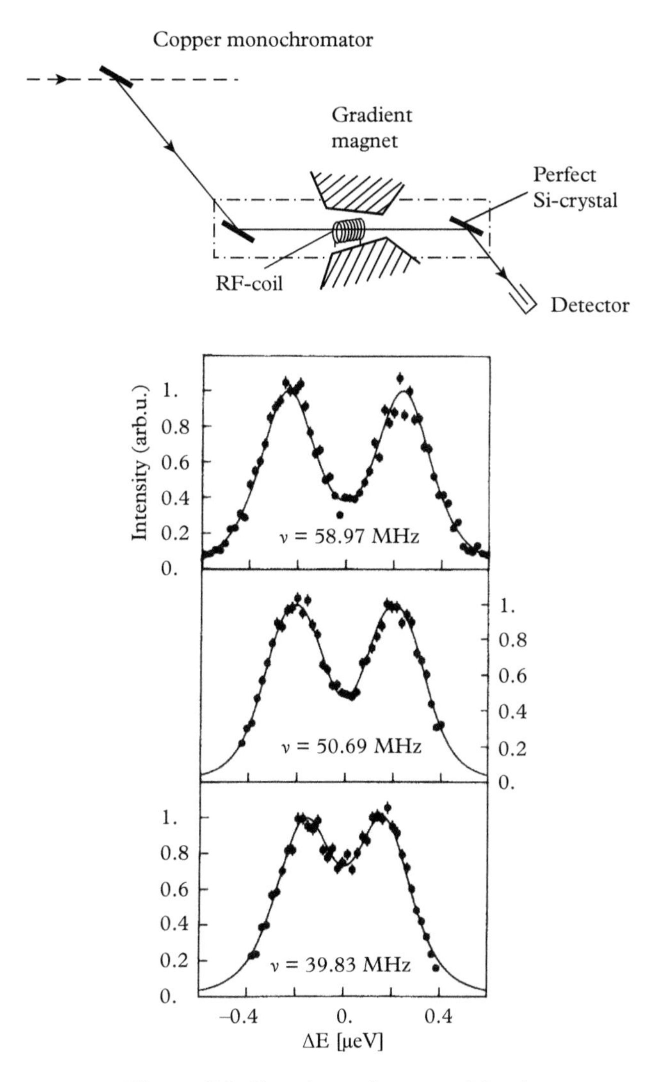

Figure 5.8 *Experimental setup (above) and verification of an energy transfer between the neutron and the resonator system during a Rabi spin-flip (below; Weinfurter et al. 1988)*

and Nieto 1968, Gerhardt et al. 1974), it follows that the presence or absence of a single photon $\Delta N = 1$ cannot be detected when the phase of the field $\Delta\phi$ is known, thereby allowing one to observe the interference pattern simultaneously. This also becomes understandable when the mean number of photons for the flipper field is considered ($\overline{N} \sim 10^{22}\text{cm}^{-3}$, see comment following Eq. 5.27). Although the phase–number uncertainty relation used in this context is not correct in a rigorous sense, it can be used as an approximation in this limit. Namiki et al. (1987) gave an explanation of this epistemological question on the basis of the Machida–Namiki (1980, 1986) theory of measurement in a manner that avoids the use of the not fully accepted particle number–phase uncertainty relations. Similar constraints arise when one tries to measure the change of velocity behind the guide field of the neutrons which have passed through the resonance coil ($\Delta v_r = 2\mu B/mv$; Alefeld et al. 1981b). For a spectroscopic measurement, the velocity spread of the beam Δv must be smaller than Δv_r and when measuring by a time-of-flight method, the time channels must fulfill the condition $\Delta t > \Delta\ell/\Delta v_r$ to accumulate the neutrons with the correct polarization phase in the correct time channels, which, in turn, must fulfill the condition $\Delta t < 2\pi/\omega_r$ to allow the observation of the interference pattern discussed earlier. This is incompatible with the momentum–position uncertainty relation $\Delta k\Delta\ell > ½$, where $\Delta\ell$ is the distance of the detector behind the guide field.

The coupling of the neutron quantum system to micromaser detector systems have been discussed in detail by Scully and Walther (1989). They found that classical coherent maser states preserve spin coherence but masers prepared in number states destroy it in agreement with the discussion above. This result was also obtained by Leggett (1986) when analyzing these experiments in terms of complementarity. Therefore, no mystery concerning a virtual beam path detection remains. It is shown that a loss of coherence in measurements on quantum systems can always be traced to correlations between the measuring apparatus and the systems being observed (Eq. 4.37). Whether there exists an explanation of such complementary experiments without using the uncertainty relation is still under discussion (Scully et al. 1991, Storey et al. 1994, Wiseman et al. 1995, Wagh 1999c). Similar experiments with atoms use internal degrees of freedom for beam path labeling which makes the uncertainty-related explanation less straightforward, but still feasible (Duerr et al. 1998). Erhart et al. (2012) demonstrated in a neutron optic experiment that an error–disturbance uncertainty relation must be used instead of the standard Heisenberg uncertainty relation.

5.4 Double-Coil Experiments and the Magnetic Josephson Effect

The question arises as to whether an energy change is equivalent to a measuring process and what happens if a resonance spin turn occurs in both beam paths (Dewdney et al. 1984). When resonance coils are placed into both beam paths an energy exchange $\hbar\omega_r$ occurs with certainty and the polarization of the outgoing beam becomes inverted (Fig. 5.9). According to our previous considerations the change of the wave function for different modes of operation when the resonance flippers are tuned for resonance can be categorized into four cases. We take the initial state to be $|z>$):

(a) Both flippers are operated synchronously without a phase shift between the flipper fields

$$\psi_0 \rightarrow e^{i(\omega-\omega_r)t}\,|-z> + e^{i\chi}\,e^{i(\omega-\omega_r)t}\,|-z>, \tag{5.14}$$

which gives an intensity modulation of

$$I_0 \propto 1 + \cos\chi, \tag{5.15}$$

independent from the flipper fields.

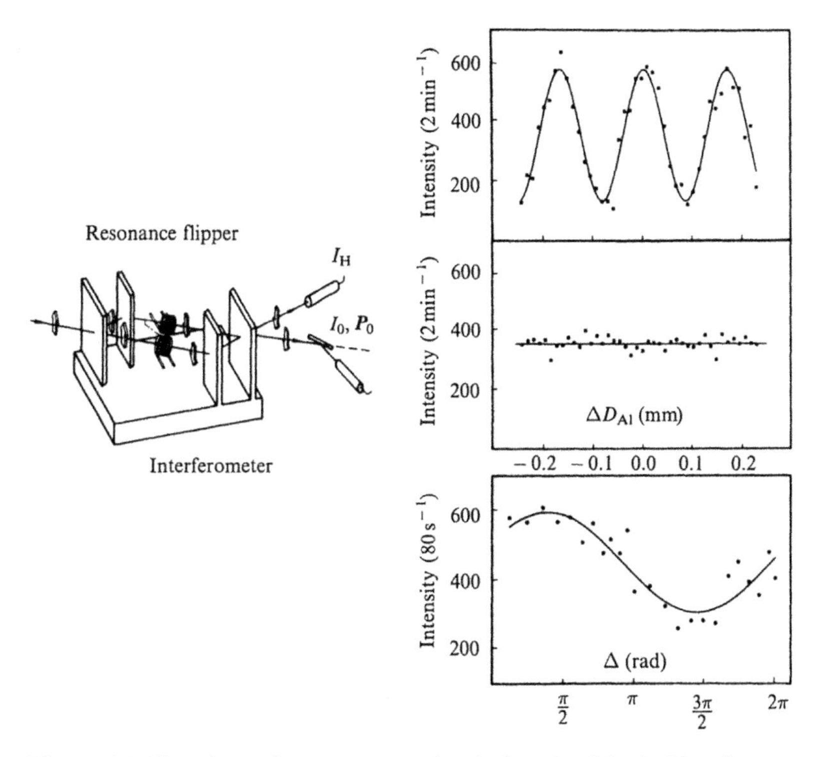

Figure 5.9 *Experimental arrangement and typical results of the double-coil experiment when both flipper fields are operated synchronously (top), non-synchronously (middle), and with a distinct phase shift (bottom) (Badurek et al. 1986)*

(b) Both flippers are operated with statistically fluctuating phase differences $\Delta(t)$, which average out during the measuring interval

$$I_0 \propto \text{const.} \tag{5.16}$$

However, it has been noted that even in this case coherence phenomena can be observed if a stroboscopic investigation is performed ($I_0 = I_0(\Delta(t))$).

(c) Both flippers are operated synchronously with a distinct phase relation Δ, such that the wave function in the 0-beam is

$$\psi_0 \rightarrow e^{-i(\omega-\omega_r)t} \left|-z > +e^{-\chi} e^{i\Delta} \cdot e^{i(\omega-\omega_r)t}\right| z >. \tag{5.17}$$

The intensity in the 0-beam is then

$$I_0 \propto 1 + \cos(\chi + \Delta), \tag{5.18}$$

which is dependent on the phase shift Δ of the resonance fields. The related experiments have shown this behavior (Fig. 5.9; Badurek et al. 1986). The dependence on the phase difference of the resonance fields indicates that the interference pattern is also sensitive to the longitude line on the Poincaré sphere on which the neutron spin is rotated from the north pole $|z>$ to the south pole $|-z>$. This influence of the topological phase will be discussed in Section 6.3.

(d) Both flippers are operated synchronously but with slightly different resonance frequencies ω_{r1}, ω_{r2}

$$\psi \rightarrow e^{i(\omega-\omega_{r1})t}\,|z> + e^{i\chi}\,e^{i(\omega-\omega_{r2})t}\,|z>. \tag{5.19}$$

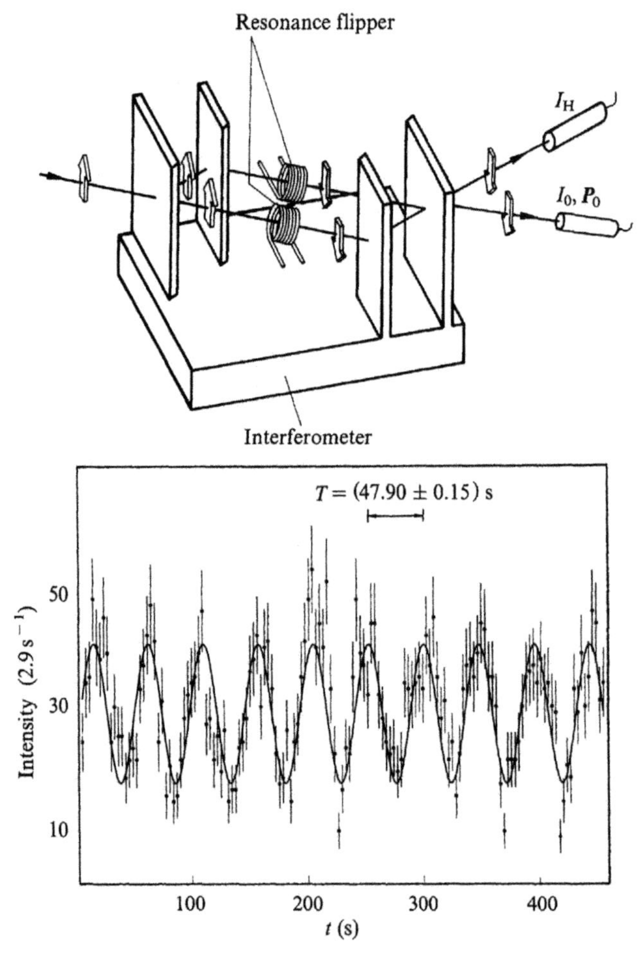

Figure 5.10 *Experimental arrangement and typical results of the double-coil experiment when the resonance frequencies of both flipper coils are chosen slightly different ($\Delta f = 0.02$ Hz; Badurek et al. 1986)*

Therefore, the intensity behind the interferometer exhibits a typical quantum beat effect, caused by an extremely small energy difference between both beams ($\Delta E = \hbar(\omega_{r1} - \omega_{r2})$):

$$I = 1 + \cos[\chi + (\omega_{r1} - \omega_{r2})t]. \quad (5.20)$$

The related experiments have shown complete agreement with this theoretical prediction (Badurek et al. 1986). The dependence of the interference pattern on the phase shift of the oscillating resonance fields shows again that the wave function behind the interferometer carries information about the physical situation of the history along both beam paths and that for definite phase shifts every neutron "knows" to enter the forward or the deflected beam. This verifies coherence remains even when an energy exchange occurs with a probability near to unity and that energy exchange is not automatically a measuring and/or localization process. Certainly, no measuring process has taken place as long as the energy transfer due to the resonance fields is smaller than the energy width of the beam. In this case, absorption (or emission) of a photon by coil I and coil II can coexits objectively (Unnerstall 1990).

Here we display the result of the quantum beat measurement (Fig. 5.10) where the intensity behind the interferometer oscillates between the forward beam (0-beam) and deviated beam (H-beam) without any apparent change inside the interferometer. The time constant of this modulation can reach a macroscopic scale which is again correlated with an uncertainty relation $\Delta E \Delta t \leq \hbar/2$. The periodicity of the intensity modulation $T = 2\pi/(\omega_{r1} - \omega_{r2})$ has a value of $T = 47.90 \pm 0.15$ s caused by a frequency difference of about 0.02 Hz. This corresponds to a mean difference of the energy transfer to the two beams of $\Delta E = 8.6 \times 10^{-17}$ eV and to an energy sensitivity of 2.7×10^{-19} eV, which is many orders of magnitude greater than that of other advanced spectroscopic methods. This high resolution is essentially independent of the monochromaticity of the neutron beam. In this case the monochromaticity was $\Delta E \cong 5.5 \times 10^{-4}$eV centered around the mean energy of the beam $E_0 = 0.023$ eV.

This experiment gives additional possibilities for a comparison of the Bohr–Heisenberg and the Einstein–de Broglie view of quantum mechanics (Vigier 1985, 1988). The experimental results agree with the predicted outcome of the quantum-mechanical formalism and can, therefore, help for interpretational questions, but only indirectly. An adequate discussion of the above experiment in terms of a modernized Copenhagen measurement theory has been given by Omnès (1994).

The resonance spin flippers used for the time-dependent spinor superposition (Section 5.3) and for the double-coil experiment described in this section have been operated with comparable strengths of the guide field and of the amplitude of the oscillating field ($B_0 \sim B_1$). In this case the Larmor frequency $\omega_L = 2|\mu|B_0/\hbar$ deviates from the resonance frequency ω_r due to the so-called Bloch–Siegert shift (Bloch and Siegert 1940, Greene 1978, Pendlebury et al. 2004). For the case $B_1 \leq B_0$ the frequency shift can be written as $\omega r = \omega_L \left(1 + B_1^2/16B_0^2\right)$—see Eq. (2.42). The oscillatory field must be considered to consist of two counter-rotating fields where only one is at resonance while the other contribute to a pseudo-guide field-shifting of the resonance frequency. The Bloch–Siegert shift can be calculated using the classical description of the rf field and the Bloch equation (Eq. 2.27), but the dressed atom approach allows a more global analysis to be made. In this case the rf field is treated as a quantum entity as well. The results show also the correlation of all the resonance phenomena with the properties of the energy diagram, which is created by a related Hamiltonian (Cohen-Tannoudji et al. 1992, Pendlebury et al. 2004).

A Japanese group performed such a double-coil resonance experiment with a spin-echo arrangement (Section 2.4), putting two resonance coils behind each other and operating them with a frequency difference of 20 μHz. This corresponds to an energy difference of 8.27×10^{-20}eV and a sensitivity near to 10^{-22}eV (Ebisawa et al. 1998a, Yamazaki et al. 1998). The arrangement and typical results are shown in Fig. 5.11. These high sensitivities could be used for the search of new

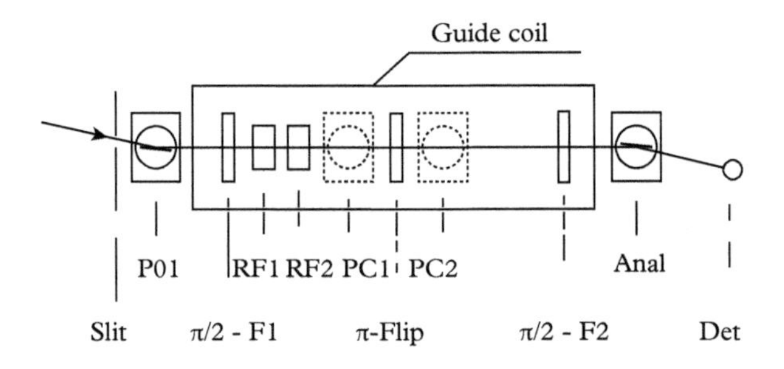

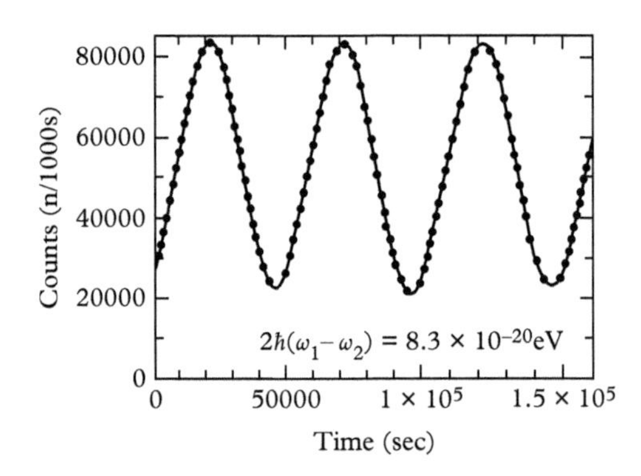

Figure 5.11 *Magnetic Josephson experiment with a spin-echo arrangement and two flipper coils operated at slightly different frequencies. Reprinted with permission from Ebisawa et al. 1998a, copyright by Physical Society of Japan.*

effects in the neutron–gravity interaction, an electric dipole moment of the neutron, and perhaps other exotic phenomena.

The quantum beat effect can also be interpreted as a magnetic analog of the Josephson effect known for superconducting tunnel junctions (Josephson 1974). In that case, the common boson phase of the Cooper pairs in both superconductors vary when a voltage V is applied at the tunnel junction as

$$\frac{\partial}{\partial t}(\phi_2 - \phi_1) = \frac{1}{\hbar}(E_1 - E_2) = \frac{2e}{\hbar}V, \tag{5.21}$$

which yields

$$\phi = \phi_2 - \phi_1 = \frac{2\text{ eV}}{\hbar}t. \tag{5.22}$$

This produces a rapidly oscillating tunnel current

$$I_s = I_{\max}\sin\phi. \tag{5.23}$$

In the neutron case (Eq. 5.19) we can formulate the different energy transfer in both beams as a time-dependent phase shift (Rauch 1991)

$$\Delta = \Delta_2 - \Delta_1 = (\omega_{r2} - \omega_{r1})\, t = \frac{2\mu\Delta B_0}{\hbar} t, \tag{5.24}$$

which obeys an equation analogous to Eq. (5.21). That is,

$$\frac{\partial}{\partial t}(\Delta_2 - \Delta_1) = \omega_{r2} - \omega_{r1} = \frac{1}{\hbar}\, 2\mu\Delta B_0, \tag{5.25}$$

which yields again Eq. (5.20) but now in the form

$$I \propto (1 + \cos\Delta(t)), \tag{5.26}$$

where ΔB_0 denotes the difference of the strength of the guide fields in both flippers. Whereas the ordinary Josephson effect is driven by an electrical force, the neutron analog is driven by a magnetic force. Both are phase-sensitive phenomena and are therefore extremely sensitive to any additional driving forces. The slowly varying phase difference occurring in that experiment can also be interpreted as a time-varying geometrical phase (Wagh and Rakhecha 1990; see Section 6.3).

The more complete treatment of neutrons in a static field, B_0, and in an oscillating field, B_1, must start with the second quantized Jaynes–Cummings Hamiltonian (Cohen-Tannoudji and Haroche 1969, Cohen-Tannoudji et al. 1992)

$$\mathcal{H} = \mu B_0 \sigma_z + \hbar\omega a^+ a^- + \hbar \frac{B_1}{\sqrt{\overline{N}}} (a^+ \sigma_- + \sigma_+ a^-), \tag{5.27}$$

where the first term denotes the Zeeman term, the second one adds the energy of N photons of the oscillating field, and the last term describes the coupling between the photons and the particle's magnetic moment. Here $\sigma_\pm = \sigma_x \pm i\sigma_y$, where σ_i are the Pauli matrices of the neutrons, a^+ and a^- are the creation and annihilation operators and $\overline{N} = a^+a^-$ denotes the mean number of photons in the field, which is related to the amplitude of the oscillating field by $\overline{N}\hbar\omega = B_1^2 V/2\mu_0$ (Cohen-Tannoudji and Haroche 1969, Schmidt et al. 1993). In the case discussed in Sections 5.3 and 5.4, $\overline{N}$ is on the order of $10^{22}\mathrm{cm}^{-3}$. The Hamiltonian (Eq. 5.27) has been used for the description of a two-level atom in a radiation field which is homomorphic to a spin-½ particle precessing in a magnetic field (Feynman et al. 1957). It correctly describes the appearance of 1-, 3-, 5-, etc. quanta in spin resonance transitions. The particle number–phase uncertainty relation in the symmetrical form reads as

$$(\Delta N)^2 \frac{(\Delta S)^2 + (\Delta C)^2}{\langle S\rangle^2 + \langle C\rangle^2} \geq 1/4, \tag{5.28}$$

where S and C are the Hermitian cosine and sine operators which can be expressed by $C = (a^+ + a^-)/2(N+1)^{1/2}$ and $S = (a^+ - a^-)/2i(N+1)^{1/2}$ and whose matrix elements couple coherent Glauber states (Carruthers and Nieto 1968). C and S permit the definition of unitary phase operators $U_c = \exp(i\phi_c)$ and $U_s = \exp(i\phi_s)$ giving the uncertainty of a phase measurement in the form

$$(\Delta\phi)^2 = (\Delta S)^2 + (\Delta C)^2. \tag{5.29}$$

For $N \to \infty$, the uncertainty relation can be simplified to $\Delta N \Delta \phi \geq \frac{1}{2}$ where ϕ is the phase of the oscillating field, and for a coherent state system, where the number states are Poisson distributed ($(\Delta N)^2 = N$) one gets $\Delta \phi = (2N)^{-1/2}$—see Section 4.4.3. Although the existence of a phase operator in quantum mechanics is still under discussion, the relevance of the related uncertainty relations has been proven in many optical investigations. The definition of the phase operator with phase-dependent states seems to be a useful concept (e.g., Pegg and Barnett 1988). The coupling term of the Hamiltonian (Eq. 5.27) causes the appearance of level-crossing and anticrossing in the related energy-level diagram (Muskat et al. 1987). In analogy to the well-known "dressed-atom" phenomena (Cohen-Tannoudji and Haroche 1969) these effects can be classified as "dressed-neutron" phenomena. Related experiments with polarized slow neutrons have demonstrated these interesting effects (Muskat et al. 1987, Dubbers 1989). The Jaynes–Cummings (1963) model also describes the dephasing and the later revival of the coherent spin precession which means that the spin precession in one component vanishes and appears in another component and sweeps back again. Related experiments have verified this behavior (Schmidt et al. 1993). Spontaneous polarization effects for an unpolarized beam are also predicted but seem to be rather small for feasible parameters. This kind of "hidden coherence" is another form of interference in phase space as discussed in Section 5.5 (Gea-Baracloche 1990).

The coupling of a polarized neutron beam to two resonator cavities in series where an initial spin flip in the first cavity can be undone in the second one is given by Scully et al. (1989). It is shown again that spin coherence is preserved if the cavities are prepared in coherent state modes where no measurable correlation is imposed on the resonator system due to the neutron–resonator interaction. The mutual exclusivity of observing interference and path information has also been verified for entangled photon pairs when the partner photon of the interfering one is intended to deliver additional information (Herzog et al. 1995). By Ramsey interferometry with Rydberg atoms it has been shown that superposition states between atoms and cavity states can be produced (Brune et al. 1996). But it has also been shown that a classical behavior is expected even in the case of small quantum number excitations and a dissipative cavity subsystem (Kim et al. 1999).

Rauch and Vigier (1990) proposed to repeat the double-coil experiment under the condition of strong magnetic fields causing energy shifts larger than the energy widths of the beam ($\Delta E_{\text{hf}} = 2\mu B_0 = \hbar\omega_r > \delta$E). It can be shown that in this case the neutrons become partly labeled because they are shifted out of the original phase space. It can easily be shown that the phase Δ now varies significantly during the coherence time of the beam ($\Delta t_c = \Delta x_c/v = \hbar/2\delta E$), causing a kind of phase-chopping effect, which might be interesting by itself, but it reduces the contrast of the interference pattern considerably. Thus, the plane wave formalism of Section 5.3 and of this section must be replaced by a wave packet formalism if Δt_c approaches $\Delta t_{\text{hf}} = h/2\mu B_0$. It can be shown that there is a smooth transition between the classical situation where the phase shift (total energy change) can be described by the different Zeeman energy changes the particle experiences at the entrance and exit of the time-dependent field

$$E = E_0 + \boldsymbol{\mu}[B(t_{\text{ex}}) - B(t_{\text{en}})] \tag{5.30}$$

and the quantum behavior where the particle ends up in a superposition state

$$E = \sum a_n \left| E_0 + n\hbar\omega \right>, \tag{5.31}$$

where n are integers, ω is the frequency of the oscillating field, and a_n are the transition amplitudes (see Section 5.5). Which of the pictures must be taken depends crucially upon the coherence time of the wave packet (Summhammer 1996). This effect is somehow analogous to the (spatial) lattice diffraction effect or the diffraction-in-time effect from a very fast chopper, as discussed in Section 4.3.2. For the labeled neutrons ($n \neq 0$) only a beam trajectory can be assigned. These neutrons do not contribute to the interference pattern, leaving open the Einstein–Bohr controversy concerning the completeness of quantum mechanics (Unnerstall 1990).

Various schemes have been discussed for simultaneous interference and beam path detection, such as Einstein's recoiling slit (Jammer 1974), Feynman's light microscope (Feynman et al. 1965), and Scully's quantum eraser system (Scully et al. 1991). For the understanding of the phenomenon of interference or which path detection, it is irrelevant whether a momentum (or energy) disturbance occurs locally or non-locally. For a double-slit situation this has been visualized by means of Wigner functions (Wiseman et al. 1997). Analogies to the non-locality phenomena in Aharonov–Bohm situations exist as well (Section 6.2). Depending on what we choose to measure, the nature of the related disturbance can sometimes be interpreted as a momentum change or a localization of the particle. In the former case, the interference pattern persists but is shifted, whereas in the latter, the interference pattern becomes featureless due to the coupling with the apparatus. This coupling does not imply that the alteration of the system is uncontrolled unless information about the detector state is discarded. Thus, an entangled state of the system and the path detector exists and the probability density of the detector states influences the interference pattern causing gradual or complete loss of coherence (Tan and Walls 1993). In the case of a two-level detector model, each subclass of particles which left the detector in the state $|0>$ or $|1>$ form an interference pattern with unit visibility but with the minima and maxima interchanged. When all the particles are considered together, however, no interference is observed. This brings us back to post-selection experiments as they were discussed in Section 4.5.

5.5 Multiphoton Exchange Experiments

In the previous section, the strength and the frequency of the time-dependent magnetic field were tuned to resonance which caused a spin-flip probability near to unity and a photon exchange probability near to unity, as well. The feature of simultaneous spin flip and energy change has first been demonstrated by Alefeld et al. (1981b) at a high-resolution backscattering instrument and in more detail by Weinfurter et al. (1988) with a perfect crystal diffraction camera (Fig. 5.8). In these cases the energy resolution of the apparatus and the energy change has been measured directly. Here we will focus on multiphoton exchanges where we will obtain information about the energy exchanges from the related phase changes.

When the resonance conditions are not fulfilled multiphoton exchange become possible with different probability amplitudes. Here a close analogy to the dressed neutron phenomenon exists (Muskat et al. 1987, Dubbers 1989, Dubbers and Ströckmann 2013). Thus for a neutron in the $|+z>$-state and with an energy $\hbar\omega_0$ we expect a wave function beyond the time-dependent field as

$$|\psi_f\rangle = \sum_{j=-\infty}^{+\infty} [a_j|+z\rangle + b_j|-z\rangle]|\omega_j\rangle, \tag{5.32}$$

where $\omega_j = \omega_0 + j\omega$. In our case we assume a static and an oscillating magnetic field existing in a certain region of space where they are taken to be spatially homogeneous. Summhammer (1993) performed related calculations for various field configurations. In the so-called parallel field configuration

$$B(t) = \begin{pmatrix} 0 \\ 0 \\ B_0 + B_1 \cos\omega t \end{pmatrix}. \tag{5.33}$$

The solution can be taken from the solution for an oscillating step potential (Haarvig and Reifenberger 1982), because the two spin components are decoupled and no spin flip can occur ($b_j \equiv 0$). The interaction of neutrons with oscillating surfaces of material slabs is another example of a time-dependent interaction with a step potential causing the exchange of quanta between the neutron and the oscillating system (Felber et al. 1966). The situation discussed here is also rather similar to that used in the measurement of the scalar Aharonov–Bohm effect (Section 6.2; Allman et al. 1992, Badurek et al. 1993). The transition amplitudes can be written in terms of ordinary Bessel functions of order j:

$$a_j = \mathcal{J}_j(\alpha B_1) \tag{5.34}$$

with

$$\alpha = \frac{\mu}{\hbar\omega}\sin(\omega\tau/2),$$

and τ is the time-of-flight through the field region. The absorption and emission probabilities of a given number of photons are the same, independent of whether an even or an odd number of photons are exchanged. The transition amplitudes do not depend on the static component of the magnetic field, which contribute only to the phases, in analogy to the 4π-symmetry experiments (Section 5.1).

In the so-called perpendicular field configuration the magnetic field is

$$B(t) = \begin{pmatrix} B_0 \\ 0 \\ B_1 \cos\omega t \end{pmatrix}, \tag{5.35}$$

which causes a coupling of position and time coordinates in the neutron wave function. When the time-of-flight through the magnetic field can be assumed to be the same for both spin components, a separation of spatial and time coordinates becomes possible (Shirley 1965). For thermal and cold neutrons and magnetic fields which are not too strong or turned on too long, this is reasonably well satisfied. The exchange of an odd number of quanta will result in a spin flip, whereas the exchange of an even number of quanta leaves the neutron spin state unaffected. The amplitudes for emission and absorption for odd numbers of photons are in general very different from each other (Summhammer 1993). This changes the occupation number of different levels. The resulting transition amplitudes depend on the strength of the static and oscillating field and on the time-of-flight through the field. Any depolarization of the neutron beam when entering and leaving the perpendicular fields must be avoided.

Experiments related to the above discussion have been performed at the MURR using a skew-symmetric perfect crystal interferometer and an oscillating field coil in one of the coherent beams (Fig. 5.12; Summhammer et al. 1995). A static spin flipper in the second beam was used to measure the spin-flip amplitudes when the perpendicular field configuration was investigated. The intensity at detector C3 is given by the overlap of the undisturbed (or spin-flip) wave function of beam path I and the wave function exposed to the oscillating field (Eq. 5.33),

$$I \propto \left||+z\rangle + e^{i\chi}|\psi_f\rangle\right|^2 = 1 + \left|\sum_{j=-\infty}^{+\infty}(a_j|+z\rangle + b_j|-z\rangle)\, e^{-ij\omega t}\right|^2 + 2\sum_{j=-\infty}^{+\infty}|a_j|\cos(\phi_j + \chi - j\omega t), \tag{5.36}$$

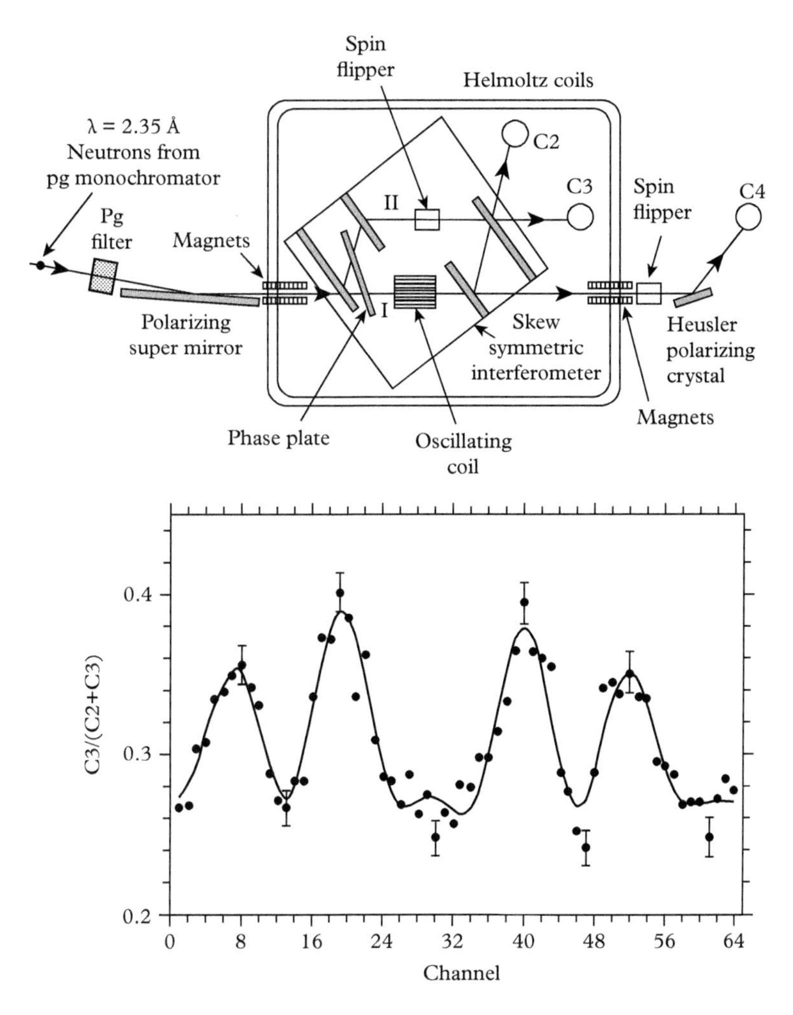

Figure 5.12 *Experimental arrangement and typical results obtained when an oscillating magnetic field in one beam path causes multiphoton exchange processes. Reprinted with permission from Summhammer et al. 1995, copyright 1995 by the American Physical Society.*

where χ is the phase shift induced by the phase shifter plates, and ϕ_j are the phases of a_j (without static spin flip) and b_j (with static spin flip): The counting rate was stroboscopically measured with the phase of the oscillating field, whose frequency was chosen as 7534 Hz, corresponding to a photon energy of 3.24×10^{-11}eV. A typical data set is shown in Fig. 5.12. In Fig. 5.13 the photon exchange amplitudes obtained with the parallel field configuration are shown as a function of the oscillating field strength. The full lines are a least squares fit of the theoretically expected Bessel functions. The general consistency between observed and predicted values confirms that in this configuration an even as well as an odd number of photons can be absorbed or emitted

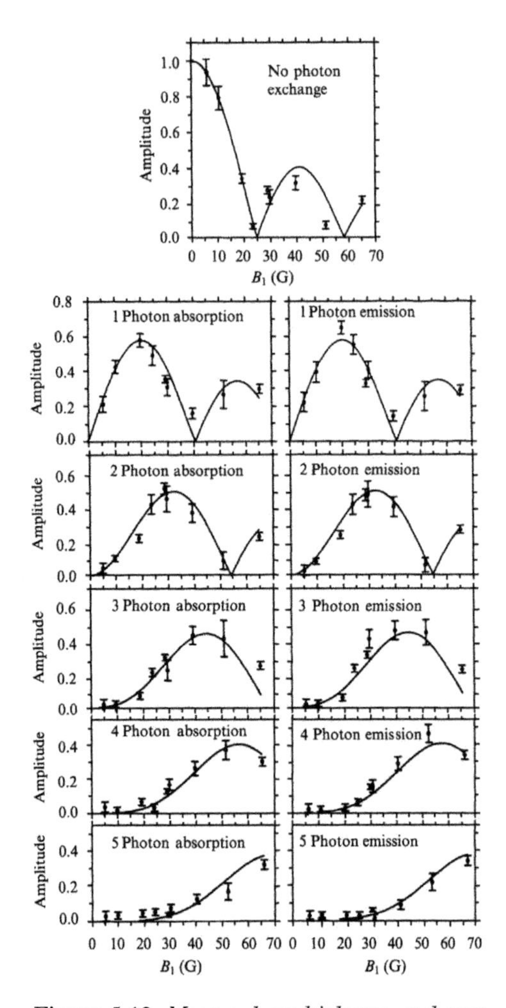

Figure 5.13 *Measured multiphoton exchange amplitudes in the parallel field configuration of an oscillating field. Reprinted with permission from Summhammer et al. 1995, copyright 1995 by the American Physical Society.*

by the neutron, despite the fact that no neutron spin flip occurs. In the case of an odd photon number exchange, a part of the whole angular momentum must be absorbed by the magnetic field. Much more experiments with off-resonance magnetic fields have been done in connection with dephasing (decoherence) experiments described in Section 4.6.2 (Sulyok et al. 2012).

Measurement with the perpendicular field configuration described by Eq. (5.35) has been performed as well (Summhammer et al. 1995). It has been confirmed that in the case of an even number of exchanged photons (no spin flip) the amplitudes of absorption and emission are equal, whereas for an odd number of photon exchange (spin-flip processes) the amplitudes for emission and absorption become different.

The experiments described in this section can only approximately be explained in a quasi-static manner without energy exchange by using a time-dependent phase acquired by the two spin components inside the oscillating field (Eqs. 4.152–4.155). This simplified description relies on the rotation of the polarization vector in the oscillating field and does not account for the whole physical situation as discussed by Zhang et al. (1994) and Summhammer (1996). A direct measurement of the energy change would require oscillation frequencies of the field higher than 100 kHz. These rather large changes of the kinetic energy come into the scope of the sensitivity of perfect crystal small-angle scattering cameras. On the other hand, this interferometric method represents an alternative method of measuring small changes of energy without the constraint that the energy change must be larger than the energy resolution of a spectrometer. This provides the basis for novel Fourier methods in condensed matter physics, as discussed in Section 9.3 (Rauch 1995b). When oscillating nuclear phase shifters are used, similar multiphoton exchange processes are expected (Littrell et al. 1996). In this case the phase shifts can be calculated on the basis of this multiphoton exchange process and it can be related to the neutron interaction in non-inertial reference frames as discussed in Sections 8.3 and 8.5. How multiphoton exchange contributes also to dephasing and decoherence effects is discussed in Section 4.6.2, and how this may be used for beam cooling is described in Section 10.16.

6

Topological and Geometric Phases

Topological and geometrical effects appear in the solution of the Schrödinger equation due to special geometrical forms of the interaction, or due to interactions which do not result from conservative forces acting on the system (see Section 1.2). A conscientious reader may point out that all these effects could be entirely explained by only using the Schrödinger equation. However, the geometric phase formalism is conceptually useful, because it often plays a role in understanding fundamental quantum phenomena. It shows that additional phase factors exist which are easily overlooked in making the interpretation of measured effects more mysterious than necessary. Thus, our understanding of quantum mechanics has been deepened since the existence of geometric phases was explicitly revealed. This occurred due to the seminal works of Pancharatnan (1957) and Berry (1984). Subsequent papers by Anandan (1992), Bhandari (1997), Shapere and Wilczek (1989), and Manoukin (2006) are important in the history of this subject. It also shows that a wave function often carries much more information than is usually extracted in a standard experiment. A typical example is the assumption of adiabaticity where the magnetic moment of the neutron produces a dynamical phase shift proportional to the path integral $\int B ds$ (inside a slowly varying magnetic field), but when the direction of the field varies slowly an additional, i.e., a geometric phase appears as well. The back reaction of a quantum measurement can also be described as a kind of an induced topological phase (Aharonov et al. 1998). A review summarizing the Aharonov–Casher effect experiment in the next section and the scalar Aharonov–Bohm effect experiment discussed in Section 6.2 is given by Werner and Klein (2010).

6.1 Aharonov–Casher Topological Phase

Aharonov and Bohm (1959) identified two important effects in their famous paper: the so-called magnetic (or vector) AB effect and the scalar interaction effect of an electron interacting with the vector potential and with a purely time-dependent electric potential. Both effects have counterparts for neutrons carrying a magnetic dipole moment moving around an electric line charge or when they are exposed to a pure time-dependent magnetic field (Fig. 6.1). In both cases the canonical momentum $\boldsymbol{p}$ changes but the kinematical momentum $m\boldsymbol{v}$ does not. Therefore, these effects also carry topological features. Both effects have been observed first by neutron interferometry. The magnetic Aharonov–Bohm effect for electrons found many experimental realizations using various solenoids and various magnetic shielding including superconductors (Chambers 1960, Boersch et al. 1961, Möllenstedt and Bayh 1962, Tonomura et al. 1986, Hasselbach 2010). Three of the Aharonov–Bohm effects mentioned earlier have been verified. The scalar Aharonov–Bohm effect for charged particles (electrons) has not found experimental verification so far.

Neutron Interferometry. Second Edition. Helmut Rauch and Samuel A. Werner.

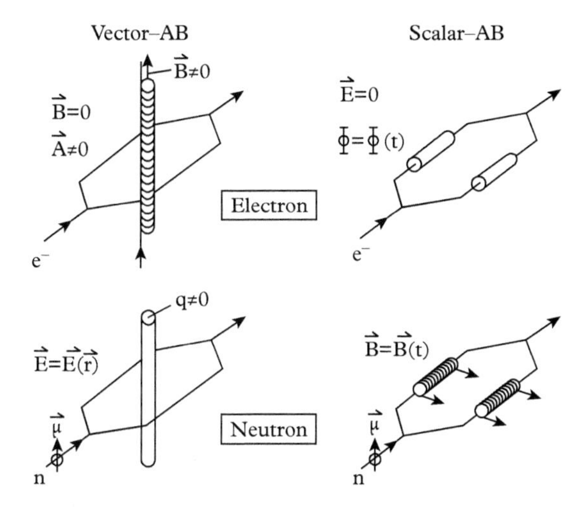

Figure 6.1 *Schematic view of the various electron and neutron Aharonov–Bohm situations. Reprinted with permission from Badurek et al. 1993, copyright 1993 by the American Physical Society.*

The difficulty is the rather high velocity of the electrons inside an electron interferometer which requires a very fast switching of the electrical potential in one of the Faraday cages, which would imply temporal fields and other disturbances. A steady state version of the electric Aharonov–Bohm effect using two oppositely charged metallic plates has been reported by Matteucci and Pozzi (1985), but it was challenged by Boyer (1987b). A follow-up experiment by Hilbert et al. (2011) showed indeed a time delay of the packet passing through such a double-wire arrangement and therefore disproving a forceless interaction for this geometry also. A verification of the electric Aharonov–Bohm effect has also been achieved by means of a Ramsey–Bordé atom interferometer where the time-dependent interaction has been generated by a pulsed laser field creating an induced atomic dipole (Mueller et al. 1995).

The interaction of the neutron magnetic moment with an electric field is given by the Schwinger interaction as formulated in Chapters 1 and 3 (Eqs. 1.21 and 3.42)

$$V_s = \frac{\hbar}{mc} \mu \boldsymbol{\sigma} \cdot (\boldsymbol{E} \times \boldsymbol{k}), \tag{6.1}$$

which permits the definition of an effective magnetic field since the neutron moves relative to the electric field:

$$\boldsymbol{B}' = \frac{\hbar}{mc} (\boldsymbol{E} \times \boldsymbol{k}). \tag{6.2}$$

Thus $\boldsymbol{B}'$ is the residue of the Lorentz transformation of $\boldsymbol{E}$ to the rest frame of the neutron. The reader will recognize the Schwinger interaction above as nearly identical to the spin–orbit interaction in atomic physics. For atoms there is a factors of ½ which comes from Thomas precession in the accelerated (circular) frame of the electron in an atom. From the analogy of the movement of a charged particle (electron) around a magnetic flux tube and the movement of a magnetic dipole (neutron) around a line of electric charge, Aharonov and Casher (1984) introduced a gauge-invariant Lagrangian which yields for the two polarization states an additional phase shift. It can be calculated from a proper time integral or the path integral of the canonical momentum, namely (Eq. 1.38)

$$\Delta_{AC} = \oint \frac{\boldsymbol{\mu} \cdot \boldsymbol{B}'}{\hbar} \mathrm{d}t = \oint \frac{\boldsymbol{\mu} \cdot \boldsymbol{B}'}{\hbar v} \mathrm{d}s = \pm \frac{2\mu}{\hbar c} E\ell. \tag{6.3}$$

The magnetic field $\boldsymbol{B}'$ appears in the frame of the moving neutron due to the existence of the $\boldsymbol{E}$ field in the laboratory frame over a distance ℓ. In this case, the canonical momentum is

$$\boldsymbol{p} = m\boldsymbol{v} + \frac{\boldsymbol{\mu}}{c} \times \boldsymbol{E}, \tag{6.4}$$

which also yields the same formula for the phase shift of Eq. (6.3). The Aharonov–Casher phase shift has also been obtained when a mean potential due to spin–orbit coupling is assumed (Rauch 1979, Anandan 1982).

The important experiments related to these topological effects have been carried out at the University of Missouri Research Reactor with an electrical field applied to the coherent beams of a perfect crystal interferometer. The field strength was 30 kV/mm and the effective length of this field was 2.53 cm, which yields for the arrangement shown in Fig. 6.2 an expected phase a shift of only 1.5 mrad (Kaiser et al. 1988, 1991; Cimmino et al. 1989). It was not necessary to use polarized neutrons if an additional spin-independent phase shift is judiciously introduced and adjusted (see Eq. 5.8). The authors used the gravitationally induced COW phase shift (Section 8.1) by tilting the interferometer slightly about the incident beam. In order to achieve the highest sensitivity an additional magnetic phase shifter was applied and adjusted to the maximum positive or the maximum negative slope of the interference pattern. In this setting, the relative intensity variations

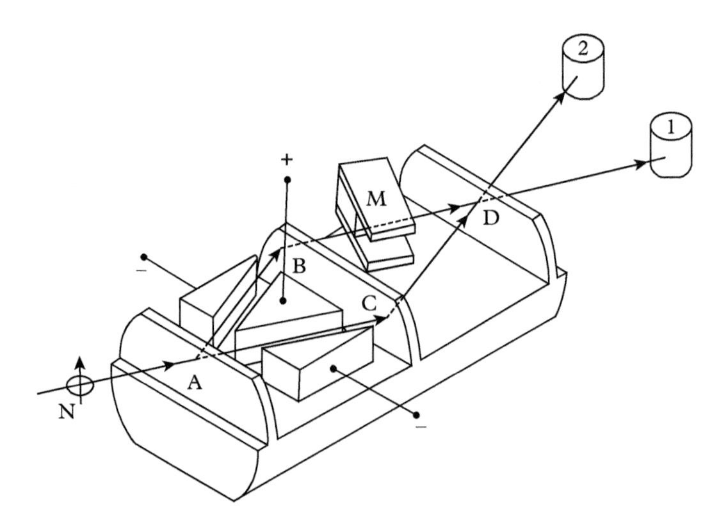

Figure 6.2 *Schematic diagram of the experimental arrangement for the measurement of the Aharonov–Casher effect (Klein and Werner 1991)*

in the counters behind the interferometer become linearly proportional to the Aharonov–Casher phase shift as the electric field is switched alternatively into the positive and negative directions. One finds that

$$\frac{I^+ - I^-}{I^+ + I^-} \cong V|\Delta\phi_{AC}|, \tag{6.5}$$

where + and – denote the different directions of the electrical field which was changed periodically and V denotes the visibility of the interference fringes (Eqs. 4.46 and 4.84). In a long-term experiment extended over 2 years, Cimmino et al. (1989) obtained a value $\Delta\phi_{AC} = 2.19(52)$mrad, which for the first time verified this interesting effect. Sangster et al. (1993, 1995) observed the Aharonov–Casher phase without a spatial beam separation using a Ramsey–Bordé arrangement for molecular beams (Bordé 1989). Strictly speaking the Sangster experiment did not verify the essential topological aspect of the Aharonov–Casher effect since their beam did not encircle the charge. See Ramsey (1993) for a discussion of this point.

The Aharonov–Casher effect can also be interpreted as a kind of a Schwinger spin–orbit term (Eq. 1.21), where in the semi-classical picture the phase shift is caused by a force acting on the neutron (Klein 1986, Boyer 1987a). The discussion, whether there is a force acting on the neutron in their experiments, is still continuing (Boyer 1987a, Aharonov et al. 1988, Anandan 1989, Goldhaber 1989, Casella and Werner 1992, Anandan and Hagen 1994, Wagh and Rakhecha 1997, Peshkin 1999, Cappelletti 2012). In the frame of the successful current loop model of the neutrons' magnetic dipole moment, the retro-action of an external field on the internal mechanical momentum must be included to keep the model relativistically invariant (e.g., Mezei 1988). This appears to cause a force $\boldsymbol{F} = \boldsymbol{\mu} \cdot \nabla(\boldsymbol{E} \times \boldsymbol{v})/c$, which would act at the entrance and exit of the field. This is, in fact, not the case as we now demonstrate for neutrons moving around a line charge of density Λ extending along the z-axis, for which the electric field is

$$\boldsymbol{E} = \frac{2\Lambda}{r}\,\hat{r}, \tag{6.6}$$

which is a radial vector in the xy plane. The Hamiltonian for the moving neutron then consists of two terms:

$$\mathcal{H} = \frac{p^2}{2m} - \frac{1}{mc}\,\boldsymbol{\mu} \cdot (\boldsymbol{E}\, x\, \boldsymbol{p}). \tag{6.7}$$

The canonical momentum $\boldsymbol{p}$, as given by Eq. (6.4), is obtained from the first of Hamilton's equations

$$\dot{r} = \frac{\partial\mathcal{H}}{\partial\boldsymbol{p}} = \frac{\boldsymbol{p}}{m} - \frac{\boldsymbol{\mu} \times \boldsymbol{E}}{mc}. \tag{6.8}$$

For neutrons polarized along the z-axis ($\boldsymbol{\mu} = \mu\sigma\hat{z}$, $\sigma = \pm 1$), we see that $\boldsymbol{\mu} \times \boldsymbol{E}/c$ is a solenoidal vector field (taking the place of $e\boldsymbol{A}/c$ in the canonical momentum for an electron moving around a tube of magnetic flux). The AC phase shift for the geometry of Fig. 6.1 is given by the line integral

$$\Delta\Phi_{AC} = \frac{1}{\hbar c}\oint \boldsymbol{\mu} \times \boldsymbol{E} \cdot \mathrm{d}\boldsymbol{s}, \tag{6.9}$$

where the line element ds, in cylindrical coordinates (r, φ, z), is

$$ds = dr\,\hat{r} + r d\varphi\,\hat{\varphi}\,dz\,\hat{z}. \tag{6.10}$$

But

$$\boldsymbol{\mu} \times \boldsymbol{E} = \frac{2\sigma\mu\Lambda}{\nu}\,\hat{\varphi}. \tag{6.11}$$

Thus, carrying out the integral in Eq. (6.9) we have

$$\Delta\Phi_{\mathrm{AC}} = \sigma\,\frac{4\pi\,\mu\Lambda}{\hbar c}. \tag{6.12}$$

This derivation and result clearly display the topological nature of the AC effect in that the phase shift, $\Delta\Phi_{\mathrm{AC}}$, is independent of where the line change penetrates (threads) the loop of the neutron beams in the xy plane of the interferometer. The phase shift depends only upon the net charge Λ per unit z, thus allowing the use of an extended electrode (sum of many line charges on its surfaces) as in the experiment of Cimmino et al. (1989). Equation (6.12) is then seen to be identical to Eq. (6.3) if we use Gauss' law, from which we obtain

$$\Lambda = 2V\ell/4\pi D, \tag{6.13}$$

where the electric field $E = V/D$ along the beam paths is given by the electric potential difference V across the gaps of width D (see Fig. 6.2).

We now establish that there is no kinematical effect (to order $1/c$) of the electric field on the neutron's trajectories (classical paths), i.e., that the acceleration, $\boldsymbol{a}$, is zero. Differentiating Eq. (6.8) with respect to time, we obtain

$$m\boldsymbol{a} = \dot{\boldsymbol{p}} - \frac{1}{c}\frac{d}{dt}(\boldsymbol{\mu} \times \boldsymbol{E}). \tag{6.14}$$

But $\mathbf{E}$ is a static field here, so that the total derivative is

$$\frac{\mathrm{d}}{\mathrm{d}t} = \boldsymbol{v} \cdot \boldsymbol{\nabla}, \tag{6.15}$$

where $\mathbf{v} = \dot{\mathbf{r}}$ is the neutron's group velocity. The time derivation of the canonical momentum is given by the second of Hamilton's equations:

$$\dot{\boldsymbol{p}} = -\frac{\partial \mathcal{H}}{\partial \boldsymbol{r}} = \frac{1}{mc}\boldsymbol{\nabla}\,[(\boldsymbol{\mu} \times \boldsymbol{E}) \cdot \boldsymbol{p}]. \tag{6.16}$$

Therefore, using Eqs. (6.15) and (6.16) in Eq. (6.14) we obtain the equation of motion

$$m\boldsymbol{a} = \frac{1}{mc}\{\boldsymbol{\nabla}[(\boldsymbol{\mu} \times \boldsymbol{E}) \cdot \boldsymbol{p}] - m\boldsymbol{v} \cdot \boldsymbol{\nabla}\,(\boldsymbol{\mu} \times \boldsymbol{E})\}. \tag{6.17}$$

Using rules of elementary vector analysis, and keeping terms up to order $1/c$, we find that

$$m\boldsymbol{a} = -\frac{1}{c}\,(\boldsymbol{\mu} \cdot \boldsymbol{\nabla}).\,(\boldsymbol{v} \times \boldsymbol{E}). \tag{6.18}$$

This is the result first clearly derived by Aharonov, Pearle, and Vaidman (1988).

For neutrons polarized along the z-axis, we have

$$\boldsymbol{\mu} \cdot \nabla = \sigma \mu \frac{\partial}{\partial z}, \tag{6.19}$$

but for a line charge (or distribution of line charge, as in the experiment of Cimmino et al. (1989)), the electric field $\boldsymbol{E} = \boldsymbol{E}(x, y)$ is independent of z; thus Eq. (6.18) requires

$$m\boldsymbol{a} = 0, \tag{6.20}$$

as originally stated by Aharonov and Casher (1984). The fact that there is no force on the neutron upon entering or leaving the region of the electric field has caused a fair amount of skepticism. The paper by Cappelletti (2012) has addressed these concerns in a transparent and elegant manner but see also the book by Aharonov and Rohrlich (2005) for further discussions.

There remains a subtlety in applying this result to the experiment of Cimmino et al. (1989) since the experiment was done with unpolarized neutrons. It was realized during the course of the experiment that the axis of spin quantization (axis of polarization) must also be the z-axis. In a somewhat fortuitous way, this quantization axis was provided by the stray magnetic fields from the DC magnet in path II of the interferometer, used to set the spin-dependent phase, at the location of the electrode assembly. A discussion of these matters is given in the paper by Kaiser et al. (1991). The exact equivalence of the Aharonov–Bohm effect and the Aharonov–Casher effect for relativistic spin-½ particles has been shown by an analysis based on the Dirac equation (Hagen 1990). The different views may be linked together and may have their justification if the axis of quantization is included in the discussion. In any case, it remains an interesting debate because the force—if existing at all—is a non-conservative one due to its velocity dependence. Zeilinger et al. (1991) and Peshkin (1999) claimed the Aharonov–Casher effect is purely topological in nature. A comprehensive discussion of this effect has been given by Aharonov and Rohrlich (2005) and by Werner and Klein (2010). A crucial test would be whether there is a Zeeman energy shift within the effective magnetic field $\boldsymbol{B}'$ given in Eq. (6.2). According to experiments which are related to a search for an electric dipole moment of the neutron, it is very likely that the effective field $\boldsymbol{B}'$ acts as a real magnetic field causing an additional energy shift and additional Larmor rotation (Golub and Pendlebury 1972). The paper by Cappelletti (2012) addresses the question of the Zeeman splitting in the motional magnetic field $\boldsymbol{B}'$. He concludes that the field is indeed real in the frame of the neutron, and that the spin-up and spin-down states are split in energy. The energy for this splitting comes from the electromagnetic field as the neutron enters the region containing the $\boldsymbol{E}$ field and gives it back to the field upon leaving the field region. The analogous situation describing the motion of an electric dipole within a magnetic field has been elucidated by Spavieri (1999), but seems to be accessible for atom optics only. The Aharonov–Casher effect also must be considered in accurate electric dipole moment measurements of the neutron (Pendlebury et al. 2004). A new experiment by the Toulouse atom interferometer group has observed the AC effect using Li atoms in a separated atom interferometer (Gillot et al. 2014). This experiment is very similar geometrically to the neutron experiment.

6.2 Scalar Aharonov–Bohm Effects

Consider now the case in which a purely time-dependent magnetic field is applied when the neutron is inside a long solenoid, where there is no gradient of the field. The magnetic field pulse gives rise to a phase shift (Zeilinger 1984)

$$\Delta\phi_{\mathrm{SAB}} = \pm(\mu/\hbar)\int B(t)\,\mathrm{d}t, \tag{6.21}$$

which is called the scalar Aharonav–Bohm (SAB) effect. The experiment has been done with unpolarized neutrons, with a stationary bias coil and a pulsed coil in the coherent beams (Allman et al. 1992; Fig. 6.3). Neutrons were registered into a multichannel scaler and synchronized with the pulsed field. Thus, one could select those neutrons which were in the coil when the magnetic field pulse was turned on. The bias coil was necessary to impose a different dynamical phase shift ($\alpha/2 = \pm\mu B\ell/\hbar v$; Section 5.1) to both spin states which make them distinguishable for the data analysis. The difference counting rate for a positive and a negative magnetic pulse gives the SAB phase directly:

$$(I_+ - I_-) \propto \sin(\Delta\phi_{\mathrm{SAB}}). \tag{6.22}$$

The related results for both beams behind the interferometer are shown in Fig. 6.3. A special accuracy estimate based upon a maximum likelihood prescription determined the error bars in the data analysis (Opat 1991).

An essential feature of the Aharonov–Bohm effects is their non-dispersivity and, therefore, they show up only as an overall phase factor of the spatially non-shifted wave packet. They are only observable in interference experiments. Because of non-dispersivity the spatial coherence function is not affected (Section 3.2.2), and Aharonov–Bohm phase shifts can be much larger than a phase shift corresponding to the coherence length of the beam can occur. This non-dispersive feature has been demonstrated by Badurek et al. (1993) on the basis of a time-dependent spin-echo arrangement using a rather polychromatic beam. A pulsed magnetic field was applied when the neutrons were inside the precession field and high-order spin rotations, which are washed out in the case of a static precession field (due to its dispersive action), have been observed.

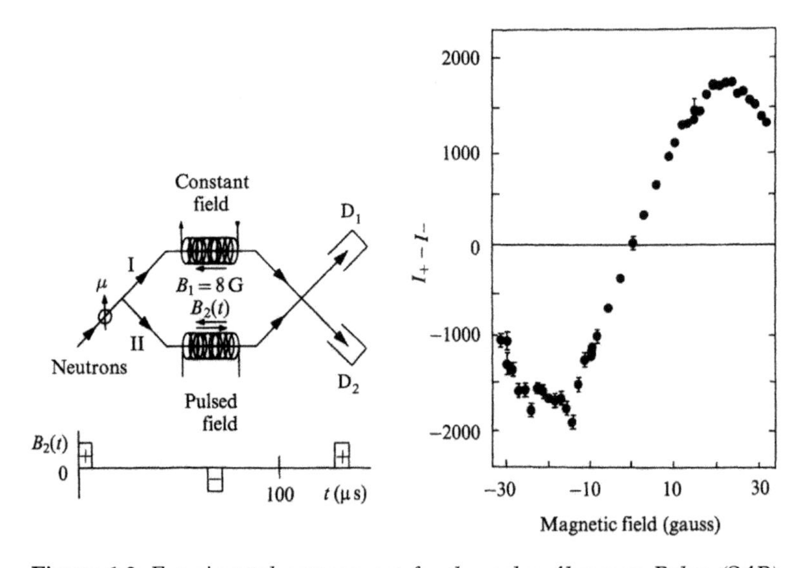

Figure 6.3 *Experimental arrangement for the scalar Aharonov–Bohm (SAB) experiment using unpolarized neutrons. The time dependence of the magnetic field is shown at the bottom. The difference counting rate is shown on the right when positive and negative magnetic fields are applied (Allman et al. 1992)*

In the discussion about the topological nature of this SAB effect it should be remembered that the relevant interaction is still given by the Hamiltonian

$$H_m = -\mu\boldsymbol{\sigma} \cdot \boldsymbol{B}(t). \tag{6.23}$$

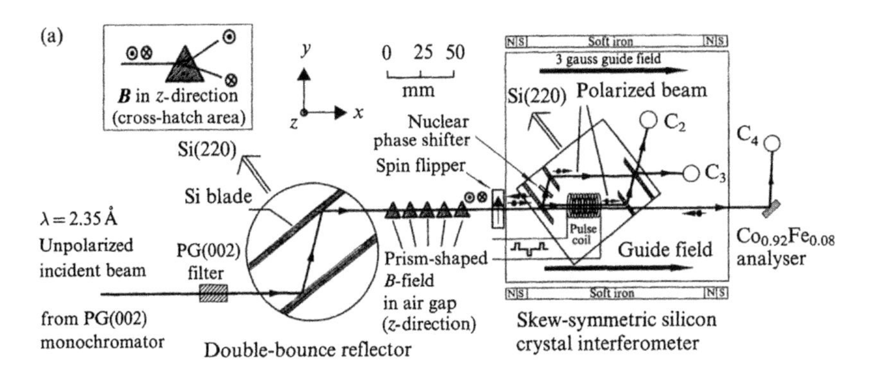

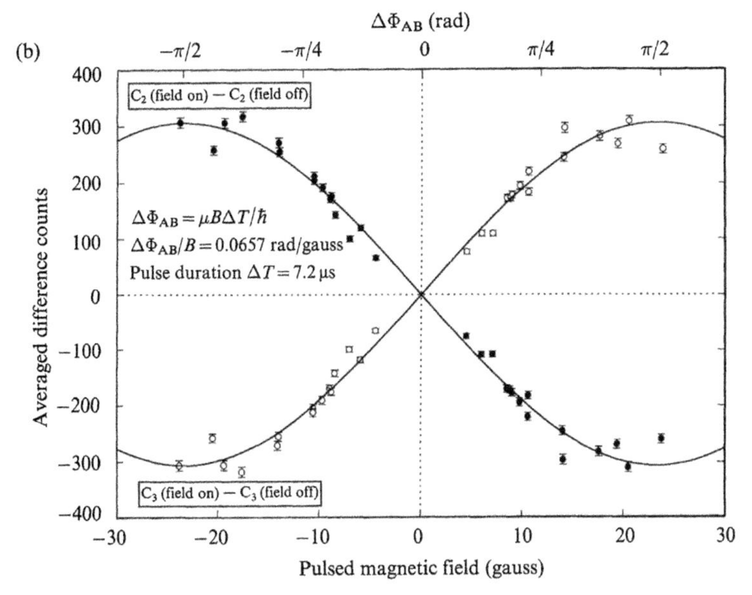

Figure 6.4 *(a) Experimental setup for observing the scalar AB effect with longitudinally polarized neutrons and characteristic results showing the different counts of positive and negative field pulses of a duration of 7.2 μs (Lee et al. 1998). The incident neutron beam is split into spin-up and spin-down states with triangular wedge-shaped fields in the air gaps of Sm-Co permanent magnets by the Stern–Gerlach effect. A π/2 DC spin flipper turns the neutron polarization into the longitudinal direction before entering the Si interferometer. Rotating the interferometer against the double-bounce Si crystal allows one to pick out one of the two incident spin states. (b) Photo of the crystal interferometer used*

Even for polarized beams there exist interactions along all three axes due to the spinor features of the Pauli matrices $\boldsymbol{\sigma}$ (Eq. 3.44). The mean values of the related torque angle can become zero for certain directions but the fluctuation does not vanish. This can be described by spin-autocorrelation functions which give a more local description of the SAB effect (Peshkin 1995).

In an experiment performed by Lee et al. (1998) longitudinally polarized neutrons were used to observe the SAB effect (Fig. 6.4). The pulsed magnetic field was used in a fashion similar to those in Fig. 6.3. In this case neither a classical force nor a classical torque on the neutrons can be expected. The interference can be explained only by the quantum-mechanical SAB effect, having no classical analog. But when the spinor properties of the magnetic interaction are accepted, a semi-classical explanation is possible as discussed above. A rather detailed description of this experiment and of its quantum-mechanical interpretation is given by Allman et al. (1999) and by Comay (2000).

An experiment aimed at directly demonstrating the essential non-dispersivity of the SAB effect has been carried out by van der Zouw et al. (2000) using the cold-neutron phase-grating interferometer at the ILL (Fig. 2.16). The magnitude of the Aharonov–Bohm phase shift remains observable far beyond the limit of dispersive phase shifts, which are determined by the coherence lengths of the interfering beams. A schematic diagram of the experiment is shown in Fig. 6.5a. Two antiparallel solenoids of rectangular cross-sections are used, which have a length-to-width ratio of 60:1, thus ensuring a field homogeneity of better than 1% over a large part of their length. Due to the large wavelength spread of the incident beam ($\lambda = 92 \pm 17$ Å), the neutrons that were in the field region during the time width of the pulsed magnetic could not be *post-selected* as in the experiments by Allman et al. (1992) and Lee et al. (1998), but had to be *pre-selected* using a chopper placed in front of the solenoids. The magnetic field pulse was applied 0.5ms after the chopper had opened for 2.6 ms, thus ensuring that the neutrons leaving the chopper had reached the homogeneous region of magnetic field within each solenoid. The field pulse was quite short (0.84 ms), thus ensuring that no neutron left the homogeneous region of pulsed magnetic field before the end of the pulse. Adequate timing allowed a duty cycle of 0.4 (open-to-closed ratio) of the chopper.
The AB phase shift should be independent of the wavelength of the interfering particles, and therefore no positional or wave packet spreading effects are observable. Thus, in the case of the pulsed operation the AB phase shifts remain observable far beyond the limit of dispersive phase shifts which are determined by the coherence length of the interfering beams (Fig. 6.5c); this is in contrast to a stationary field within the coils (Fig. 6.5b). Because this experiment (like the Allman et al. 1992 experiment) was carried out with unpolarized neutrons, the spin-independent phase had to be adjusted to $n\pi$ ($n = 0, 1, 2, \ldots$) to make the spin-dependent AB fringes maximally visible. This was achieved by translating the first grating. Two experiments were carried out: one with DC current applied to the solenoids (dispersive) and the second with pulsed current applied to the solenoids (non-dispersive). The results are shown in Fig. 6.5b. When a DC current is applied, the neutron accelerates when entering the field region and decelerates when leaving it (or vice versa). The interference pattern is clearly washed out after a few fringes (Fig. 6.5b), showing the dispersive character of the wave packet associated with the wavelength spectrum of Fig. 6.5c. In that experiment the magnetic field is pulsed on and off. The counting rate is plotted against the current pulse integral. The interference contrast remains constant over the entire range. Far more fringes are observed than in the static field case. These two experiments using the same interferometer beautifully demonstrate the robustness of the non-dispersive character of AB phase shifts.

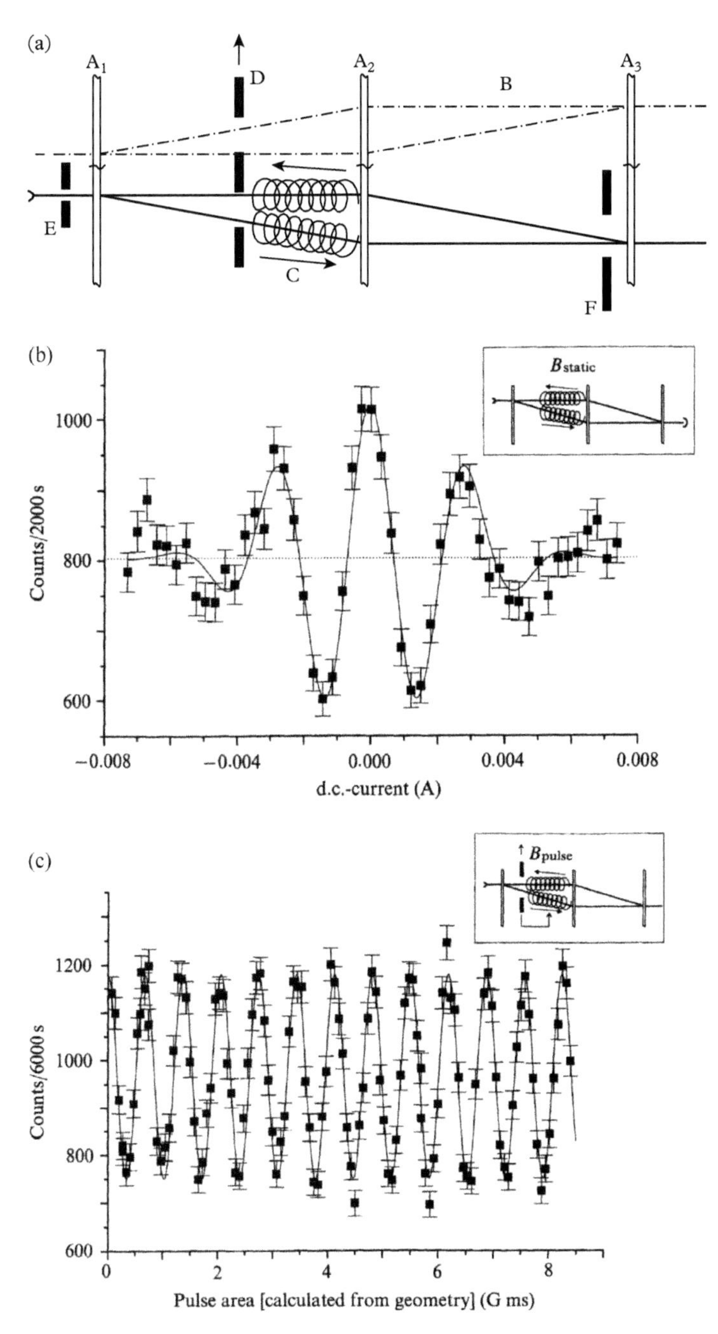

Figure 6.5 *(a) Schematic view of the grating interferometer. The chopper (D) produces pulses synchronized with the field pulses within the coils (C). (b) Measured interference pattern in a stationary situation as a function of the DC coil current, indicating the dispersivity of the phase shift. (c) The measured intensity as a function of the field pulse internal over time, demonstrating the non-dispersivity of the phase shift in the case of a pulsed field operation. Reprinted from van der Zouw et al. 2000, copyright 2000, with permission from Elsevier.*

6.3 Geometric Phases: Theoretical Background

The concepts of geometric and topological phases are related to the concept of anholonomy which cause global changes without local ones. Consider a vector transported along a geodesic of a sphere in such a way that the angle between the geodesis and the vector is held constant. Due to the curvature of the sphere the direction of the vector will be shifted when it returns to the initial position. The transport of the polarization of light within a fiber bundle can be seen as an example (Simon 1983, Berry 1988). The general connection of geometric phases to classical phenomena was given by Hanney (1989).

The quantum analog of a geometric phase was first formulated by Pancharatnan (1957), but it was implicitly included earlier in many textbooks (e.g., Schiff 1955). Nevertheless, only the seminal work of Berry (1984) and Simon (1983) made the general character clear and stimulated a lot of work in this direction (e.g., Anandan 1992, Bhandari 1997). The constraint of adiabaticity originally used by Berry (1984) was subsequently removed by Aharonov and Anandan (1987) to include any cyclic evaluation which is governed by the time-dependent Schrödinger equation (Eq. 1.2). In this case one can write

$$\left|\psi(t)\right\rangle = e^{i\Phi(t)}\left|n(\boldsymbol{R}(t))\right\rangle, \tag{6.24}$$

where $|n(\boldsymbol{R}(t))\rangle$ denotes the eigenstate of the instantaneous Hamiltonian $H(\boldsymbol{R}(t))\,|n(\boldsymbol{R}(t))\rangle = E_n(t)|n(\boldsymbol{R}(t))\rangle$ and $\Phi(t)$ is a generalized phase. Inserting this equation into the time-dependent Schrödinger equation (Eq. 1.2) and integrating over a closed path C in parameter space such that $|\psi(\boldsymbol{R}(T))\rangle = |\psi(\boldsymbol{R}(0))\rangle$ one gets a separation into a dynamical phase (δ) accumulating the energy (momentum) change along the loop and a geometric phase (γ) which is independent of energy and is gauge invariant (Berry 1984). For a constant and uniform magnetic field only a dynamical phase exists, as discussed in the previous sections. The total phase is in general made up of two parts

$$\begin{aligned}\Phi(T) = \arg < \psi(T)|\psi(0) > &= -\frac{1}{\hbar}\int_0^T < \psi(t)|\mathrm{H}|\psi(t) > \mathrm{d}t + i\int < \phi(t)|\frac{\mathrm{d}}{\mathrm{d}t}|\phi(t) > \mathrm{d}t \\ &= \int_0^T E_n(\boldsymbol{R}(t))\mathrm{d}t + i\oint \mathrm{d}\boldsymbol{R} < n(\bar{R})|\nabla_{\boldsymbol{R}}|n(\boldsymbol{R}) > = \delta + \gamma.\end{aligned} \tag{6.25}$$

In a constant magnetic field the Hamiltonian is $\mathcal{H}_{mag} = -\boldsymbol{\mu}_n\boldsymbol{B}(\boldsymbol{r})$ and the spin state can be written as

$$|\psi> = \cos\frac{\theta}{2}|\uparrow> + e^{i\phi}\sin\frac{\theta}{2}|\downarrow>. \tag{6.26}$$

Here θ denotes the polar angle and ϕ the azimuthal angle when plotted on a Bloch sphere. The states develop as time evolves ($t = r/v$)

$$|\psi(t)> = e^{\mathrm{iH}t/\hbar}|\psi(0)> = e^{-\mathrm{i}\mu\mathrm{B}t/\hbar}|\psi(0)>, \tag{6.27}$$

which verifies Eq. (5.2) and shows that the dynamical phase δ is ½ the precession angle α, that is $\delta = \alpha/2$ (Section 5.1). In the case of a slow change of the Hamiltonian (magnetic field) which corresponds to an adiabatic evolution the neutron spin which will be pinned to the direction of the magnetic field

$$\boldsymbol{B}(t) = B\boldsymbol{n}(t) = B\begin{pmatrix} \cos\phi(t)\sin\Theta(t) \\ \sin\phi(t)\sin\Theta(t) \\ \cos\Theta(t) \end{pmatrix}, \tag{6.28}$$

with the eigenvectors

$$\left|\psi_{\uparrow}(\Theta,\phi)\right\rangle = \begin{pmatrix} \cos\dfrac{\Theta(t)}{2} \\ e^{i\varphi(t)}\sin\dfrac{\Theta(t)}{2} \end{pmatrix}, \tag{6.29}$$

$$\left|\psi_{\downarrow}(\Theta,\phi)\right\rangle = \begin{pmatrix} \sin\dfrac{\Theta(t)}{2} \\ -e^{i\phi(t)}\cos\dfrac{\Theta(t)}{2} \end{pmatrix}. \tag{6.30}$$

This evolution yields

$$< \psi_{\uparrow}\left|\frac{\partial}{\partial\phi}\right|\psi_{\uparrow} > = \frac{i}{2}\big(1-\cos\Theta(t)\big). \tag{6.31}$$

When we move along a latitude circle ($\phi =$ constant) we obtain the well-known Berry phase, namely

$$\gamma^{\uparrow} = \mathrm{i}\int_{0}^{2\pi}\frac{\mathrm{i}}{2}\,(1-\cos\Theta(t))\,d\phi = -\pi\big(1-\cos\Theta\big) = -\Omega/2, \tag{6.32}$$

i.e., the geometric phase is just half of the solid angle Ω enclosed by the path. For example, if $\theta = \pi/2$, a walk along the equatorial line gives $\gamma^{\uparrow} = -\pi$ and the encompassed angle as seen from the degeneracy point $B = 0$ is half the total solid angle of a sphere, namely $\Omega = 2\pi$. Such a rotation produces a sign change of a fermionic wave function, which is equivalent to a sign change of a spinor undergoing a SU(2) rotation (Sections 5.1 and 6.2). A neutron interferometric experiment has been performed by Wagh et al. (1997) with Larmor precession coils directed with opposite sense in the two coherent beams. This experiment is discussed in the next section. Complete agreement between theoretical prediction and experiment has been achieved. In an experiment with ultra-cold neutrons a spin-echo method has been used to compensate the dynamical phase and to measure the Berry phase rather accurately (Filipp et al. 2009). They obtained a value of $\gamma^{\uparrow} = -0.51(1)\Omega$ in good agreement with Eq. (6.32).

Aharonov and Anandan (1987) generalized this approach to any cyclic evolution of a quantum system. In this case we can add additional phases to any point of the curve

$$|\Psi(T)> = e^{-if(t)}\,|\psi(t)> \tag{6.33}$$

with $|\phi(T)> = |\phi(0)>$ and $f(T) - f(0) = \Phi$. This shows that any excursion curve (C) in Hilbert space having the same projections onto P has the same geometric phase modulus 2π (Fig. 6.6). That is,

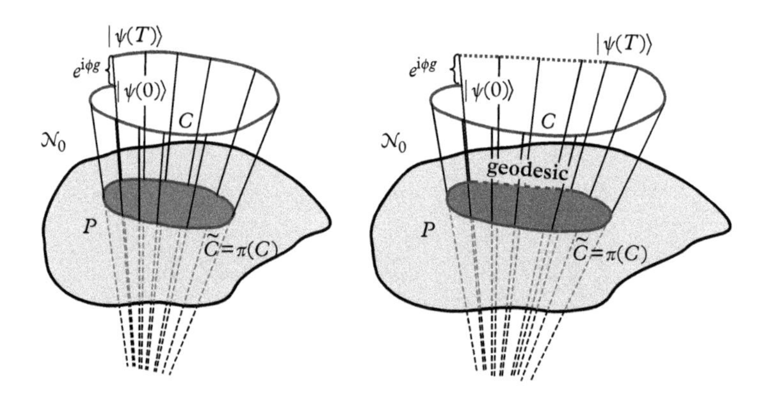

Figure 6.6 *Non-adiabatic (left) and a non-adiabatic plus non-cyclic (right) phase evolution*

$$\gamma = \mathrm{i}\int_0^T < \Psi(t)|\frac{\mathrm{d}}{\mathrm{d}t}|\psi(t) > \mathrm{d}t = 2\pi n + \mathrm{i}\int_0^T \langle\psi(t)\,|\frac{\mathrm{d}}{\mathrm{d}t}|\,\psi(t)\rangle\,\mathrm{d}t. \tag{6.34}$$

Thus, neutrons moving in a magnetic field pick also up a dynamical phase according to ($\delta = \mu BT\cos\theta/\hbar = (\alpha\cos\theta)/2$) due to its state component parallel to the field and a geometric phase ($\gamma = \pi(1-\cos\theta)$) according to the state rotation around (or with) the field direction.

This shows that the character of the phase shift varies from a dynamical phase when the polarization is parallel ($\theta = 0$) or antiparallel ($\theta = \pi$) to a geometric phase when the polarization is perpendicular ($\theta = \pi/2$) to the magnetic field. The total phase in all cases is $\phi = \alpha/2$, which agrees with Eq. (5.3) and shows that the 4π-symmetry of spinor wave functions can be measured with polarized and unpolarized neutrons as well (Section 5.1). Measurements with polarized neutrons allowed the extraction of the cyclic dynamical phase, although this was not mentioned explicitly in original paper (Badurek et al. 1988; Fig. 3.15).

The next important generalization of the geometric phase concept was given by Samuel and Bhandari (1988), who extended the treatment to non-cyclic evolutions, where these evolutions are geodesically closed to make a comparison of states possible (see also Shapere and Wilczek 1989, Aitchison and Wanelik 1992, Mukuda and Semion 1993, Bhandari 1997). This circumvented the closure "trick" and gave a solution for any closed or open path. The starting point is that the geometric phase is the total phase minus the accumulation of local phase changes. The local phase changes between $|\psi(t)>$ and $|\psi(t+\delta t)>$ are given by

$$\delta\phi \cong -\mathrm{i} < \psi(t)\frac{\mathrm{d}}{\mathrm{d}t}|\psi(t) > \delta t, \tag{6.35}$$

which follows from a Taylor series expansion of Eq. (6.25). Thus, one gets a rather general formula for the geometric phase

$$\gamma = \arg(<\psi(0)|\psi(t)> = +\mathrm{i}\int_0^t <\psi(t')\,|\frac{\mathrm{d}}{\mathrm{d}t'}|\psi(t')>\mathrm{d}t', \tag{6.36}$$

which reduces to Eq. (6.25) in the case of cyclic evolutions. In this connection, it can then be shown that the dynamical phase takes the general form

$$\delta = -\int_0^t <\psi(t')|\frac{\mathrm{d}}{\mathrm{d}t'}|\psi(t')>\mathrm{d}t' = -\frac{1}{\hbar}\int_0^t <\psi(t')|\mathcal{H}|\psi(t')>\mathrm{d}t'. \tag{6.37}$$

This equation shows that the dynamical phase is associated with an averaged action of the Hamiltonian. The dynamical phase approaches zero (when neutrons polarized perpendicular to a field are considered). Although there exist Zeeman energy shifts $+\mu B$ for the spin-up state and $-\mu B$ for the spin-down state, the net (average) effect becomes zero. Geometric phases in non-relativistic quantum theory are not Galilean invariant. But because in most of the cases discussed in this book the spin is involved, the results are immune against Galilean transformations (Sjöqvist et al. 1997). This is a result of the frame-independent coupling of the spin to the magnetic field.

Adiabatical or topological and geometric phases are predicted for fermions *and* bosons as well. Verifications for photons were accomplished by means of optical fibers (Tomita and Chiao 1986) and by means of various laser interferometers (Bhandari and Samuel 1988, Chiao et al. 1988, Bhandari 1997). Wagh and Rakhecha (1990) interpreted the double-coil resonance experiments discussed in Section 5.4 (Figs. 5.9 and 5.10) as a result of time-varying geometric phases. Hasegawa and Badurek (1999) demonstrated by spin polarimetry experiments how the non-commuting behavior of spinor rotations are related to the intrinsic phases of the wave functions when they are commuted. These phase shifts contain both geometric and dynamical contributions, yielding the familiar anti-commuting relation of Pauli spin matrices (Eq. 3.44), which we will discuss in the next section.

The geometric phase is a new signature of a quantum system when it evolves and then returns to its initial state. In this case it acquires a memory of its motion which has consequences in a wide range of physical systems and in the interpretation of quantum phenomena (Anandan 1992, Anandan et al. 1997).

6.4 Interferometric Measurement of the Berry Phase

It has been emphasized that the geometrical phase factor is predicted to occur if the magnetic field of constant strength varies adiabatically in its direction and the neutron remains in an eigenstate. The wave function of a neutron in a magnetic field will take the general form (Eqs. 6.27 and 6.33)

$$\psi \rightarrow e^{-\mathrm{i}\Omega/2}e^{-\mathrm{i}\boldsymbol{\sigma}\cdot\boldsymbol{\alpha}}\psi_0, \tag{6.38}$$

where Ω is the solid angle subtended by the closed curve as seen from the origin of the spin sphere. One gets immediately an intensity modulation given by

$$I \propto 1 + \cos\frac{\alpha + \Omega}{2}. \tag{6.39}$$

The sign of Ω depends in the same manner on the spin component along the magnetic field and, therefore, this equation is valid for polarized and unpolarized neutrons as well. In a related experiment the dynamical phase α (Eq. 5.3) and, therefore, $\int Bdt$ taken over both beam paths should be zero (i.e., $\alpha = 0$). If the magnetic field is arranged to rotate the spin adiabatically and it turns back to the z-direction with an apex angle θ of the related excursion curve, the predicted intensity variation should be independent of $|B|$ and of the magnetic moment of the neutron. It is

$$I(\theta) \propto 1 + \cos[\pi(1 \pm \cos\theta)], \tag{6.40}$$

where the $\pm$ signs depend upon the rotation direction of the helical field B_1 and $\cos\theta = B_z/(B_1^2 + B_z^2)^{1/2}$, where B_z is the field along the axis of a helix. The topological Berry phase appeared early in a theoretical treatment of the phase shift of a neutron in a helical field (Eder and Zeilinger 1976). The adiabaticity condition for neutrons requires a field variation frequency ($\nu_f \sim v/l$) smaller than the Lamor frequency ($\nu_L = 2\mu B/h$). For thermal neutrons this condition can be written as $Bl >> 8$ (B in millitesla and the length l of the helical coil in meters). This constraint causes difficulties for the realization of such an experiment with today's perfect crystal interferometers. The clear verification was first realized along the lines of a proposal made by Wagh and Rakhecha (1990). The experiment was carried out at beam port C at MURR. We discuss now the concept and results of this experiment (Allman et al. 1997, Wagh et al. 1997).

Consider first the idealized experiment with polarized neutrons shown in Fig. 6.7. The interferometer contains a pair of π spin-flip coils, one in each leg of the interferometer. The interferometer is situated in the xy plane, and the incident neutrons are polarized in the z-direction.

We envision the coils as having a rectangular cross-section, and producing a magnetic field $\boldsymbol{B}$ at an angle $+\Delta\beta/2$ with respect to the y-axis in the upper path II, and a magnetic field $\boldsymbol{B}$ at an angle $-\Delta\beta/2$ with respect to the y-axis in the lower path I, as shown. The strength of these fields is just enough to allow the incident neutron to precess from up (along $+z$-axis) to down (along the $-z$-axis); i.e., the coils act as spin flippers. Since the coils are exactly the same, the phase shift due to the spin flips will be equal in each of the two legs of the interferometer, and cancel each other in the phase difference, if it were not for the relative angle $\Delta\beta$ of the magnetic field in path II with respect to that in path I. Under this condition, the *dynamical phase* shifts in the two legs of the interferometer are equal and hence their difference is zero, but the *geometric phase* difference $\Delta\Phi$ is equal to $\Delta\beta$ (Eq. 6.32). This is the Berry phase, given by $-\frac{1}{2}$ the solid angle $\Delta\Omega$ subtended by the enclosed paths on the spin sphere, as also shown in Fig. 6.7.

The neutron spin vector $\boldsymbol{S}_{\text{II}}$ on path II precesses from the north (N) pole to the south (S) pole of the spin sphere along the path shown, while the spin vector $\boldsymbol{S}_{\text{I}}$ on path I precesses down along the other path. One-half the solid angle subtended at the origin of this spin sphere by these paths is the geometric phase, as mentioned earlier. That is

$$\Delta\Phi_{\text{geom}} = -\frac{1}{2}\Delta\Omega = -\Delta\beta. \tag{6.41}$$

It is only the difference in phases of the neutron wave function accumulated on beam path II relative to that on beam path I that is physically measurable.

But why is it that there is an additional phase shift given by Eq. (6.41) related to the non-collinear precession axes? Let's see. The interaction of a neutron having magnetic dipole moment $\boldsymbol{\mu}$ with a magnetic field $\boldsymbol{B}$ is given by

$$V = -\boldsymbol{\mu} \cdot \boldsymbol{B} = -\mu\boldsymbol{\sigma} \cdot \boldsymbol{B} = -\mu \begin{bmatrix} B_z & B_x - iB_y \\ B_x + iB_y & -B_z \end{bmatrix}, \tag{6.42}$$

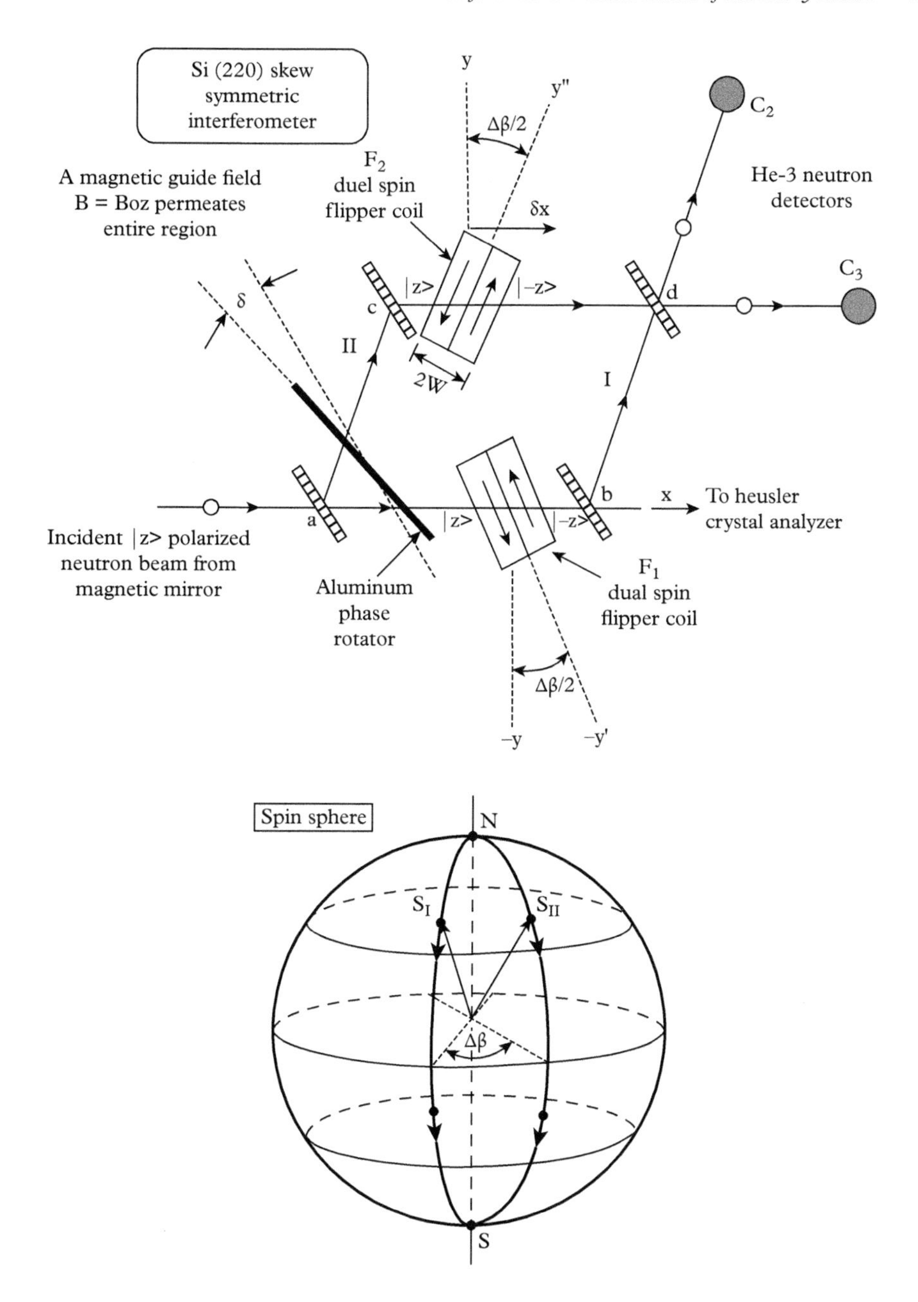

Figure 6.7 *Schematic of an idealized interferometer experiment designed to measure the geometric phase independent of the dynamical phase and diagram of the trajectories of the spin vector on the spin sphere on path II and path I (Allman et al. 1997)*

where we have used the Pauli spin matrices given in Eq. (3.44). In case of $B_z = 0$ one gets

$$\begin{aligned} B_x - \mathrm{i}B_y &= B\,[\sin(\Delta\beta/2) - \mathrm{i}\cos(\Delta\beta/2)] = -\mathrm{i}Be^{\mathrm{i}\Delta\beta/2} \\ B_x + \mathrm{i}B_y &= B\,[\sin(\Delta\beta/2) + \mathrm{i}\cos(\Delta\beta/2)] = \mathrm{i}Be^{-\mathrm{i}\Delta\beta/2}. \end{aligned} \tag{6.43}$$

Thus, the interaction matrix of a neutron with the spin-flipper coil on path II is

$$V = -\mu B \begin{bmatrix} 0 & -\mathrm{i}e^{\mathrm{i}\Delta\beta/2} \\ \mathrm{i}e^{-\mathrm{i}\Delta\beta/2} & 0 \end{bmatrix}. \tag{6.44}$$

So we see that the spin flipper in path II adds a phase of $-\Delta\beta/2$ to the spin-flip process. Likewise, the spin flipper in path I adds a phase $+\Delta\beta/2$ to its spin-flip process. Therefore, the interferometric phase difference is

$$\Delta\Phi = \Phi_{\mathrm{I}} - \Phi_{\mathrm{II}} = -\Delta\beta = -\Delta\Omega/2, \tag{6.45}$$

as stated earlier. Consequently, we understand now that the geometric phase shift comes from the off-diagonal matrix elements, which arise from the non-collinear precession fields in path II relative to path I. If B_z were not zero, then we would have an additional phase shift. This would be a dynamical phase shift. It comes from the diagonal elements in the interaction potential in Eq. (6.42). One can measure the geometric phase difference as a function of $\Delta\beta$ by carrying out a phase flag (phase rotator) scan for each $\Delta\beta$. The shift of these interferograms will be $\Delta\Phi = -\Delta\beta$. This geometric phase difference does not depend upon the Hamiltonian, or the rate of precession. Of course the reason that the spin vectors $\mathbf{S}_{\mathrm{I}}$ and $\mathbf{S}_{\mathrm{II}}$ rotate from the north pole to the south pole of the spin sphere is the torque $\boldsymbol{\mu} \times \boldsymbol{B}$ on the neutron's magnetic dipole moment $\boldsymbol{\mu}$ created by the magnetic fields $\boldsymbol{B}$ in each leg of the interferometer.

This is all quite interesting, of course, but an even more remarkable experimental idea is the one we will now discuss, involving the construction and use of *dual spin flippers*, and the direct demarcation of the geometric phase from the dynamical phase. The use of dual flippers is necessary because there is always a z-component of the magnetic field which defines the axis of quantization of the incident and transmitted beams in Fig. 6.7. This field must be added vectorially to the horizontal field within the spin flipper. Thus, to get a perfect spin flip, the horizontal flipper field should be tipped down to make the total field horizontal. This requirement is inconvenient experimentally. Furthermore, the use of dual flippers allows us to measure the dynamical phase separately from the geometric phase as we shall see later.

The single coil flippers shown in Fig. 6.7 were replaced with dual-coil spin flippers as shown schematically in Fig. 6.8. On path II, the first coil of the pair produces a field B_0 in the negative y''-direction, and the second coil produces a field B_0 in the positive y''-direction. An identical dual flipper on path I produces a similar pair of oppositely directed magnetic fields. The angle between the flipper fields on path II relative to path I is $\Delta\beta$, in a manner similar to the situation shown in Fig. 6.7. A uniform field of magnitude $B_0 \approx 30$ gauss in the vertical ($+z$-direction) was provided over the entire experimental region of the interferometer by a pair of water-cooled and temperature-stabilized Cu-wire Helmholtz coils.

The field B_0 from the Helmholtz pair is in the z-direction, and thus adds vectorially to the fields created by each of the dual flipper coils such that the net magnetic field that the neutron feels in its passage through each coil is directed at 45° to the horizontal plane and also 45° to the z-axis. It has a magnitude $\sqrt{2}B_0$ in the $\boldsymbol{q}$ direction in the first coil and it has the direction

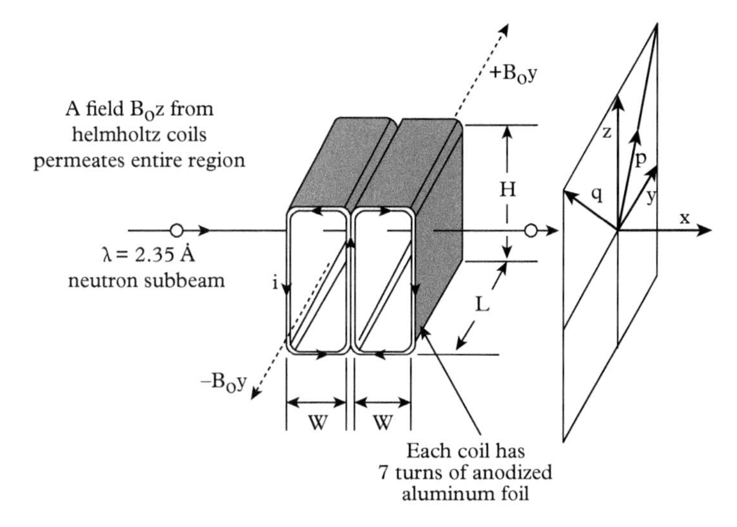

Figure 6.8 *Isometric view of the dual flipper assembly showing the neutron trajectory in the* x*-direction through the aluminum foil windings (Allman et al. 1997)*

$\boldsymbol{p}$ in the second coil (see Fig. 6.8). The magnitude of the field B_0 was set at 30 gauss, which is appropriate for the incident neutron polarized in the z-direction to precess down to the horizontal plane in the first coil and then on down to the south pole of the spin sphere in the second coil. With the aid of a little bit of spherical trigonometry, one can show that the solid angle $\Delta\Omega$ that this closed cycle evolution subtends at the origin of the spin sphere is $2\Delta\beta$. The experimental strategy for measuring the spin-dependent phase shift was to rotate the 1.05-mm-thick aluminum phase flag with the flippers turned off, and then with the flippers turned on. It is this phase difference between these two interferograms that is of interest to us here.

To measure the geometric phase, the dual spin flippers were individually rotated in opposite directions to $+\Delta\beta/2$ and to $-\Delta\beta/2$. This was done so that the phase shifts acquired in the neutron's passage through the flipper materials would cancel out in the difference phase $\Phi_{II} - \Phi_I$. We show in Fig. 6.9 a series of interferograms taken at various angular settings $\Delta\beta$ for the flippers on, but always adjusted in phase angle for the flipper-off condition for each relative rotation angle $\Delta\beta$. A summary of the phase angles of these interferograms plotted versus the relative angles of the two flippers is shown in Fig. 6.9.

A pure geometric phase can also be obtained without physically rotating the flippers at all, by simply reversing the current (and hence the magnetic field) in one of the dual flippers. This is equivalent to producing a 180° rotation of the precession axes without any physical motion. This ensures that the original offset phases remain unchanged. The dual flipper produces two successive π precessions, first about the $\boldsymbol{p}$-direction and then about the $\boldsymbol{q}$-direction shown in Fig. 6.8. These successive operations bring the $|+z\rangle$ state into the $|-z\rangle$ state. It is represented by two successive operations of two orthogonal Pauli spin matrices, first σ_p and then σ_q. Reversing the current in one of the dual flippers reverses the order of this precession to first σ_q and then σ_p, which causes a 180° phase shift, since the loop around the enclosed solid angle ($\Delta\Omega = 2\pi$) is reversed. There are four such field reversal conditions. Interferograms obtained under each of

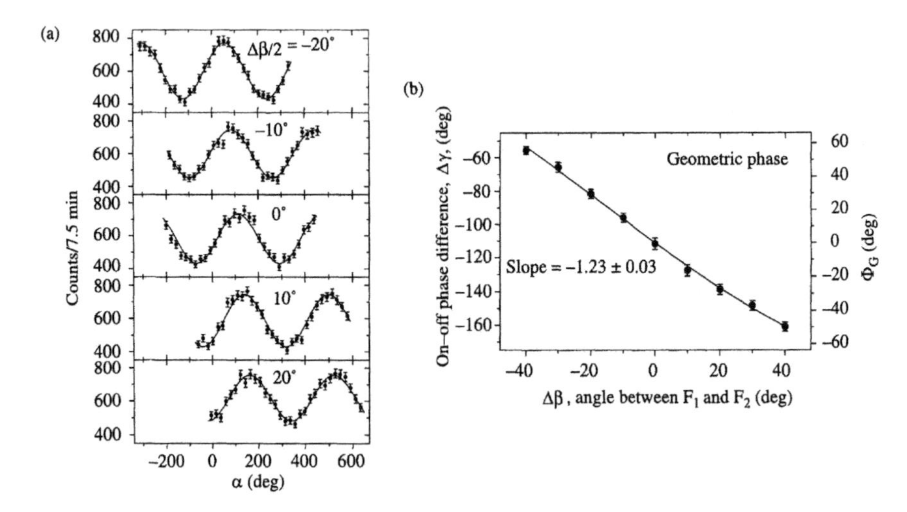

Figure 6.9 *Interferograms when the nuclear phase shifter is rotated according to Fig. 6.7 for different relative angular settings of the two flippers (left) and the measured geometric phase shift (right, Allman et al. 1997)*

these conditions are shown in Fig. 6.10. This result is perhaps the first direct observation of the anticommutation rule for the orthogonal Pauli spin matrices, namely

$$\sigma_p\sigma_q = -\sigma_q\sigma_p. \tag{6.46}$$

This sign change manifests itself in phase shifts of the neutron interferograms of approximately 180°, as shown in Fig. 6.10. The average phase shift from the top plot to the bottom plot is 180° ± 2.4°. It is thus correct to say that the anti-commutation relation of Eq. (6.46) has been confirmed to an accuracy of 1.3%. A neutron polarimetric proof has been reported by Hasegawa and Badurek (1999).

Finally, we now discuss the dynamical phase. A linear translation of one of the spin flippers along its sub-beam path results in the neutron spending more time in one spin orientation than in the other. The total energy E_0 of the neutron is fixed (the Hamiltonian is time independent here). This requires the neutron to have a slightly greater kinetic energy when its magnetic moment is pointing up than when it is pointing down (Eq. 2.29):

$$E_0 = \frac{\hbar^2 k_0^2}{2m} = \frac{\hbar^2 k_\uparrow^2}{2m} - \mu B = \frac{\hbar^2 k_\downarrow^2}{2m} + \mu B. \tag{6.47}$$

The phase accumulated along the path depends upon the line integral of the k-vector. Thus, if the spin flipper coil is translated along its sub-beam by an amount δx, there will be a phase shift of $(k_\uparrow - k_\downarrow)\delta x$, which is the dynamical phase shift. It can be written as

$$\Delta\Phi_{\text{dynam}} = -2\mu B_0 \delta x/\hbar v_0. \tag{6.48}$$

Here B_0 is the magnetic field from the Helmholtz coils (≈30 gauss), and v_0 is the mean velocity (≈ 1700 m/s) for the 2.35 Å neutrons in the incident beam. For each translation position of one

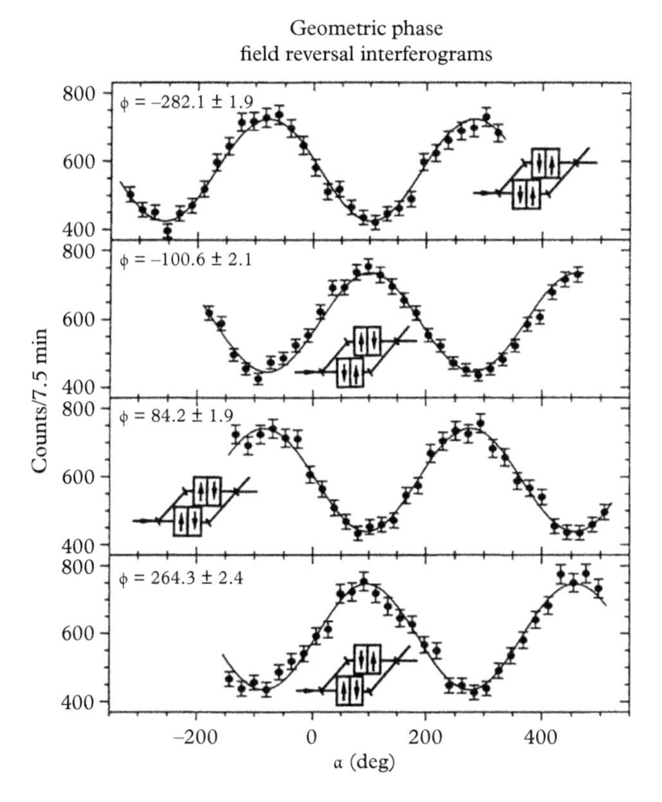

Figure 6.10 *This is a series of field reversal interferograms obtained by scanning the aluminum phase flag for various field directions in the coils of the dual flipper. π phase shifts between each of these scans accurately verifies the anti-commutation relation rules for the Pauli spin matrices (Allman et al. 1997)*

of the dual flipper coils along the beam path, a phase flag interferogram was taken. A summary of these dynamical phase shifts is shown in Fig. 6.11 for various translation positions of flipper F_2 along beam path II. The slope of the data in this figure agrees with the formula Eq. (6.48) to about 3%.

It is probably correct to say that these polarized neutron interferometric data represent the *first and only experiments in physics* where a clean and obvious separation and demarcation of the geometric phase and dynamical phase has been made. The geometric phase comes from a relative rotation of the spin precession axes of the spin flippers, while the dynamical phase comes from a translation of the spin flippers relative to the skew symmetric interferometer. A photograph of the mechanism for rotation and translation of the dual spin flippers within the interferometer is shown in Fig. 6.12 (Werner 2012).

In another interferometer experiment, Mezei et al. (2000) and Ioffe and Mezei (2001) used a slowly, but oppositely rotating field in both beam paths, which produces real space geometric

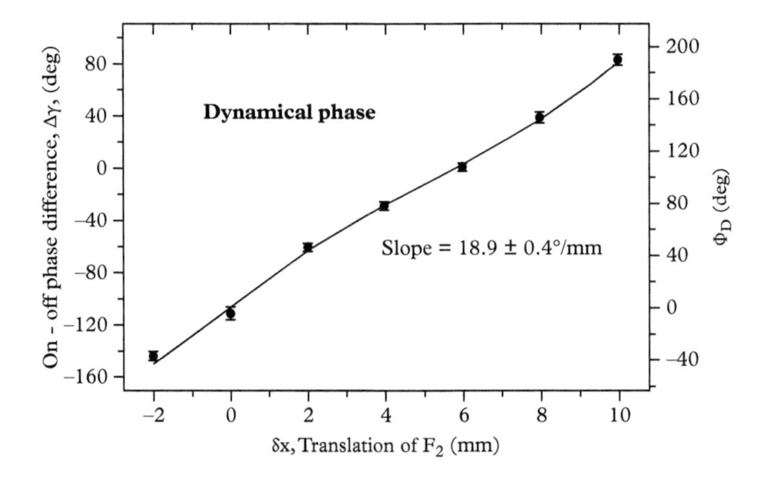

Figure 6.11 *A plot of the on–off difference count rate as a function of the translation of F_2 along beam path II, measured relative to a reference position from the first blade of the interferometer. The right-hand ordinate shows the equivalent dynamical phase (Allman et al. 1997)*

Figure 6.12 *Photograph of the precision translation–rotation mechanism and the water-cooled heat-sink blocks containing the rectangular, aluminum foil dual flipper coils. This device was fabricated by Mr. Cliff Holmes in the Missouri Physics Machine Shop (Werner 2012)*

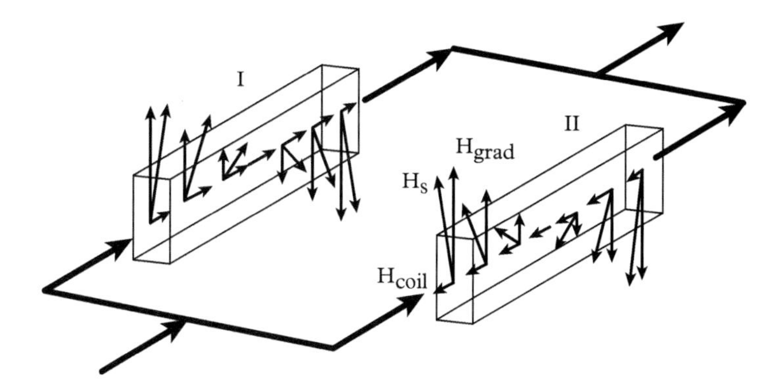

Figure 6.13 *Sketch of the adiabatic spin rotation by means of a helical field. Reprinted from Ioffe and Mezei 2001, copyright 2001, with permission from Elsevier.*

rotations and acts as a spin flipper (Fig. 6.13). For a 360° rotation they found a spinor phase shift of $\gamma = \alpha/2 = (180.4 \pm 4.0 \pm 1.2)$, where the uncertainties represent systematical and statistical ones.

6.5 Non-cyclic Berry Phases

In a later experiment Wagh et al. (1998) showed with polarized neutrons that by observing the phase shifts and the amplitudes of interferograms, that non-cyclic phases are measurable in dedicated interference experiments (see Eqs. 6.49 and 6.50). In an interferometer experiment the phases become apparent due to superposition with a reference beam. When a scalar (nuclear) phase shift χ is included one gets

$$I(\chi,\alpha) = \left|\psi_0(0,0) + \psi_0(\chi,\alpha\,)\right|^2 \propto D + \cos\chi\cos\frac{\alpha}{2} + \sin\chi\sin\frac{\alpha}{2}\;\cos\Theta = D + A\cos(\chi + \phi), \quad (6.49)$$

with the varying amplitude A given by

$$A = \sqrt{1 - \sin^2\Theta\,\sin^2\frac{\alpha}{2}}. \quad (6.50)$$

Here Θ is the polar angle of the spin on the spin sphere and the azimuthal angle α is the precession angle. This provided the basis for the observation of a non-cyclic phase by Wagh et al. (1998; see Section 6.4).

Another experiment verifying non-adiabatic and non-cyclic phases has been performed with a double-loop interferometer where two phase shifters (PS) and an absorber (A) permit quite peculiar state excursions (Fig. 6.14; Filipp et al. 2005). The upper beam $|\psi_t{}^0>$ of the first loop is used as a reference beam with adjustable phase η. As the incident wave $|p>$ for loop 2 becomes attenuated ($T = \exp(-\sigma_t ND)$) and phase shifted (χ_2), the orthogonal beam, $|p^\perp>$ becomes phase shifted by χ_1. This gives an overlap of the reference beam $|\psi_{\mathrm{ref}}>$ and the loop 2 beam $|\psi_2>$,

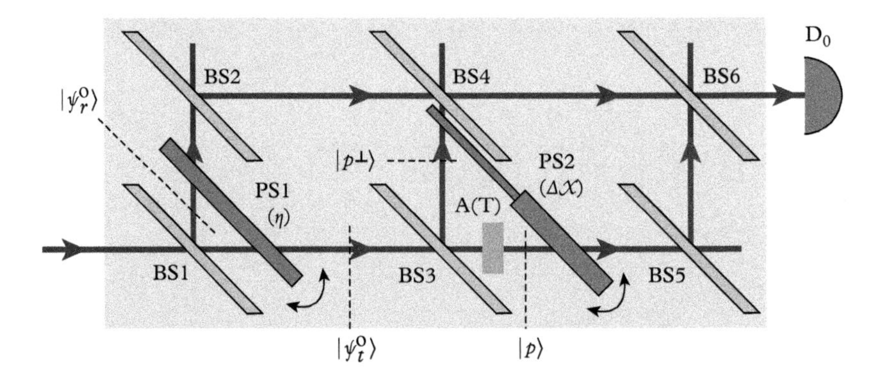

Figure 6.14 *Sketch of a two-loop interferometer (above) and two examples of quantum evolutions for measuring the geometric phase (left) and a non-cyclic phase (right; Filipp et all. 2005)*

where

$$|\psi_{\text{ref}}> \propto (|p> + |p^{\perp}>) = |q>, \tag{6.51}$$

and

$$|\psi_2> \propto (e^{i\chi_1} + \sqrt{T}\, e^{i\chi_2})|q>, \tag{6.52}$$

such that the phase angle of the interference patterns is

$$\phi_{\text{d}} = \arg < \psi_{\text{ref}}|\psi_2> = \frac{\chi_1 + \chi_2}{2} \operatorname{arctg}\left[\frac{\chi_2 - \chi_1}{2}\left(\frac{1-\sqrt{T}}{1+\sqrt{T}}\right)\right]. \tag{6.53}$$

This dynamical phase can be written as

$$\phi_{\text{d}} = \frac{\chi_1 + T\chi_2}{1+T}, \tag{6.54}$$

which becomes a constant when

$$\chi_1 + T\,\chi_2 = \text{const}. \tag{6.55}$$

A proper manipulation of phase shifters and the absorbers permit cyclic and non-cyclic evolutions on the Bloch sphere where the north pole and the south pole correspond to well-defined paths along the upper, $|p^{\perp}>$, and the lower, $|p>$, beam paths within the second interferometer loop. The absorber determines the latitude where the evolution driven by the phase shifter PS2 takes place. This allows us to write the transmission in the form

$$T = \tan^2\frac{\Theta}{2}. \tag{6.56}$$

The geometric phase can be measured when closed cycles at different latitudes are chosen. Non-cyclic evolutions occur when such a rotation is stopped before a cycle is complete and then this endpoint must be connected by a geodesis line to the equator. Figure 6.15 shows the results of such measurements. They clearly define the geometric phase and the non-cyclic phase for situations shown in Fig. 6.6b. Nearly all excursion on the Bloch sphere can be realized and contain information about the history of a quantum system. This experiment also confirms the validity of the geometric phase concept in the case when spatial degrees of freedom are involved and an absorber (thus a SU(1) system) is used to define distinct excursion paths on a Bloch sphere. The symmetry of this system is SU(1). This settles a dispute with Wagh (1999) and Sjöqvist (2001), who denied

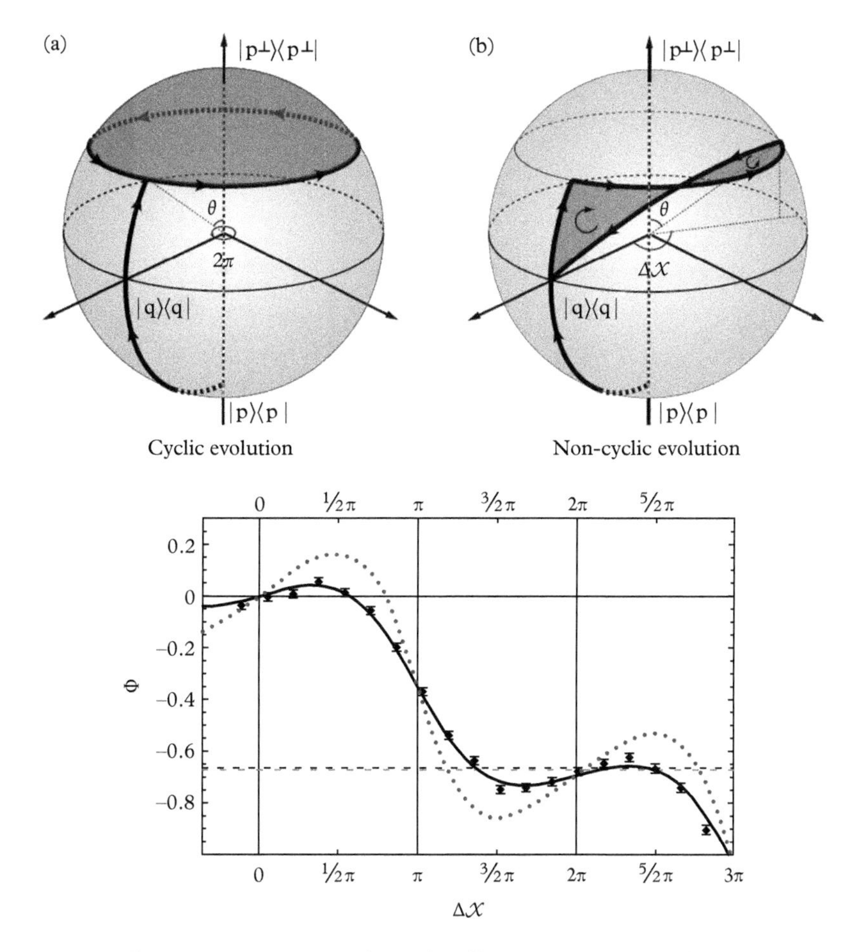

Figure 6.15 *Representations of cyclic and non-cyclic evolutions on the Bloch sphere (above) and results of the non-cyclic phase experiment according to non-cyclic excursion (below; Filipp et al. 2005)*

the applicability of SU(2) descriptions to spatial evolutions. The SU(2) symmetry appears in the two beam paths, and also in experiments with unpolarized neutrons.

The results in this section indicate that geometric phases are well defined and clearly measurable quantities. They may become even more important in the future since they seem to be less sensitive to any fluctuation and dissipative effects of external parameters (De Chiara and Palma 2003). The expected robustness of the geometric phase against any kind of disturbances has been demonstrated recently with bottled ultra-cold neutrons (Filipp et al. 2009). This robustness of the geometric phase against any kind of fluctuations may become an important aspect for future quantum communication techniques. Similar results have now been obtained in a solid state qubit system (Leek et al. 2007).

6.6 Polarization Rotation Experiments

Before the direct interferometric observation of geometric phases described in Section 6.4, Bitter and Dubbers (1987) found an alternative way to observe Berry's phase by means of a neutron spin rotation experiment. It is basically also an interference experiment (Section 2.4; Mezei 1988). The change of polarization was measured when neutrons pass through a helical magnetic field of a single turn (Fig. 6.16). From the solution of the Bloch equation

$$\frac{\mathrm{d}\boldsymbol{P}}{\mathrm{d}t} = \gamma(\boldsymbol{P} \times \boldsymbol{B}) \tag{6.57}$$

$$(\gamma = 2\mu/\hbar \ldots \text{ gyromagnetic ratio}),$$

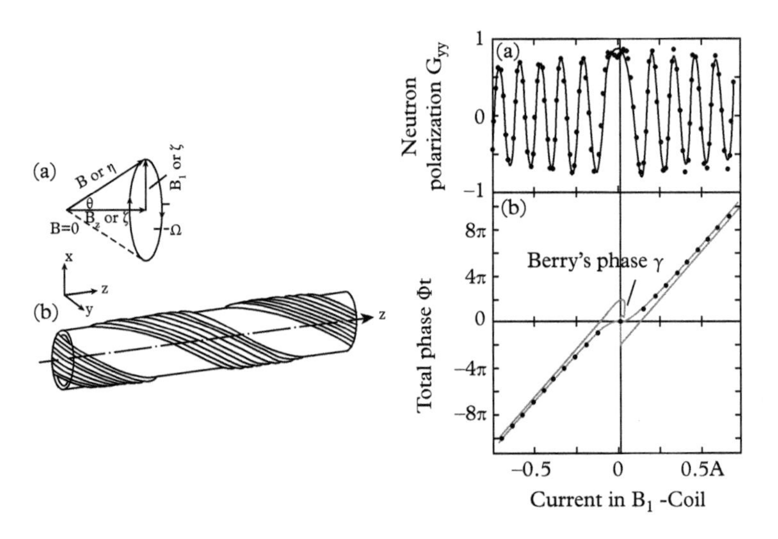

Figure 6.16 *Sketch of the experimental arrangement (left) and measured polarization and extracted total phase when the cone angle of the precession field has been changed (right). Reprinted with permission from Bitter and Dubbers 1987, copyright 1987 by the American Physical Society.*

one gets the motion of the polarization vector within any arbitrary magnetic field. The change of the y-component within the helical field reads as (e.g., Dubbers 1976)

$$G_{yy} = \cos\left(2\pi\sqrt{1+\zeta^2} - 2\pi\right), \qquad (6.58)$$

where the extra term -2π ensures that the total rotation angle becomes zero when there is no field. In Eq. (6.58), ζ denotes the Larmor rotation angle around the helical field of strength B_1 and length $L(\zeta = \gamma B_1 L/v)$. The extra phase ($2\pi$) is manifested by a non-equidistant spin rotation pattern and by a shift of the adiabatic limits for large positive and negative B_1 values (Fig. 6.16). When an additional field B_z is applied along the beam path the geometric phase becomes $\Omega = 2\pi\left(1 - B_z\big/\sqrt{B_1{}^2 + B_z{}^2}\right)$. The expected behavior has also been demonstrated by measuring the change of the z-component G_{zz} (Bitter and Dubbers 1987). This result must be compared to that expected in an interference experiment.

A rather similar demonstration of Berry's phase has been done with ultra-cold neutrons by Richardson et al. (1988). The magnetic field inside a neutron storage volume was slowly (adiabatically) varied as a function of time and the resulting spin rotations were measured and extrapolated to zero field, thus allowing the Berry phase to be extracted. The effects of multiple excursions in parameter space and of elliptical paths were also investigated. The appearance of the topological phase for non-cyclic spinor evolutions has been demonstrated by Weinfurter and Badurek (1990) using a neutron polarimetric method where neutrons inside a Rabi flipper were rotated incompletely and the resulting polarization was measured in three dimensions.

6.7 Spin-Echo Version of the Geometric Phase

An adiabatic resonance spin-echo system inherently produces a geometric phase. Each arm of the spin-echo system consists of two adiabatic resonance spin flippers as described in Sections 2.4 and 5.4 (Figs. 2.19 and 2.23). As mentioned in the previous sections a polarized neutron beam flying through a twisted or rotating field accumulates a geometric phase (Bitter and Dubbers 1987, Richardson et al. 1988, Wagh 1990, Weinfurter and Badurek 1990, Hasegawa et al. 1997).

A resonance spin-echo system as shown in Fig. 6.17 with inclined fields has been used to measure the geometric phase in an alternative way (Kraan et al. 2010). Wavelength-dependent measurements become feasible with a chopper and a time-of-flight analysis. The phase shift in the first arm consists of a wavelength-dependent dynamical phase and a wavelength-independent geometric phase $\theta_1 = \Theta_1\lambda + \theta_{1g}$. In general the precession in the second arm will not cancel this phase shift. Therefore, an auxiliary DC coil which produces a phase shift $c_{DC}B_{DC}\lambda$ is used. Let the dynamical phase cancel for a distinct field B_{DC0}, i.e., $\Theta_1 - \Theta_2 + c_{DC}B_{DC} = 0$. When one now varies B_{DC} by an amount ΔB, one gets a net phase shift of $\theta_{1g} - \theta_{2g} + \lambda\Delta B = \theta_g + \lambda\Delta B$ and a measurable polarization:

$$P(\Delta B, \lambda) = \cos\left(\theta_g - c_{dc}\lambda\Delta B\right). \qquad (6.59)$$

This then permits an accurate determination of the geometric phase θ_g. Different rotations within the resonance flippers produce different geometric phases which are monitored by a phase and amplitude control of the flippers. The λ and ΔB maps show good agreement between

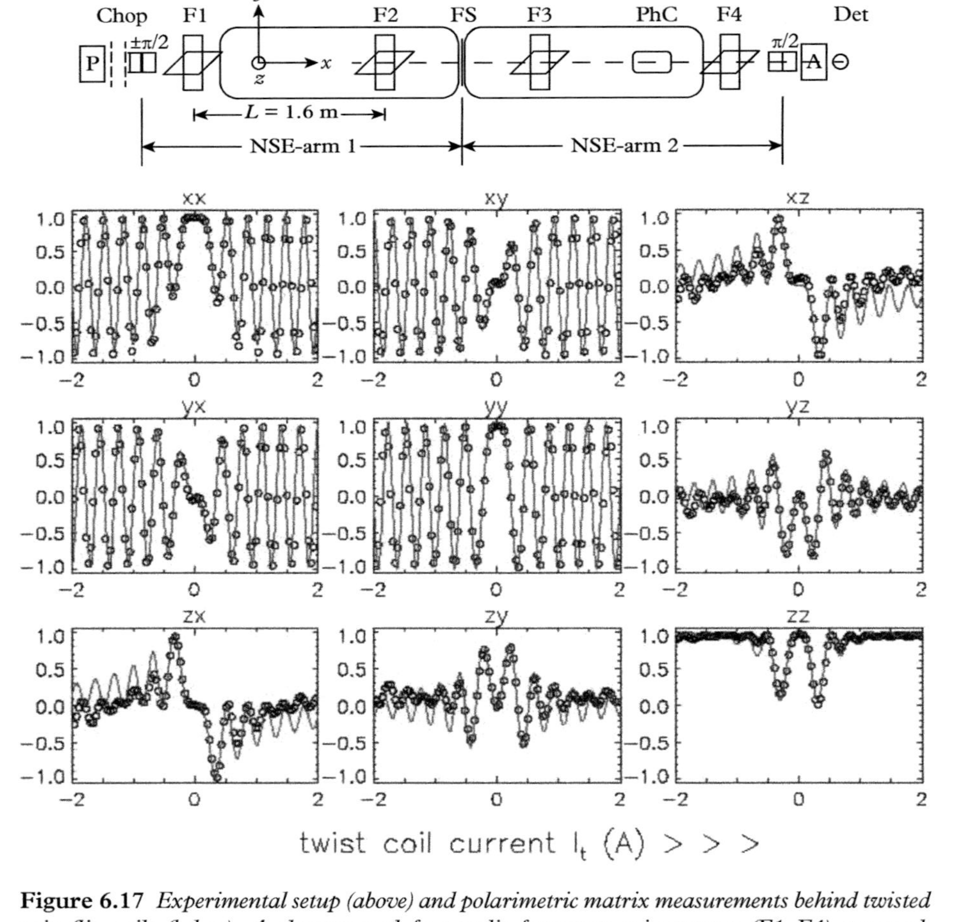

Figure 6.17 *Experimental setup (above) and polarimetric matrix measurements behind twisted spin-flip coils (below). A chopper and four radio-frequency spin-turners (F1–F4) are used. Reprinted with permission from Kraan et al. 2010, copyright 2010 by the American Physical Society.*

calculated and measured geometric phase values. In addition, a twisted DC flipper was used to reproduce the results of a previous experiment of Bitter and Dubbers (1987) described in Section 6.6.

6.8 Absorption Analog

In the previous sections it has been shown that spinor rotation of a fermion in a homogeneous magnetic field results in a dynamical and a geometric phase (Berry 1984, Aharonov and Anandan 1987). Spin-up and spin-down states in a constant magnetic field are good examples of the cyclic evolution of a quantum system in which the system can return to the original state by various anholonomy transformations. The system returns to the original state with a specific phase shift ϕ_{tot} when the difference between the phases of the spin-up and spin-down state becomes equal to an integral multiple of 2π. As is known from the magnetic interaction, the spin-up component accumulates a phase $\phi_{\uparrow} = \mu Bt/\hbar$ and the spin-down component $\phi_{\downarrow} = -\mu Bt/\hbar$. This spinor rotation is periodic with a period $\phi_{\uparrow} - \phi_{\downarrow} = 2\pi$. Please note that $\phi_{\uparrow}$ and $\phi_{\downarrow}$ denote just half of the related Larmor precession angle (Eq. 5.3).

Very similar situations exist within a split beam interferometer when phase shifters and beam attenuators are inserted (Fig. 6.18). The two basic states are given by the two split beams and represent eigenstates in analogy to the spin-up and spin-down state discussed earlier. According to Eq. (6.25) the geometric phase is given in a more general notation as

$$\phi_g = \phi_{tot} - \phi_d. \tag{6.60}$$

In the spin-state situation the dynamical phase is given as $\phi_d = \alpha/2 = \pm \mu Bt/\hbar$. In the split-beam situation the dynamical phase can be calculated as well. Considering that the normalization of the wave functions is sometimes altered by a beam attenuator the dynamical phase becomes (Eqs. 1.33 and 6.25)

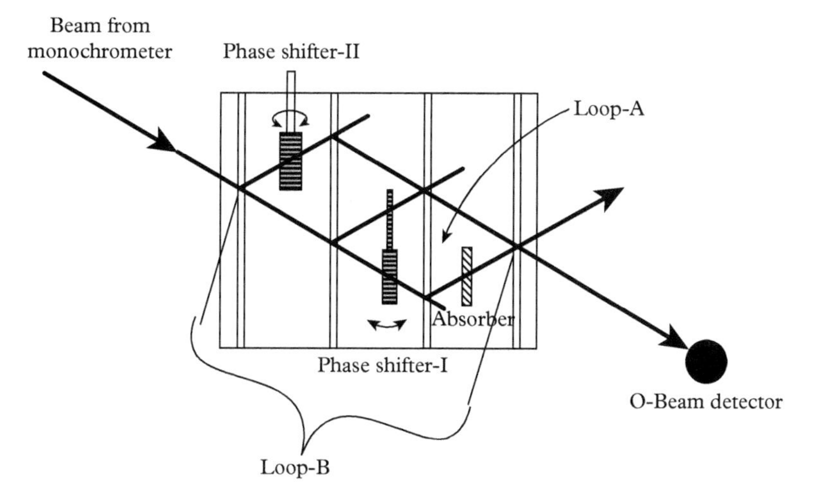

Figure 6.18 *A four-plate interferometer with coupled interference loops A and B.*

$$\phi_{\mathrm{d}} = \frac{\oint < \psi\,|\Delta \boldsymbol{k} \cdot d\boldsymbol{s}|\,\psi >}{< \psi|\psi >}$$

$$= \frac{\int < \psi_{\mathrm{I}}\,|\psi_{\mathrm{I}} > \,\Delta \boldsymbol{k}_{\mathrm{I}}\ \mathrm{d}\boldsymbol{s}_{\mathrm{I}}}{< \psi\,|\psi >} + \frac{\int < \psi_{\mathrm{II}}\,|\psi_{\mathrm{II}} > \,\Delta \boldsymbol{k}_{\mathrm{II}}\ \mathrm{d}\boldsymbol{s}_{\mathrm{II}}}{< \psi\,|\psi >}, \tag{6.61}$$

where ψ_i, $\mathrm{d}\boldsymbol{s}_i$, and $\boldsymbol{k}_i$ denote the two wave functions, beam paths, and canonical momenta, respectively, along the two beam paths through the interferometer. In this respect, the experiment described in this section is related to the partial beam path detection experiments discussed in Section 4.3. When the real part of the phase shift of the beam attenuator is included in the phase shift χ_i of the pure phase shifters, we can re-arrange Eq. (6.61) in the form

$$\phi_{\mathrm{d}} = \frac{I_{\mathrm{I}}}{I_{\mathrm{I}} + I_{\mathrm{II}}}\chi_{\mathrm{I}} + \frac{I_{\mathrm{II}}}{I_{\mathrm{I}} + I_{\mathrm{II}}}\ \chi_{\mathrm{II}} = \left(\frac{1}{1+a}\right)(\chi_{\mathrm{I}} + a\,\chi_{\mathrm{II}}), \tag{6.62}$$

where I_i represents the intensities of the two beams and a the transmission probability of the beam attenuator which is assumed to be inserted into beam path II (see Section 4.3). From this equation it becomes obvious that the dynamical phase shift becomes zero when

$$\chi_{\mathrm{I}} + a\chi_{\mathrm{II}} = 0, \tag{6.63}$$

which can be controlled by a phase flag inserted into both beams (Fig. 6.15).

The situation can be visualized by means of the Poincaré sphere, which was used earlier to describe the specific particle and wave features of the neutron inside an interferometer (Fig. 4.15). The related presentation by means of such Poincaré spheres is shown in Fig. 6.19 for the case of equal beam intensities ($I_{\mathrm{I}} = I_{\mathrm{II}}$) and for unequal intensities due to the absorber in arm II of the

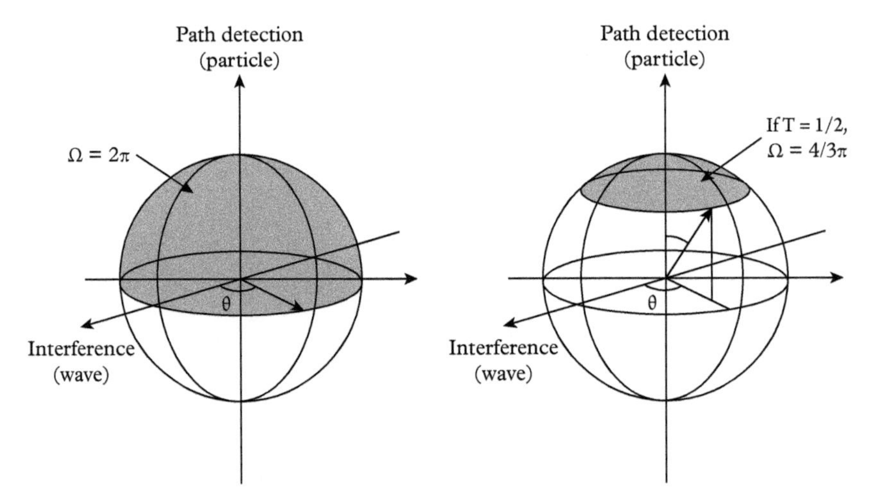

Figure 6.19 *Poincaré sphere description of a split beam experiment without (left) and with an absorber providing a = ½ (right; Hasegawa et al. 1996)*

interferometer ($I_{\mathrm{I}} > I_{\mathrm{II}}$). The vertical axis represents the relative intensity of the two beams and the polar points represent situations where either beam path I or II is closed completely and, therefore, perfect path information exists. This occurs when shifting the relative phase χ and the polar points represent situations where either beam path I or II is closed completely. When shifting the relative phase ($\chi = \chi_{\mathrm{I}} - \chi_{\mathrm{II}}$), all possible states trace on the equator or on a latitude circle depending, on the ratio between the intensities of the two beams. From these spheres the solid angle which subtends the traced curve seen from the origin is given as

$$\Omega(C) = \pi(1 \pm \cos\vartheta) = 2\varphi_g = 2\varpi\left(1 - \frac{1-a}{1+a}\right) = 4\pi\frac{a}{1+a}. \tag{6.64}$$

This shows the direct connection of the geometric phase to the transmission probability.

This geometric phase is imparted to the outgoing beam of a standard three-plate interferometer but does not show up in the intensities because the square of the wave functions determines the intensities. This geometric phase can only be detected by an additional reference beam which can be provided by a multiplate interferometer (Section 4.5.6). This method is similar to a homodyne detection known in laser optics (e.g., Walls and Milburn 1994). The neutronic version of this method has been used by Hasegawa et al. (1996) to detect the geometric phase caused by the joint action of a phase shifter and a beam attenuator within interferometer loop A and a reference beam arising from interferometer loop B (Fig. 6.19). The beam attenuators were gold plates providing transmission probabilities of $a = 0.492(4)$ and $a = 0.212(5)$, respectively. The phase shifter within this loop had different thicknesses for the left and the right beam in order to fulfill Eq. (6.55), i.e., to make the dynamical phase shift zero during the evolution. These phase shifters were adjusted to three intensity peaks (peaks 1, 2, 3) counted in the forward (0) detector when a thick absorber was inserted which blocked the beam with phase shifter II. Then phase shifter II within interferometer loop B was rotated and the interference pattern shown in Fig. 6.20 was obtained. These curves show a different shift for the various absorbers. The peak positions were determined as a function of the phase of the oscillation, which is expected to depend on the solid angle subtended by the closed curve of the cyclic evolution (Fig. 6.21 and Eq. 6.34). It shows the solid angle calculated according to Eq. (6.32) and the measured geometric phase shift. The agreement between

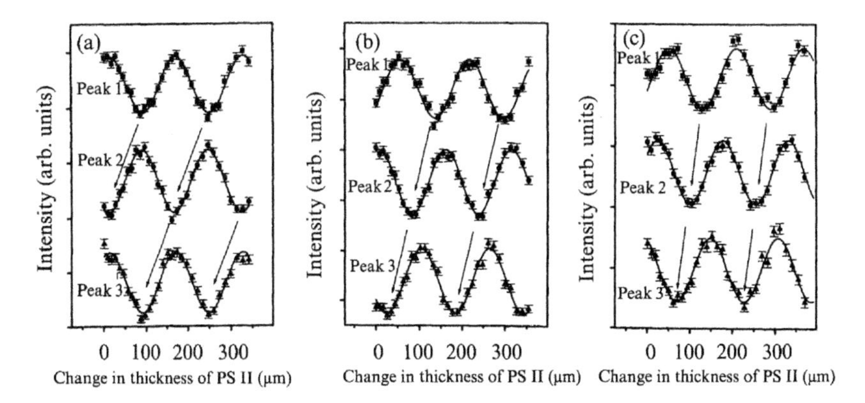

Figure 6.20 *Measured shifts of the interference pattern when phase shifter II was rotated and phase shifter I was turned to the peak intensity of loop A (Hasegawa et al. 1996)*

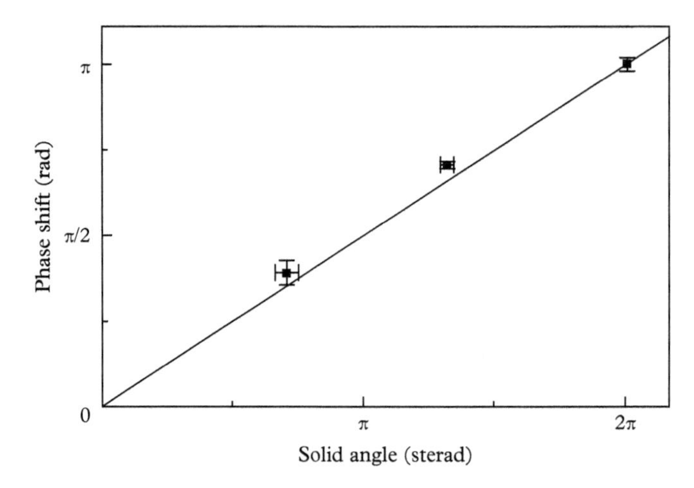

Figure 6.21 *Experimental results and theoretical predictions of the geometric phase shift as a function of the solid angle where the state traces around the axis of the Poincaré sphere (Hasegawa et al. 1996)*

measured and calculated values is fairly good, which demonstrates again that the concept of geometric phase is a rather general one. The analogy of the spin rotation experiment as discussed in Section 6.4 and of the split-beam absorption experiment discussed here is based on the fact that both situations exhibit SU(2) symmetry. This is also why it is frame independent in analogy to the spin rotation experiments discussed earlier (Sjöqvist et al. 1997). A Galilean invariant structure of the geometric phase has been constructed out of the difference between the geometric phases for two different paths in configuration space which have common initial and final points (de Polavieja 1997, Sjöqvist 2001). Thus, the two beam paths represent the SU(2) symmetry in the absorption case, which underlines the basic feature of the geometric phase as a property related to the configuration space instead of only to the ordinary space.

The analogy between the absorption and the spin rotation experiments which has been shown in this section has been criticized by Wagh (1999a, 1999b), claiming that geometric phases can only be associated with SU(2) symmetries and a real ray-space evolution. But the formal analogy shows that ray-space evolution in phase space is equally relevant. In the analogy discussed here, the different directions of the coherent beams constitute a geometric-like phase evolution. This has been verified in a dedicated experiment using a double-loop interferometer (Filipp et al. 2005; Section 6.4).

6.9 Confinement-Induced Topological Quantum Phase

Another interesting quantum phase appears when neutrons travel inside a tube or channel. In this case the traverse momentum becomes quantized according to the dimension of the tube and consequently the longitudinal component also changes. This causes a related phase shift (Lévy-Leblond 1987, Greenberger 1988). Very narrow tubes or slits must be used, and all parasitic phenomena must be identified and excluded. The confinement causes a momentum quantization

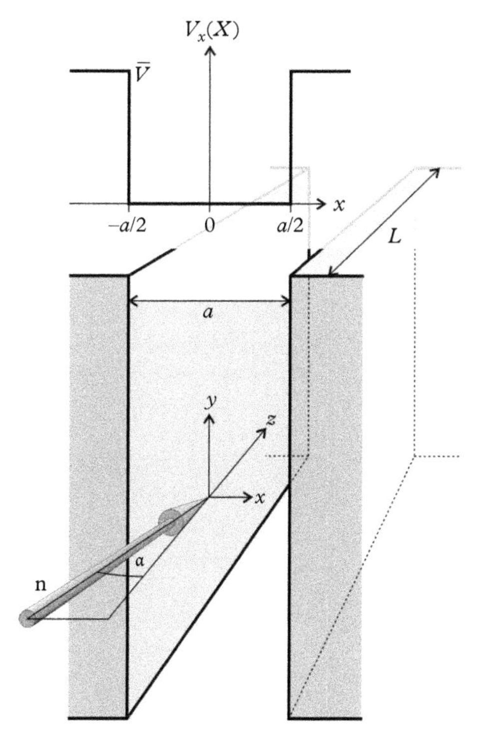

Figure 6.22 *Sketch of the slit structure and the confining wall potential*

in the transverse direction, which changes the transverse coherence properties and the longitudinal k-component as well (Section 4.2.2).

A tube of length L causes a phase shift $\Delta\varphi_{LL} \cong L\Delta k$, where Δk results from the transverse momentum quantization. This can be calculated from energy (momentum) quantization (see Fig. 6.22). For a tube with a square cross-section $a \times a$ and infinitely steep walls one gets

$$E = \frac{k_z^2\hbar^2}{2m} + (n_x^2 + n_y^2)\,\frac{\pi^2\hbar^2}{2ma^2}, \tag{6.65}$$

where n_x and n_y are integer quantum numbers. The axis of the tube is along z. The related energy levels are shown in Table 10.1. Slight changes to smaller values occur when a finite height of the wall potentials is taken into account (e.g., for the first level from 5.113 to 4.677 × 10^{-13} eV). Energy conservation causes the momentum along the tubes to change as well. For the basic mode ($n_x = n_y = 1$) one obtains (Razavy 1989, Griffin et al. 1996)

$$k' = k(1 - \pi^2/k^2a^2) \tag{6.66}$$

and from that we get the phase shift

$$\Delta\varphi_{11} = \frac{L\pi^2}{k\,a^2}, \tag{6.67}$$

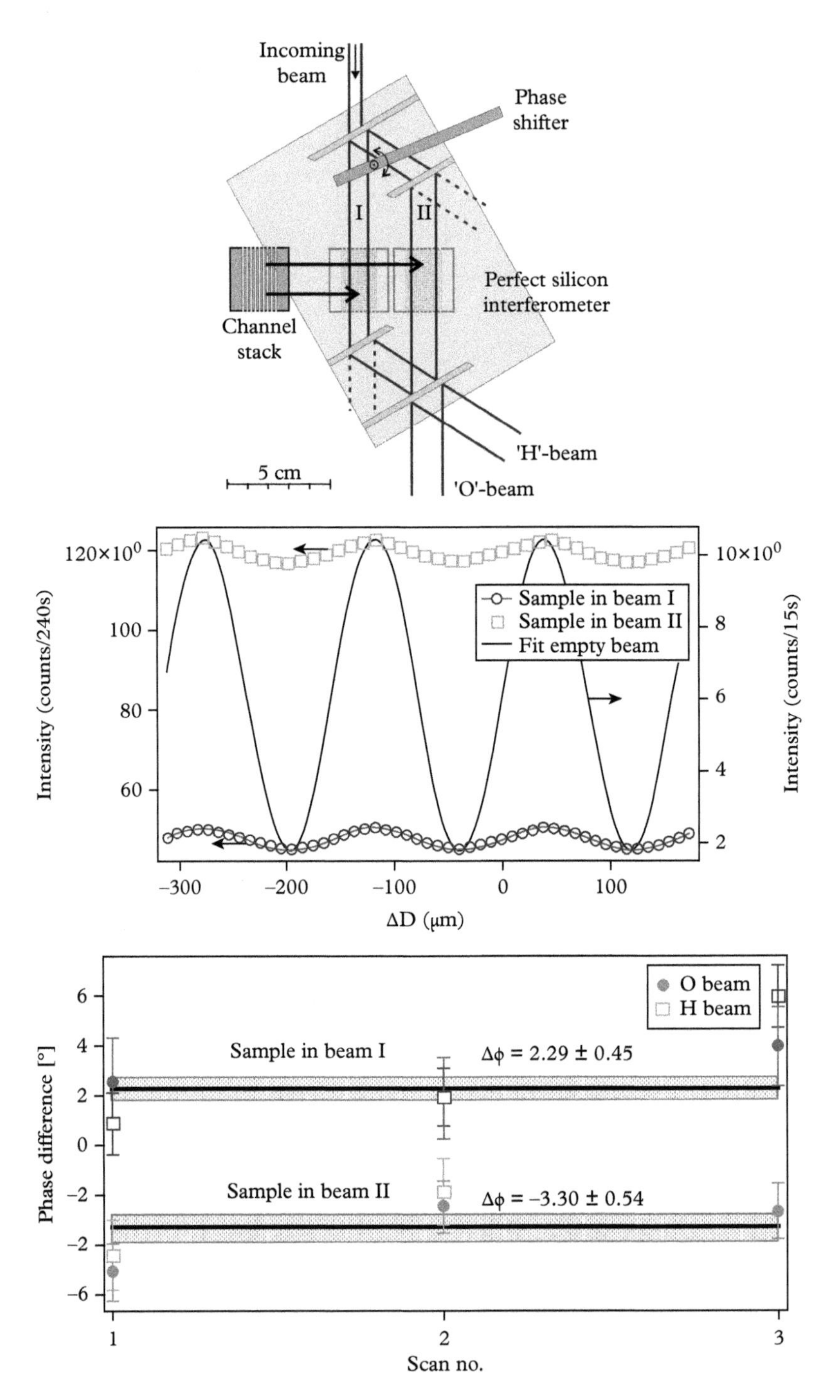

Figure 6.23 *Experimental arrangement (above) and results of the neutron confinement experiment (middle and below) (Rauch et al. 2002)*

where L is the length and a^2 the cross-section of the tube. For capillary dimensions of 20 μm and a length of 30 mm a phase shift of 0.11 rad is expected. This is a measurable effect. However, due to intensity considerations a multichannel system must be used and a high stability of the internal interferometer phase is required (Fig. 2.10). A treatment of this phenomenon using the quantum potential approach is also known (Chapter 12, Kastner 1993). An optical analog experiment of this idea has been carried out by Allman et al. (1999). The situation is rather similar to the transport of electrons through open ballistic microstructures where periodic and chaotic conductance fluctuations have been observed (e.g., Ferry et al. 1995, Wirtz et al. 1997). The advantage of neutrons in this case is that they do not interact via van der Waals or Casimir forces (Casimir and Polder 1948, Haroche and Raimond 1993, Bordag et al. 2001), which complicates the situation using atoms.

The experiment described here has been performed with narrow silicon wafer slits with a width of $a = 22.1\ \mu m$ and a length of $L = 2$ cm (Rauch et al. 2002; Fig. 6.22). Within the potential there are about 360 bound states whose excitation probability P_n depends on the angle of incidence α. For a completely parallel beam the excited energy levels and their excitation probabilities are $E_0 = 0.4172$ peV, $p_0(0) = 0.406$; $E_2 = 1.669$ peV, $p_0(0) = 0.045$; $E_4 = 6.676$ peV, $p_4(0) = 0.016$, etc. A beam component deviating 18 s from being parallel with the walls excites mainly energy levels around $n = 19$. The energy change determines the change of the longitudinal momentum as

$$k_{||,n} = \sqrt{2m(E_{\text{in}} - E_n)/\hbar^2}, \tag{6.68}$$

where E_{in} denotes the energy of the incident neutron beam. The related phase shift due to level n is given as

$$\Delta\phi_n = L(k_{||,\text{in}} - k_{||,n}) \tag{6.69}$$

The related experiment has been done with a multislit system using a sweep method as indicated in Fig. 6.23 (Rauch et al. 2002). The results show a phase shift $\Delta\phi_{\text{exp}} = 2.8(4)°$ in rough agreement with the theory ($\Delta\phi_{\text{th}} \cong 2.5°$). The intensity and the contrast is strongly reduced since the low-lying levels contribute only to a net phase shift and most of the beam becomes totally reflected from the walls and falls outside the diffraction width of the interferometer crystals. The experiment has been repeated several times with various differing setups, and in all cases the experimental values lie above the theoretical ones. This is not yet understood (see also Section 10.11).

7

Contextuality and Kochen–Specker Phenomena

The question of whether quantum phenomena can be explained by classical models with hidden variables is a subject of a long-lasting debate (e.g., Einstein et al. 1935, Bell 1964, Haroche and Raymond 2006). In 1964, John Bell showed that certain types of classical models can be ruled out by measuring quantities which determine an inequality, thereby demarcating a classical system from a quantum system. Bell's paper demonstrated that non-locality is a basic feature of quantum mechanics (e.g., Aspect et al. 1982, Weihs et al. 1998, Gröblacher et al. 2007), whereas any intuitive feature of a classical system is that a measurement has a value independent of other compatible measurements, and this predetermined value, perhaps determined by a hidden variable, therefore, shows local realism. However, a theorem derived by Kochen and Specker (1967) shows that this classical (non-contextuality) feature is in conflict with quantum mechanics and an event-by-event description of quantum phenomena is impossible. The proofs of this so-called "no-go theorem" can be converted into experimentally measurable inequalities. In this respect this idea has become testable during the past decade with photons, atoms, ions, and especially neutrons when a kind of entanglement between various degrees of freedom is introduced (Hasegawa et al. 2003, Huang et al. 2003, Moehring et al. 2004, Bartosik et al. 2009, Kirchmair et al. 2009). It has been shown that the concept of contextuality is more general than the entanglement issue. Neutrons are convenient guinea pigs for the kind of delicate experiments needed to investigate intrinsic aspects of quantum physics. It should be mentioned that the debate whether all kinds of hidden variables and even local realism must be aborted is still continuing (e.g., Penrose 1994, 2005; Nieuwenhuizen et al. 2007; Khrennikov 2009; Allahverdyan et al. 2013).

7.1 Quantum Contextuality

Entangled systems provide a new basis for related investigations and, therefore, many experiments have started to deal with these questions. Entanglement of pairs of photons or material particles is a well-known phenomenon, which produces a non-classical quantum state. It means that the state of two coupled systems cannot be separated into a product state (Einstein et al. 1935; Bell 1964; Clauser and Shimony 1978; Bertlmann and Zeilinger 2002; see also Gilder's (2008) book *The Age of Entanglement: When Quantum Physics Was Reborn*). Experiments have verified the non-local feature of quantum mechanics (Aspect 1981, 1982; Weihs et al. 1998; Moehring et al. 2004) and have addressed the question of local realism (Weihs 2007, Scheidl et al. 2009). In a more general sense entanglement is not restricted to two-particle systems but can be extended to an entanglement

Neutron Interferometry. Second Edition. Helmut Rauch and Samuel A. Werner.

of different degrees of freedom where internal and external degrees of freedom can be of concern. The falling cat which always lands on its legs is a macroscopic analog to this effect (e.g., Montgomery 1990). Thus, entanglement can also exist between different degrees of freedom of a single-particle system which yields to quantum contextuality (Kochen and Specker 1967; Mermin 1990, 1993; Peres 1993). A significant development is the formulation of a scheme for testing non-contextuality models on the basis of the quantum-mechanical violation of an inequality obtained from the Peres–Mermin proof of the no-go theorem (Cabello et al. 2008). These inequalities can be considered as generalizations of the well-known Bell inequality and can be observed for any quantum system and not just for entangled states. Pan and Home (2009) showed how contextuality is manifested within quantum mechanics when the related proofs are based on mean values obtained from sub-ensembles.

Contextuality implies that the outcome of a measurement depends on the experimental context, i.e., the outcome of a previous or simultaneous experiment of another compatible observable (Roy and Singh 1993, Basu et al. 2001). In this respect it is a more stringent demand than non-locality. In related neutron experiments (Hasegawa et al. 2003, 2006) the commuting observables of the spin path (s) on the Poincaré sphere and the beam path (p) in real space through the interferometer act as two independent degrees of freedom where the following entangled state can be produced:

$$\Psi = \left(|\uparrow>_{\mathrm{s}}|I>_{\mathrm{p}} - |\downarrow>_{\mathrm{s}}|II>_{\mathrm{p}}\right)/\sqrt{2}, \tag{7.1}$$

as we have already seen in Chapter 5. The spin state and the beam path state represent two-level systems. They each can be described by Pauli spin matrices with the commutation relations

$$\begin{aligned} [\sigma_j^{\mathrm{s}}, \sigma_k^{\mathrm{p}}] &= 0 \quad \text{for } \{j,k\} = \{x,y\}, \\ [\sigma_x^{\mathrm{s}}\sigma_y^{\mathrm{p}}, \sigma_y^{\mathrm{s}}\sigma_x^{\mathrm{p}}] &= 0. \end{aligned} \tag{7.2}$$

When applied to a Bell-like state of Eq. (7.1) the eigenvalue equations become

$$\sigma_i^{\mathrm{s}}\sigma_i^{\mathrm{p}}|\psi> = -|\psi>, i = x, y, \tag{7.4}$$

$$(\sigma_x^{\mathrm{s}}\sigma_y^{\mathrm{p}})(\sigma_y^{\mathrm{s}}\sigma_x^{\mathrm{p}})|\psi> = -|\psi>. \tag{7.5}$$

The related Bell state can be produced within the interferometer when a polarized incident beam is split coherently into two beam paths (I and II) and the spin in one beam path is rotated by Larmor precession to the $+y$-direction and in the other beam path to the $-y$-direction (Fig. 7.1).

Then various entangled Bell states can be produced

$$|\psi> = |\rightarrow>\otimes|I> \pm |\leftarrow> \otimes|II>, \tag{7.6}$$

and three others (separable and non-separable ones) can be formulated in a similar fashion. The notation $|\rightarrow\rangle \otimes |I\rangle$ means $+y$ polarization within a guide field in the z-direction along beam path I and $|\leftarrow\rangle \otimes |II\rangle$ means $-y$ polarization along beam path II.

In a separate analysis these Bell-like states have been measured in a quantum tomographical manner as shown in Fig. 7.2 (Hasegawa et al. 2006). This shows spin–path entanglement in spin–path joint measurements. In all these cases Bell-like inequalities can be formulated to demarcate a quantum world from a classical one. The phase shift χ between the beams and the spin rotation angle α are used as path and spin parameters and Bell-like inequalities can be formulated (Hasegawa et al. 2003, Klepp et al. 2014) as follows:

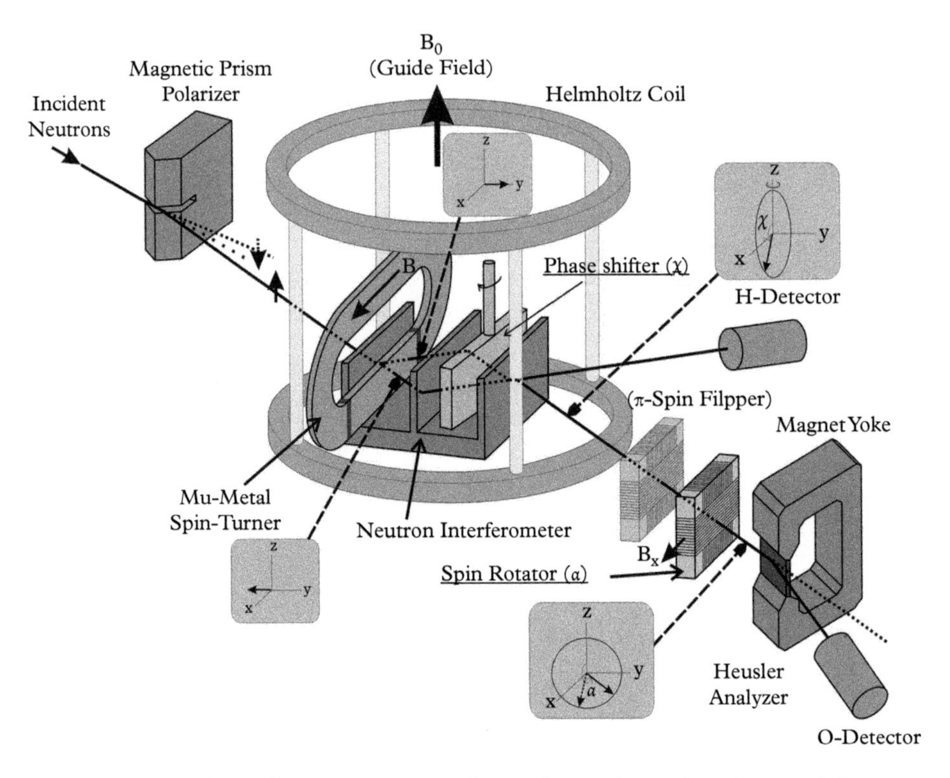

Figure 7.1 *Experimental arrangement to produce and to analyze spin-path entangled neutron states (Hasegawa et al. 2003). Here the y-axis is normal to the crystal plates*

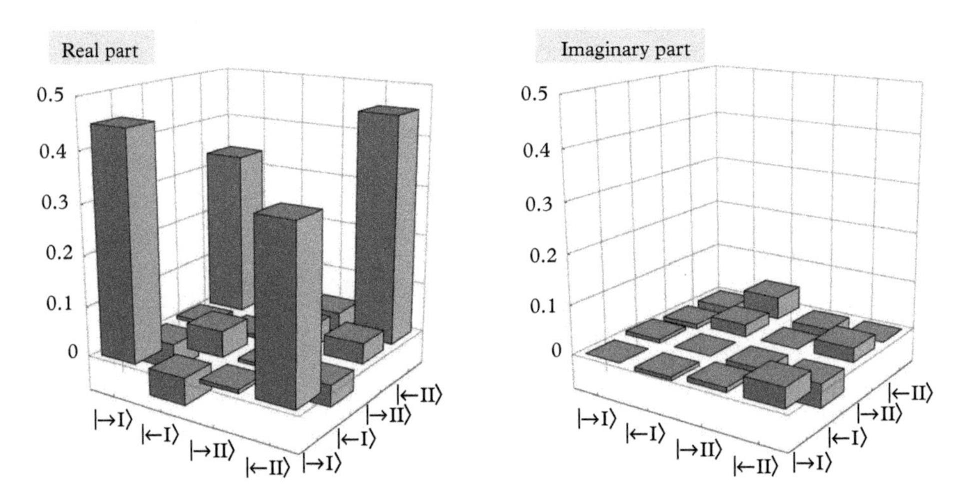

Figure 7.2 *Result of the spin-state reconstruction of an entangled neutron state (Hasegawa et al. 2006)*

$$-2 \leq S \leq 2 \text{ (classical)}, \tag{7.7}$$

$$-2\sqrt{2} \leq S \leq 2\sqrt{2} \text{ (quantum)}, \tag{7.8}$$

$$S = E(\alpha_1, \chi_1) + E(\alpha_1, \chi_2) - E(\alpha_2, \chi_1) + E(\alpha_2, \chi_2), \tag{7.9}$$

$$E(\alpha, \chi) = \frac{N(\alpha, \chi) + N(\alpha + \pi, \chi + \pi) - N(\alpha, \chi + \pi) - N(\alpha + \pi, \chi)}{N(\alpha, \chi) + N(\alpha + \pi, \chi + \pi) + N(\alpha, \chi + \pi) + N(\alpha + \pi, \chi)}. \tag{7.10}$$

This combination of counting rates $N(\alpha, \chi)$ depends on the spin rotation angle α and the nuclear phase shift χ, which occur in each beam path, as it is formulated by analogy to Bell-like states. The maximal violation of classical predictions, and moving toward the quantum-mechanical description happens for the following parameters: $\alpha = 0$, $\alpha_2 = \pi/2$, $\chi_2 = \pi/4$, and $\chi_2 = -\pi/4$. Typical results are shown in Fig. 7.3.

Careful data analysis gave a value of $S = 2.051 \pm 0.019$, i.e., beyond the classical prediction of $S < 2$. The reason why this value is considerably below $2\sqrt{2}$ (the quantum prediction) lies in imperfections of the setup. The contrast of the interference was about 74%. It was mainly limited by stray fields of the spin rotator, and the degree of polarization was 95%. Nevertheless, quantum contextuality has been demonstrated here, indicating an intrinsic correlation between the spin

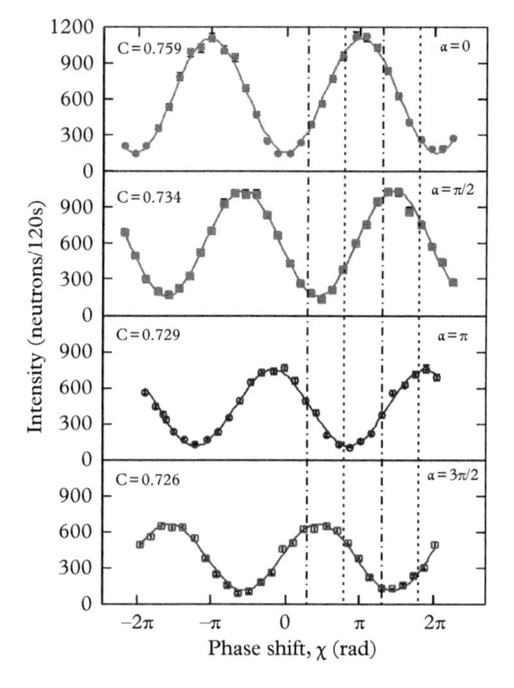

Figure 7.3 *Experimental results of the contextuality experiment with spin-path entangled neutron states (Hasegawa et al. 2003)*

and the momentum (path) variables. These results have been commented on by Weihs (2007) in relation to the general understanding of reality in quantum physics. A more precise value has been obtained by Hasegawa and Erdösi (2011) who used a Mu-metal tube to vary the path integral, resulting in opposite polarization states. Their Bell parameter was $S = 2.202 \pm 0.007$, evidently considerably above 2.

In a continuation of these experiments Hasegawa et al. (2006) dealt with the Kochen–Specker theorem (Kochen and Specker 1967) and the Mermin inequalities (Mermin 1990, 1993), where even stronger violations of classical hidden variable theories are predicted. A related test of the Kochen–Specker theorem was formulated by Simon et al. (2000) and realized for photons by Huang et al. (2003). For neutron matter-waves a related proposal came from Basu et al. (2001) and a related experiment has been performed using a setup similar to that shown in Fig. 7.1 but with the additional feature that the beam paths could be closed alternatively by means of an absorber sheet (Hasegawa et al. 2006). The measurement of the product observable $(\sigma_x^s\sigma_y^p)\cdot(\sigma_y^s\sigma_x^p)$ was done by measuring $(\sigma_z^s\sigma_z^p)$ and using a priori the non-contextuality relation. The measurable quantity is defined by a sum of product observables

$$C = \hat{I} - \sigma_x^s\sigma_x^p - \sigma_y^s\sigma_y^p - (\sigma_x^s\sigma_y^p)\cdot(\sigma_y^s\sigma_x^p). \tag{7.11}$$

In any experiment only expectation values can be measured, which gives rise to a discussion whether experimental verifications are feasible at all. For non-contextual models the last term can be separated as follows:

$$<(\sigma_x^s\sigma_y^p)><(\sigma_y^s\sigma_y^p)> = <\sigma_x^s><\sigma_y^p><\sigma_y^s><\sigma_x^p>, \tag{7.12}$$

which gives for non-contextuality theories

$$C_{\text{non–contextual}} = \pm 2, \tag{7.13}$$

whereas quantum mechanics predicts

$$C_{\text{quantum}} = 4. \tag{7.14}$$

The measured value was

$$C_{\text{exp}} = 3.138 \pm 0.0115, \tag{7.15}$$

which is well above the non-contextuality (classical) limit of 2 and provides an all-versus-nothing type of contradiction. It provides a Peres–Mermin proof of quantum mechanics against non-contextual hidden variable theories.

A debate in literature (Simon et al. 2000, Cinelli et al. 2005) criticized the a priori use of the non-contextuality relation $(\sigma_x^s\sigma_y^p)\cdot(\sigma_y^s\sigma_x^p) = (\sigma_z^s\sigma_z^p)$ and in this connection the use of an absorber to measure this quantity. A follow-up proposal of Cabello et al. (2008) required the measurement of this quantity in the same context as the measurement of the other observables. This has been achieved with a setup shown in Fig. 7.4 (Bartosik et al. 2009).

The maximally entangled state is generated in the first part of the interferometer. The second part serves together with a phase shifter as a path measurement apparatus and a spin-analysis system allows the selection of neutrons with certain spin properties. The spin flippers in both beam paths are required for the measurement of the product observable $\langle\sigma_x^s\sigma_y^p\cdot\sigma_y^s\sigma_x^p\rangle$ (Bartosik

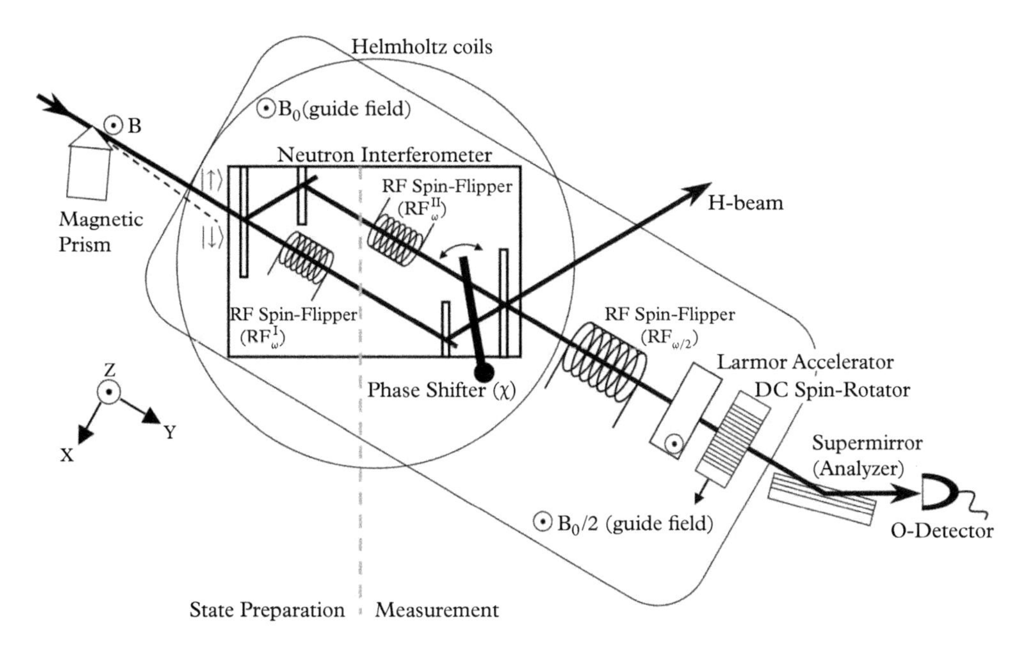

Figure 7.4 *Experimental setup for the loophole-free verification of quantum contextuality (Bartosik et al. 2009)*

et al. 2009). The previous result of Hasegawa et al. (2003) has been verified and a numerically higher violation factor has been achieved ($C_{\text{exp}} = 3.291 \pm 0.008$). In this case a quantum erasure method has been used instead of an absorber and, therefore, all quantities required for Eq. (7.11) could be measured within the same context as required by the theory. Nearly simultaneously an experimental test of quantum contextuality has been performed with trapped ions which yielded compatible results (Kirchmair et al. 2009, Zhang et al. 2013) and with single photons (Simon et al. 2000, Amselem et al. 2009). A series of non-contextuality hidden variable theories (e.g., Bell 1987, Gühne et al. 2010) are shown to be non-valuable by these experiments. However, the discussion whether a loophole-free verification has been achieved is still continuing (e.g., Amselem et al. 2013).

Years ago it was shown that in the case of neutron resonance the Zeeman energy ($\hbar\omega = 2\mu B_0$) is exchanged between the neutron and a resonance coil coherently (Alefeld et al. 1981). This provides the basis for triple-entanglement experiments using spin–path–energy as independent degrees of freedom (Sponar et al. 2008, Hasegawa et al. 2010). These GHZ-like states are a new tool for basic neutron quantum optics experiments and may entangle new components of quantum computing elements like NOTNOT gates. The energy states have a geometric nature and may be rather robust under dissipative effects as shown in Section 4.6.

The production of triply entangled states in a single-neutron system within the neutron interferometer containing a time-dependent flipper is used to provide an energy exchange of $\hbar\omega_1 = 2\mu \text{B}_{01}$ (Section 5.3), which gives a rotating spin behind the interferometer whose rotation is stopped by means of another time-dependent flipper providing half the energy exchange of the first one ($\hbar\omega_2 = 2\mu B_{02}$ with $B_{02} = B_{01}/2$; Fig. 7.4). In this case the three subspaces are spanned by orthogonal two-bases, i.e., two-level systems:

$$
\begin{aligned}
|\psi_{\text{spin}}> &= \{|\uparrow>, |\downarrow>\} \\
|\psi_{\text{path}}> &= \{|I>, |II>\} \\
|\psi_{\text{energy}}> &= \{|E_0 - \hbar\omega>, |E_0>\}.
\end{aligned} \tag{7.16}
$$

In consequence, one can generate the state of neutrons in a triply entangled GHZ-like state (Greenberger et al. 1989)

$$
|\Psi_n^{\text{GHZ}}> = \tfrac{1}{2}\left\{|\uparrow> \otimes |I> \otimes |E_0> + |\downarrow> \otimes |II> \otimes |E_0 - \hbar\varpi>\right\}. \tag{7.17}
$$

Mermin (1990) derived an inequality suitable for experimental tests to distinguish between predictions by quantum mechanics and by local realism theories:

$$
M = E[\sigma_x^{\text{p}}\sigma_x^{\text{s}}\sigma_x^{\text{e}}] - E[\sigma_x^{\text{p}}\sigma_y^{\text{s}}\sigma_y^{\text{e}}] - E[\sigma_y^{\text{p}}\sigma_x^{\text{s}}\sigma_y^{\text{e}}] - E[\sigma_y^{\text{p}}\sigma_y^{\text{s}}\sigma_x^{\text{e}}]. \tag{7.18}
$$

Non-contextual theories set a strict limit for a maximum possible value of $M = 2$, whereas quantum theory predicts a maximum value of $M = 4$. The expectation values in Eq. (7.18) can be measured for certain values of the path phase shift χ, the spin phase α, and the energy phase γ. χ is manipulated by an auxiliary phase shifter, α by an accelerator DC coil, and $\gamma (= 2\omega_1 L/v)$ by the zero-field precession between the two flipper coils located a distance L apart (Golub et al. 1994; Sponar et al. 2008). The experimental result after a careful data analysis was

$$
M = 2.558 \pm 0.004, \tag{7.19}
$$

exhibiting a clear violation of $M = 2$ resulting from non-contextual theories (Hasegawa et al. 2010, Sponar et al. 2010). Even higher degrees of multiple entanglement and higher degrees of fidelity of such states have been demonstrated by Erdösi et al. (2011) and summarized by Klepp et al. (2014). The results demonstrate that the sequence (context) of measurements of different subspaces is essential and an intrinsic feature of quantum mechanics, since the energy entanglement can be extended to multi-entanglements (up to 1000) by an energy manipulation scheme with multiple frequency systems in serial.

7.2 Quantum Cheshire Cat

All interferometer experiments suggest that a quantum system "feels" the physical situations along both beam paths and all physical features should be transported in both beams. This is also of relevance for all degrees of freedom which should exist always in both beams. Is it then possible that one feature is transported in one beam and the other feature in the other beam, respectively? This yields to a Cheshire cat situation where, i.e., the particle properties are transported in one beam and the spin becomes transported in the other beam, as indicated in Fig. 7.5. The nomenclature follows the children's novel *Alice's Adventures in Wonderland.* Actual experiments are related to pre- and post-selection experiments, as described in Section 4.5, and to absorber experiments, as described in Section 4.3. In combination this permits "weak measurements" which can visualize various quantum paradoxes (Aharonov et al. 1988, Aharonov and Vaidman 1991). In this connection the quantum Cheshire cat has attracted attention (Aharonov et al. 2013, Matzkin and Pan 2013, Yu and Oh 2014) and a dedicated neutron experiment has been done by Denkmayr et al. (2014). In this case the beam splitter not only splits the quantum wave describing the neutron itself but also its spin, and a special analyzer system selects in both beams separately the neutron particle features (absorber) and the spin feature (spin rotator) (Fig. 7.5).

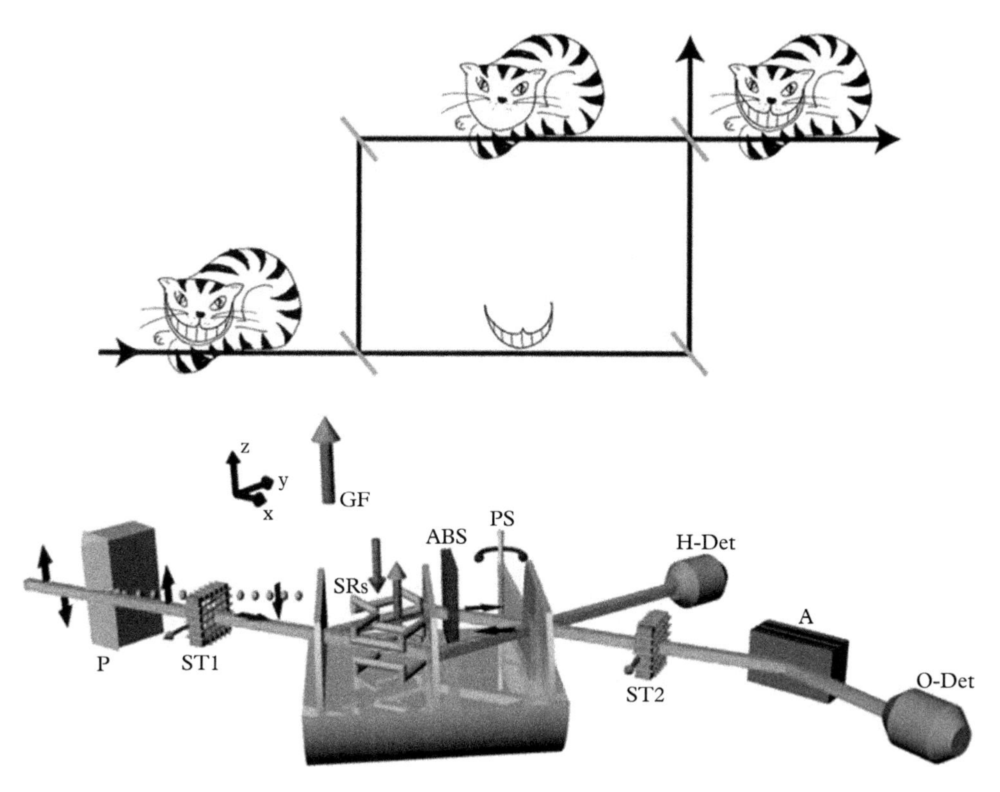

Figure 7.5 *Sketch (above) of a Cheshire cat neutron interferometer experiment and its realization (below). STx are $\pi/2$-spin turners and SRx are spin-rotators use for weak measurements and ABS denotes the absorber (Denkmayr et al. 2014). Reprinted with permission of Denkmayr et al. (2013) and Nature Publ. Group*

In the dark position of the O-detector the insertion of the absorber and the activation of the spin rotator cause distinct neutron counting rates. These intensities are different when the absorber is set into beam I or beam II, indicating a weak particle path measurement. In a second step an additional magnetic field is applied in one or the other beam path which causes also distinct spin rotations ($\sim 20°$) and distinct neutron counting rates related to a weak spin measurement. The results indicate significantly higher count rates when the absorber is put into beam path II and higher count rates when the additional spin rotation is applied to beam I ($N_a(II)/N_a(I) = 6.95 \pm 0.52$; $N_s(II)/N_a(I) = 0.17 \pm 0.39$). The obtained results exhibit the characteristics of a Cheshire cat in a matter-wave interferometer.

The neutrons behave as if particle and spin properties are spatially separated while traveling through the interferometer. These statements can be given only for the system behind the analyzer, but such statements are generally questionable because they cannot be proven or disproven. The situation is rather similar to that in the spin-superposition experiment discussed in Section 5.2 where a post-selected spin-up state after the superposition can or cannot be attributed to the spin-up state before superposition (Fig. 5.5). The observation of Cheshire cat situations is another example of paradoxical phenomena found within the framework of quantum mechanics (see e.g. Klepp et al. 2014).

8

Gravitational, Inertial, and Motional Effects

The fundamental dual nature of neutrons—sometimes a particle (when detected) and sometimes a wave (when traversing the interferometer)—is wonderfully manifested by the non-local nature of the neutron interferometry experiments discussed in this chapter. We begin with a discussion of the series of experiments, called gravitationally induced quantum interference, in Section 8.1. Then in Section 8.2, the effect of the Earth's rotation on the phase of a neutron de Broglie wave is described. This is the quantum-mechanical analog of the Michelson, Gale, Pearson (1925) experiment with light. An experiment in which this Sagnac phase shift was observed due to rotation of the interferometer on a turntable within the laboratory frame was verified (Atwood et al. 1984). Such experiments prove the universality of free fall by comparing the acceleration of neutrons to that of classically freely falling objects (Wolf et al. 2011). In this respect the neutron interferometer can be seen as a quantum gravimeter. An experiment related to the small effective mass of the neutron propagating in a crystal under Bragg reflecting conditions and its deflection due to the Coriolis force is discussed in Section 8.2.3 (Raum et al. 1995). Acceleration-induced interference, which is related to the gravity experiments by the principle of equivalence, is discussed in Section 8.3 (Bonse and Wroblewski 1983). It is natural to ask about the connection of these neutron interference experiments to similar ones carried out with photons. Section 8.4 is devoted to this question. Phase shifts caused by the motion of matter within the interferometer are related to the optical Fizeau effect. Several neutron Fizeau-type experiments are discussed in Section 8.5.

8.1 Gravitationally Induced Quantum Interference

Neutrons, like all matter, are subject to Newton's (1686) universal gravitational force. This fact has been demonstrated by verifying that neutrons fall on a parabolic trajectory in the Earth's gravitational field (Dabbs et al. 1965, Koester 1976). This is a consequence of classical mechanics and is expected from the principle of equivalence. For a review of the principle of equivalence see Hughes (1993). However, gravity and quantum mechanics do not simultaneously play an important role in most phenomena experimentally accessible in terrestrial physics. In this section we describe a series of neutron interferometry experiments for which the outcome necessarily depends upon both the gravitational constant G and Planck's constant h. The first observation of the phase shift of a neutron de Broglie wave induced by the Earth's gravity was made in an experiment carried out at the 2 MW University of Michigan Reactor by Colella, Overhauser,

Neutron Interferometry. Second Edition. Helmut Rauch and Samuel A. Werner.

Table 8.1 *Contributions to the Earth's Gravitational Acceleration.*

Effect	**Oder of magnitude (m/s^2)**
Earth sphere gravity	10
Ellipsoidal and Sagnac effect	10^{-3}
Landscape	10^{-4}
Local density variations	10^{-5}
Tidal forces	10^{-7}
General relativity theory	10^{-9}

and Werner (1975). A series of increasingly sophisticated and precise experiments, now collectively called COW (Colella–Overhauser–Werner) experiments, were pursued at the University of Missouri Research Reactor (MURR; Staudenmann et al. 1980; Werner et al. 1988; Jacobson 1993; Littrell 1996, 1997). For reviews of related experiments with electrons and atoms see Hasselbach (2010) and Cronin et al. (2009), respectively.

The experiments deal with a measurement of the local acceleration g due to the Earth, which is determined by the local gravity and small additional effects collected in Table 8.1. The ordinary gravity and the Sagnac (Coriolis) contributions have been verified with neutrons. A general relativistic treatment of the COW experiment yields, in first order, the same gravity-induced phase shifts as those obtained by the Newtonian theory of gravity (Varju and Ryder 2000, Wolf et al. 2011, Greenberger et al. 2012). More details are found in Section 8.5.6.

8.1.1 Geometry of the COW Experiment

A schematic diagram of the overall apparatus used in the experiments carried out at MURR is shown in Fig. 2.12. The double-crystal monochromator provides a selectable and continuously variable wavelength ($\lambda \sim 0.8$ to 2.4 Å) incident beam directed along the local north–south axis of the Earth, a fact which we will see is important in these experiments. In the earlier experiments, the monochromator crystals were pyrolytic graphite, PG (004); the more recent experiments have utilized a set of vertically focusing copper, Cu(220), crystals. The nominally monochromatic ($\Delta\lambda/\lambda \sim 0.005$), collimated ($\Delta\theta \sim 0.4°$) beam is directed through a series of slits, and is incident upon the interferometer along the horizontal line SA shown in Fig. 8.1. Tilting the interferometer about this incident beam direction through an angle α, while maintaining the Bragg condition, requires the neutron wave packets on the sub-beam paths I and II to mix and recombine in the third crystal plate near point D, which is higher above the Earth by an amount $H(\alpha) = H_0 \sin(\alpha)$ than the entrant point A. The phase accumulated along the rising segment AC on path II is equal to the phase accumulated along the rising segment BD on path I. However, due to conservation of energy, the momentum, $p = \hbar k$, of the wave packet along the upper, horizontal beam segment CD is less than the momentum, $p_0 = \hbar k_0$, along the lower beam segment AB, that is (similar to Eq. 2.29)

$$E_0 = \frac{\hbar^2 k_0^2}{2m} = \frac{\hbar^2 k^2}{2m} + mgH(\alpha), \tag{8.1}$$

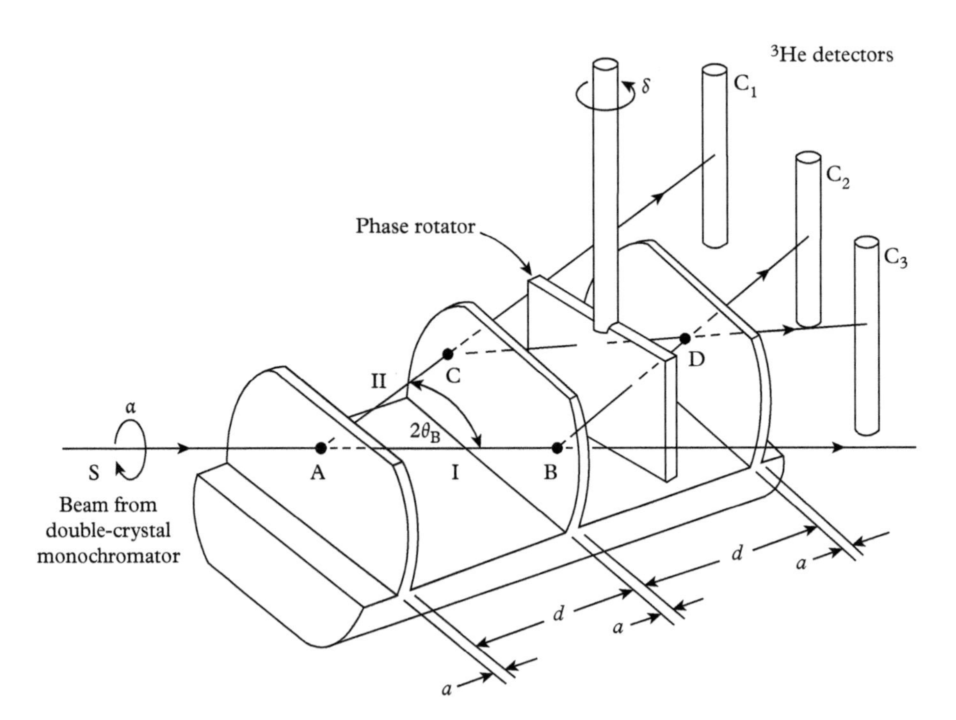

Figure 8.1 *Schematic diagram of the LLL interferometer used in the gravitationally induced quantum interference experiments. The interferometer crystal, along with the 1.2-cm-diameter 20-atm* ^3He *neutron detectors is tilted about the incident beamline SA through angles* α*, while maintaining the Bragg condition for wavelength* λ *and Bragg angle* θ_B*. It was machined by Mr. Clifford Holmes in the Physics Department Machine Shop at the University of Missouri*

where g is the acceleration due to gravity. Since the gravitational potential energy difference is small ($mg \sim 1.003$ neV/cm) compared to the kinetic energy, $E_0 \sim 20$ meV, of the incident neutron, the wave vector difference according to Eq. (8.1) is given approximately by

$$\Delta k = (k - k_0) \cong -\frac{m^2 gH}{\hbar^2 k_0} \sin\alpha, \tag{8.2}$$

where $\alpha = 0$ corresponds to the horizontally level interferometer situation, with both beam paths (I and II) parallel to the Earth's surface. The phase difference of the neutron traversing path II relative to path I is therefore given by

$$\Delta\Phi_{\text{COW}} = \Phi_{\text{II}} - \Phi_{\text{I}} = \Delta k S, \tag{8.3}$$

where S is the path length of the segments AB and CD. Writing $k_0 = 2\pi/\lambda$ we have

$$\Delta\Phi_{\text{COW}} = -2\pi\lambda \frac{m^2}{h^2} g A_0 \sin\alpha, \tag{8.4}$$

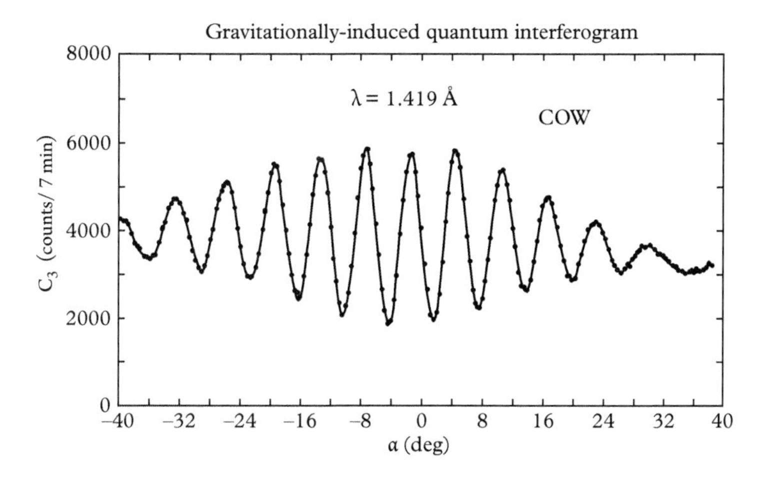

Figure 8.2 *Gravitationally induced quantum interferograms taken with the LLL interferometer shown in Fig. 8.1, for which d = 34.518 ± 0.002 mm and a = 2.464 ± 0.002 mm (Staudenmann et al. 1980)*

where $A_0 = H_0S$ is the area of the parallelogram enclosed by beam paths I and II. An alternative derivation of this formula has been given by Mannheim (1998). Many graduate students study the derivation of this formula in standard quantum-mechanics textbooks such as the one by Sakurai (1994). We see that this phase shift is proportional to the neutron wavelength λ and depends explicitly on Planck's constant h and the acceleration g due to the Earth's gravity. The experimental procedure used to measure this phase shift involves tilting the interferometer about the incident beam direction through various angles α and observing the variation in counting rates of the two outgoing interfering beams in the two ^{3}He detectors, labeled C_2 (H-beam) and C_3 (0-beam) in Fig. 8.1. An example of such a gravity-induced interferogram (Staudenmann et al. 1980) is shown in Fig. 8.2. The frequency of the oscillations of this interferogram differs slightly from that predicted by the formula (8.4), due to subtle dynamical diffraction effects, the Sagnac effect, and the bending of the interferometric under its own weight. We will discuss these effects in some detail in Section 8.1.4.

8.1.2 Formal Derivations of the COW and Sagnac Phase Shifts

Since the experiment was carried out in a frame fixed to our rotating Earth, a non-inertial frame of reference, the Lagrangian $\mathcal{L}$ governing the neutron's motion as it traverses the interferometer involves the Earth's rotation frequency, $\mathbf{\Omega}$. Transforming the Lagrangian in an inertial frame ($m_i v^2/2 + Gm_g M/r$) to this rotating frame, it is a straightforward matter to show that

$$\mathcal{L}(\boldsymbol{r}, \boldsymbol{v}) = \frac{1}{2}m_i v^2 + G\frac{m_g M}{r} + m_i(\mathbf{\Omega} \times \boldsymbol{r}) \cdot \boldsymbol{v} + \frac{1}{2}m_i|\mathbf{\Omega} \times \boldsymbol{r}|^2, \tag{8.5}$$

where G is Newton's universal gravitational constant, m_i and m_g are the inertial and gravitational masses of the neutrons, M is the Earth's mass, and $\boldsymbol{r}$ and $\boldsymbol{v}$ are the neutron's position and velocity in the rotating frame of the interferometer. The Hamiltonian $\mathcal{H}$ is obtained from $\mathcal{L}$ using the Legendre transformation (Eq. 1.32), and is found to be (Landau and Lifshitz 1969)

$$\mathcal{H}(\boldsymbol{r},\boldsymbol{p}) = \frac{\mathrm{p}^2}{2m_\mathrm{i}} - G\frac{m_\mathrm{g}M}{r} - \boldsymbol{\Omega}\cdot\boldsymbol{L}, \tag{8.6}$$

where $\boldsymbol{L} = \boldsymbol{r}\times\boldsymbol{p}$ is the angular momentum of the neutron's motion about the center of the Earth ($\boldsymbol{r} = 0$). The neutron's canonical momentum is

$$\boldsymbol{p} = \frac{\partial\mathcal{L}}{\partial\boldsymbol{v}} = m_\mathrm{i}\boldsymbol{v} + m_\mathrm{i}\boldsymbol{\Omega}\times\boldsymbol{r}. \tag{8.7}$$

Using Hamilton's equations

$$\dot{\boldsymbol{r}} = \frac{\partial\mathcal{H}}{\partial\boldsymbol{p}} \text{ and } \dot{\boldsymbol{p}} = -\frac{\partial\mathcal{H}}{\partial\boldsymbol{r}}, \tag{8.8}$$

it is easy to obtain the well-known equation of motion for a particle in a rotating frame of reference, that is

$$m_\mathrm{i}\ddot{\boldsymbol{r}} = m_\mathrm{g}\boldsymbol{g}_0 - m_\mathrm{i}\boldsymbol{\Omega}\times(\boldsymbol{\Omega}\times\boldsymbol{r}) - 2m_\mathrm{i}\boldsymbol{\Omega}\times\dot{\boldsymbol{r}}. \tag{8.9}$$

The acceleration due to gravity is

$$\boldsymbol{g}_0(\boldsymbol{r}) = -\frac{GM}{r^2}\hat{\boldsymbol{r}}. \tag{8.10}$$

Thus, we see that the term $-\boldsymbol{\Omega}\cdot\boldsymbol{L}$ in the Hamiltonian gives rise to both the centrifugal force and the Coriolis force. Since we are only interested in the neutron's motion over a distance corresponding to the dimensions of the interferometer, which are very small compared to the Earth's radius R, we define an effective gravitational acceleration in the usual manner, namely

$$\boldsymbol{g} = \boldsymbol{g}_0(\boldsymbol{R}) + \left(\frac{m_\mathrm{i}}{m_\mathrm{g}}\right)\boldsymbol{\Omega}\times(\boldsymbol{\Omega}\times\boldsymbol{R}), \tag{8.11}$$

which is independent of the neutron's instantaneous position. Under this assumption, the solution of the equation of motion (8.9) to leading order in Ω is

$$\boldsymbol{r}(t) = \boldsymbol{r}_0 + \boldsymbol{v}_0 t + \frac{1}{2}\boldsymbol{g}t^2 + \frac{1}{3}t^3\boldsymbol{\Omega}\times\boldsymbol{g}. \tag{8.12}$$

Based upon a sidereal day of 23 h 56 m, one gets $\Omega = 7.29\times10^{-5}\ \mathrm{s}^{-1}$. The transit time for thermal neutrons (v ~ 2 mm/μs) across a Si-crystal interferometer ($l \sim 10$ cm) is about 50 μs. Thus, the term in Eq. (8.12) involving Ω is smaller than $gt^2/2$ by a factor of about 10^{-9}. Consequently, the effect of the Coriolis force on the trajectory over these small distances is negligible. However, its effect on the neutron phase is not negligible, as we shall see in Section 8.2. Between the crystal blades of the interferometer the neutron wave packets move on a parabola. But over the small distances involved within the interferometer the curvature of the trajectories is very small. In fact, the angular deviation from a straight line is only about 0.25 μrad ~ 0.05 arc sec. This is well within the Darwin width (~2 arc sec) for reflection by the Si-crystals (see Chapter 11), which is necessary (and fortunate) for the neutrons to satisfy the Bragg conditions in the second and third crystals of the interferometer. For experiments carried out at MURR described in this section, the centrifugal acceleration of the neutron was 0.0265 m/s^2 and the Coriolis acceleration was 0.41 m/s^2 for 1.4-Å neutrons.

The discussion so far in this section has been based upon classical mechanics. In order to calculate the quantum mechanical phase shift $\Delta\Phi$ we must associate with the canonical momentum $\boldsymbol{p}$ the operator $-i\hbar\nabla$. In the spirit of the WKB approximation the momentum is then $\boldsymbol{p} = \hbar\boldsymbol{k}$, and the phase shift is obtained by the integration of the action along the unperturbed (by gravity or rotation) trajectories according to the Feynman–Dirac prescription, namely

$$\Delta\Phi = \frac{1}{\hbar}\int_{\mathrm{ACD}} \boldsymbol{p}\cdot\mathrm{d}\boldsymbol{r} - \frac{1}{\hbar}\int_{\mathrm{ABD}} \boldsymbol{p}\cdot\mathrm{d}\boldsymbol{r} = \frac{1}{\hbar}\oint \boldsymbol{p}\cdot\mathrm{d}\boldsymbol{r} \tag{8.13}$$

The momentum appearing in this line integral on the closed path ACDBA along the trajectories of the neutron within the interferometer is given by Eq. (8.7). Thus, the phase shift involves two terms,

$$\Delta\Phi = \frac{m_{\mathrm{i}}}{\hbar}\oint \boldsymbol{v}\cdot\mathrm{d}\boldsymbol{r} + \frac{m_{\mathrm{i}}}{\hbar}\oint (\boldsymbol{\Omega}\times\boldsymbol{r})\cdot\mathrm{d}\boldsymbol{r}. \tag{8.14}$$

To first order in g, the first term yields $\Delta\Phi_{\mathrm{COW}}$ as given by Eq. (8.4), where the area $A_0 = (2d^2 + 2ad)\tan\theta_{\mathrm{B}}$. Here the crystal blade thickness is a and the blade separations are d. However, the square of the neutron mass should be replaced by the product of the neutron's inertial and gravitational masses, so that we can write

$$\Delta\Phi_{\mathrm{COW}} = -q_{\mathrm{COW}}\sin\alpha, \tag{8.15a}$$

where

$$q_{\mathrm{COW}} = 2\pi\lambda\frac{m_{\mathrm{i}}m_{\mathrm{g}}}{h^2}gA_0. \tag{8.15b}$$

The neutron wavelength λ is given by its laboratory velocity $v(\lambda = h/m_{\mathrm{i}}v)$. Measuring this phase shift induced by the Earth's gravity can therefore be regarded as a test of the principle of equivalence ($m_{\mathrm{i}} = m_{\mathrm{g}}$) in the quantum limit.

The second term in Eq. (8.14), which we call $\Delta\Phi_{\mathrm{Sagnac}}$, is due to the Earth's rotation. Using vector calculus, it can be written as a surface integral, yielding

$$\Delta\Phi_{Sagnac} = \frac{2m_{\mathrm{i}}}{\hbar}\boldsymbol{\Omega}\cdot\boldsymbol{A}_0. \tag{8.16}$$

This formula was obtained by Page (1975) using wave-optical arguments, and by Anandan (1977) and Stodolsky (1979, 1979a) within the framework of general relativity. A very readable general relativistic treatment of these COW and Sagnac effect experiments is given by Varju and Ryder (2000). An interesting derivation has been given by Dresden and Yang (1979) in which the phase shift for either a rotating neutron or optical interferometer is derived from the point of view of the Doppler shift of waves from the moving mirrors (crystals) relative to an inertial frame of reference. A number of important observations and conclusions are made by Dresden and Yang (1979). First, the reflection from the moving mirrors also changes the beam displacements in first order, but these displacements affect the phase shifts in second order because of Fermat's principle for reflection. The importance of integrating the wave vector along the classical unperturbed beam paths enters many other calculations in this book (e.g., Opat 1995). Second, the phase shift is calculated at a given instant of time. Therefore, only the $\boldsymbol{k} \cdot \boldsymbol{x}$ term enters the phase shift; the ωt part of the phase plays no role. This is true of all phase shifts where the Hamiltonian is independent of time, so that the energy $E = \hbar\omega$ is a constant of the motion. The Dresden–Yang derivation is also valid for any optical or atom interferometer. Sakurai (1979) pointed out that this phase shift can be regarded as a topological effect due to the flux of rotation penetrating the loop of the neutron wave packet's motion around the interferometer. This is analogous to the topological Aharonov–Bohm effect for electrons encircling a long tube of magnetic flux. Sakurai (1994) notes that the Coriolis force $2m\boldsymbol{v} \times \boldsymbol{\Omega}$ is of the same form as the Lorentz force on a charged particle in a magnetic field, namely,$(e/c)\, \boldsymbol{v} \times \boldsymbol{B}$ if one identifies $(e/c)\, \boldsymbol{B}$ with $2m\boldsymbol{\Omega}$.

For a horizontally directed incident beam, the dot product in Eq. (8.16) is easily evaluated; the result is

$$\Delta\Phi_{\text{Sagnac}}(\alpha) = \frac{2m_{\text{i}}}{\hbar}\Omega A_0 (\cos\theta_{\text{L}} \cos\alpha + \sin\gamma \sin\theta_{\text{L}} \sin\alpha), \tag{8.17}$$

where θ_{L} is the colatitude angle at the point on the Earth's surface where the experiment is carried out, and γ is the angle of the incident neutron beam west of due south. In the experiment at beam port B at MURR, γ is nearly exactly 0 and $\theta_{\text{L}} = 51.37°$. Therefore, the Sagnac phase shift for the horizontally directed beam experiment can be written as

$$\Delta\Phi_{\text{Sagnac}}(\alpha) = q_{\text{Sagnac}} \cos\alpha, \tag{8.18}$$

where

$$q_{\text{Sagnac}} = \frac{4\pi\, m_{\text{i}}}{h}\Omega A_0 \cos\theta_{\text{L}}. \tag{8.19}$$

Additional terms which are orders of magnitude smaller than these two leading terms appear when a relativistic treatment and a spin–gravity coupling is taken into account (Mashhoon 1988, Varju and Ryder 2000).

An alternative approach to understanding the effect of Earth's gravity on a phase of matter waves in terms of a gravitational redshift has been put forward by Müller et al. (2010), who analyzed previous quantum gravity experiments done with atoms (Peters et al. 1999), while emphasizing the Compton frequency as an internal clock of particles ($\omega_{\text{C}} = mc^2/\hbar \approx 10^{25}$ Hz). They use this method to claim a very precise determination of the gravitational redshift. The same type of analysis can be done for the COW neutron interferometer experiment. There is an ongoing discussion in the literature of whether this view is correct. In any case it shows that interference experiments provide a pathway to achieve extremely high sensitivities and a deeper understanding of quantum physics (Hohensee et al. 2011, Sinha and Samuel 2011, Wolf et al. 2011, Greenberger et al. 2012, Lan et al. 2013). In Section 8.5.6 we discuss the importance of proper time τ and general relativistic considerations in understanding these inherently non-relativistic matter wave interference experiments in the gravitational field of our rotating Earth.

8.1.3 The Total Phase Shift Including Bending

As we have already pointed out, there is an additional effect on the measured phase shift $\Delta\Phi(\alpha)$ resulting from bending (or warping) of the silicon crystal interferometer under its own weight. Since the interferometer is tilted about the incident beam direction, which is not an axis of elastic symmetry of the device, the effect of bending cannot be reliably modeled and calculated.

To circumvent this problem the phase shift due to bending has been separately measured with X-rays. The procedure involves using molybdenum K_α X-rays ($\lambda = 0.71$ Å) directed along the same incident beam direction (SA in Fig. 8.1) and observing the interfering X-ray beams with an X-ray sensitive proportional argon gas-filled detector as a function of tilt angle α. The effect of gravity (gravitational redshift, see Section 8.5.6) on the X-ray photon is negligible over the distances involved in the interferometer. It has been found that the phase shift due to bending is proportional to sin α. This is understandable since the path length difference, $\Delta L(\alpha) = L_{II}(\alpha) - L_I(\alpha)$, is expected to increase with increasing interferometer tilt, such that

$$\Delta\Phi_{bend}(\alpha) = -k \cdot \Delta L(\alpha) = -(2\pi/\lambda)\Delta L_0 \sin\alpha \equiv -q_{bend}\sin\alpha, \tag{8.20}$$

where ΔL_0 depends upon the elastic properties of the Si interferometer and its mounting. The sign of this phase shift is experimentally found to be the same as that due to gravity.

Thus, the total phase shift is composed of three terms,

$$\begin{aligned}\Delta\Phi(\alpha) &= \Delta\Phi_{grav}(\alpha) + \Delta\Phi_{bend}(\alpha) + \Delta\Phi_{Sagnac}(\alpha)\\ &= -q_{grav}\sin\alpha - q_{bend}\sin\alpha + q_{Sagnac}\cos\alpha\\ &= q\sin(\alpha - \alpha_0)\ ,\end{aligned} \tag{8.21}$$

where the frequency of the interferogram oscillations is given by

$$q = \left[\left(q_{grav} + q_{bend}\right)^2 + \left(q_{Sagnac}\right)^2\right]^{1/2}, \tag{8.22}$$

and

$$\alpha_0 = \tan^{-1}\left[\frac{q_{Sagnac}}{q_{grav} + q_{bend}}\right]. \tag{8.23}$$

Here q_{grav} differs from q_{COW} by a small correction factor discussed in the next section. The ratio q_{Sagnac}/q_{grav} is 0.025 for λ = 1.4 – Å neutrons. Typically, q_{bend}/q_{grav} is about 0.04. Since q_{Sagnac} enters Eq. (8.22) in quadrature, its contribution to the total frequency of oscillation q is small (of order 3 parts in 10^4). It should be noted that the three contributions $\Delta\Phi_{grav}$, $\Delta\Phi_{bend}$, and $\Delta\Phi_{Sagnac}$ to the total phase shift depend upon the neutron wavelength as λ^1, λ^{-1}, and λ^0, respectively. This different dependence on wavelength for the three contributions to the total phase shift was exploited in the "two-wavelength" experiment discussed in Section 8.1.5. When the contribution of bending to the phase shift is experimentally measured with X-rays, the scaling of q_{bend} with λ^{-1} is taken into account.

8.1.4 Neutron-X-Ray Difference Experiments

Various techniques for mounting the interferometer in its V-shaped cradle have been utilized. The method that has proven most successful in minimizing $\Delta\Phi_{bend}$ is to use a type of pliable double-sided sticky-back tape between the interferometer and the cradle. We show in Fig. 8.3 full 360°-tilt-interferograms as recorded in the C_2 and C_3 detectors. The oscillations in the two detectors are 180° out of phase with each other, as expected. The loss of contrast as the interferometer tilt angle approaches 90°, and then the recovery of contrast as the interferometer is turned upside down (α = 180°), are due to a dynamical diffraction effect, and not to bending of the interferometer, as first pointed out by Horne (1986). The three-crystal LLL interferometer is not a simple two-path device, but really an eight-path interferometer, as shown in Fig. 8.4. This

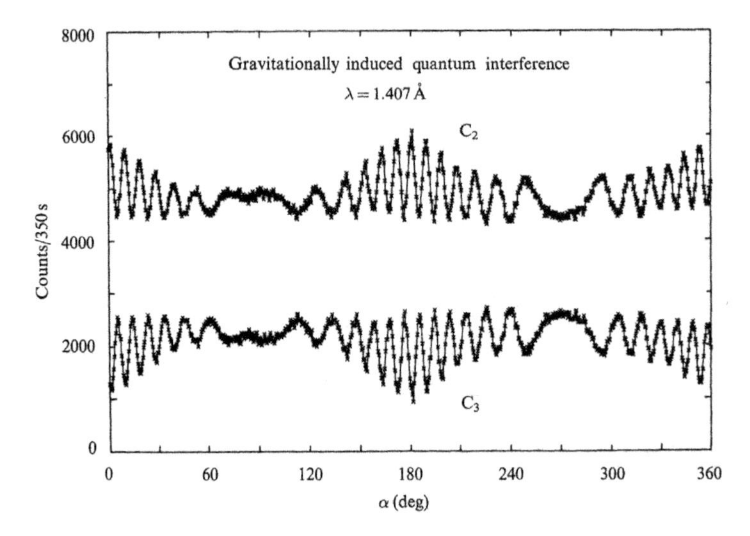

Figure 8.3 *Full-rotation, gravitationally induced quantum interferogram taken with the interferometer shown in Fig. 8.1. Here the Si-interferometer is attached to its cradle with double-sticky-back tape. One should note the 180° phase difference between the C_2 and C_3 data (Werner et al. 1986)*

fact is discussed in detail in Chapter 11. It has the effect that the single interferometer area A_0 appearing in the COW phase shift formula, Eq. (8.15), should be replaced by a dynamical diffraction intensity-weighted average over three areas: A_0, $A_0 + \delta A$, and $A_0 - \delta A$, where $\delta A/A_0 = |\Gamma|\, a/d$, and $|\Gamma|$ is a factor (less than unity) dependent upon the misset angle $\Delta\theta$ of a given incident ray from the exact Bragg angle (see Chapter 11). Since the frequency of oscillation of the gravity-induced interferogram depends on the interferometer area, this intensity-weighting process over the spectrum of areas results in a loss of contrast in much the same way as a wave packet is formed from a spectrum of plane waves of various frequencies. Furthermore, there is a shift of the central frequency of the interferogram oscillations from q_{COW} to

$$q_{\text{grav}} = q_{\text{cow}}\ (1 + \varepsilon(\alpha)), \tag{8.24}$$

where the correction factor $\varepsilon(\alpha) \approx 2a/3d$, independent of α to leading order in a/d. For the interferometer used to obtain the data in Fig. 8.3, $\varepsilon = 0.0476$. These dynamical diffraction corrections were first noted by Bonse and Wroblewski (1983, 1984) in their analysis of the acceleration-induced interferometry experiments discussed in Section 8.3. A detailed analysis of these effects has been given by Littrell et al. (1998).

In the early experiments of Colella, Overhauser, and Werner (1975) and of Staudenmann et al. (1980) the frequency of oscillation q of the gravity-induced interferogram was obtained by Fourier transformation of the data. In the more recent experiments of Werner et al. (1988) and Littrell et al. (1996, 1997; see Section 8.1.5) the following procedure was used to obtain q, q_{grav}, and q_{bend}: The phase shift $\Delta\Phi(\alpha)$ is measured directly by first setting $\alpha = 0$, and rotating the phase flag (typically, a 2-mm-thick, polished aluminum plate) through successive angles δ. The phase of the resulting interferogram is obtained by standard least-squares fitting methods. The interferometer is then tilted to some other angle α and another phase-rotator interferogram is recorded. The difference in phase between these two interferograms is $\Delta\Phi(\alpha)$. This phase difference is then plotted versus

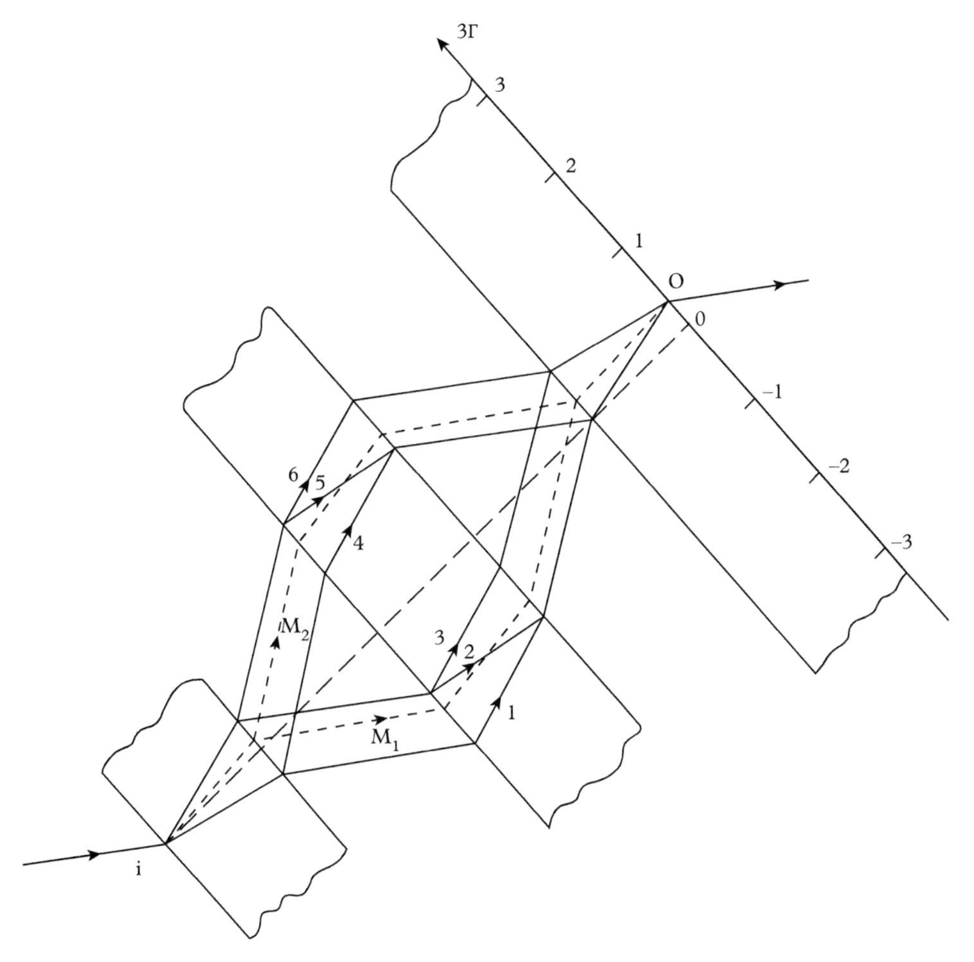

Figure 8.4 *This diagram shows the 8 beam trajectories in the LLL interferometer. The solid lines give the "primary" trajectories and the dashed lines are the "maverick" trajectories (see Section 11.7.2). Reprinted from Horne 1986, copyright 1986, with permission from Elsevier.*

sin $(\alpha - \alpha_o)$, as shown in Fig. 8.5 (Werner et al. 1988). The slope of these plots is denoted as q_{exp}. An analogous experiment was then carried out with X-rays, where the phase rotator is a thin plate of plastic. A plot of the phase $\Delta\Phi_{\text{X-rays}}$ versus sin α gives the phase shift due to bending as shown in Fig. 8.5. The slope of this plot is q_{bend} appropriate to the X-ray wavelength (λ = 0.711 Å). Scaling this value of q_{bend} to the wavelength of the neutron experiment (λ = 1.417 Å) then gives the bending correction of the neutron data, that is

$$q_{\text{grav}} = \left(q_{\text{exp}}^2 - q_{\text{Sagnac}}^2\right)^{1/2} - q_{\text{bend}}. \tag{8.25}$$

In the X-ray experiment, the Si(440) Bragg reflection was used, while in the neutron experiment the Si(220) Bragg reflection was used. Consequently, the X-rays and neutrons follow the same

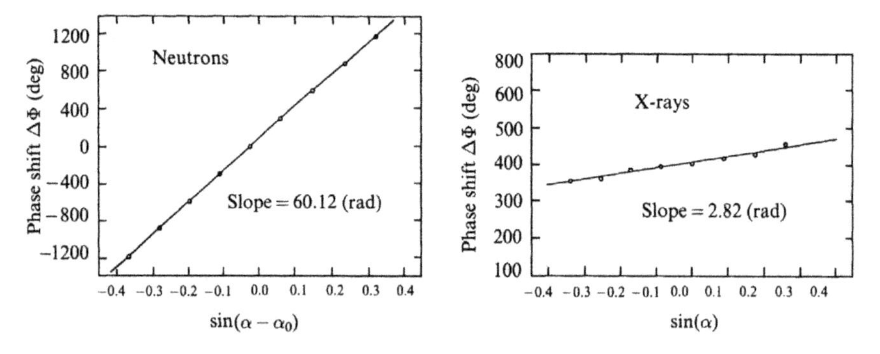

Figure 8.5 *Phase shift as a function of the tilt angle* α *measured with neutrons (left) and with X-rays (right). The angle* $\alpha_0 = 1.4°$ *is due to the Sagnac effect (Werner et al. 1988)*

trajectories through the interferometer. Numerically Eq. (8.25) reads

$$q_{\text{grav}}(\text{exp}) = (60.12^2 - 1.45^2)^{1/2} - 1.42 = 58.72 \pm 0.03 \;\; \text{rad}, \tag{8.26}$$

for the data shown in Figs. 8.5 and 8.6. Theory predicts that

$$q_{\text{grav}}(\text{theory}) = (1 + \varepsilon) q_{\text{COW}} = 59.2 \pm 0.1 \;\; \text{rad}, \tag{8.27}$$

where the stated uncertainty is due to the precision in the measured neutron wavelength. Thus, the observed value of the gravity-induced phase shift from these data is 0.8% lower than theory predicts.

8.1.5 Neutron Two-Wavelength Difference Experiment

Subsequent to the publication of the above result, Layer and Greene (1991) suggested that the small discrepancy between theory and experiment might be due to the fact that X-rays interrogate a somewhat different and smaller region of the interferometer crystal blades than neutrons. This is the result of the much stronger absorption of X-rays by the Si crystals which leads to an incomplete filling of the Borrmann fans (see Chapter 11) as the X-rays traverse the interferometer. Since X-rays experience substantial absorption in silicon, the rays traverse the interferometer via the anomalous transmission effect and do not spread out very much. This is shown schematically in Fig. 8.6. Detailed elastic deformation calculations using finite element analysis indicated that such an effect would lead to an over-correction of the bending contribution to the phase shift of about the right magnitude. Figure 8.7 shows the type of distortion that the interferometer undergoes.

A series of X-ray experiments were undertaken by Arif et al. (1994) on the same interferometer used in the gravity experiments, in which the interferometer was translated in small steps across the incident beam. The phase shifts due to bending were found to be dependent upon the location in the first crystal blade where the X-rays are Bragg diffracted, and were observed to be a non-linear function of position. Consequently, it was found to be somewhat difficult to model the bending phase shifts contributing to the neutron experiments using the X-ray data. Nevertheless, it was noted that the observed phase shifts with X-rays could be made very small (less than 1% of $\Delta\Phi_{\text{grav}}$) by carefully modifying the interferometer mounting.

Since the phase shift induced by gravity is proportional to λ, and the phase shift due to bending is proportional to λ^{-1}, a simultaneous measurement with two neutron wavelengths, say λ_1 and λ_2,

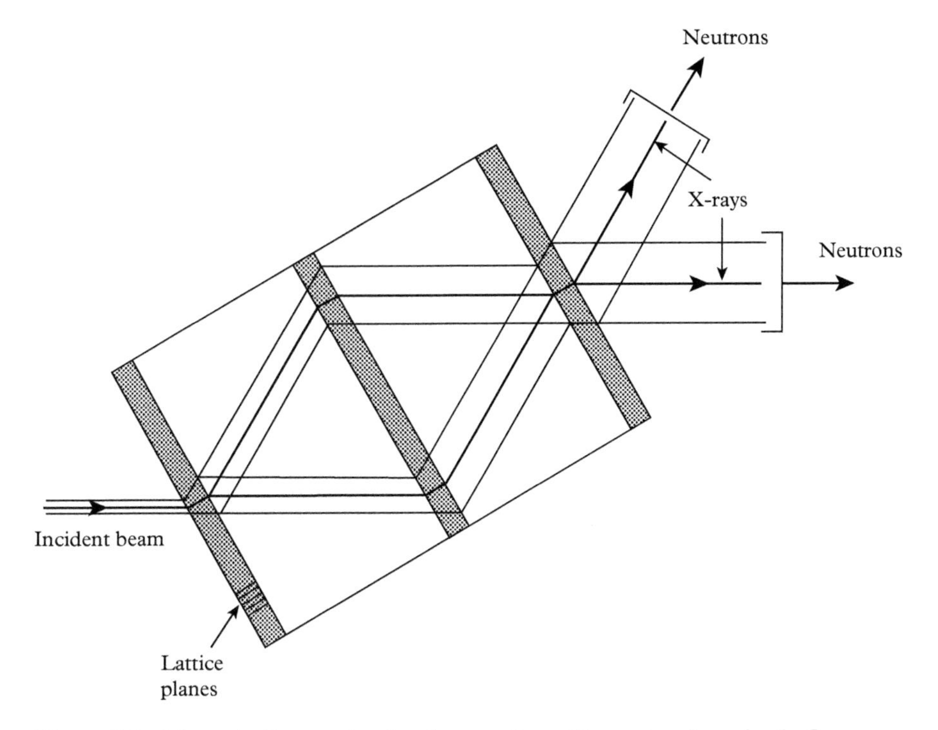

Figure 8.6 *Schematic diagram showing the spreading of the neutron beam by the Borrmann fans in each crystal plate and less spreading in the X-ray case*

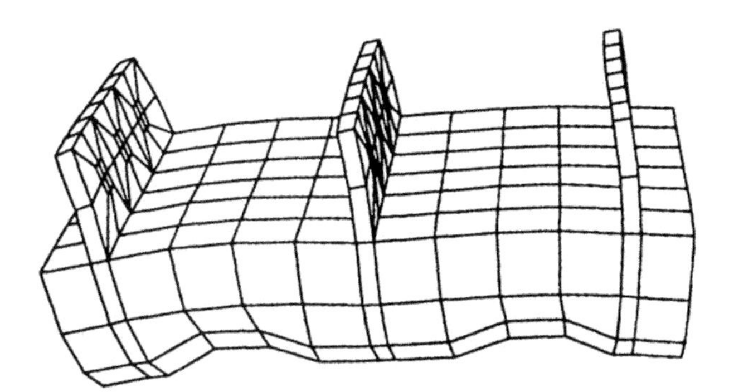

Figure 8.7 *This diagram shows the bending or warping of the interferometer under its own weight. The Si is taken to be elastically isotropic in a finite element calculation. The distortions are magnified by a factor of 10^6 Arif et al. 1994, copyright 1994 Society of Photo Optical Instrumentation Engineers.*

Table 8.2 *Dimensions of Interferometers*

LLL 1	$d = 35$ mm, $a = 2$ mm
LLL 2	$d = 34.518(2)$ mm, $a = 2.464(2)$ mm
Large symmetric LLL	$d = 50.404(3)$ mm, $a = 3.077(3)$ mm
Skew symmetric	$d_1 = 16.172(3)$ mm, $d_2 = 49.449(3)$ mm, $a = 2.621(3)$ mm

can, in principle, be used to determine both contributions. However, it is imperative that the geometry of the neutron paths, that is the Bragg angles (θ_{B1} and θ_{B2}) and the Borrmann fans in all three crystal plates, for both wavelengths be the same. This requirement is met if λ_2 is the perfect second harmonic of λ_1 (i.e., $\lambda_2/\lambda_1 = 2/1$) and neutrons of wavelength λ_1 are diffracted by the (440) lattice planes, while those of wavelength λ_2 are diffracted by the (220) lattice planes in the silicon interferometer. This idea was pursued by Littrell et al. (1996, 1997) on two interferometers, a large symmetric interferometer, and a skew symmetric interferometer (see Table 8.2).

The geometry of the beam trajectories for the skew symmetric interferometer experiment is shown in Fig. 8.8. The wavelength of the neutrons participating in the interference was measured using a pyrolytic graphite (PG) crystal with the (002) planes nominally parallel to its surface attached to a shaft perpendicular to the scattering plane of the interferometer in beam path II,

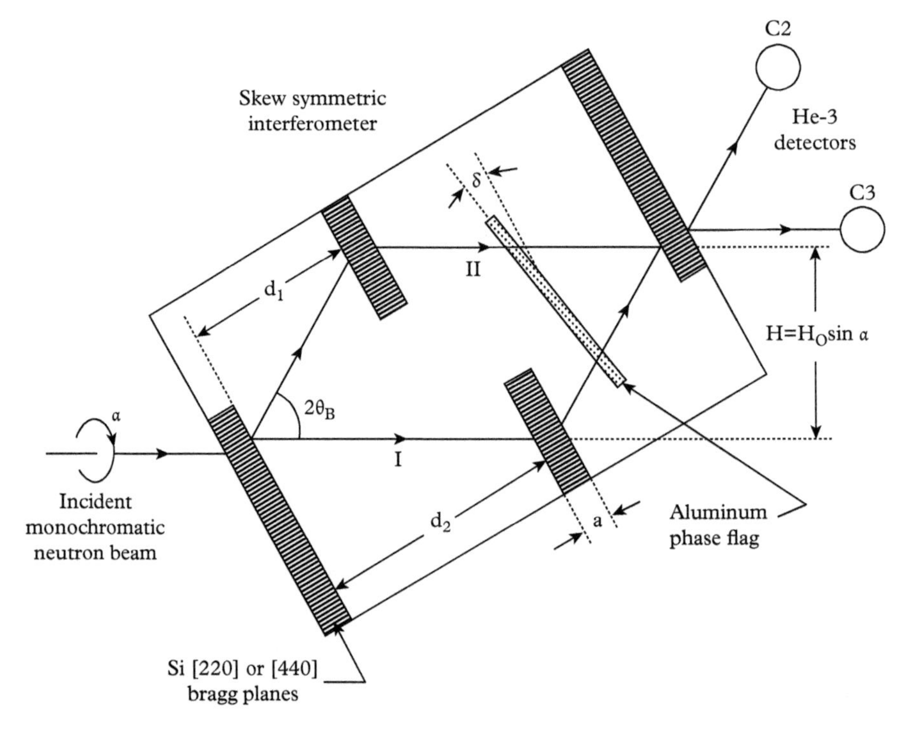

Figure 8.8 *Schematic diagram of the gravity experiment using a skew symmetric interferometer (Littrell et al. 1997)*

with beam path I blocked with a piece of B_4C in epoxy resin. Determination of each wavelength was obtained by measuring the separation between the sharp transmission minima of the sum of the C_2 and C_3 counting rates as the PG crystal was rotated through its first- and second-order Bragg reflections, and averaging the results. The neutron wavelengths were selected on the basis of being compatible with the geometry and performance of each interferometer, and accessible with the double-crystal Cu monochromator assembly. The mean wavelengths used with the skew symmetric interferometer were $\lambda_1 = 1.0780(6)$ Å and $\lambda_2 = 2.1440(4)$ Å, corresponding to the Bragg angles $\theta_{B1} = 34.15°$ and $\theta_{B2} = 33.94°$ with respect to the Si(440) and Si(220) lattice planes, respectively. The wavelengths for the experiment with the large symmetric interferometer were $\lambda_1 = 0.9464\,(6)$ Å and $\lambda_2 = 1.8796\,(3)$ Å, with corresponding Bragg angles $\theta_{B1} = 29.504°$ and $\theta_{B2} = 29.304°$. The fact that these measured /wavelengths were not precise harmonics of each other was taken into account in the data analysis, but was not important in altering the requirement that the geometry be the same for both wavelengths.

For each wavelength, the phase difference $\Delta\Phi(\lambda, \alpha)$ was obtained using the phase rotator interferogram technique, and referred to the phase shift measured for the level interferometer, i.e., at $\alpha = 0°$. The Sagnac phase differences, taken relative to $\alpha = 0°$, namely $\Delta\Phi_{\mathrm{Sagnac}}(\alpha) = (m_i/\hbar)A_0(1 - \cos\alpha)$, are very small for the range of tilt angles spanned in the experiments; they were calculated and subtracted from the phase data. Therefore, for each tilt angle there are two measured phase shifts, namely

$$\Delta\Phi(\lambda_1, \alpha) = -A_g(\lambda_1 \sin\alpha)F_g(\lambda_1, \alpha) - A_b(\lambda_1^{-1} \sin\alpha)F_b(\lambda_1), \tag{8.28}$$

and

$$\Delta\Phi(\lambda_2, \alpha) = -A_g(\lambda_2 \sin\alpha)F_g(\lambda_2, \alpha) - A_b(\lambda_2^{-1} \sin\alpha)F_b(\lambda_2). \tag{8.29}$$

Here

$$F_g(\lambda, \alpha) = (1 + \varepsilon(\lambda, \alpha)) \tan\theta_B(\lambda), \tag{8.30}$$

and

$$F_b(\lambda) = \sin^2\theta_B(\lambda). \tag{8.31}$$

The dependence of the bending effect on the Bragg angle is inferred from the X-ray data of Staudenmann et al. (1980). Since $\lambda_2/\lambda_1 \approx 2/1$, the Bragg angles for the two wavelengths are nearly equal as stated already, so that $F_b(\lambda)$ is essentially a constant, independent of λ. In the dynamical diffraction theory, the correction factor $\varepsilon(\lambda, \alpha)$ actually depends upon the variables $(a/d_i)\, q_{\mathrm{COW}} \sin\alpha = (\text{constant})\lambda \sin\alpha$. For small α, the leading term is $2a/3d$ for the symmetric interferometer, and $4a/3(d_1 + d_2)$ for the skew symmetric interferometer. The parameters A_g and A_b characterize the gravity- and the bending-induced phase shift, respectively. We therefore have two equations (8.28 and 8.29) for these two unknown, for each tilt angle. If the theory is correct A_g and A_b should be independent of α. Theory predicts that $A_g = q_{\mathrm{COW}}/\lambda \tan\theta_B = 2\pi\,(m_i m_g/h^2)A_0/\tan\theta_B$, independent of λ.

Figure 8.9 shows a series of phase-rotator interferograms taken with the skew symmetric interferometer for tilt angle near $\alpha = 0°$, for both wavelengths λ_1 and λ_2. The incident beam was defined by a circular Cd aperture, 6 mm in diameter, placed immediately in front of the

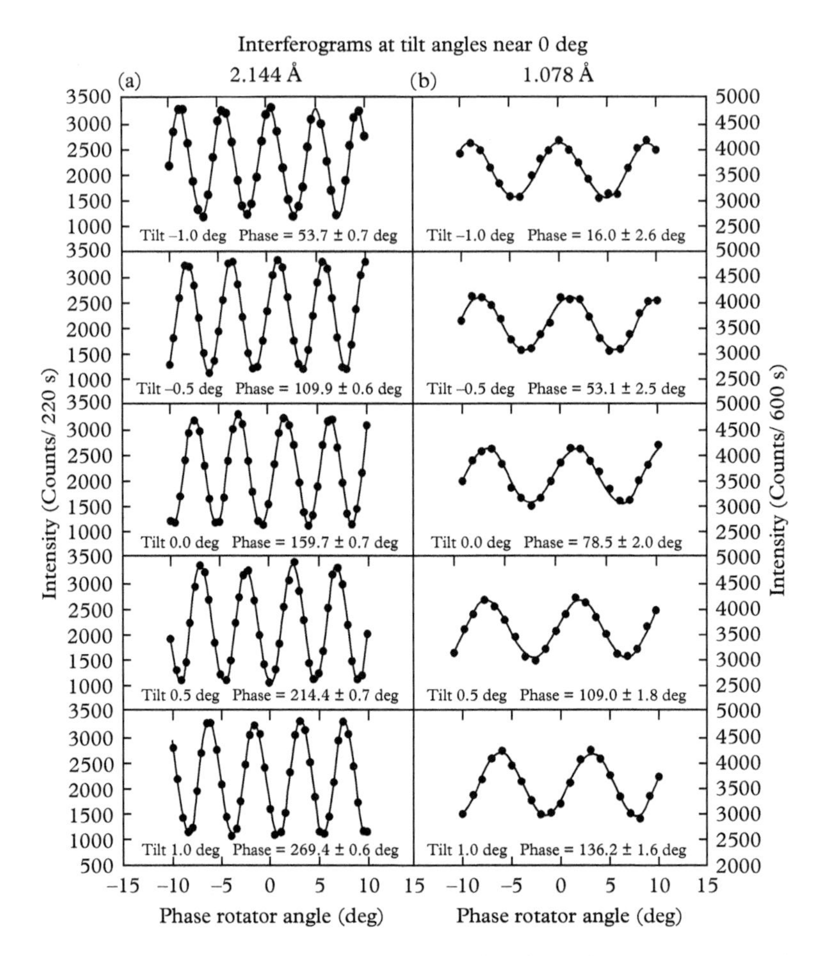

Figure 8.9 *A series of phase rotator scans taken for various values of the tilt angle α using wavelengths (a) 2.1440 Å and (b) 1.0780 Å in the skew symmetric interferometer. The phase advances by almost the same amount with each step and nearly twice as much for the 2.1440-Å data as for the 1.0780-Å data (Littrell et al. 1997)*

interferometer. Fitting each of the interferograms of this type taken for a wide range of tilt angles gives a series of phase differences $\Delta\Phi(\lambda, \alpha)$ which are shown in Fig. 8.10 for both interferometers. The values of A_g and A_b as a function of tilt angle are then obtained from these data. Using Eqs. (8.28) and (8.29) the mean values of these parameters give the solid lines in Fig. 8.10. For tilt angles $|\alpha|$ less than about 11°, these parameters are nearly independent of α as theory predicts. Clearly, for larger tilt angles, this independence of α is only approximately satisfied, indicating the presence of other effects. Table 8.3 contains the numerical values for q_{COW} and q_{bend} for both interferometers. The restricted range data mean $|\alpha| = 11°$. Figure 8.11a shows interferograms obtained directly by tilting the skew symmetric interferometer, using 2.1440-Å neutrons. A theoretical calculation of these interferograms, taking into account the detailed dynamical

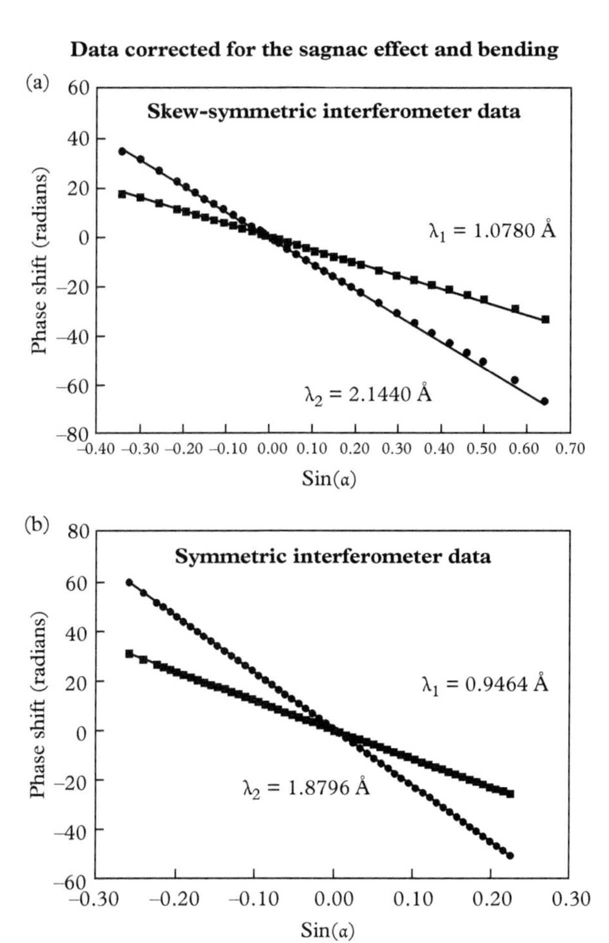

Figure 8.10 *Graphical representation of the phase shift data using (a) the skew symmetric interferometer and (b) the symmetric interferometer with the Sagnac effect phase shift (calculated) subtracted. The solid lines are phase shifts calculated from theory (Littrell et al. 1997)*

diffraction effects discussed earlier (and in Chapter 11), is shown in Fig. 8.11b. It is evident that the observed loss of contrast as a function of tilt angle is predicted reasonably well.

An historical summary of these gravitationally induced quantum interference experiments is given in Table 8.3. In all cases, the experimentally determined value of the frequency of oscillation of the gravity-induced interferogram, q_{COW}, is less than predicted theoretically using $g = 980.0$ cm/s^2, which is the proper value for Columbia, Missouri. The latest data of Littrell et al. (1997) taken with the large symmetric interferometer are probably the most reliable, yet they still yield a discrepancy with theory of order 1%. The importance of these experiments and their connection to relativity and gravitation theory is discussed in many papers (e.g., Audretsch

Table 8.3 *History of Gravity-Induced Interference Experiments.*

Authors	Interferometer	λ [Å]	A_0 [cm^2]	θ_B [°]	q_{COW} (theory) [rad]	q_{COW} (exp) [rad]	q_{bend} [rad]	Agreement with theory (%)
Colella et al. (1995)	Symmetric LLL #1	1.445(2)	10.52(2)	22.10(5)	59.8(1)	54.3(2.0)		12
Staudenmann et al. (1980)	Symmetric LLL #2	1.419(2)	10.152(4)	21.68(1)	56.7(1)	54.2(1)	3.30(5)	4.4
		1.060(2)	7.332(4)	16.02(1)	30.6(1)	28.4(1)	2.48(5)	7.3
Werner et al. (1988)	Symmetric LLL #2	1.417(1)	10.132(4)	21.65(1)	56.50(5)	56.03(3)	1.41(1)	0.8
Jacobson et al. (1993)	Symmetric LLL #2	1.422(1)	10.177(4)	21.73(1)	56.94(5)	54.7(2)	1.6(1)	3.9
Littrell et al. (1997)	Skew symmetric							
(440) reflection	Full range data	1.078(6)	12.016(3)	314.15(1)	50.97(5)	49.45(5)	2.15(4)	3.0
	Restricted range data	1.078(6)	12.016(3)	314.15(1)	50.97(5)	50.18(5)	2.03(4)	1.5
(220) reflection	Full range data	2.1440(4)	11.921(3)	33.94(1)	100.57(10)	97.58(10)	1.07(2)	3.0
	Restricted range data	2.1440(4)	11.921(3)	33.94(1)	100.57(10)	99.02(10)	1.01(2)	1.5
Littrell et al. (1997)	Large symmetric LLL							
(440) reflection	Full range data	0.9464(5)	30.50(1)	29.50(1)	113.60(10)	112.89(15)	8.09(6)	0.6
	Restricted range data	0.9464(5)	30.50(1)	29.50(1)	113.60(10)	112.62(15)	8.36(6)	0.9
(220) reflection	Full range data	1.8796(10)	30.26(1)	29.30(1)	223.80(10)	222.38(30)	4.02(3)	0.6
	Restricted range data	1.8796(10)	30.26(1)	29.30(1)	223.80(10)	221.85(30)	4.15(3)	0.9

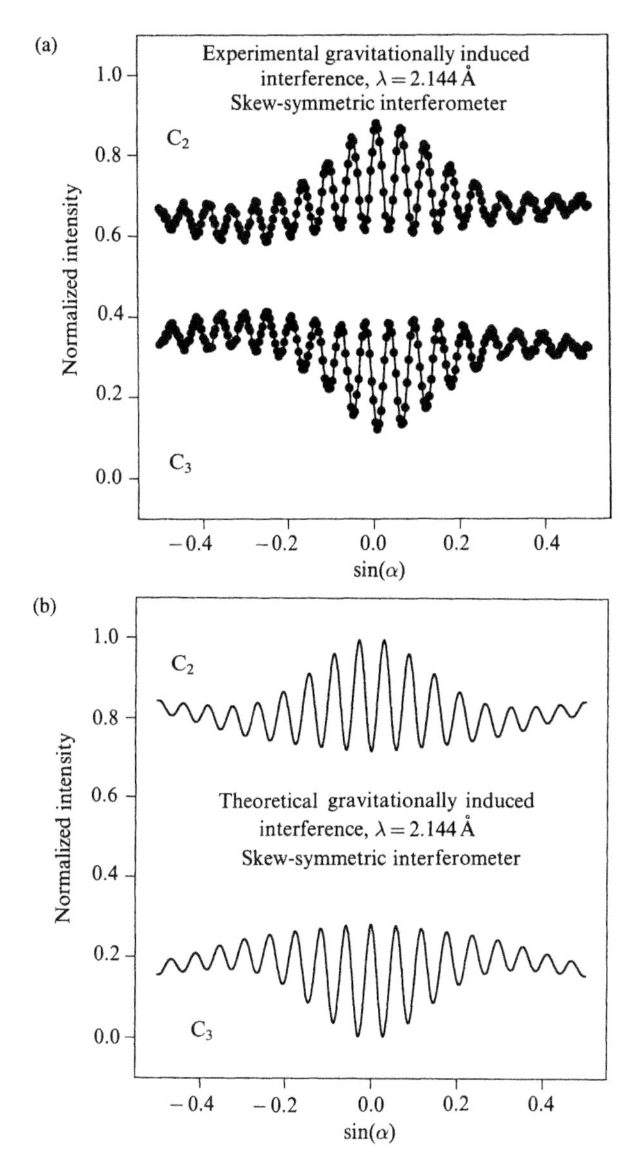

Figure 8.11 *(a) Gravity-induced interferogram obtained with the skew symmetric interferometer (Werner 1996). (b) Full dynamical diffraction calculation of the gravity-induced interferogram for the skew symmetric interferometer (Littrell 1997)*

and Lammerzahl 1983, Fabri and Picasso 1986, Takahashi 1987, Brown 1996, Viola and Onofrio 1997, Camacho and Camacho-Galvin 2007). A proposal of Kaiser et al. (2007) suggests the use of a floating interferometer crystal to improve the accuracy. Following the first demonstrations of atom interferometers in the early 1990s (Carnal and Mlynek 1991; Keith et al. 1991; Riehle et al. 1991; Kasevich and Chu 1991, 1992) interest in repeating or extending some of the fundamental physics experiments first carried out by neutron interferometry ensued. Substantial, and perhaps unexpected, success has been achieved in gravitationally induced quantum interference by the Stanford group (Peters et al. 1999, 2001). Their technique uses cesium atoms in an atomic fountain of laser-cooled atoms. Their atom optics is based upon stimulated Raman transitions, in which transitions are induced between stable hyperfine ground states (F-levels). Upon spatial separation of the wave packets corresponding to each of the two superposed atomic states following the absorption of one unit of momentum ($\hbar k$) from an optical pulse, this phase difference is affected along a vertical line by the Earth's gravity. The wave packets are put back together by an optical pulse of opposite momentum ($-\hbar k$), and the interferometric phase shift is measured by detecting the number of atoms in, say, the upper state. Accuracies in the measurement of g on the order of 5×10^{-9} have been achieved.

Metrological applications of neutron interferometry were never an original driving motivation. However, it is apparent that atom interferometry may become competitive with other techniques, such as the Michelson interferometer with a falling corner cube, as a high precision gravimeter (Cronin et al. 2010).

The results of a gravitationally induced quantum interference experiment using the very cold neutrons (VCN) grating interferometer at the ILL have been reported by van der Zouw et al. (2000). This interferometer is described in Sections 2.4 and 6.3. It uses VCNs of mean wavelength $\lambda \approx 9.5$ Å with a wavelength spread of $\Delta\lambda \approx 3$ Å. The incident flux is about 1000 n/cm^2/s, yielding

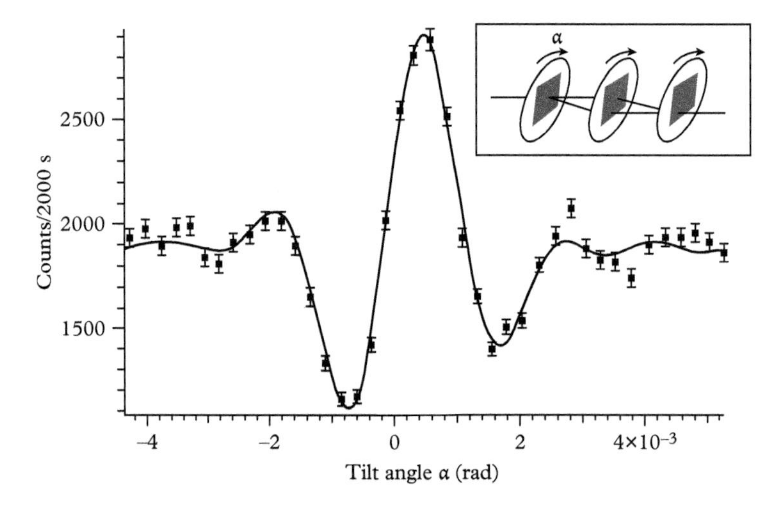

Figure 8.12 *Gravitationally induced interferograms use the VCN grating interferometer at the ILL. The solid line is the result of a least squares fit to the data using the measured incident neutron wavelength spectrum and averaging the interferogram over this spectrum. Reprinted from van der Zouw et al. 2000, copyright 2000, with permission from Elsevier.*

typical counting rates in the interfering beams of about 0.6 n/s when the total length of the device is 102 cm. Even though the beam separation angle is only 5.4 mrad (corresponding to a grating period of 2 μm), the interferometer is very sensitive to the Earth's gravity because the $\Delta\phi_{COW}$ is proportional to the wavelength (Eq. 8.4)

The intensity variation of the interfering beams as a function of tilt angle α is shown in Fig. 8.12. The rapid washing out of the interference oscillation is due to the wavelength spread $I(\alpha) = \langle I_0 [1 + c \cdot \cos \Delta\Phi_{COW}(\alpha)] \rangle_\lambda$. Using the measured incident spectrum and fitting the observed interferogram yields the solid line in Fig. 8.12. This fitting procedure gives a mean frequency of oscillations $q_0 = 4.012(36) \times 10^{13}$ rad^{2}s^{2}m^{-4}. This is to be compared to the theoretical value $q_0 = 2\pi m_i m_g / h^2 = 4.0148 \times 10^{13}$ rad^{2}s^{2}m^{-4}. The accuracy and level of agreement with theory is smaller (0.08%) than the perfect Si crystal interferometry measurement of Littrell et al. (1997), but the correction factor due to the asymmetric wave length distribution is quite large. Special note should be taken of the very small tilt angles ($\approx$2.2 mrad) necessary to cause a 2π change in the gravity-induced phase in this experiment.

8.2 Sagnac Effect

The French scientist M. G. Sagnac (1913) demonstrated that angular rotation can be detected by means of an optical interferometer having an enclosed area, such as the Mach–Zehnder interferometer. This effect is now used routinely in navigation and is the basis of the ring-laser gyroscope. For an excellent early review of the ideas, instrumentation, and literature of the Sagnac effect the reader is referred to the paper by Post (1967). A review, including a description of the large-scale Sagnac gyroscope in Christchurch, New Zealand, is given by Stedman (1997). In 1925 Michelson, Gale, and Pearson carried out a heroic experiment (for that time) in which they constructed an interferometer in the form of a rectangle of size 2010 ft $\times$ 1113 ft, and were able to detect the retardation of light due to the Earth's rotation corresponding to about $\frac{1}{4}$ of a fringe in agreement to theory. In view of the inertial and coordinate transformation differences between light waves and matter waves, it cannot be taken for granted that directly analogous quantum-mechanical phase shifts should exist for matter waves, especially for neutrons. The first Sagnac effect experiment with matter waves was carried out with Cooper pairs (Zimmerman and Mercereau 1965) in a superconducting Josephson junction electron interferometer. It has also been observed with free electrons (Hasselbach and Nicklaus 1993) and with atoms (Riehle et al. 1991; Lenef et al. 1997; Gustavson et al. 1997, 2000; Hasselbach 2010).

8.2.1 Earth's Rotation

Since the gravitationally induced interference experiments discussed in Section 8.1 were carried out on the surface of our rotating Earth, which is thereby a non-inertial frame, the neutron's Hamiltonian involves a third term ($-\boldsymbol{\Omega} \cdot \boldsymbol{L}$), giving rise to the phase shift $(2m_i/\hbar)\boldsymbol{\Omega} \cdot \mathrm{A}$ (Eq. 8.16). The effect of this phase shift is quite small for the experiments described in Section 8.1 using a horizontally directed beam. However, its dependence on the interferometer orientation with respect to verticality and to the local north–south axis of the Earth is quite different.

In an experiment carried out by Werner et al. (1979), an incident beam directed vertically was utilized, as shown in Fig. 8.13. The phase shift was measured, using the phase-rotator technique, as a function of the interferometer orientation angle α about the vertical (plumb-line) axis. From symmetry, it is clear that there is no α-dependent gravity-induced phase shift in this geometry. However, the angle between the axis of the Earth's rotation $\boldsymbol{\Omega}$ and the interferometer's normal

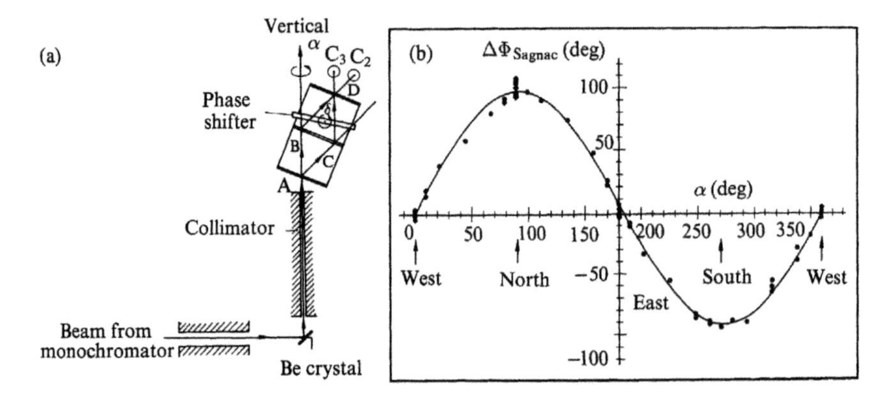

Figure 8.13 *(a) Schematic diagram of the vertical-beam geometry used in observing the neutron Sagnac effect. (b) Phase shift due to the Earth's rotation. The angle α specifies the orientation of the interferometer normal area vector with respect to the local N–S axes of the Earth (Werner et al. 1979)*

area vector A is varied as the interferometer's orientation α is changed. In terms of the colatitude angle θ_L at the point on the Earth's surface where the experiment was carried out (Columbia, MO, USA), and the angle α, the Sagnac phase shift for this geometry is given by

$$\Delta\Phi_{\text{Sagnac}}(\alpha) = (2m_i/\hbar)\Omega A \sin\theta_L \sin\alpha. \tag{8.32}$$

The experimental results are shown in Fig. 8.13 as well. The phase shift is zero when the normal area vector A points west or east, but when it points north or south it is +95° and −95°, respectively. Small deviations can be explained by the action of the Coriolis force on the beam paths inside the interferometer crystals (Littrell 2007). The results are in reasonable agreement with Eq. (8.32), which predicts that it should be +92° and −92° for the north and south orientations, respectively. It is interesting to note that the results of this experiment depend upon the inertial mass of the neutron, m_i, whereas the results of the gravity experiment depend upon the product $m_i m_g$. Consequently, one can interpret the combination of these two experiments as independent measurements of the inertial and gravitational neutron masses obtained from quantum-mechanical interference. The energy shift corresponding to the Sagnac phase shift is exceedingly small compared to the neutron energy $E \approx 60$ meV. It is $\hbar\Omega \approx 6 \times 10^{-17}$ meV, corresponding to the Doppler shift of the moving "mirrors" due to the Earth's rotation. That is $\Delta E/E \approx 10^{-18}$. This high energy resolution manifests itself also in a correspondingly high angular sensitivity of about 10^{-6} arc sec, which may be used for precise measurements of the neutron electron interaction, for the search of additional gravitational terms at small distances or the precise measurement of the Earth Coriolis force (Zawisky et al. 2010).

8.2.2 Turnable Rotation

The neutron interferometry analog of Sagnac's original experiment was carried out by Atwood et al. (1984). The experiment was done with a two-crystal LL interferometer mounted on a turntable. The phase shift was observed as the interferometer was oscillated back and forth through the Bragg reflecting condition within a somewhat divergent beam. The results of this experiment are shown in Fig. 8.14, in which the phase shift due to rotation is plotted as a function of

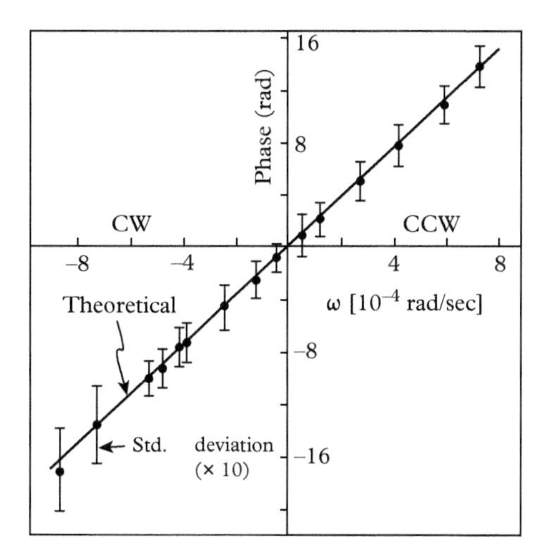

Figure 8.14 *Phase shift induced by turntable rotation of a LL interferometer as a function of angular velocity. Reprinted with permission from Atwood et al. 1984, copyright 1984 by the American Physical Society.*

rotation speed. The neutron detection is done in a phase-sensitive manner, such that the neutrons arriving at the detector leave the interferometer when the angular velocity Ω of the oscillating interferometer is maximum (angular acceleration is zero). The linear dependence of the phase shift on Ω is nicely observed. The arrow on the abscissa gives the angular rotation velocity of the Earth ($\Omega_{\text{Earth}} = 0.727 \times 10^{-4}$ rad/s).

8.2.3 Effective Mass and the Coriolis Force

Neutrons propagating in a perfect single crystal under Bragg reflecting conditions exhibit many interesting and unexpected phenomena. Some of these effects are discussed in Section 11.3. The analogy between the band gaps in the ε versus k spectrum of electrons in metals at the Brillouin zone boundary caused by Bragg reflection and the two-beam theory of dynamical diffraction of neutrons is indeed very close. In particular, the curvature of the dispersion relation $\varepsilon(k)$ is large near the energy gaps at the Brillouin zone boundaries, and the effective mass m^* is inversely proportional to this curvature. If an external force is applied to the neutron, such as may be implemented by the gravitational field of the Earth ($\boldsymbol{F} = m\boldsymbol{g}$) or a magnetic field gradient ($\boldsymbol{F} = \nabla(\boldsymbol{\mu} \cdot \boldsymbol{B})$), the deflection of a neutron along its trajectory can be quite large. The acceleration a_μ is related to the force F_ν by the effective mass tensor

$$a_\mu = \left(\frac{1}{m^*}\right)_{\mu\nu} F_\nu, \tag{8.33}$$

where

$$\left(\frac{1}{m^*}\right)_{\mu\nu} = \frac{1}{\hbar^2}\frac{\partial^2\varepsilon(\boldsymbol{k})}{\partial k_\mu \partial k_\nu}. \tag{8.34}$$

This is well known from solid state physics. The effective mass of neutrons, m^*, propagating under diffraction conditions inside a perfect Si crystal is reduced by a factor of more than 10^5 from its rest mass. Of course, this effective mass is its inertial mass. The deflection trajectories were studied in detail by Werner (1980), and these calculations were recast in an effective mass formalism by Zeilinger et al. (1986).

The experimental verifications of this small effective mass has been demonstrated in three experiments: (a) deflection of a neutron by a gradient magnetic field (Finkelstein et al. 1986, Zeilinger et al. 1986), (b) deflection by the gravitational field (Raum et al. 1995), and (c) deflection by the Coriolis force engendered within a slowly rotating crystal (Raum et al. 1995). The experimental ideas are similar in these three cases. We discuss here the experiment (c) showing the deflection of the neutron inside a perfect crystal due to the Coriolis force. A schematic diagram of the experiment is shown in Fig. 8.15.

Assuming that neutrons are traveling close to the Bragg angle θ_B for a single set of lattice planes characterized by the reciprocal lattice vector $\boldsymbol{G}$ (220 in the experiment), it is sufficient to use the superposition of two plane wave states (the incident beam and the Bragg reflected beam) as discussed in Chapter 11. Calculation of the probability current leads to an explicit formula for the effective mass in terms of the periodic crystal potential $V(\boldsymbol{G})$ for the neutron interacting with the nuclei. The expression for the acceleration $\boldsymbol{a}$ in terms of the external force $\boldsymbol{F}$ is

$$\boldsymbol{a} = \frac{\boldsymbol{F}}{m} \pm (1-\Gamma^2)^{3/2}\left(\frac{\hbar^2}{4m^2 V(\boldsymbol{G})}\right)(\boldsymbol{G}\cdot\boldsymbol{F})\,\boldsymbol{G}, \tag{8.35}$$

where the dimensionless parameter $\Gamma = \tan\Omega/\tan\theta_B$ (Eq. 11.46) characterizes the slope of the trajectory at any point inside the crystal relative to the Bragg planes, such that $-\theta_B \le \Omega \le \theta_B$ (see Fig. 8.15). This Γ parameter is related to the deviation of the local wave vector from the exact Bragg condition; that is, to the y-parameter (Eq. 11.28). The first term in Eq. (8.35) is the acceleration of a free neutron. It can be neglected for $|\Gamma| \ll 1$, that is, near the exact Bragg condition, where $\Gamma = 0$. The two possible signs for the acceleration correspond to the two independent solutions of the Schrödinger equation. The (+) sign corresponds to, say, the α-branch of the dispersion relation, and the (–) sign corresponds to the β-branch (see Chapter 11). The two different effective mass states are precisely analogous to electrons and holes in a semiconductor; that is, to states above and below the energy gap at the Brillouin zone boundary.

From Eq. (8.35) we see that only the component of force $\boldsymbol{F}$ parallel to $\boldsymbol{G}$ (perpendicular to the lattice planes) is relevant. The Coriolis force ($2m\mathbf{v} \times \boldsymbol{\omega}$) acting on a neutron inside of a crystal rotating around an axis perpendicular to both $\boldsymbol{G}$ and $\boldsymbol{v}$ (i.e., normal to the plane of Fig. 8.15) is given by

$$F = 2mv_d\omega, \tag{8.36}$$

where ω is the angular velocity and $v_d = v\cos\theta_B$ is the drift velocity of the neutrons inside the crystal (Shull et al. 1980). The fact that the "effective," or drift velocity has a cos θ_B factor in

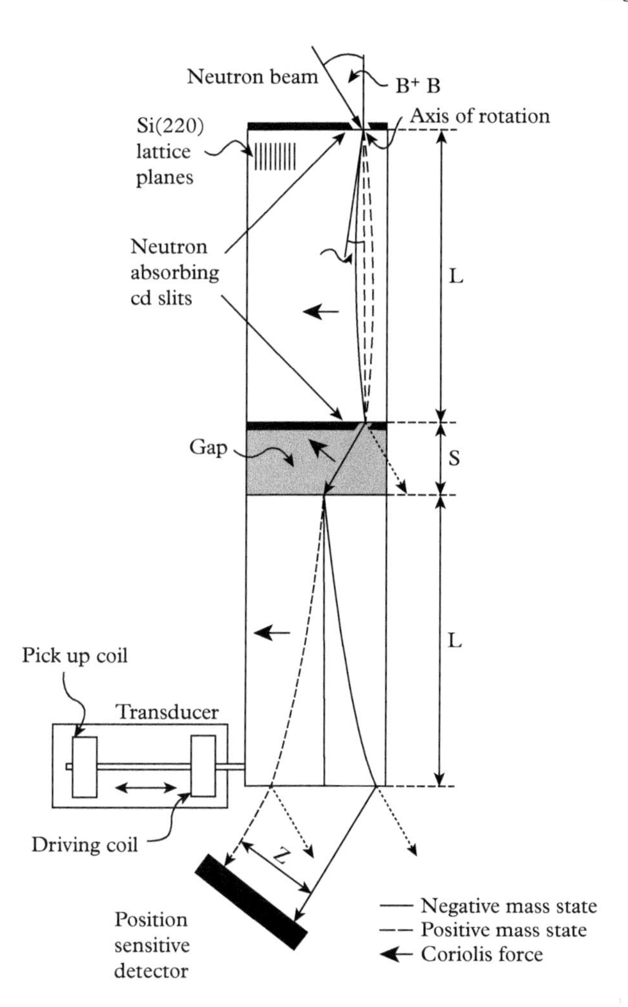

Figure 8.15 *Principle of the experimental setup viewed from above. The fine dashed lines are the neutron trajectories without external force; the fat line trajectories are shown for one sense of rotation. The dimensions of the crystals are* L = *52.3 mm and* S = *9.6 mm. Reprinted with permission from Raum et al. 1995, copyright 1995 by the American Physical Society.*

it comes about because of the zigzag character of the trajectories on the scale of the unit cell (Chapter 11, Fig. 11.15). This results in an acceleration for small Γ given by

$$a = 2\,\frac{m}{m^*}\,\omega v_{\mathrm{d}}, \tag{8.37}$$

directed along G as indicated by the arrows in Fig. 8.15, and with

$$m^* = \pm m\left(\frac{2V(G)}{\hbar^2 G^2/2m}\right). \tag{8.38}$$

The deflection caused by this force was observed using the two-crystal arrangement shown in Fig. 8.15. The crystal consists of two sections with a small gap between them and a common base. The first crystal acts as a collimator that selects only those neutrons which can pass the entrance and exit slits. When the crystal is at rest, the Coriolis force is zero, and the selected neutrons are those that pass through the first crystal section parallel to the (220) lattice planes. For a slowly rotating crystal, $F \neq 0$, and neutrons experience an effective acceleration in the first crystal such that only those with a slight deviation $\delta\theta$ from the exact Bragg condition will make it through the second slit, following the curved trajectories shown. The action of the Coriolis force on the neutron within the second crystal results in a further bending of the trajectories. The separation of the positive and negative effective mass states when leaving the second crystal section is readily calculated to be

$$Z = 4\frac{\omega m}{v_{\mathrm{d}} m^*} \cos\theta_{\mathrm{B}}\, L^2 \left(1 + \frac{S}{L}\right), \tag{8.39}$$

where L is the length of each crystal section and S is their separation. The factor $(1 + S/L)$ is a correction factor coming from the fact that the Coriolis force also acts on the neutron within the gap between the two crystals. This expression is valid not only in the rest frame of the crystal but also in the laboratory frame.

Figure 8.16 shows the neutron counts as a function of position along the exit face of the second crystal section for three frequencies, including $\omega = 0$, where the peak separation $Z = 0$. A summary of the peak separations, Z, as a function of angular velocity ω is shown in Fig. 8.16b. The

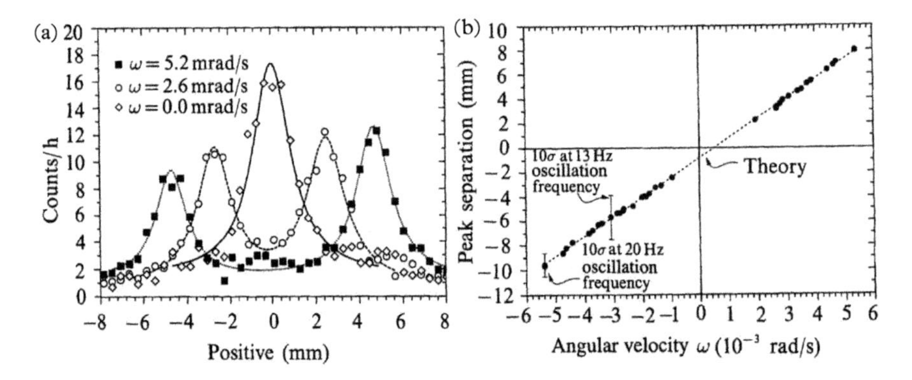

Figure 8.16 *(a) Neutron counts as a function of the horizontal position across the exit beams for different angular velocities. The lines shown result from a Lorentzian fit to the data. (b) Distance of the peaks to the two effective mass states as a function of angular velocity. The diagram shows all measured data points at oscillation frequencies of 13 and 20 Hz. The deflection for counterclockwise rotation of the crystal is defined negative. The dashed line has the slope predicted by theory and an offset resulting from a linear fit. Reprinted with permission from Raum et al. 1995, copyright 1995 by the American Physical Society.*

expected linear dependence is observed with a slope $(dZ/d\omega)_o = 1.609 \pm 0.014$ m/rad/s. This agrees nicely with the predicted slope (slope $(dZ/d\omega)_p = 1.614 \pm 0.008$ m/rad/s. Although the maximum acceleration of any part of the crystal assembly was only 0.1 m/s^2, it was necessary to take care that there were no effects due to bending of the crystal itself due to this acceleration.

It is truly remarkable that such small angular rotation frequencies utilized in this experiment can have such enormous amplification effects on the bending of neutron's trajectories. It is fair to say that the periodic potential of the crystal lattice creates a violation of the free-space principle of equivalence by a factor of 10^5.

8.3 Acceleration-Induced Interference

In order to truly verify the principle of equivalence in the quantum limit for neutrons propagating in free space, one should carry out two experiments—one in which there is the presence of a gravitational field $\boldsymbol{g}$ with no acceleration $\boldsymbol{a}$, and then one in which the effects of gravity are zero but the entire experimental setup is in a state of uniform acceleration. Bonse and Wroblewski (1983, 1984) carried out this second experiment at the ILL in Grenoble (Section 2.2.1). A schematic diagram of their experimental setup is shown in Fig. 8.17.

The LLL interferometer was fastened to a specially designed horizontal traverse. This traverse utilized leaf-spring guidance for smooth and practically frictionless movement under sinusoidal, forced-oscillation conditions parallel to the three slabs of the LLL interferometer. The sinusoidal oscillation was facilitated with a pair of standard loudspeaker magnets. The leaf-spring mechanism was designed in a way to minimize any deleterious rotational motion which would lead to phase shifts from the Sagnac effect discussed earlier in Section 8.2, and the large effects of the Coriolis force within the crystal blades as described in Section 8.2.3. This requirement on the mechanical structure of the traverse was severe indeed.

The intensity of the outgoing 0-beam was monitored stroboscopically in a manner to record neutrons leaving the interferometer at the inversion points of the oscillation, i.e., when the

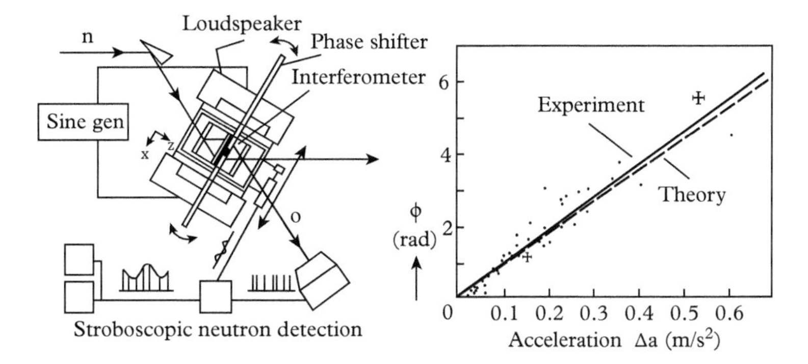

Figure 8.17 *Experimental setup for the observation of an acceleration-induced phase shift (left). The observed phase shift as a function of the acceleration (right). Reprinted with permission from Bonse and Wroblewski 1983, copyright 1983 by the American Physical Society.*

momentary acceleration $a_\pm$ corresponds to $\pm x_0\omega^2$, where x_0 is the amplitude and ω is the frequency of the oscillation. A complete array of interference fringes for both a_+ and a_- was obtained by repeated measurements at various settings of a stationary aluminum phase shifter, in a manner analogous to the measurement of the gravity-induced phases described in Section 8.1. The amplitudes of oscillation were in the range of 10 to 45 μm, and frequencies ranging from 3 to 19 Hz were used. The neutron wavelength was 1.831 Å.

A summary of the results is given in Fig. 8.17 where the phase shift $\Delta\Phi$ is plotted as a function of the mean acceleration, $a = \frac{1}{2}(a_+ - a_-)$. The Schrödinger equation for neutrons in the non-inertial (accelerating) frame of the interferometer is

$$\left[-\frac{\hbar^2}{2m_\mathrm{i}}\nabla^2 + V(\boldsymbol{r}) + m_\mathrm{i}\,\boldsymbol{a}\cdot\boldsymbol{r}\right]\psi(\boldsymbol{r}) = E\psi(\boldsymbol{r}), \tag{8.40}$$

which can be obtained by transforming the wave equation according to the laws of the macroscopic, classical principle of equivalence. The Schrödinger equation in the gravitational field $\boldsymbol{g}$ of the Earth, but under conditions of zero acceleration, is

$$\left[-\frac{\hbar^2}{2m_\mathrm{i}}\nabla^2 + V(\boldsymbol{r}) - m_\mathrm{g}\,\boldsymbol{g}\cdot\boldsymbol{r}\right]\psi(\boldsymbol{r}) = E\psi(\boldsymbol{r}). \tag{8.41}$$

In these equations $V(\boldsymbol{r})$ is the crystal potential responsible for the diffraction of neutrons in each of the crystal plates of the Si interferometer. Here m_i and m_g are again the inertial and gravitational masses, respectively. These two Schrödinger equations are equivalent only when $\boldsymbol{a} = -\boldsymbol{g}$, and the principle of equivalence $m_\mathrm{i} = m_\mathrm{g}$ in the quantum limit is strictly correct. The transformation of the wave equation into an accelerated frame was placed on firmer theoretical grounds by Klink (1997).

Effects due to dynamical diffraction, analogous to the gravity experiments, are also important in these acceleration-induced experiments. The difference between the solid line and the dashed line in Fig. 8.17b is the result of these dynamical diffraction effects. The final agreement of the data with theory, after all known corrections are made, is at the 4% level.

8.4 Connections with Photons

Here we make a few comments on the Sagnac effect for light and also on the intriguing possibility of observing the gravity-induced phase shift for light. Dresden and Yang (1979) derived the Sagnac phase shift formula in a way that makes it clear that it applies equally well to Schrödinger matter waves and to Maxwell electromagnetic waves. They write the Sagnac phase shift as

$$\Delta\Phi_\mathrm{Sagnac} = 2\frac{k}{v_0}\,\boldsymbol{\Omega}\cdot\boldsymbol{A}, \tag{8.42}$$

where k is the wave vector of the radiation and v_0 is the group velocity. Since for thermal neutrons $mv_0 = \hbar k$, one sees that $k/v_0 = m/\hbar$, and we get the formula we derived by integrating the canonical momentum in a rotating frame around the closed loop of the interferometer (Eq. 8.16). For light (photons) $v_0 = c$. Thus, using the same Si crystal interferometer, but with X-ray photons of say $\lambda = 2$ Å, we have $\Delta\Phi_\mathrm{Sagnac}(\text{X}-\text{rays}) = (v/c)\,\Delta\Phi_\mathrm{Sagnac}(\text{neutrons}) \approx 1.5\times10^{-5}$rad for the interferometer (area $A = 15.6$ cm^2) used for the original neutron Sagnac experiment. Even for an interferometer 10 times larger in linear dimension (100 times larger in area), which appears feasible today, the Sagnac phase shift for X-rays would only be 1.5 mrad. This would be an experimental challenge to observe. However, if one could devise a method for the X-ray to go around a ring many times, as in a ring-laser gyroscope, the conclusion would be very different. The perfect crystal resonant cavities as described by Rostomyan et al. (1989) provide an example of a possible scheme.

If one replaces the neutron mass by its Compton frequency according $mc^2 = \hbar\omega$, and uses the de Broglie relation $\hbar k = mv$, the COW formula (Eq. 8.15) can be written in the form

$$\Delta\Phi_{\mathrm{COW}} = -\frac{g_0 A\omega}{vc^2}\sin\alpha. \tag{8.43}$$

It is interesting to ask what the result of such a COW-type experiment would be with photons. It was supposed earlier that the gravitational effect on the X-rays used in the measurement of the bending effect was negligible. The Earth's gravitational field can be viewed as a region of space with an index of refraction for photons, namely (see also Section 8.5.5)

$$n(r) = \frac{k}{k_\infty} = 1 + \frac{2GM}{rc^2} \cong 1 + \frac{2GM}{R_0 c^2} - \frac{2g_0 z}{c^2}. \tag{8.44}$$

Here k_∞ is the photon wave vector far from the Earth (see Cheng 2010). The Schwarzschild radius $R_S = 2GM/c^2 = 8.8$ mm for the Earth. For $z = \mathrm{H}$, it is now straight-forward to work out the phase shift for a COW-type experiment with photons. The result is

$$\Delta\Phi_{\mathrm{COW}}(\text{photons}) = -\frac{2g_0 A\omega}{c^3}\sin\alpha. \tag{8.45}$$

One notices that simply replacing the velocity v in Eq. (8.43) by c, the resulting phase shift is ½ the correct general relativistic formula (8.45). This result is reminiscent of the factor of 2 in the correct general relativistic formula for the bending of light by the Sun.

This result was obtained earlier by Cohen and Mashhoon (1993) by deriving the index of refraction n in the exterior field of a spherically symmetric distribution of matter, in which n is only a function of the radial coordinate of the exterior Schwarzschild geometry which is asymptotically flat. Thus, for X-ray photons of the same wavelength as the neutrons, say 2 Å, we have $\Delta\Phi_{\mathrm{grav}}$ (X-rays) $= 2(v_0/c)^2\Delta\Phi_{\mathrm{grav}}$ (neutrons). We have $(v_0/c)^2 \approx 0.44 \times 10^{-10}$, so that observing the COW effect for X-ray photons would be quite challenging indeed, but not totally out of the question. Suppose we used 0.1-Å X-rays and had an interferometer $10 \times 10\ \mathrm{m}^2$, then $\Delta\Phi_{\mathrm{grav}}$ (X-rays) = 0.6 mrad. Clearly, such an experiment would be a tour de force, but it would be a wonderful, laboratory-based, general relativity experiment. A laboratory-based experiment using an optical fiber interferometer was proposed by Tanaka (1983). The estimated phase shift is about 1 μrad. The paper by Zych et al. (2012) provides a very useful discussion on the predictions of general relativity on the effect of gravity on the flow of time in various experiments, including the Pound–Rebka (1960) and Haferle–Keating (1972) experiments. They give a detailed analysis of a possible COW experiment with photons and a clear derivation of the phase shift starting with a Schwarzschild metric (see also Section 8.5.6).

Ahluwalia and Burgard (1996) have investigated the effects of gravity-induced quantum phase shifts in neutrino oscillations. In the neighborhood of a 1.4 solar-mass neutron star they predict that the gravity-induced phases are roughly 20% of their kinematical counterparts.

8.5 Neutron Fizeau Effects

The effect of motion of transparent matter on the phase of transmitted light waves was the subject of the historic Fizeau 1851 and 1859 experiments. The observation of the "dragging" effect of the propagating light waves by the moving medium was initially considered to be a triumph of Fizeau's ideas of the dragging of the ether by the moving matter. However, Einstein (1905) showed that the observed effects follow directly from the special theory of relativity. Subsequently, a number of more precise experiments were carried out by Zeeman (1914, 1915, 1927), and in more detail by Macek et al. (1964).

Analogous experiments have been carried out with neutrons and have led to some interesting conclusions. Ordinarily, the phase shift that arises from the motion of a medium introduced into one leg of a neutron interferometer turns out to depend only on the motion of the boundaries of the medium and not on the motion of the bulk. This somewhat surprising conclusion, though not at all obvious, depends upon the specific form of the dispersion relation, $\omega(\boldsymbol{k})$, for neutrons in most materials. If a slab of material of thickness L is moving with velocity $\boldsymbol{w}$, as shown in Fig. 8.18, there will, in general, be a shift of both the wave vector $\boldsymbol{k}$ and the frequency ω of the neutron de Broglie waves inside the medium. This leads to an accumulated phase shift when the neutron traverses the moving medium, over and above the phase shift that would occur in the same medium at rest. To first order in the optical potential, V_0, of the medium, this motion-induced phase shift is given by (see Eq. 8.58)

$$\Delta\Phi_{\text{Fizeau}} = -\frac{\alpha\beta\, k_x L}{2(1-\alpha)}, \tag{8.46}$$

where α is the ratio of the velocity of the boundary ($w_x = \boldsymbol{w}\cdot\hat{x}$) to the x-component of the neutron velocity and β is the ratio of the optical potential to the x-part of the incident neutron kinetic energy $E_{x0} = \hbar^2 k_x^2/2m$, namely

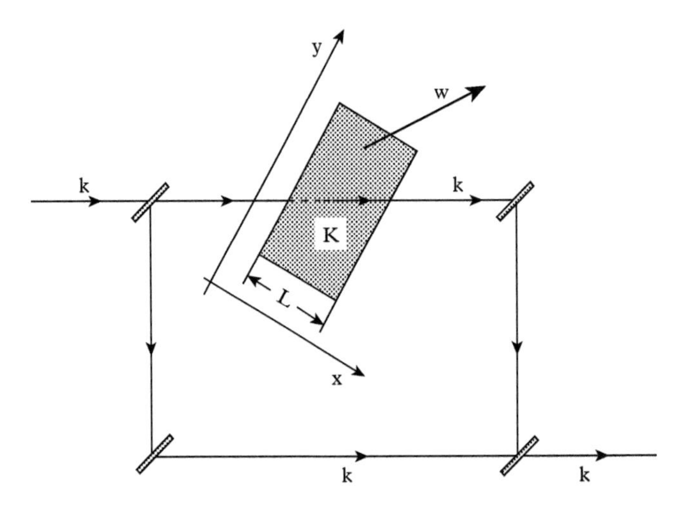

Figure 8.18 *Phase plate of thickness L moving with velocity w in one beam path of an interferometer. The x–y coordinate system (frame F) is at rest, but x = 0 coincides with the plate's entrant surface at time t = 0*

$$\alpha \equiv w_x/v_x = mw_x/\hbar k_x \quad \text{and} \quad \beta = V_0/E_{x0}. \tag{8.47}$$

We will derive this expression for the motion-induced phase shift in Section 8.5.2, following the arguments of Horne et al. (1983). We begin our discussion with a description of the experiment of Klein et al. (1981a) in which this effect was first observed using a rotating square quartz rod within a wave-front division neutron interferometer. We note that if the motion of the medium is parallel to its boundaries, $w_x = 0$, Eq. (8.46) predicts that there should be no motion-induced phase shift. An experiment verifying this "null" Fizeau effect was carried out by Arif et al. (1985) using a flat rotating quartz disk within a perfect silicon crystal LLL interferometer. This experiment is described in Section 8.5.3. For neutrons having an energy near a nuclear resonance, the optical potential becomes dependent upon the neutron wave vector, such that $V_0 = V_0\,(\boldsymbol{k})$, and a motion-induced phase shift is expected, even if $w_x = 0$. Such an experiment has been carried out by Arif et al. (1989) using a rotating aluminum disk with Sm metal foils attached to its surface, as described in Section 8.5.5. Finally, a very difficult experiment, carried out by Bonse and Rumpf (1986), is described in Section 8.5.4. They were successful in observing the motion-induced phase shift resulting from an aluminum propeller rotating about an axis perpendicular to the plane of a square LLL perfect silicon crystal interferometer.

8.5.1 Rotating Quartz Rod Experiment

The experiment of Klein et al. (1981a) was carried out at the optical bench at the ILL, as shown in Fig. 1.2, where the object was replaced by a rotating quartz rod (Fig. 8.19). The parameters of the setup were as follows: The incoming neutrons were monochromated by prism refraction and entered the apparatus through a 20-μm-wide entrance slit. The central diffraction peak of the slit coherently illuminated a double-slit assembly located 5 m downstream. The double slit assembly consisted of a single slit of 146 μm width, in the center of which a boron wire of 102 μm diameter was mounted, to form two slits, each of 22 μm nominal width. Downstream 5 m, the diffracted intensity profile was scanned with a 20-μm-wide exit slit, and counted with a

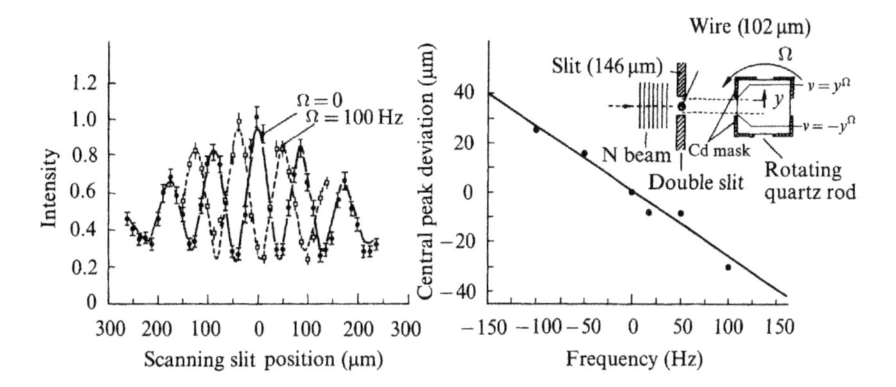

Figure 8.19 *(left) Double-slit interference pattern when the quartz rod was at rest (full lines) and when it was rotated with a frequency of 100 Hz (dashed lines). (right) Shift of the interference pattern as a function of the rotation frequency. Reprinted with permission from Klein et al. 1981a, copyright 1981 by the American Physical Society.*

BF_3 detector. The neutron mean wavelength was 18.45 Å with a spread of ±1.40 Å as measured by time-of-flight techniques.

A quartz rod with a square cross-section was placed directly behind the double slit and rotated about a vertical axis. As the Fizeau phase shift, Eq. (8.46), depends only on the velocity difference of the phase-shifting material in the two beams of the interferometer, the centering of the rod relative to the double-slit assembly is not critical. Therefore, the velocity components of the material traversed by the two beams can be taken to be $w_x = \pm\Omega_y$, as shown in Fig. 8.19. Here Ω is the angular frequency of rotation of the quartz rod and $2y$ is the slit separation. Thus, there is a continuous lateral phase gradient introduced by the rotating quartz rod, which manifests itself as an angular deflection, $\delta\theta$ of the whole interference pattern given by

$$\delta\theta = k_0^{-1} d\,(\Delta\Phi)/dy = (LV_0/2E_0 v_0)\,\Omega, \tag{8.48}$$

where $E_0 = \hbar^2 k_0^2/2m = mv_0^2/2$ is the kinetic energy of the incident neutrons and L is the thickness of the quartz rod. Hence, the rotating rod acts like a refracting wedge. The corners of the rod were covered with cadmium foil to define the neutron paths. The rod was approximately 10 mm square made of "optosil" quality quartz and characterized by a neutron index of refraction $n = 1 - V_0/2E_0 = 1 - 1.88 \times 10^{-4}$ for $\lambda = 18.45$ Å neutrons. Thus, Eq. (8.48) predicts an expected fringe displacement of 28.05 μm (at 5 m) at 100 Hz.

Figure 8.19 also gives the experimental result of the two-slit diffraction pattern as measured with the quartz rod at rest together with one obtained with a rotation frequency of 100 Hz. The displacement of the fringe structure due to the rotation is clearly observed. The solid line is a straight line fit to the data, giving a slope of 27.8 ± 2.5 mm/100 Hz, which compares nicely with the theoretically expected value. It is the first experimental verification that the neutron Fizeau effect comes from the motion of the boundary surfaces of moving matter.

An experiment inducing a time-dependent Fizeau effect is proposed in Section 10.16.

8.5.2 Theoretical Origin of the Neutron Fizeau Phase Shift

Motivated by the motion-induced phase shift experiment discussed above, Horne et al. (1983) carried out a detailed theoretical analysis of neutron de Broglie waves traversing uniformly moving media. We follow their treatment here.

8.5.2.1 Relativity considerations

Consider again the arrangement shown in Fig. 8.18 where a parallel-faced slab of thickness L and index of refraction $n(\boldsymbol{k})$ is inserted into one beam of an interferometer. If the slab is at rest, the relative phase of the two interfering beams upon recombination is

$$\phi(\boldsymbol{k}) = (K_x - k_x)\,L, \tag{8.49}$$

where $K_x = n(\boldsymbol{k})\,k_x$ is the x-component of the wave vector $\boldsymbol{K}$ inside the slab. Since the y-component of the internal wave vector $\boldsymbol{K}$ must be equal to the y-component of the external wave vector $\boldsymbol{k}$ due to the phase matching at the entrant boundary, this expression can be rewritten as

$$\phi(\boldsymbol{k}) = \left\{\sqrt{n^2(\boldsymbol{k})k^2 - k_y^2} - k_x\right\}L \tag{8.50}$$

Now suppose that the slab is set in motion with a laboratory velocity $\boldsymbol{w}$ as shown in Fig. 8.18. Then, in the rest frame of the slab, the relative phase will be $\phi(\boldsymbol{k}')$, where $\boldsymbol{k}'$ is the wave vector in this (moving) frame, and ϕ is the function given in Eq. (8.50). Since for uniform motion, the relative phase is a relativistic invariant quantity, $\phi(\boldsymbol{k}')$ must be the relative phase in all inertial frames, including the laboratory frame. Therefore, the phase shift induced by the motion of the slab is

$$\Delta\phi = \phi(\boldsymbol{k}') - \phi(\boldsymbol{k}). \tag{8.51}$$

This is the Fizeau effect phase shift. An evaluation of $\Delta\phi$ depends first upon the transformation $\boldsymbol{k}'(\boldsymbol{k})$, and secondly upon the specific form of $n(\boldsymbol{k})$.

From Einstein's special relativity, the transformation of the momentum–energy four-vector for a particle of rest mass m_0 and momentum $\boldsymbol{p} = \hbar\boldsymbol{k}$ in the laboratory frame gives the wave vector in the frame of the moving slab

$$\boldsymbol{k}' = \boldsymbol{k} - \left\{(1-\gamma)\frac{\boldsymbol{k}\cdot\boldsymbol{w}}{w} + \gamma\frac{w}{c}\sqrt{k + \frac{m_0^2c^2}{\hbar^2}}\right\}\frac{\boldsymbol{w}}{w}, \tag{8.52}$$

where $\gamma \equiv \sqrt{1 - w^2/c^2}$. With this transformation rule, Eq. (8.51) applies equally well to photons (where $m_0 = 0$) and to neutrons. Of course, the specific form of the index of refraction $n(\boldsymbol{k})$ depends upon the type of radiation and the material of the moving slab.

For thermal or cold neutrons, and for low slab velocities ($w/c << 1$), Eq. (8.52) reduces to the Galilean transformation, namely

$$\boldsymbol{k}' = \boldsymbol{k} - \boldsymbol{q} \tag{8.53}$$

where the wave vector $\boldsymbol{q} = m\boldsymbol{w}/\hbar$. The appropriate index of refraction for the slab at rest (obtained from conservation of energy) is

$$n(\boldsymbol{k}) = \frac{K_x}{k_x} = \sqrt{1 - \frac{2mV_0\left(|\boldsymbol{k}|\right)}{\hbar^2k^2}}, \tag{8.54}$$

where $V_0\left(|\boldsymbol{k}|\right)$ is the neutron optical potential of the medium for neutrons of kinetic energy $E = \hbar^2k^2/2m$. It is only the x-part of the momentum and energy that plays a role here. Using the expressions Eqs. (8.53) and (8.54) in Eqs. (8.50) and (8.51) we get the motion-induced phase shift for slow neutrons:

$$\Delta\phi = \left\{\sqrt{(1-\alpha)^2 - \beta'} + \alpha - \sqrt{1-\beta}\right\}k_xL, \tag{8.55}$$

where α and β are defined by Eq. (8.47), such that $\beta' \equiv V_0\left(|\boldsymbol{k}'|\right)/\left(\hbar^2k_x^2/2m\right)$, which is the ratio of the optical potential appropriate to the frame of the moving slab to the x-part of the neutron kinetic energy $E_x = mv_x^2/2$, in the laboratory frame, and $\beta = V_0\left(|\boldsymbol{k}|\right)/E_x$.

In the thermal or sub-thermal energy regions, the neutron optical potential for most materials is independent of k, so that $\beta = \beta'$. Assuming this to be the case, the Fizeau phase shift $\Delta\phi$ to first order in β is easily obtained from (Eq. 8.57). The result is

$$\Delta\phi = -\frac{\alpha\beta k_xL}{2(1-\alpha)} = -\frac{w_x}{v_x}\left(\frac{V_0}{E_x}\right)\frac{kL}{2\left(1 - w_x/v_x\right)}. \tag{8.56}$$

For very cold neutrons and/or for glancing angles of incidence, the denominator in the expression can approach zero, making the phase shift extremely sensitive to motion. For the rotating quartz rod experiment described in Section 8.5.1, this was not the case. The surface velocity, $w_x = 3.9 \times 10^{-2}$m/s at 100 Hz, and the neutron velocity was $v_x = 214$ m/s, such that $\alpha = 1.8 \times 10^{-4}$. Likewise, the value of β was small, namely $3.8 \times 10^{-4}(V_0 = 9.0 \times 10^{-8}\text{eV},\ E_x = 2.4 \times 10^{-4}\text{ eV})$.

8.5.2.2 Neutron Wave Functions in the Laboratory Frame

Since the Schrödinger equation is Galilean invariant, the neutron wave function in the laboratory (interferometer) frame, say F, should be related to the neutron wave function in the frame of the uniformly moving slab, say F', by a Galilean $(\omega, \boldsymbol{k})$ 4-vector transformation. We now explicitly show that this is, in fact, the case. From the discussion in the previous section, this is fundamentally a one-dimensional problem, involving only the x-component Ω_x of the velocity of the moving slab. Because of phase matching across the entrant and exit boundaries, the y-components of the external an internal wave vector are equal in both frames, i.e., $k_y = K_y$ and $k_y' = K_y'$. Thus, in the following analysis, we will omit the y-dependence of the wave function for simplicity. The y-component of the wave vector does matter in obtaining a value for the optical potential for media in which the incident neutron energy is near a nuclear resonance. Of course, the relevant neutron energy is its incident kinetic energy in the rest frame of the moving slab, i.e., $\varepsilon_{ix} = \hbar^2 k'^2 / 2m = \hbar^2 (\boldsymbol{k} - \boldsymbol{q})^2 / 2m$, such that $V_0 = V_0(\varepsilon')$. The experiment discussed in Section 8.5.5, involving a rotating disk with Sm metal foils, requires a knowledge of $V_0(\varepsilon')$ near the resonance at $\varepsilon' = 97.3$ meV in ^{149}Sm.

The calculation of the neutron wave function traversing and being partially reflected by the moving slab as shown in Fig. 8.18 is therefore identical to the quantum-mechanical one-dimensional problem of a moving square barrier shown in Fig. 8.20, having three regions. The solution of the time-dependent Schrödinger equation appropriate to the three regions, labeled 1, 2, and 3, are

$$\Psi_1(x,t) = Ae^{ik_{ix}x - i\omega_{ix}t} + Be^{-ik_{rx}x - i\omega_{rx}t}; \quad x \le w_x t, \tag{8.57a}$$

$$\Psi_2(x,t) = Ce^{iK_{ix}x - i\Omega_{ix}t} + De^{-iK_{rx}x - i\Omega_{rx}t}; \quad w_x t \le x \le w_x t + L, \tag{8.57b}$$

$$\Psi_3(x,t) = Fe^{ik_{tx}x - i\omega_{tx}t}; \quad w_x t + L \le x. \tag{8.57c}$$

Across the entrant boundary at $x = w_x t$, the continuity of the wave function $\Psi_1 = \Psi_2$ gives for the amplitudes

$$A + B = C + D, \tag{8.58}$$

provided that the phases match for all times t, which in turn requires that

$$k_{ix}w_x - \omega_{ix} = -k_{rx}w_x - \omega_{rx} = K_{ix}w_x - \Omega_{ix} = K_{rx}w_x - \Omega_{rx}. \tag{8.59}$$

In order for the incident and reflected waves in region 1 to satisfy the Schrödinger equation ($i\hbar\dot{\Psi} = H\Psi$), we must have

$$E_{ix} = \hbar\omega_{ix} = \hbar^2 k_{ix}^2 / 2m$$

and

$$E_{rx} = \hbar\omega_{rx} = \hbar^2 k_{rx}^2 / 2m, \tag{8.60}$$

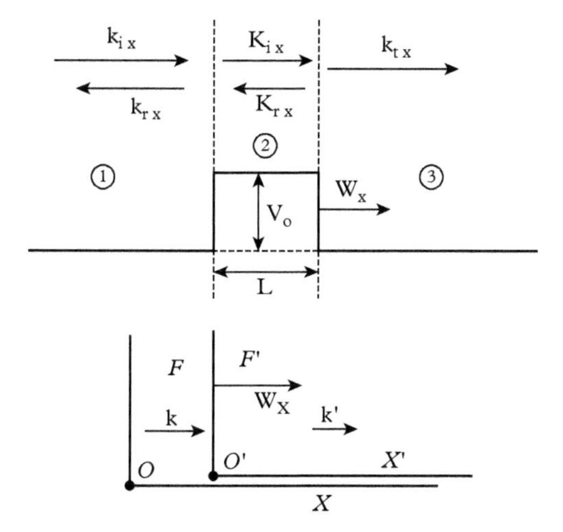

Figure 8.20 *One-dimensional moving square barrier as viewed in the laboratory frame F. The frame F' is attached to the moving barrier having a laboratory velocity W_x in the positive x-direction. The wave incident on the barrier has a laboratory wave vector k_{ix}. Part of this wave is reflected by the barrier, and Doppler-shifted down to a wave vector k_{rx}. There is a transmitted wave having a wave vector k_{tx} equal to k_{ix}. Within the barrier there are waves in the incident and reflected directions, having wave vectors K_{ix} and K_{rx}, respectively*

for the energies, ε_{ix} and ε_{rx}, of the incident and reflected waves, respectively. Using our definition of the wave vector $q_x \equiv mw_x/\hbar$ we see that the first part of Eq. (8.59) along with Eq. (8.60) requires that

$$k_{ix}q_{ix} - \frac{1}{2}k_{ix}^2 = -k_{rx}q_{rx} - \frac{1}{2}k_{rx}^2 \tag{8.61}$$

Thus we find that the shift of the wave vector $(k_{ix} - k_{rx})$ and the energy $(E_{ix} - E_{rx})$ due to reflection from the moving entrant boundary are

$$k_{ix} - k_{rx} = 2q_x \tag{8.62}$$

and

$$E_{ix} - E_{rx} = 4\frac{\hbar^2}{2m}\left(k_{ix}q_x - q_x^2\right). \tag{8.63}$$

The last of Eq. (8.59) similarly requires that

$$\frac{\hbar^2}{2m}K_{ix}q_{ix} - E_{ix} = -\frac{\hbar^2}{2m}K_{rx}q_{rx} - E_{rx}, \tag{8.64}$$

where, the energies of the internal incident and reflected waves are

$$E_{ix} \equiv \hbar\Omega_{ix} = \frac{\hbar^2 K_{ix}^2}{2m} + V_0\left(|\boldsymbol{k}_i - \boldsymbol{q}|\right) \tag{8.65a}$$

and

$$E_{rx} \equiv \hbar\Omega_{rx} = \frac{\hbar^2 K_{rx}^2}{2m} + V_0\left(|\boldsymbol{k}_i - \boldsymbol{q}|\right), \tag{8.65b}$$

respectively. Thus, Eq. (8.64) requires that the difference of the internal wave vectors and energies due to motion are

$$K_{ix} - K_{rx} = 2q_x \tag{8.66}$$

and

$$E_{ix} - E_{rx} = 4\frac{\hbar^2}{2m}\left(K_{ix}q_x - q_x^2\right) \tag{8.67}$$

These equations are analogous to those for the external wave vectors and energies given by Eqs. (8.62) and (8.63). It should be noted that the optical potential is dependent upon the total incident energy, $E_i' = E_{ix}' + E_{ry}' = \frac{\hbar^2 k_i'^2}{2m}$ in the rest frame of the moving slab (barrier), or equivalently upon the magnitude of $\boldsymbol{k}_i' = \boldsymbol{k}_i - \boldsymbol{q}$, as already pointed out.

In order to relate the internal wave vector K_{ix} to the external incident wave vector k_{ix}, we need to have the motion-dependent index of refraction $n = K_{ix}/k_{ix}$. From Eq. (8.59) again, we have

$$\frac{\hbar^2}{2m}k_{ix}q_{ix} - \hbar\omega_{ix} = \frac{\hbar^2}{2m}K_{ix}q_{ix} - \hbar\Omega_{ix}. \tag{8.68}$$

Using Eqs. (8.60) and (8.65a) for the energies $\varepsilon_{ix} = \hbar\omega_{ix}$ and $E_{ix} = \hbar\Omega_{ix}$, we have an equation that relates K_{ix} to k_{ix}, namely

$$\frac{\hbar^2}{2m}k_{ix}q_x - \frac{\hbar^2}{2m}k_{ix}^2 = \frac{\hbar^2}{2m}K_{ix}q_x - \frac{\hbar^2}{2m}K_{ix}^2 - V_0\left(|\boldsymbol{k}_i - \boldsymbol{q}|\right). \tag{8.69}$$

In terms of α and β this equation is easily solved for the ratio $n = K_{ix}/k_{ix}$. The result is the motion-independent index of refraction

$$n\left[\alpha, \beta(\boldsymbol{q})\right] = \alpha + \sqrt{(1-\alpha)^2 - \beta(\boldsymbol{q})}. \tag{8.70}$$

Thus, the motion-dependent part of the phase shift of the incident wave traversing the moving barrier is

$$\Delta\phi = \phi(\boldsymbol{q}) - \phi(0) = \left\{n\left[\alpha, \beta(\boldsymbol{q})\right] - n\left[0, \beta(0)\right]\right\}k_i L \tag{8.71}$$

which is identical to Eq. (8.55), where $\beta' = \beta(\boldsymbol{q})$ and $\beta = \beta(0)$. From Eq. (8.68) we note that the energy difference between the incident external wave, ε_{ix}, and the incident internal wave, E_{ix}, is

$$\varepsilon_i - E_i = \hbar\left(\omega_{ix} - \Omega_{ix}\right) = \frac{\hbar^2}{2m}q_x k_{ix}\left(1 - n\left(\alpha, \beta\right)\right). \tag{8.72}$$

Therefore, the fractional energy shift due to motion is

$$\frac{\varepsilon_{\mathrm{i}} - E_{\mathrm{i}}}{\varepsilon_{\mathrm{i}}} = \frac{\omega_{\mathrm{ix}} - \Omega_{\mathrm{ix}}}{\omega_{\mathrm{ix}}} = 2\alpha\left(1 - \alpha - \sqrt{(1-\alpha)^2 - \beta(\boldsymbol{q})}\right), \tag{8.73}$$

which equals $\alpha\beta(\boldsymbol{q})$ for small $\beta(\boldsymbol{q})$.

All of the above results, namely the expressions for the wave vectors shifts, the energy shifts, and the motion-dependent index of refraction, have been obtained from the single condition of continuity of the wave function in region 1 with the wave function in region 2. We have not yet used the condition of continuity of the derivative of Ψ across the boundary. At $x = w_x t$, the continuity of neutron current would seem to require $(\partial\Psi_1/\partial x = \partial\Psi_2/\partial x)$, which gives

$$k_{\mathrm{ix}}A - (k_{\mathrm{ix}} - 2q_x)B = K_{\mathrm{ix}}\,C - (K_{\mathrm{ix}} - 2q_x)D, \tag{8.74}$$

where we have used Eqs. (8.62) and (8.66).

At the exit boundary, for which $x = w_x t + L$, we require that $\Psi_2 = \Psi_3$, that is,

$$Ce^{\mathrm{i}\left(K_{\mathrm{ix}}w_{\mathrm{ix}}t + K_{\mathrm{ix}}L - \Omega_{\mathrm{ix}}t\right)} + De^{\mathrm{i}(-K_{\mathrm{rx}}w_{\mathrm{rx}}t - K_{\mathrm{rx}}L - \Omega_{\mathrm{rx}}t)} = Fe^{\mathrm{i}(k_{\mathrm{tx}}w_x t + k_{\mathrm{tx}}L - \omega_{\mathrm{tx}}t)}. \tag{8.75}$$

For this to be true for all values of time, we must have

$$K_{\mathrm{ix}}w_x - \Omega_{\mathrm{ix}} = -K_{\mathrm{rx}}w_x - \Omega_{\mathrm{rx}} = k_{\mathrm{tx}}w_x - \omega_{\mathrm{tx}}. \tag{8.76}$$

Together with Eq. (8.59), and the fact that the transmitted wave must satisfy the Schrödinger equation in free space, we find that

$$k_{\mathrm{tx}} = k_{\mathrm{ix}} \quad \text{and} \quad \varepsilon_{\mathrm{tx}} = \varepsilon_{\mathrm{ix}}. \tag{8.77}$$

That is, the momentum and energy of the transmitted wave are the same as those of the incident wave. Using Eq. (8.66) we see that the continuity condition (Eq. 8.76) across the exit boundary requires that

$$Ce^{\mathrm{i}K_{\mathrm{ix}}L} + De^{\mathrm{i}2q_x L}e^{-\mathrm{i}K_{\mathrm{ix}}L} = Fe^{\mathrm{i}k_{\mathrm{ix}}L}. \tag{8.78}$$

Continuity of the neutron current across the exit boundary requires continuity of the derivative, $\partial\Psi_2/\partial x = \partial\Psi_3/\partial x$, at $x = w_x t + L$. Using $K_{ix} - K_{rx} = 2q_x$ again, we see that our final continuity condition can be written as

$$K_{\mathrm{ix}}Ce^{\mathrm{i}K_{\mathrm{ix}}L} - (K_{\mathrm{ix}} - 2q_x)\,De^{\mathrm{i}2q_x L}e^{-\mathrm{i}K_{\mathrm{ix}}L} = k_{\mathrm{ix}}Fe^{\mathrm{i}k_{\mathrm{ix}}L}. \tag{8.79}$$

We now have four equations (Eqs. 8.58, 8.74, 8.78, 8.79) for four unknown amplitude ratios B/A, C/A, D/A, and F/A. Solving these equations is a bit tedious, but straightforward. The solution for the transmission coefficient F/A and reflection coefficient B/A are

$$\left(\frac{F}{A}\right) = e^{-\mathrm{i}\varphi(1-\alpha)}\frac{1}{\cos\theta - \mathrm{i}\left\{\left[(n-\alpha)^2 + (1-\alpha)^2\right]/\left[2(1-\alpha)(n-\alpha)\right]\right\}\sin\theta} \tag{8.80}$$

and

$$\left(\frac{B}{A}\right) = -\mathrm{i}\frac{\{[(1-\alpha)^2-(n-\alpha)^2]/[2(1-\alpha)(n-\alpha)]\}\sin\theta}{\cos\theta - \mathrm{i}\{[(n-\alpha)^2+(1-\alpha)^2]/[2(1-\alpha)(n-\alpha)]\}\sin\theta}, \tag{8.81}$$

where $n = n[\alpha, \beta(q)]$ is the motion-dependent index of refraction given by Eq. (8.70), $\varphi = k_{ix}L$ and $\theta = (n-\alpha)\varphi$.

The above expressions for the transmission and reflection coefficients appear to be fairly complicated. The reason for this is that they include all multiple reflection paths for the incident wave to give rise to a transmitted or reflected wave; that is, they include all Feynman paths. For example, the transmitted wave arises from the sum of a wave transmitted through the entrant surface then transmitted through the exit surface, plus a wave transmitted through the entrant surface, reflected back to the entrant surface, reflected back to the exit surface where it is transmitted, and so on. However, this is the same physics of waves transmitted through a static barrier, that is, the above problem when viewed in the moving frame F' in which the slab is at rest. Furthermore, since $|F|^2$ and $|B|^2$ are the neutron densities, which are invariant under Galilean transformation the above results must be the same as those obtained in the frame F. To see that this is in fact the case, we note that the external and internal wave vectors in F' are related to those in F by

$$k'_{ix} = k_{ix} - q = (1-\alpha)k_{ix} \tag{8.82a}$$

and

$$K'_{ix} = K_{ix} - q = (n-\alpha)k_{ix}. \tag{8.82b}$$

Thus, the transmission coefficient of Eq. (8.80) can be written in terms of the wave vectors in the moving frame F' as

$$\left(\frac{F}{A}\right) = e^{-\mathrm{i}k'_{ix}L}\frac{1}{\cos(K'_{ix}L) - \mathrm{i}[(K'^2_{ix}+k'^2_{ix})/2k'_{ix}K'_{ix}]\sin(K'_{ix}L)}. \tag{8.83}$$

And the reflection coefficient of Eq. (8.81) is

$$\left(\frac{B}{A}\right) = -\mathrm{i}\frac{[(k'^2_{ix}-K'^2_{ix})/2k'_{ix}K'_{ix}]\sin(K'_{ix}L)}{\cos(K'_{ix}L) - \mathrm{i}[(K'^2_{ix}+k'_{ix}K'_{ix})/2k'_{ix}K'_{ix}]\sin(K'_{ix}L)}. \tag{8.84}$$

These are the standard textbook results. Since the velocity of the neutrons in the incident, reflected, and transmitted beams are all equal in this F' frame (i.e., $v'_x = \hbar k'_{ix}/m = \hbar k'_{rx}/m = \hbar k'_{tx}/m$), the transmission T' and the reflectivity R' sum to unity. That is

$$T' + R' = \left|\frac{F}{A}\right|^2 + \left|\frac{B}{A}\right|^2 = 1, \tag{8.85}$$

as can be easily verified using the above expressions.

Since energy is conserved in this frame (the Hamiltonian is time independent) we know that

$$\varepsilon'_{ix} = \frac{\hbar^2 k'^2_{ix}}{2m} = \frac{\hbar^2 K'^2_{ix}}{2m} + V_0(|\boldsymbol{k}_i|), \tag{8.86}$$

and the transmission probability T' can then be written in the well-known form (e.g., Born and Wolf 1975)

$$T'\left(\varepsilon'_{ix}\right) = \frac{1}{1 + \left[V_0^2/4\varepsilon'_{ix}(\varepsilon'_{ix} - V_0)\right]\sin\left(\sqrt{(2m/\hbar^2)\left(\varepsilon'_{ix} - V_0\right)L^2}\right)}. \tag{8.87}$$

Clearly when α and β are small, $B/A \approx 0$, and

$$\left(\frac{F}{A}\right) \approx e^{-i\varphi(1-\alpha)}e^{i\theta} = e^{i(n-1)k_{ix}L} = e^{-i\phi(q)}, \tag{8.88}$$

so that the phase shift of the incident wave traversing the moving slab is

$$\phi(q) = \left[\alpha + \sqrt{(1-\alpha)^2 - \beta(q)}\right]k_{ix}L, \tag{8.89}$$

in agreement with Eq. (8.71). However, when α and β are not small the motion-induced phase shift is the phase of (F/A) given by Eq. (8.83). This regime has not yet been investigated experimentally.

8.5.2.3 Neutron Currents in the Laboratory Frame

It may appear to the casual reader that the complicated analysis of the previous section could have been avoided since the results for the transmissivity T' and the reflectivity R' in the frame F' attached to the moving slab are fairly easily obtained from standard quantum mechanics textbooks. It would appear that the transmissivity and reflectivity, T and R, in the laboratory frame, F, would simply be given by

$$T = \left(\frac{k_{tx}}{k_{ix}}\right)\left|\frac{F}{A}\right|^2 = \left|\frac{F}{A}\right|^2 \tag{8.90}$$

and

$$R = \left(\frac{k_{rx}}{k_{ix}}\right)\left|\frac{B}{A}\right|^2. \tag{8.91}$$

Thus, $T = T'$, which is correct; but $R + T$ cannot then be equal to unity, since $k_{rx}/k_{ix} = \left(k'_{ix} - q\right)/\left(k'_{ix} + q\right) \neq 1$. What is the problem here? The answer is that there is an interference current between the incident wave and the reflected wave in both region 1 and region 2 when viewed in the laboratory frame, F. We demonstrate this for region 1. The current density in region 1 is given by

$$\mathcal{J}_{1x}(x,t) = \frac{\hbar}{2mi}\left[\Psi_1^*\frac{\partial\Psi_1}{\partial x} - \Psi_1\frac{\partial\Psi^*_1}{\partial x}\right], \tag{8.92}$$

where $\Psi_1 = \Psi_1(x,t)$ is given by Eq. (8.57a). Working this out, one gets

$$\mathcal{J}_{1x}(x,t) = \frac{\hbar}{m}\left[k_{ix}|A|^2 - k_{rx}|B|^2\right] + \frac{\hbar}{m}q_x\left[AB^*e^{i\chi(x,t)} + A^*Be^{-i\chi(x,t)}\right], \tag{8.93}$$

where the phase angle

$$\chi(x,t) = (k_{ix} + k_{rx})\,x - (\omega_{ix} - \omega_{rx})\,t. \tag{8.94}$$

Thus, the current density in region 1 is not simply the incident current density, say $\mathcal{J}_A$, minus the reflected current density, say $\mathcal{J}_B$, but it contains an interference current density $\mathcal{J}_{AB}$, such that

$$\mathcal{J}_{1x} = \mathcal{J}_A + \mathcal{J}_B + \mathcal{J}_{AB}. \tag{8.95}$$

In the frame F', $k'_{ix} = k'_{rx}$ $(q'_x = 0)$, the current density $\mathcal{J}'_{AB}$ vanishes. At the entrant boundary, $x = w_x t$, the phase angle $\chi = 0$, as can be seen from Eqs. (8.62) and (8.66). Thus the current which crosses the boundary involves the phase of $B = |B|\,e^{ib}$ relative to the phase of $A = |A|\,e^{ia}$. This current density then must be equal to the current density in region 3, namely,

$$\mathcal{J}_{3x} = \frac{\hbar k_i}{m}|F|^2, \tag{8.96}$$

where we have used the fact that $k_t = k_i$. Thus,

$$k_{ix}|A|^2 - k_{rx}|B|^2 + 2q_x\,|AB|\cos(b-a) = k_{ix}|F|^2. \tag{8.97}$$

But since $k_{rx} = k_{ix} - 2q_x$ and $|A|^2 = |B|^2 + |F|^2$, we have the requirement that

$$\cos(b-a) = -\left|\frac{B}{A}\right| \tag{8.98}$$

Thus, the interference current term in Eq. (8.97) is $-2q_x|B|^2$, precisely accounting for the discrepancy of $+2q_x|B|^2$ in the reflected current density $\mathcal{J}_B$. We thus see that the result of Eqs. (8.83) and (8.84) for F/A and B/A are correct aside from a phase factor. The boundary condition of continuity of the derivatives of Ψ at the moving entrant and exit surfaces is consistent with the continuity of the current density, and is correct only up to a constant phase factor dependent on the velocity $w_x = \hbar q_x/m$.

8.5.3 Null Fizeau Effect Experiment

Arif et al. (1985) carried out an experiment designed to verify the prediction that the phase shift of a neutron wave traversing a slab of matter moving with a velocity $\boldsymbol{w} = w_y\hat{y}$ that is parallel to its boundaries is independent of w_y. The experimental setup is shown in Fig. 8.21. A 5-cm-radius disk of 1-cm-thick fused quartz was rotated at various speeds within an LLL interferometer. The disk intercepted both coherent beams. The rotor, supported by gyro-compass bearings, was enclosed in a partially evacuated box and was driven by integrally mounted turbines supplied with jets of air from the ambient, thermally controlled enclosure. In order to minimize systematic effects, the fused quartz disk was sawn off as shown in the diagram. In this manner, sectors of quartz alternated with empty sectors, and the neutron counts were synchronously gated into separate counting channels, with suitable intervals within the LLL interferometer. In this way, two interleaving sets of data points were obtained when an aluminum phase plate was scanned in angular position inside the interferometer, thereby obtaining two neutron interferograms simultaneously for each rotation speed. Typical interferograms taken at a rotation frequency of 25 Hz are shown in Fig. 8.21B.

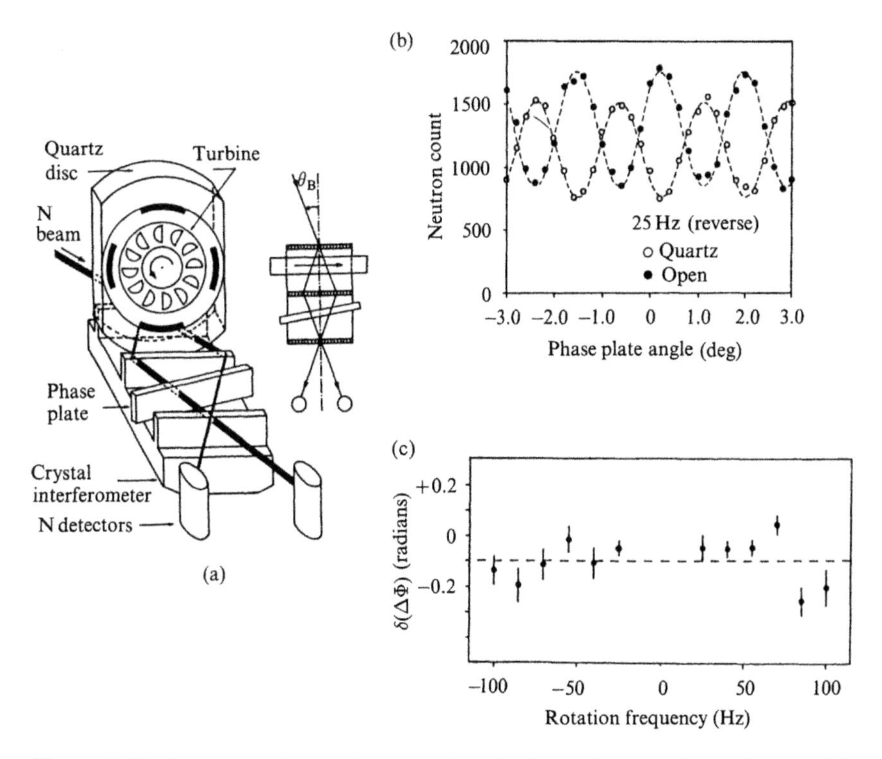

Figure 8.21 *Experimental setup (a), typical results (b), and motion-induced phase shift (c) for the null Fizeau effect experiment (Arif et al. 1985)*

The motion-induced phase shift for small α and β is given by Eqs. (8.70) and (8.71),

$$\Delta\phi = \phi(\boldsymbol{w}) - \phi(0) = -\frac{1}{2}\big(\beta(\boldsymbol{q}) - \beta(0)\big)k_{\mathrm{ix}}L, \tag{8.99}$$

where

$$\beta(\boldsymbol{q}) = V_0\left(\frac{\hbar^2}{2m}(\boldsymbol{k}-\boldsymbol{q})^2\right)\bigg/\left(\frac{\hbar^2 k_{\mathrm{ix}}{}^2}{2m}\right). \tag{8.100}$$

In this case $\boldsymbol{q} = \left(m w_y/\hbar\right)\hat{y}$, so

$$\beta(\boldsymbol{q}) \approx V_0\left(\varepsilon_{\mathrm{i}} - \hbar k_{\mathrm{iy}} w_y\right)/\varepsilon_{\mathrm{ix}}, \tag{8.101}$$

where we have neglected the quadratic term $\hbar^2 q^2/2m$ in the argument of V_0. The contribution of the term $\hbar k_{\mathrm{iy}} w_y$ to the energy argument is equal in both beams I and II in the interferometer, but of opposite signs. Furthermore, for the geometry shown in Fig. 8.21 we have $\varepsilon_{\mathrm{ix}} = \varepsilon_{\mathrm{i}}\cos^2\theta_{\mathrm{B}}$. Expanding V_0 to first order in w_y, we have

$$V_0\left(\varepsilon_{\rm i} - \hbar k_{\rm iy} w_y\right) \approx V(\varepsilon_{\rm i}) - \left[\frac{\partial V_0}{\partial \varepsilon}\right]_{\varepsilon_{\rm i}} \cdot \hbar k_{\rm iy} w_{\rm i}. \tag{8.102}$$

We have for the velocity-dependent phase difference

$$\delta(\Delta\phi) = \Delta\phi_{\rm II} - \Delta\phi_{\rm I} = -2\left(m w_y/\hbar\right) L \tan\theta_B \left[\frac{\partial V_0}{\partial \varepsilon}\right]_{\varepsilon_{\rm i}}. \tag{8.103}$$

Thus if the optical potential V_0 is independent of the energy, near ε_1, this phase shift is expected to be 0.

A summary of the measurements of $\delta(\Delta\phi)$ is shown in Fig. 8.21C as a function of frequency for both clockwise and counterclockwise rotations. In view of the fact that the phase shift of the $\lambda = 1.268$ Å neutrons traversing the $L = 1$-cm-thick quartz slab is about 500 rad, the deviations of $\delta(\Delta\phi)$ from 0 are indeed very small. A fit to the data of Fig. 8.21C indicates a variation of $\delta(\Delta\phi)$ with a rotation frequency of $(0.8 \pm 5.9) \times 10^{-4}$ rad/Hz, which within experimental error agrees with the theoretical prediction of a null variation of the motion-induced phase shift with w_y. This result places an upper limit of the energy variation of the optical potential V_0 for quartz of

$$\left|\left(\frac{\partial V}{\partial \varepsilon}\right)\right|_{\varepsilon_i} \leq 2.1 \times 10^{-8} \tag{8.104}$$

for $\varepsilon_i = 50.8$ meV neutrons.

8.5.4 Rotating Aluminum Propeller Experiment

Bonse and Rumpf (1986) carried out a Fizeau effect experiment, using a four-winged propeller rotating within a specially prepared square interferometer. A diagram of the setup is shown in Fig. 8.22a. Each of the four 10-mm-thick blades engages the coherent beams simultaneously as the propeller rotates about an axis normal to the plane of the interferometer. The experiment was carried out at the S18 neutron interferometer station at the ILL using $\lambda = 1.92$ Å ($v = 2060$ m/s) neutrons (see Section 2.2.1). This wavelength has a Bragg angle $\theta_{\rm B} = \pi/4$ for the (400) reflection in silicon, thus allowing the fabrication of the square interferometer. The propeller is encapsulated within a 6.5 cm × 6.5 cm aluminum box fitted with Mylar windows at the entrance and exit points of the neutron beam.

Having four symmetrically situated blades on the propeller serves two important purposes. The first is that it provides a balanced rotator, and therefore mitigates the difficulties associated with transmitting deleterious vibrations or fluctuating rotations to the silicon interferometer crystal. Secondly, the expected motion-induced phase shifts are multiplied by a factor of 4. The propeller was rotated up to angular speeds of 120 Hz corresponding to motion along the beams of $w_x = 240$ cm/s ($\alpha = 0.00116$). A summary of the 147 measured motion-induced phase shifts is shown in Fig. 8.22b. Each phase shift was obtained from an interferogram using the tantalum phase flag. The typical contrast was 37%. Data were obtained stroboscopically with a multichannel analyzer. The pertinent channels correspond to the time slices when each of the propeller wings is nearly normal to the corresponding beam. For data analysis purposes the collected counts in the angular interval [−17.6°, +17.6°] were used. The measured phase shifts as a function of frequency were compared to the phase shift obtained for quasi-static rotations ($f \approx 0$). The straight line fit to the data of Fig. 8.22b is shown; it has a slope 0.8102(0.01) rad/[m/s]. This is to be compared to the theoretical slope of 0.8171 rad/[m/s] based upon values for the neutron optical potential

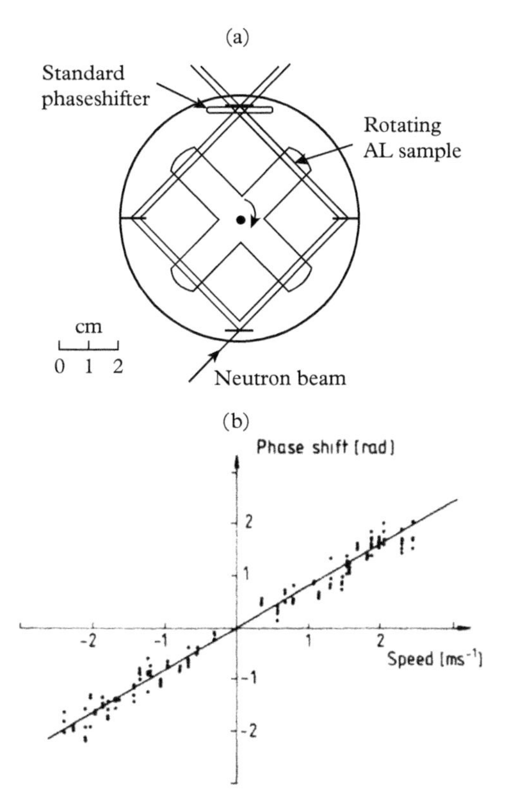

Figure 8.22 *Experimental setup (a) and typical results (b) of the rotating propeller experiment. Reprinted with permission from Bonse and Rumpf 1986, copyright 1986 by the American Physical Society.*

of aluminum, the measured neutron wavelength, the geometry of the propeller, and the proper averaging over the angular factors for the selected angular interval above. Thus, the agreement between theory and experiment is slightly better than 1%.

8.5.5 Rotating Samarium Disk Experiment

As we have already discussed in some detail, there should be no motion-induced phase shift in neutron interferometry if the matter moves parallel to its boundary. This prediction was established quite accurately in the null Fizeau effect experiment. However, if the neutron–nuclear optical potential $V_0 = 2\pi\hbar^2 Nb/m$ depends upon the incident energy ε' in the frame F' of the moving slab of matter, there will be a motion-induced phase shift expected as given by Eq. (8.103).

Arif et al. (1989) carried out an experiment with a rotating disk of aluminum upon which were attached Sm foils of thickness 33 ± 2 μm in sectors, as shown in Fig. 8.23. The isotope ^{149}Sm has a nuclear absorption resonance at $\varepsilon' = 97.3$ meV, and occurs with abundance $a = 13.9\%$ in natural samarium. In the vicinity of this resonance the neutron–nuclear scattering length varies

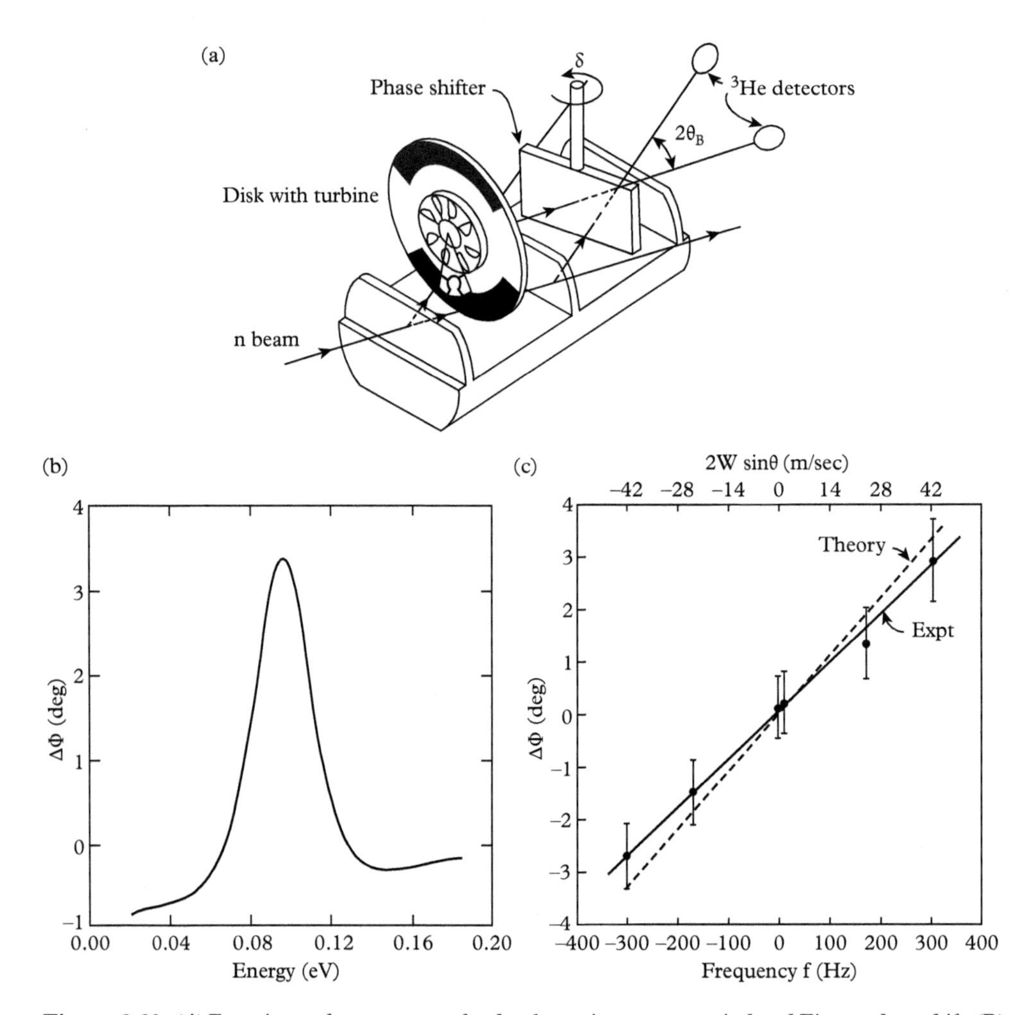

Figure 8.23 *(A) Experimental arrangement for the absorption resonance-induced Fizeau phase shift, (B) absorption resonance, and (C) a comparison between theoretical and experimental values. The shadowed segments of the disk indicate the Sm foils (Arif et al. 1989)*

drastically. For a foil of thickness L mounted on the disk of radius r = 4.6 cm, and rotating with frequency f, Eq. (8.103) can be written as

$$\delta\,(\Delta\phi) = 8\pi^2\hbar^2 aLrf \tan\theta_B \cos\frac{\Omega}{2}\left(\frac{\partial b}{\partial\varepsilon}\right)_{\varepsilon_i}. \tag{8.105}$$

Here Ω is the angle subtended by the two beams, I and II, in the interferometer at the center of the rotating disk. Using the known resonance parameters of ^{149}Sm, a calculation of the expected motion-induced phase shift as a function of ε' is also shown in Fig. 8.23.

Two simultaneous interferograms (counts versus δ) were obtained stroboscopically for the neutrons passing through the pure aluminum sector and then through the Sm-foil–Al composite. The difference of the phase shift as a function of the frequency was then compared to this difference obtained quasi-statically ($f \approx 0$). The resulting phase shifts obtained at three frequencies, $f = 3$, 170, and 303 Hz, for both clockwise and counterclockwise rotations is shown in Fig. 8.23C. The incident neutron energy was 95.8 meV ($\lambda = 0.924$ Å) such that the Bragg angle $\theta_B = 13.92°$ for the (220) reflection in Si. It is noted that these phase shifts are really quite small, less than a few degrees. A straight line fit to the data yields the derivative $(\partial b/\partial \varepsilon)_{95.8\,\text{meV}} = (1.90 \pm 0.34) \times 10^{-10}$ cm/eV. This is to be compared to the calculated derivative at this energy of 2.39×10^{-10}cm/eV.

Successful completion of this experiment required very precise mechanical tolerances in the apparatus and nearly perfect vibration isolation of the very fast (18,000 rpm) rotating disk from the interferometer. Furthermore, data collection times were quite long because of the large absorption cross-section of Sm near the resonance. Clearly, the averaging of the results from many runs was necessary.

8.5.6 Proper Time and General Relativity Considerations

Shortly after the initial COW experimental result was published, questions about extending the precision to include general relativistic (GR) effects on the phase were posed (Greenberger 1984; Anandan 1977; Stodolsky 1979, 1979a). With the precision achievable with modern laser-based atom interferometry, this question has recently resurfaced regarding the possibility of directly observing a gravitational redshift via matter wave interferometry techniques (Müller et al. 2010, Wolf et al. 2010, Greenberger et al. 2012). This suggestion is best characterized as quite controversial at the present time. Here we use the concept of proper time for calculating the gravitationally induced phase shift in neutron interferometry, thereby connecting time and particle mass (Lan et al. 2013), i.e., non-relativistic quantum theory to relativity phenomena.

Formally the neutron phase is given by integration over proper time τ (provided by a clock moving with the neutron):

$$\phi = -\frac{mc^2}{\hbar}\int \mathrm{d}\tau = -\frac{mc^2}{\hbar}\int \left(1 + \frac{2U}{c^2} - \frac{v^2}{c^2}\right)^{1/2} \mathrm{d}t. \tag{8.106}$$

Here $U = -GM/r$ is the gravitational potential external to our Earth of mass M. We have used the Schwarzschild metric

$$\mathrm{d}\tau = \sqrt{g_{\mu\nu}\mathrm{d}x_\mu \mathrm{d}x_\nu} \tag{8.107}$$

with $g_{00} = 1 + 2U/c^2$and $g_{ii} = -1$. Thus, in the limit of low velocity, and a weak gravitational field, the phase of the neutron de Broglie wave is

$$\phi \cong -\frac{mc^2}{\hbar}t + \frac{1}{\hbar}\int \left(\frac{1}{2}mv^2 - mU\right) dt. \tag{8.108}$$

The term with the Compton frequency $\omega_c = mc^2/\hbar$ gives an overall phase factor in an interferometer experiment, which is not observable since it appears in both beam paths.

Thus, we see that the phase shift in the interferometer can be written as integral over the non-relativistic Lagrangian $L = \left(mv^2/2 - mU\right)$, namely

$$\Delta\Phi = \frac{1}{\hbar}\int_{\text{path II}} L\mathrm{d}t - \frac{1}{\hbar}\int_{\text{path I}} L\mathrm{d}t. \tag{8.109}$$

As shown in Chapter 1, these integrals can be transformed to path integrals over the canonical momentum using a Legendre transformation, namely $L = \boldsymbol{p}\cdot\boldsymbol{v} - H$ (see Eq. 1.32). For a time-independent Hamiltonian when energy is conserved $H = H(\boldsymbol{p}, \boldsymbol{r})$, we use

$$\int L\mathrm{d}t = \int (\boldsymbol{p}\cdot\boldsymbol{v}dt - H)\mathrm{d}t = \int \overline{\boldsymbol{p}}\cdot\mathrm{d}\boldsymbol{r} - Et \tag{8.110}$$

and note again that the phase shift comes from a path integral over the canonical momentum $\boldsymbol{p}$. Carrying out the integration will give the COW formula Eq. (8.4). The integration must be carried out over the undisturbed trajectory which follows from the "golden rule" of small perturbations. It appears as a residue of relativistic effects where proper time shows up as a matter wave phase (Greenberger et al. 2012).

But what about higher order gravity effects? If one expands the square root in the integrand in Eq. (8.106) to order $1/c^2$, we get

$$\Delta\Phi_{\text{grav}} = \Delta\Phi_{\text{COW}}\left(1 + \frac{v_0^2}{2c^2} - \frac{R_S}{2R_E} - \frac{3}{2}\frac{R_S H}{R_E^2}\right). \tag{8.111}$$

where $R_S = 2GM/c^2 = 8.8$ mm is the Schwarzschild radius for the Earth of radius $R_E = 6.4 \times 10^6$ m. For 2-Å neutrons $v_0{}^2/c^2 = 4.4 \times 10^{-11}$ and $R_S/R_E = 1.4 \times 10^{-9}$. These corrections to the COW phase shift formula are indeed very small.

Using a Kerr metric Kuroiwa et al. (1993) calculated the correction to the Sagnac phase due to the Earth's rotation. The result is

$$\Delta\Phi_{\text{rotation}} = \Delta\Phi_{\text{Sagnac}}\left\{1 + \frac{2}{5}\frac{R_S}{R_E}\left[1 - \frac{3(\hat{R}\cdot\hat{A})(\hat{\Omega}\cdot\hat{R})}{\hat{\Omega}\cdot\hat{A}}\right]\right\}. \tag{8.112}$$

We have taken the Earth to be of uniform mass density to get this simplified formula. $\hat{R}$ is the outward radial unit vector, $\hat{\Omega}$ is the rotation frequency unit vector, and $\hat{A}$ is the interferometer area unit vector. This correction term is smaller than $\Delta\Phi_{\text{Sagnac}}$ by about R_S/R_E, and is due to the Lense–Thirring frame drag.

Thus, it is clear that the first *GR* terms are of order 10^{-9} of the non-relativistic phase shifts $\Delta\Phi_{\text{COW}}$ and $\Delta\Phi_{\text{Sagnac}}$. Since the Hamiltonian appropriate to the neutron moving in a laboratory fixed on the surface of our rotating Earth is time independent, the neutron's total energy E is a constant of the motion. Therefore, the frequency $\omega = E/\hbar$ of the de Broglie matter wave is a constant of the motion, and is independent of the path within the interferometer. The gravity-induced phase shift comes entirely from the line integral of the wave vector along the undisturbed trajectory. In the case of the neutron interferometer, one can say that one is indeed seeing the residues of the redshift and of the twin paradox, even though one cannot prove that it is a general relativistic observation (Greenberger et al. 2012). The situation with neutrons appears to be quite different to the situation with the Kasevich–Chu atom interferometers where the atom–laser interaction causes the beam splitting and recombination and where momentum and energy changes occur during the interaction (Schleich et al. 2013a, 2013b). Nevertheless, the Berkeley group continues to believe that there is a clock directly marking time to the particle's mass, and the atom interferometer based upon Raman resonance absorption is sensitive to the atom's Compton frequency (see Lan et al. 2013).

The paper by Zych et al. (2011) takes a very interesting point of view toward an internal clock carried by the neutron. They point out that the precessing neutron in a magnetic field is certainly a clock, and the polarization state of the neutron in either beam path is a "witness" of the proper time. One might imagine that the entire region of the interferometer could be immersed in a large, constant, and uniform magnetic field. The spin precession difference between the upper and lower path in the COW interferometer would then give the proper time difference. Unfortunately, even for a 10-T magnetic field the effect is quite small, of order 10^{-10} of the COW phase.

9

Solid State Physics Applications

Solid state and material science are fields with major applications of neutron physics. Elastic and inelastic scattering experiments give fundamental information about the static and dynamic behavior of atoms in condensed matter (van Hove 1954, Marshall and Lovesey 1971, Willis and Carlile 2009). Rather sophisticated scattering theories are required to understand these experiments. Data from neutron optics, where neutron interferometry belongs, use quantum diffraction theories based upon the Schrödinger equation directly (e.g., Sears 1989).

Most neutron interferometric experiments carried out until now have dealt with fundamental physics applications, but it was known from the beginning that the interferometric method can open new horizons for solid state physics research, as well. Samples with density or magnetic fluctuations or with decomposition effects cause inhomogeneous phase shifts which cause a measurable loss of contrast of the interference pattern. Such experiments give complementary information to small-angle scattering, critical, and depolarization experiments (Rauch and Seidl 1987). Various phase imaging and phase tomography methods for neutrons have been realized, which permits inspection of materials without absorbing or scattering neutrons from the object (Dubus et al. 2005). Various examples of related investigations will be discussed together with their future perspectives. A general Fourier method for the investigation of condensed matter correlation functions is under development and is described in Section 10.15.

9.1 Contrast Reduction due to Inhomogeneities

In the context with the discussion of the coherence function (Section 4.1.2), it has been shown that density (and thickness) variations cause a reduction of the contrast of the interference pattern (Eq. 4.48). This is caused by different phase shifts across the beam cross-section. This is generally called a loss of coherence although it is, strictly speaking, a dephasing process. This is also the basic feature for the Christiansen filter method for measuring coherent scattering lengths of irregularly shaped samples, as described in Section 3.1.2.3.

The attenuation factor of the interference contrast stems from a variation of the phase shift for various beam paths through the sample or/and variations across the volume that the neutron wave feels during its passage through the sample. For a composite material the mean overall phase shift can be written as

$$\chi = -\lambda \, D_{\mathrm{eff}} \sum_i N_i b_{ci}. \tag{9.1}$$

Neutron Interferometry. Second Edition. Helmut Rauch and Samuel A. Werner.

A phase shift of 2π defines the λ-thickness $D_\lambda = 2\pi/(\lambda \sum_i N_i b_{ci})$. Whenever inhomogeneities of the scattering length density $\sum_i N_i b_{ci}$ reach dimensions δ which are comparable to D_λ, the interference contrast becomes reduced. In most important cases, the phase shifts within individual inhomogeneities will be small compared to 2π (or $\delta << D_\lambda$). When the phase distribution function of the beam paths through the inhomogeneities is nearly Poissonian, i.e., $<\delta^2> \cong <\delta>^2$, one obtains from Eqs. (4.42) and (4.48) the damping (dephasing) factor

$$\begin{aligned} D_\mathrm{d} &= \exp\left[-(\delta N/N_0)^2 \left(\frac{<\delta>}{D_\mathrm{eff}}\right)(\Delta_0 k_0)^2/2\right] \\ &= \exp\left\{-[\delta\,(Nb_\mathrm{c})\,\lambda_0]^2 <\delta> D_\mathrm{eff}/2\right\}. \end{aligned} \tag{9.2}$$

This modifies the expected interference pattern, such that

$$I = I_0\left[A + D_\mathrm{d}\, B \cos\left(2\pi \frac{D_\mathrm{eff}}{D_\lambda} + \varphi_0\right)\right], \tag{9.3}$$

where A, B, and φ are the characteristic parameters for a sample without precipitates (Eq. 2.5). For differently shaped precipitates $<\delta>$ can be taken from the rational approximation where $<\delta> = 4V/S$, where V is the volume and S is the surface of these inhomogeneities. One notices the close analogy to the well-known neutron depolarization formalism (Halpern and Holstein 1941, Rekveldt 1973). In the case of magnetic inhomogeneities in the form of a ferromagnetic domain structure the dephasing factor reads as ($\gamma = 2\mu/\hbar$)

$$D_\mathrm{d}^m = \exp\left[-(\gamma B m \lambda_0)^2 <\delta> D_\mathrm{eff}/h^2\right], \tag{9.4}$$

where $\gamma = 2\mu/\hbar$. This equation has basically the same structure as Eq. (9.3). In both cases dephasing $(1 - D_\mathrm{d})$ increases with increasing dimensions of the inhomogeneities. The close connection between dephasing, decoherence, and depolarization has been addressed by Rauch et al. (1999). For very large inhomogeneities ($\delta >> D_\lambda$) the opposite dependence on the dimensions of the inhomogeneities is expected. A detailed treatment of the dephasing effect which takes into account the different coherence properties of the beam in the different directions and the beam deflection due to small-angle scattering effects is still missing. A smooth transition from the stochastic situation as discussed here to a deterministic situation is expected where a partial beam path detection due to beam deflection occurs (see Section 4.3.2). This small-angle scattering effect can be written down in the Guinier approximation (e.g., Kostorz 1979) as

$$\frac{d\sigma}{d\Omega} \propto [\delta\,(Nb_\mathrm{c})]^2\, e^{-Q^2<\delta>^2/3}, \tag{9.5}$$

where the momentum transfer Q is directly related to the scattering angle by $Q \cong k\cdot\Theta$. Even very small beam deflection (parts of seconds of arc) cause a labeling of neutrons which can be measured by means of a perfect crystal small-angle scattering camera (Bonse and Hart 1965, Miksovsky et al. 1992). One notices that small-angle scattering is expected when dephasing is small and vice versa.

Related interferometric measurements have been done for various metal-hydrogen (deuterium) systems where at a certain concentration a transition from the homogeneous α-phase to the inhomogeneous β-phase happens. This is a situation where hydride precipitates coexist with the α-phase-saturated bulk material. The behavior of hydrogen interstitials in metals is an interesting material science problem with considerable technological impacts as has been summarized by Alefeld and Voelkl (1978) and Wipf (1997). Such systems show incoherent, i.e., phonon-assisted, tunneling effects and, at rather low temperatures, coherent tunneling effects. Neutron diffraction and NMR investigations are the major tools for obtaining information about the dynamics of hydrogen in metals. The knowledge of the hydrogen content and of its distribution is an important aspect of such investigations. The following are standard methods for determining hydrogen content:

- Weighing the sample;
- Volumetric method;
- Electrical resistance;
- X-ray diffraction; and
- Quartz microbalance.

They have their advantages and limitations and, therefore, the development of additional methods is advised. In this connection, neutron interferometric methods have been tested to show their capabilities. The experimental results show the variation of the modulation frequency due to the variation of the λ-thickness of the composed materials (Eq. 9.1) and a reduction of the contrast when the H(D) concentration approaches the phase boundary (Fig. 9.1; Rauch et al. 1978b). The observed reduction of the contrast even in the α-phase indicates precursors of hydride

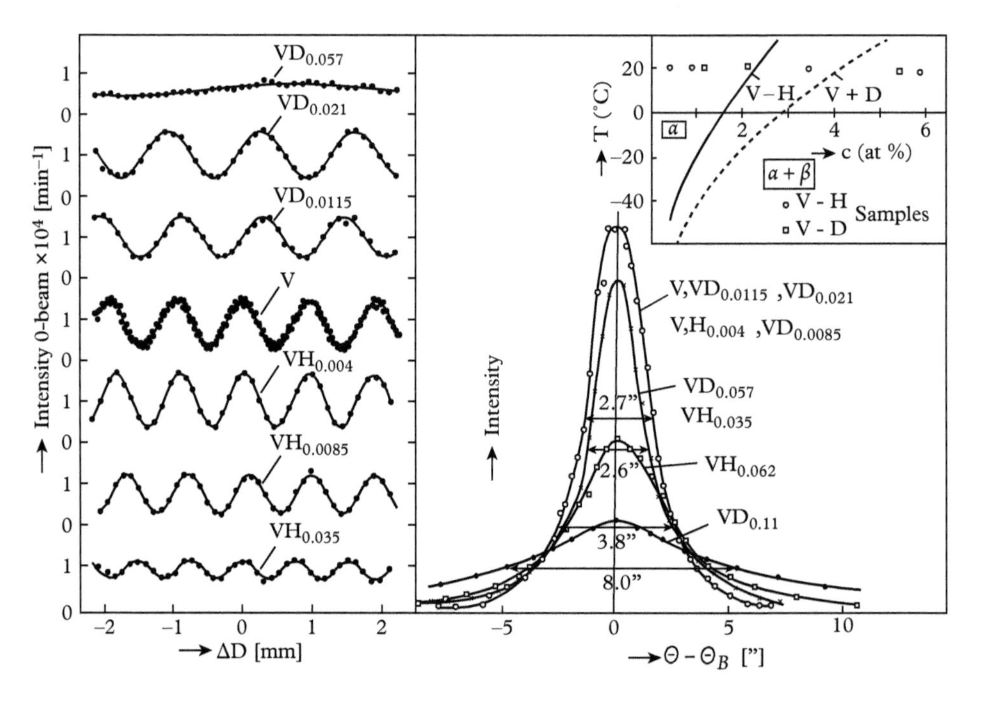

Figure 9.1 *Interference pattern (left) of various VH and VD samples near to the α- to β-phase transition (insert) and related small-angle scattering pattern (right) (Rauch and Seidl 1987)*

precipitates. The interferometric method seems to be more sensitive to this phenomenon than the ultra-small-angle scattering method where a reduction of the peak intensity and a broadening of the double-crystal rocking curve occur only for H(D) concentrations which belong to the β-phase. The accuracies achieved for determining H(D) concentration were 0.02–0.06 at.% or 6–12 ppm by weight. By using the non-dispersive measuring method described in Sections 3.1.2.1 and 4.2.2 at least one order of magnitude in sensitivity can be gained. This may become a useful method for a non-destructive hydrogen (deuterium) determination in materials.

Similar measurements (Rauch 1979) with non-annealed alloys ($Al(Zn\ Mg_2)_{0.015}$) showed stronger dephasing in the case of larger precipitates (annealed sample), which agrees with observations made by electron microscopy (Skalicky and Oppolzer 1972). Various other test measurements with samples with a known inhomogeneity structure are reported by Tuppinger et al. (1988) (Figs. 3.11 and 3.12). A special challenge would be an investigation of the contrast variation near to critical phase transitions.

The hydrogen content in an evaporated Pd film has been determined by means of a Jamin-type interferometer (see Fig. 2.16). The gap layer consists of a palladium–hydrogen system which varies the phase shift depending on the hydrogen content according to Eq. (2.17), where Nb_c must be replaced by $(Nb_c)_{Pd} + (Nb_c)_H$. The change of the layer thickness or of the layer composition causes a varying phase shift which can be written as

$$\phi = n_\ell DQ, \tag{9.6}$$

where n_ℓ and D denote the index of refraction and the thickness of the gap layer, respectively, and Q the transferred momentum $Q = 4\pi \sin \Theta/\lambda$, which is related to the lattice constant d of the diffraction layers. As the volume of the layers changes due to the absorption of hydrogen, the quantities D and N change as well ($\Delta V/V = \varepsilon \cdot c$, where c denotes the ratio H/Pd and $\varepsilon = 0.19$). The interference layer structure (MINI) was set in a holder and kept at different hydrogen pressures. The related reflection curves are shown in Fig. 9.2 (Tasaki et al. 1995). An accuracy of 5.5% for the hydrogen content determination has been achieved.

In another experiment the atomic density of a polymer film supported on a silicon substrate was investigated by means of a perfect crystal interferometer by Wallace et al. (1999). It was found that the atomic density in thin films can be a strong function of the preparation condition for many materials. In a non-dispersive arrangement (see Section 3.1.2.1) the phase shift of a deuterated polystyrene film with a thickness of 0.5210(6) μm was measured as 0.400(2) rad, which gives an atomic density of $9.339(48) \times 10^{22}$ atoms/cm^3, which is about 3% below the bulk density of that material.

The domain structure of ferromagnetic materials causes an inhomogeneous phase shift and small-angle scattering effects, as well. Both contribute to the reduction of the interference contrast. Figure 9.3 shows a related result obtained with non-annealed and annealed iron foils (Rauch 1980). The growth of ferromagnetic domains in the annealed sample (5 h at 600°C) is visible. The depolarization factor D_n of these samples has been measured separately at a proper polarized neutron facility. The loss of contrast ($D_p{}^m$) must be compared approximatively to the square of the depolarization factor ($D_n{}^2$) because the samples were placed in both beams of the interferometer. The comparison yields for the non-annealed sample

$$D_p{}^m = 0.605 \quad \text{and} \quad D_n{}^2 = 0.593$$

and for the annealed sample

$$D_p{}^m = 0.151 \quad \text{and} \quad D_n{}^2 = 0.16,$$

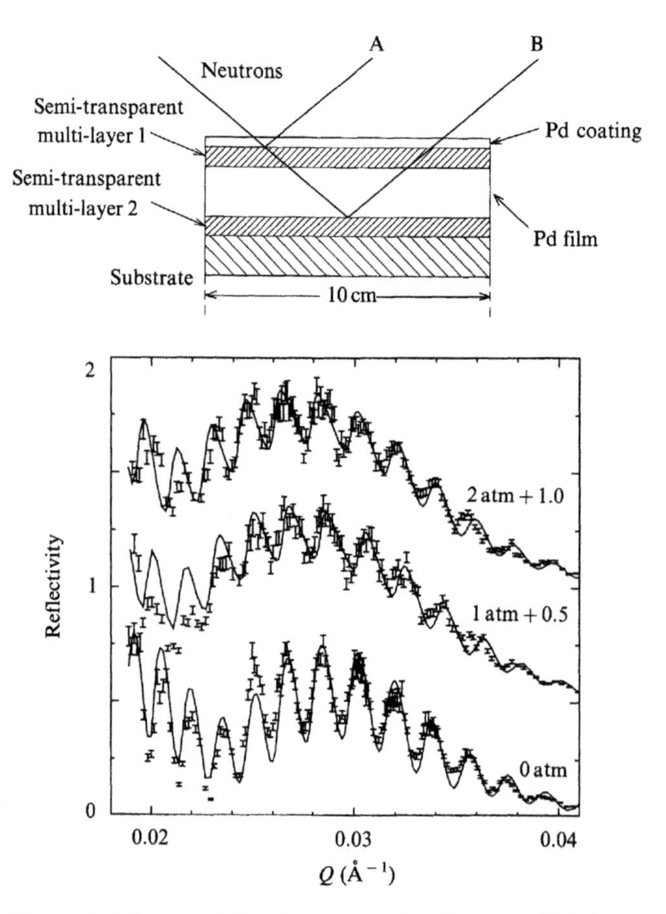

Figure 9.2 *Layout of the micro-neutron interferometer (MINI) and various reflection curves at different hydrogen content of the Pd film. Reprinted with permission from Tasaki et al. 1995. Copyright 1995, AIP Publishing LLC.*

which corresponds to a domain size 3.5 times larger than that for the non-annealed sample. Related small-angle scattering experiments showed a similar behavior.

9.2 Phase Imaging Topography and Phase Tomography

The inside of an object can be observed either by measuring the intensity transmitted through the object or by observing the phase shift caused by the material in the beam. A three-dimensional pattern of the interior structure can be reconstructed when intensity or phase patterns are taken for different orientations of the object in relation to the beam. Both techniques are well established for X-rays (Ando and Hosoya 1972, Hart 1975, Bonse et al. 1986, Kinney and Nicols 1992, Momose 1995, Momose et al. 1995, Beckmann et al. 1997). The phase contrast X-ray computed

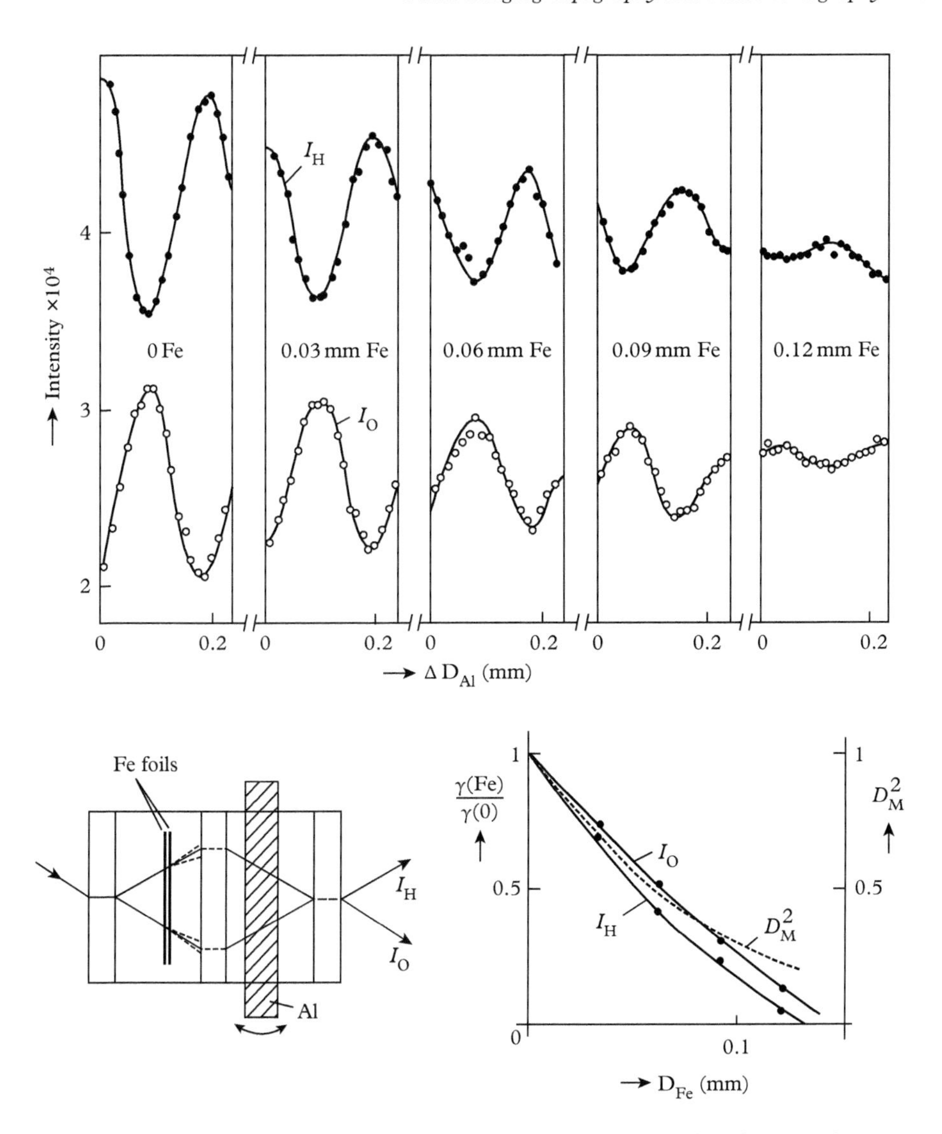

Figure 9.3 *Interference pattern (above) of various Fe foils inserted in both beams and a comparison of the dephasing and depolarization factor (right) (Rauch 1980)*

tomography performed mainly at synchrotron sources has reached a standard which allows its use for biological and medical investigations (Momose et al. 1996, Bonse et al. 1997). The advent of high-resolution position-sensitive neutron detectors makes neutron phase tomography and phase imaging possible. Consequently, in the past years several attempts have been made to detect the neutron phase variations induced by an object. These methods can be classified into double-crystal methods (Strobl et al. 2003, Treimer et al. 2003), pinhole methods (Allman et al. 2000), grating methods (Pfeiffer et al. 2006), and interferometric methods (Zawisky et al. 2004). Their basic principles and their applicability for different problems will be discussed. One of the fascinating features is that images can be produced without beam attenuation; i.e., nearly interaction-free inspection of matter becomes feasible.

9.2.1 Double-Crystal Method

Podurets et al. (1989) noticed that the rocking curves of a perfect crystal, double-crystal arrangement can be used for the reconstruction of the sample inhomogeneity structure. This method has been further developed (Chapman et al. 1997, Treimer and Feye-Treimer 1998) and uses a perfect crystal, double-crystal arrangement and the refraction broadening due to small-angle scattering within the sample (Strobl et al. 2003, Treimer et al. 2003). This broadening Δ is different for each beam path and depends on the index of refraction distribution $n(x, y)$. The beam deflection is in the direction of the gradient of the index of refraction and therefore one gets a point of a projection when rotating (θ) the sample around a vertical axis as

$$P_\theta(\Delta) = \int_{\text{path}} \nabla n(x, y) \cdot \boldsymbol{k}_\perp \mathrm{d}s. \tag{9.7}$$

The total set $P_\theta(\Delta)$ can be taken as input for a Radon transformation to obtain $n(x, y)$. Figure 9.4 shows the experimental arrangement with curved perfect crystals, which permit the measurement of the broadening by means of a position-sensitive detector. A typical image of a test sample is shown as well. One notices the advantage of the refraction method (b) in comparison with the standard attenuation method (a).

9.2.2 Pinhole Method

This method has been developed and tested for neutrons by Allman et al. (2000) and uses a non-interferometric phase recovery method based on the so-called transport of energy equation (Teague 1983, Gureyev et al. 1996)

$$\frac{2\pi}{\lambda}\frac{\partial I(\boldsymbol{r}_\perp)}{\partial z} = \nabla x\,[I(\boldsymbol{r}_\perp)\nabla\Phi(\boldsymbol{r}_\perp)], \tag{9.8}$$

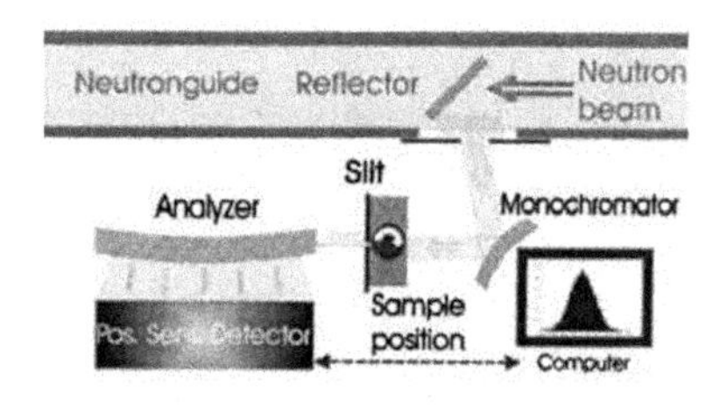

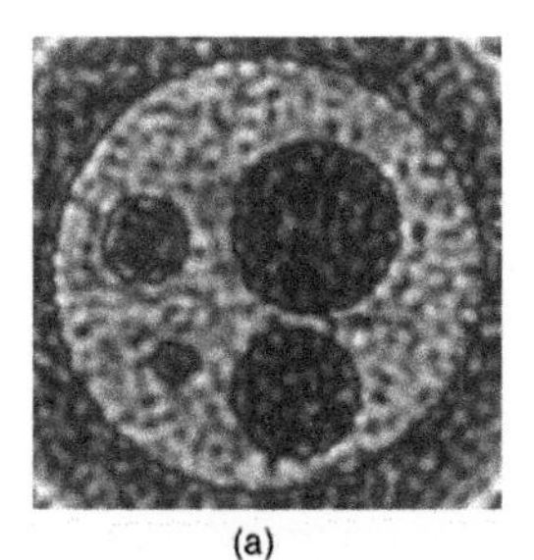

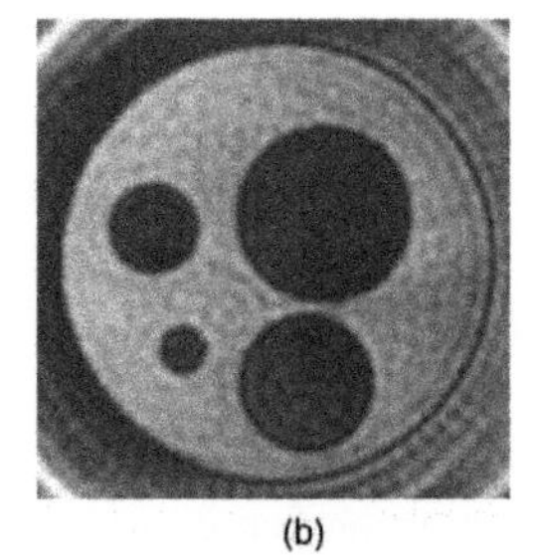

Figure 9.4 *Experimental setup and results of reconstruction of a brass sample with holes ranging from 1 to 4.3 mm. Reprinted with permission from Treimer et al. 2003. Copyright 2003, AIP Publishing LLC.*

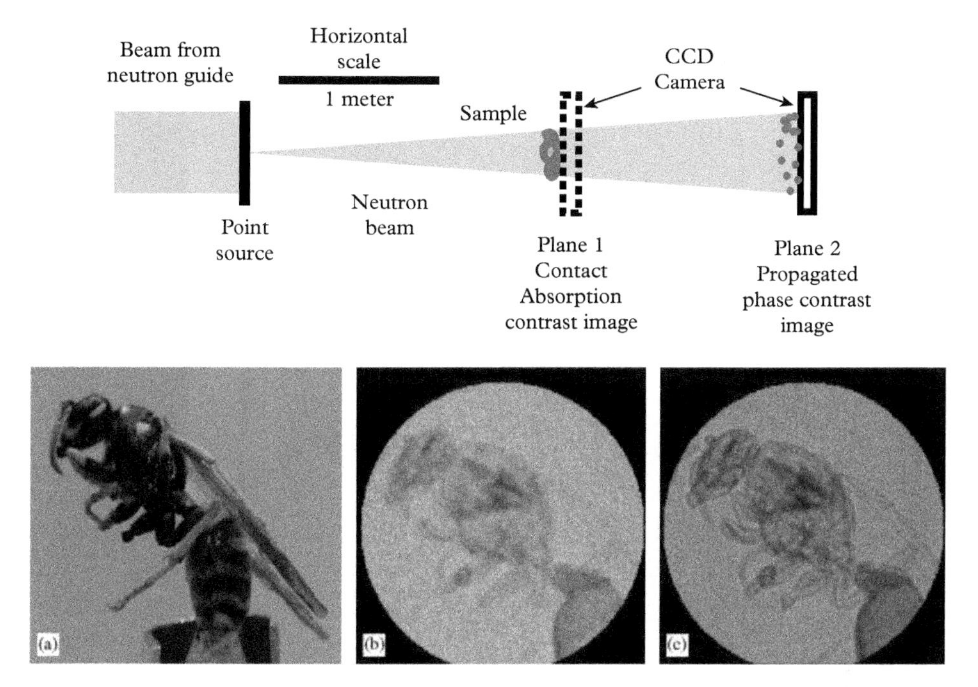

Figure 9.5 *Experimental arrangement of the pinhole system used for phase contrast radiography and typical results for a yellow-jacket wasp: (a) photograph, (b) attenuation radiograph, (c) phase contrast radiograph. Reprinted from Allman and Nugent 2006, copyright 2006, with permission from Elsevier.*

where a wave with intensity $I(\boldsymbol{r}_\perp)$ and phase $\Phi(\boldsymbol{r}_\perp)$ is described. ∇ and $\boldsymbol{r}_\perp$ denote the gradient and the position vector, respectively, in the plane perpendicular to the longitudinal optical axis behind the aperture and $\partial I(\boldsymbol{r}_\perp)/\partial z$ denotes the intensity variation along the optical axis. Thus, the phase can be obtained by measuring at two different distances behind the object (Fig. 9.5). The phase image obtained is, to a good approximation, described by a convolution of the perfect image with the intensity distribution of the effective source. A typical phase contrast image taken with a neutron wavelength of 4.43 Å, a pinhole diameter of 0.4 mm, a pinhole-sample distance of 1.8 m, and the position-sensitive detector just behind and 1.8 m behind the sample is shown in Fig. 9.5. The increased visibility of the phase contrast radiograph (c) is visible. More applications of this technique can be found in a paper by Allman and Nugent (2006).

9.2.3 Grating Method

This is an extended version of the pinhole method and uses instead of a single small pinhole different gratings, permitting a broad beam and a more effective use of neutrons (Pfeiffer et al. 2006). The setup is shown in Fig. 9.6. An absorbing source grating (G0) creates an array of individually coherent, but mutually incoherent beams. The coherence is obtained when the lattice constants p of the two lattices G0 and G2 fulfill the condition $p_0 = p_2 l/d$. The final spatial resolution depends also on the width w of the beam cross-section (wd/l). The differential phase

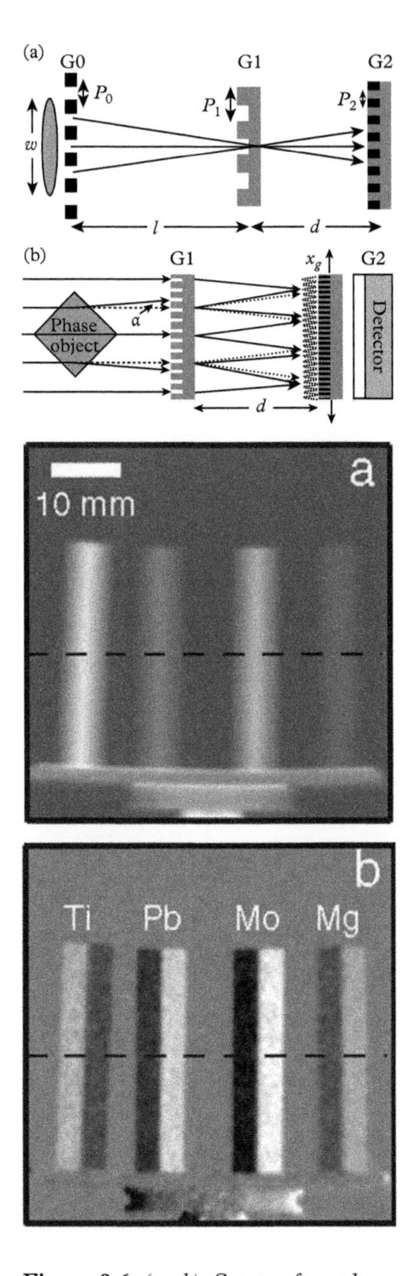

Figure 9.6 *(a, b) Setup of a phase-grating imaging system and a typical result for various quadratic metal rods (Ti, Pb, Mo, Mg), showing a conventional neutron transmission image (c) and a phase contrast image (d). Reprinted with permission from Pfeiffer et al. 2006, copyright 2006 by the American Physical Society.*

contrast image formation is achieved by the gratings G1 and G2, where G1 acts as a phase grating with a phase shift of π and as a beam splitter whose beams interfere downstream (Fig. 9.3c). Neither the period nor the lateral position of *the* interference fringes depend on the wavelength, but any perturbation due to a phase object lead to local displacements of the fringes. Scanning the grating G2 in front of a position-sensitive detector gives an intensity signal $I(x,y)$ which oscillates as a function of the G2 position. This determines the phases at that position $\Theta(x,y)$, which relates to the phase of the wave field $\chi(x) = -Nb_c\lambda D$ as $\Theta(x,y) = (\lambda d/p_2)(\delta\chi/\delta x)$. Typical parameters for such an arrangement are $p_0 = 1.08$ mm, $p_1 = 7.97$ μm, $p_2 = 4.00$ μm, $l = 5.23$ m, $d = 19.4$ mm, $w = 20$ mm, $\lambda = 4.1(9)$ Å, and a spatial resolution of the detector of 250μm. The production technique for proper high-resolution gratings is described by Kim et al. (2013).

9.2.4 Interferometer Method

In this case the perfect crystal interferometer is used as shown in Fig. 9.7 and the different phase shifts of an inhomogeneous material provide the relevant signals. Inhomogeneities of the scattering length density which are larger than the spatial resolution of the recording system can be observed similarly to X-ray phase contrast imaging and in neutron Larmor precession mapping (Schlenker and Shull 1973). A review about standard neutron tomography methods was given by Schlenker and Baruchel (1986). Phase tomography uses the interference of a coherent signal and a reference beam and produces two complementary pictures after superposition, as shown schematically in Fig. 9.7. The spatial resolution Δ_r in the case of phase tomography is given by the related coherence lengths (Δ_c) and the resolution of the imaging system (Δ_a), which can in

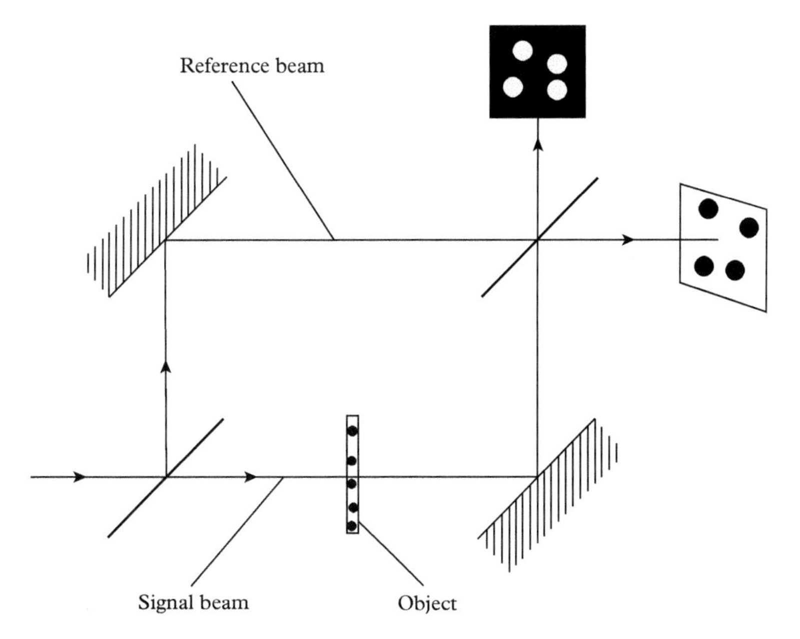

Figure 9.7 *Schematic diagram of a phase tomography setup based upon an interferometry method*

most cases be written in the form $\Delta_r^2 \cong \Delta_c^2 + \Delta_a^2$. In most cases a much higher resolution is obtained in the vertical direction where no Borrmann spreading exists. Therefore, the objects must be rotated around a horizontal axis. The tomographic picture depends on the phase shift only and, therefore, a picture of inhomogeneities can be obtained without any beam attenuation. The 2π-periodicity of the interference pattern requires a subfringe and fringe scanning analysis before the data can be used for computer tomography analysis which are based on Radon transformation methods.

Standard radiography is sensitive to the overall beam attenuation ($I = I_0 \exp[-\Sigma_t D]$), whereas phase topography to the interference term of the interference pattern reads $\cos(2\pi D/D_\lambda)$. In the case of maximum sensitivity ($D = (n + ½)D_\lambda$) and equal intensity phase tomography has an enhanced sensitivity given by (Rauch 1997, Beckmann 1998)

$$G \propto \frac{b_c \lambda}{\sigma_a + \sigma_s}, \tag{9.9}$$

which is for most substances on the order of 1,000 to 10,000. This still remains large even when the reduced intensity of an interference experiment and an interference contrast smaller than unity is taken into account.

For the neutron case ferromagnetic domains in Fe-3% Si was investigated by this method (Schlenker et al. 1980). Typical results are shown in Fig. 9.8 where an auxiliary magnetic field was applied to the reference beam to distinguish up and down magnetizations of the domains for unpolarized neutrons. The domain walls appear in the picture not only due to their inhomogeneous phase shift but also due to small-angle deflection effects. The resolution of this early phase tomograph is rather poor and this technique should be developed further before it becomes a routine method for observing magnetic domains. It contains a large amount of information on neutron optics.

Nakatami et al. (1992) measured the overall interference pattern when a Fe-3% Si magnet with 180° Bloch walls was inserted under different angles additionally to a pure nuclear (Al) phase shifter. They obtained the expected dependence of the amplitude and of the phase which verified that saturation of magnetization within the domains has been reached even without external fields. Data evaluation showed that the correct domain size can be extracted.

Three-dimensional computed tomography investigations are feasible when pictures are taken at different sample orientations. Contrast-matching methods similar to those discussed in Section 3.1.2.3 can be applied as well.

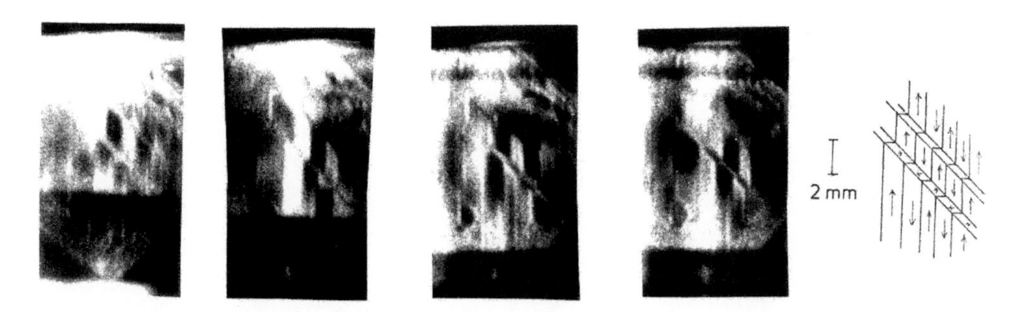

Figure 9.8 *Neutron phase tomography pattern of magnetic domains taken behind a perfect crystal neutron interferometer (Schlenker et al. 1980)*

Neutron Larmor precession transmission tomography experiments of polarized neutrons through precession coils or magnetic materials yield information about the magnetic field distribution (Rekveldt and Kraan 1987, Kraan et al. 1988, Shitnev 1989, Rosman and Rekveldt 1991, Rekveldt et al. 2006). In this case the degree of polarization plays a similar role as the coherence function in interference experiments. Therefore, Fourier spectroscopy methods become feasible for neutron Larmor precession experiments and neutron interferometer experiments as well.

For a full tomographic inspection where projections $P(\Theta, \mathbf{r})$ must be taken from an object $f(x, y)$ under different orientations, one has

$$\begin{aligned} P(\Theta, \boldsymbol{r}) &= \int f(x, y) \mathrm{d}s \\ &= \iint f(x, y) \delta(x \cos\Theta + y \sin\Theta - \boldsymbol{r}) \mathrm{d}x\, \mathrm{d}y, \end{aligned} \tag{9.10}$$

where $\boldsymbol{r}$ denotes the position at the projection screen ($\boldsymbol{r} = x\cos\Theta + y\sin\Theta$). Equation (9.7) denotes the well-known Radon transform (Radon 1917, Herman 1980) of $f(x, y)$, which can be used more easily in the frequency domain, that is

$$\begin{aligned} F_1(\Theta, \omega) &= \int P(\Theta, \boldsymbol{r}) e^{-\mathrm{i}\omega t} \mathrm{d}t \\ F_2(\mu, u) &= \iint f(x, y) e^{-\mathrm{i}(ux + vy)} \mathrm{d}x\, \mathrm{d}y, \end{aligned} \tag{9.11}$$

where

$$P(\Theta, \boldsymbol{r}) = F(\omega, \Theta) = F(u, v),$$

with $u = \omega\cos\Theta$ and $v = \omega\sin\Theta$. From these relations one obtains the object function

$$\begin{aligned} f(x, y) &= \iint F(\omega, \Theta) |\omega| e^{\mathrm{i}\omega t} \mathrm{d}\omega\, \mathrm{d}\Theta \\ &= \iint P(\Theta, \omega) |\omega| e^{\mathrm{i}(x\cos\Theta + y\sin\Theta)} \mathrm{d}\omega\, \mathrm{d}\Theta. \end{aligned} \tag{9.12}$$

The projection pattern in the case of attenuation tomography is given by the exponential attenuation law

$$P(\Theta, \boldsymbol{r}) = I_0(\Theta, r) \exp\left[-\int \sum (x, y, z) \mathrm{d}z\right], \tag{9.13}$$

where $\sum(x, y, z) = \sum_i N_i(x, y, z)\sigma_{\text{tot},i}$. In the case of phase tomography the projection pattern is given by the interference, namely

$$V(\Theta, \mathbf{r}) = I(x, y) + K(x, y)\ \cos[\varphi(x, y) + \varphi_0(x, y)]. \tag{9.14}$$

Here, the phase functions are

$$\phi(x,y) = \int \phi(x,y,z)\mathrm{d}z,$$

with

$$\phi(x,y,z) = -\lambda \sum_i N_i(x,y,z)b_{ci}.$$

In most cases the phase pattern of the empty interferometer must be substracted.

A related experiment has been done with a perfect crystal interferometer and a sample consisting of an aluminum matrix and an aluminum–magnesium screw with a small scattering length density difference only (Dubus et al. 2005). After reconstruction the screw and the air holes between the matrix and the screw become visible (Fig. 9.9).

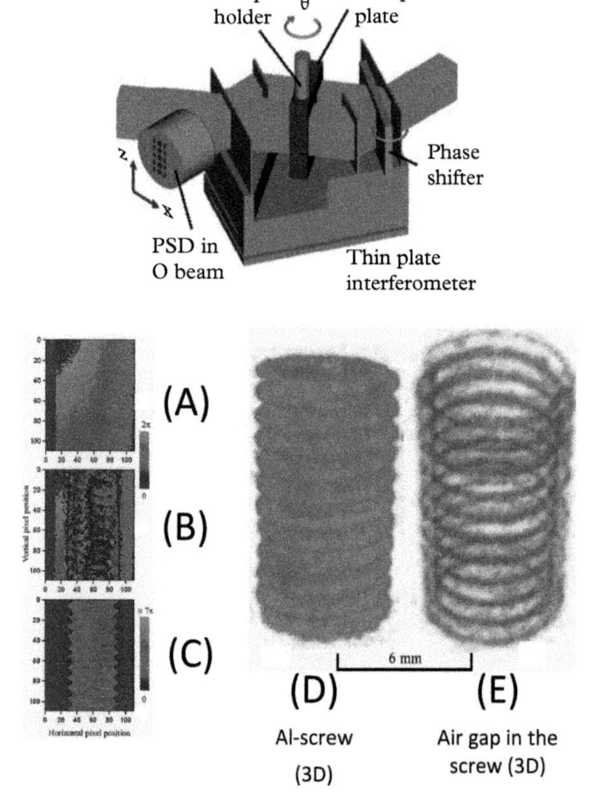

Figure 9.9 *Experimental setup and results of an interferometric phase tomography experiment inspecting a Al–Mg screw within an Al bulk material (Dubus et al. 2005)*

9.3 Observation of the Goos–Hänchen Effect

In the case of specular reflection the beam penetrates somehow into the material and the exit point becomes somewhat shifted compared to the entrance point. This was first suggested for light by Isaac Newton (1730) and experimentally verified by Goos and Hänchen (1947, 1949). The situation is shown in Fig. 9.10 where reflection from a magnetic material is considered. In this case the penetration depths for both spin components are different and the exit beams interfere with each other, which appears as a polarization variation. The neutron spin-echo reflectometer (Section 2.4; Fig. 2.20) permits a clear observation of this effect (de Haan et al. 2010). A more indirect observation has been reported by Pleshanov (1994). The superposition of the incident wave and spherically reflected waves from the scattering centers in the material can be written as

$$\Psi_{k_y}^{\pm} = \alpha^{\pm}(k_y)\left[e^{ik_y y} + \rho^{\pm}(k_y)e^{-ik_y y}\right], \tag{9.15}$$

in the y-direction. The reflection coefficients are given by

$$\rho^{\pm}(k_y) = (k_y - k_y^{c,\pm})/(k_y + k_y^{c,\pm}) \tag{9.16}$$

with wave vectors

$$k_y^{c,\pm} = \sqrt{k_y^2 - (k_c^{\pm})^2}. \tag{9.17}$$

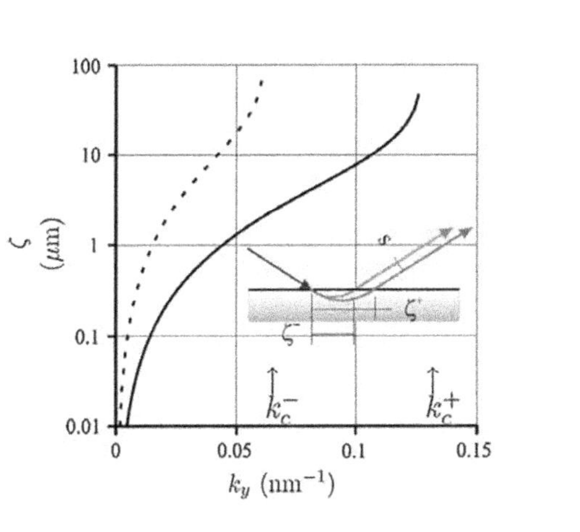

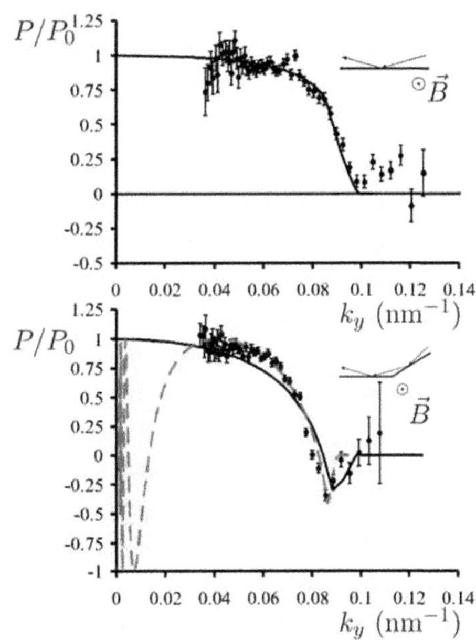

Figure 9.10 *Calculated Goos-Hänchen shift for both spin components (left) and measured polarization (right) of the exit beam as a function of the perpendicular wave vector k_y for a single and a double reflection from a magnetized Permalloy film. The insert shows the splitting of both exit beams. Reprinted with permission from de Haan et al. 2010, copyright 2010 by the American Physical Society.*

The critical wave vector is given as $(k_c^{\pm})^2 = 2m(V_n \mp \mu_n B)/\hbar^2$. From this the spatial displacement, the Goos–Hänchen shift can be calculated as (e.g., Renard 1964)

$$\xi^{\pm} = \frac{k}{(k_c^{\pm})^2} \frac{2k_y}{\sqrt{(k_c^{\pm})^2 - k_y^2}}, \tag{9.18}$$

which is shown in Fig. 9.10. Both exit beams experience a different phase shift, which appears as a k_y-dependent polarization (Fig. 9.10). This is an analog to the spin-superposition experiment described in Section 5.2.

10

Forthcoming, Proposed, and More Speculative Experiments

In this chapter we discuss some interferometric measurements which are intended to prove or disprove alternative physics theories and to search for extremely small effects where upper limits can only be anticipated. A general discussion of neutron optical searches for violations of quantum mechanics has been given by Klein (1988) and by Pokotilovky (2013). There are also experiments described here which have not been done yet. Thus, a search for speculative effects does not mean that no measurable effects can be found for all time. On the contrary, some of the phenomena described are already under investigation and a positive outcome can be expected. Some experiments are included where currently the accuracy is not high enough to make decisive conclusions. Condensed matter investigations have been done mainly by scattering experiments, but interferometric investigations are still missing. By measuring the neutron coherence function (Section 4.2), one gets direct access to the condensed matter correlation functions $G(\mathbf{r}, t)$ instead of measuring its Fourier transform $S(\mathbf{Q}, \omega)$ by spectroscopic methods. The essentials of neutron Fourier spectroscopy are formulated. Most experiments described in this chapter deal with basic properties of the neutron, with its fundamental interactions and partly with novel condensed matter research.

10.1 Non-linearity of the Schrödinger Equation

Various authors have proposed the addition of a non-linear term to the Schrödinger equation to obtain nonspreading wave packets (Bialynicki-Birula and Mycielski 1976, 1979) or manifold solutions with just one packet stationary in time (Pearle 1976, 1984a). Shimony (1979) proposed an interferometric test to observe a non-linear term which fulfills the separability condition and results in nonspreading wave packets. Such a term must have the form

$$F_{\mathrm{nl}} = b \ln(a^n |\psi|^2), \tag{10.1}$$

where a and b are constants. b must be small because of the great success of standard quantum mechanics. An absorbing phase shifter, with an attenuation factor α inserted into the coherent beam of the interferometer downstream by a distance ℓ, causes a phase shift of

$$\chi_{\mathrm{nl}} = (2\ell/v\hbar) b \ln |\alpha|, \tag{10.2}$$

in non-linear quantum mechanics. A related experiment was performed by Shull et al. (1980b) using a two-plate interferometer (Fig. 2.3, middle), placing an absorbing plate at different positions

Neutron Interferometry. Second Edition. Helmut Rauch and Samuel A. Werner.

of the Borrmann fan-broadened beam in the interferometer. From the equivalence of the net phase shifts, the authors concluded that $b \leq 3.4 \times 10^{-13}$ eV.

Non-linear terms also cause deviations from the standard diffraction pattern, because the transversal spreading of the wave packet becomes different. This yields an additional shift Y of the diffraction pattern from macroscopic objects (Fig. 1.4). At a distance Z behind the object the shift can be written as (Gähler et al. 1981)

$$Y = \frac{b}{E} \int_0^z \frac{l}{|\psi|} \frac{d\psi}{dy} (Z - z) dz, \tag{10.3}$$

which yields for the Fresnel diffraction at an absorbing edge

$$Y = \frac{2b}{E\sqrt{\lambda}} \cdot Z^{3/2} \cdot c. \tag{10.4}$$

Here c is a geometrical constant of the order 1. Y has also been calculated by a numerical method applied to the time-dependent non-linear Schrödinger equation to avoid various approximations involved in the closed-form Fresnel formula (Born and Wolf 1975). The numerical method predicts an effect about half as large as that from the Fresnel approximation (Kamesberger and Zeilinger 1988). The discrepancy is not yet completely clear and certainly further experimental effects like a partial transmission of neutrons near to the edge must be accounted for in order to compare theoretical and experimental results. The result of the related experiment, which was a diffraction rather than an interferometer experiment, yields for b an upper limit $b < 3.3 \times 10^{-15}$ eV (Gähler et al. 1981). Pearle (1984b) extracted from these measurements a mean self-reduction time $\tau_R > 5$ s for a dynamical state vector reduction, which is a model for explaining the collapse of the wave field (see Chapter 12).

10.2 Aharonov–Bohm Analog

Aharonov and Bohm (1959) predicted for charged particles a phase shift due to their coupling to the electromagnetic vector potential, i.e., in a region where no real force is acting on the particle. This phase shift is wavelength independent and given by the charge e of the particle and the magnetic flux ϕ enclosed by the coherent beam paths $\chi_{AB} = -e\phi/h$ (see Section 6.1 and Fig. 6.1). The verification of this effect is one of the most outstanding achievements of electron interferometry (Bayh 1962, Boersch and Lischke 1970, Tonomura 1998). For a neutral particle this term does not exist because the first electromagnetic non-zero coupling term is due to the magnetic moment which couples directly to the field and not to the potential, and therefore, such a particle is uneffected by a first-order gauge transformation. Assuming couplings other than the standard one makes it possible to define upper limits for such contributions. A related experiment was performed by Greenberger et al. (1981) using a two-plate interferometer (Fig. 2.3) and putting a magnetic loop crystal around one coherent beam. In this case, no magnetic field was acting on the neutron and no shift of the interference fringes occurred. Within the experimental uncertainty, no measurable Aharonov–Bohm effect acting on a neutral particle was found. The upper limit of $< 4.9 \times 10^{-12}$ compared to the charged particle Aharonov–Bohm effect was given.

An Aharonov–Bohm effect of the second kind may occur when modular momentum and modular energy are exchanged. Aharonov (1984) and Zeilinger (1984) proposed a realization by applying a purely time-dependent magnetic interaction to the coherent beams of a two-plate or a double-slit interferometer. The experimental technique corresponds to that described in Section 6.2 for the observation of the scalar Aharonov–Bohm effect and some of the expected results are described there.

In another experiment it was tested whether moving matter placed between the coherent beams has a similar effect on the neutron phase that a current (moving charge) has on the phase of electrons in the standard Aharonov–Bohm effect. The analogy exists because moving matter represents an isospin current which could couple to the neutron system. In this case, a rod of uranium was rotated between the coherent beams of a three-plate interferometer (Zeilinger et al. 1984). No influence of the rotation of the sample on the phase of the neutrons was found up to a level of 0.64°/100 rpm. Thus, one can conclude that any neutron–neutron interaction is smaller by a factor of 8.9×10^{-16} than the analogous electron–electron interaction would be in that situation. This gives also constraints to a possible isospin interaction (Wu and Yang 1975). A gravitational Aharonov–Bohm effect can be expected when particles are constrained to move in a region where the Riemann curvature tensor vanishes. It could be observed by flying an interferometer in an aircraft on a straight and level flight path and in a parabolic flight path. Reasonable phase shifts are predicted but the feasibility for such an experiment—if existing at all—may be more feasible with an atom interferometer (Ho and Morgan 1994).

Other non-zero Aharonov–Bohm effects are discussed in Chapter 6 (see Fig. 6.1).

10.3 Quaternions in Quantum Mechanics

Quaternions are hypercomplex numbers and represent a generalization of complex numbers. They have been discussed in connection with quantum mechanics to enlarge the standard complex to a quaternion number field (e.g., Kaneno 1960, Finkenstein et al. 1962, Adler 1995). Such contributions can be tested by neutron interferometry when it is assumed that the phase shifts, which are usually complex numbers (Eq. 3.18), have small quaternion components (Peres 1979). There exists a universal relationship between the six coherent scattering cross-sections of any three scatterers, taken singly and then pairwise, and there exists a non-commutativity of the phase shifts experienced by the neutron inside the interferometer (i.e., $\chi_1 + \chi_2 \neq \chi_2 + \chi_1$). A related experiment was performed by Kaiser et al. (1984) by interchanging thick aluminum and titanium phase shifters. They found that $|\chi_{Al} + \chi_{Ti} - (\chi_{Ti} + \chi_{Al})| \leq 5$ millirad and concluded that any noncommutative quaternion contribution to the scattering length of these nuclei is less than 1 part in 30,000 of the real part. In connection with the phase-echo experiments as they are described in Section 4.2.4 it has been discussed that a complete revival of the wave function behind various interaction regions become impossible in principle due to unavoidable quantum losses. Thus, more precise measurements will show that small non-commutative phase shifts occur due to standard quantum mechanics, which relates to an intrinsic irreversibility (see Fig. 4.11).

The interaction with the Al and Ti phase shifter might be collinear in the sense of quaternion interaction. Therefore, another experiment combining nuclear and gravitational interaction was started (Allman et al. 1996). In this case alternative phase shifters are introduced and the interferometer is tilted around a horizontal axis. Tests with nuclear and magnetic fields show a null effect within present experimental sensitivity.

It should be mentioned that the exchange of phase shifter ($1 \leftrightarrow 2$) results in slightly different wave functions also in standard quantum mechanics when all back-and-forth reflected waves are taken into account. This has been discussed in connection with spin-echo experiments and the question of a complete retrieval of the wave function (Section 4.2.4, Fig. 4.10).

10.4 Non-ergodic Effects

The probability of having more than one neutron in the interferometer at a given time is extremely low. Therefore, one repeats the one-particle interference experiment many times to obtain the interference pattern. One simply assumes that the experimental setup is the same for each particle, independent of how many particles had already passed the interferometer. In this case the time average should equal the ensemble average, whereas the non-ergodic interpretation predicts a difference.

The non-ergodic interpretation of quantum mechanics assumes certain memory effects in a hypothetical medium through which successive particles would interact indirectly (Buonomano 1980, 1988). The trajectory of a particle would depend on the beam path of the previous particles and, therefore, on the position of the phase shifter or if one or both beams were open at that time. In this sense it represents a hidden variable theory. Memory effects on very short time scales are not yet ruled out by experiment.

Summhammer (1985) performed a related experiment where one arm of the interferometer could be blocked by a mechanical shutter. A 50:50 probabilistic decision was made after a neutron was registered at the detector as to whether the shutter changes. The non-ergodic interpretation expects that the first particle passing the interferometer under changed conditions should essentially find the same trajectory as the previous one by a hypotized memory effect. The neutrons were registered according to their arrival time at the detectors after the shutter changed. As expected, the interference pattern only depends on the status of the shutter and not on the previous neutrons. The data analysis showed that one can reject a memory effect at a generous significance level of 10^{-6}. Thus, neutrons which experienced both beam paths open produced an interference pattern, whereas those which experienced only one open beam path did not show this phenomenon. The switching time of the shutter was about 0.15 s, which is much larger than the time-of-flight through the interferometer. Thus, diffraction in the time domain was negligible (see Section 4.5.5). A reanalysis of the data came to the same conclusion (Buonomano 1989).

In the framework of axiomatic probabilities Aerts (1986) developed a hidden variable theory which supposes memory effects inside the detector. In this case the distribution of the arrival times inside a detector should become non-Poissonian (Durt 1999). Neutron (Section 4.5.4) and atom (Lawson-Daku et al. 1996) interference experiments rule out such memory effects with characteristical times longer than 10^{-6} s.

Non-ergodicity would also mean that repeating measurements with one particle differ from successive measurements with equally prepared particles, which would indicate a difference between ensemble and time averages. Dedicated measurements with atom beams and a single atom show that ergodicity is fulfilled up to a high accuracy level (Huesmann et al. 1999).

10.5 Wheeler Delayed-Choice Experiments

Delayed choice experiments became very popular during the 1980s and 1990s in testing the question of locality and separability of quantum mechanics (Wheeler 1978). In such experiments the experimental settings must be changed quickly during the time-of-flight of the particle through the apparatus (Bohm and Aharonov 1957). If such settings have an influence on the outcome of such an experiment, the question of hidden variables would appear in a new light with many epistemological implications. To date most experiments in this field have used correlated pairs of photons (Aspect et al. 1982). They were interpreted in terms of the Einstein–Podolsky–Rosen–Bohm (1935) gedanken experiment and the "realistic local theories" (Clauser and Shimony 1978).

All the results are in agreement with the standard quantum-mechanical predictions and, therefore, confirm the quantum limit of the generalized Bell inequalities (Bell 1965). The quantum formulation of delayed-choice experiments in terms of quantum potentials has been given by Bohm et al. (1985).

Here it should be mentioned that delayed-choice experiments have been performed using an atomic Stern–Gerlach interferometer operating with metastable hydrogen atoms. It was possible to modify the operating mode of the interferometer (interference or path detection) while the atom has already entered the interferometer. The results have not shown any discrepancy between stationary and delayed-choice operating modes of the interferometer (Lawson-Daku et al. 1996).

The most direct verification of a Wheeler delayed-choice experiment for photons was done by Jacques et al. (2007), where the interferometer and which-way option was switched by an electro-optical modulator. In this case relativistically separated random number generators which decide path or interference behavior are used. For neutrons the time scale would be strongly relaxed due to the slow velocity of neutrons. Figure 10.1 shows such an arrangement which closely corresponds to the original idea of Wheeler (1978). A switchable beam splitter of a Mach–Zehnder interferometer (BS_{output}) changes the system from wave detection to a particle detection system. In the neutron case a pulsed magnetic field can switch the perfect crystal of the interferometer from a diffraction to a transmission option. This has been tested in connection with a perfect crystal resonator system (Schuster et al. 1990, Jericha et al. 1996). Such neutron delayed-choice experiments are feasible because the switching times are rather relaxed (μs instead of ns).

In an experiment with the neutron interferometer, a delayed choice must be made as to whether the interference pattern is to be observed or the neutron's route inside the interferometer is to be chosen for measurement; for example, an interference measurement behind the interferometer or a path measurement by a detector introduced into one of the coherent beams before the third interferometer plate. Due to the slow velocity of neutrons the decision process could easily be adapted to the electronic and mechanical part of the system. Various proposals for such experiments exist in the literature (Greenberger et al. 1984, Zeilinger 1984, Rauch 1984c, Bozic et al. 2011). The experiment with a time-dependent absorption in one beam as described in Section 4.3.2 (Fig. 4.15) is directly related to this subject (Rauch and Summhammer 1984b). Various aspects of delayed choice are also stressed in the spin superposition experiments discussed in Sections. 5.3 and 5.4, where the beam polarization can be used as a path identifier. The interferometer experiments with pulsed beams described in Section 4.5.5 can be extended to a delayed-choice experiment if a neutron detector can be introduced into the beam randomly and

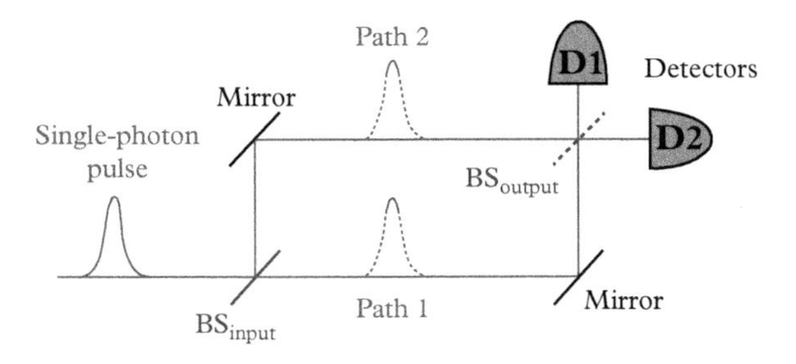

Figure 10.1 *Mach–Zehnder interferometer with a switchable beam splitter to vary between a wave and a particle detection system From Jacques et al. 2007. Reprinted with permission from AAAS.*

rapidly after the wave packet has been split at the first interferometer plate. Kawai et al. (1998a) developed a fast neutron pulsar on the basis of a Jamin-type interferometer (Fig. 2.17). In this case a pulsed π-flipper in front of the interferometer and a pulsed magnetic field acting on the second composite mirror produced the condition of whether one is measuring the interference pattern or observing the beam path. The results of these measurements show that the interference pattern can be observed regardless of whether the second composite mirror was put into the interference mode before or after the beam is split at the first composite mirror (Kawai et al. 1998b). This is a nice verification of the quantum-mechanical principle that the system evolves simultaneously in both beam paths and does not choose a priori one of the two possible paths. An experiment performed with photon pairs produced by parametric down-conversion showed the same results; i.e., the results are independent of whether the switching process between pathway or phase information takes place before or after the photon passes the first beamsplitter of the interferometer (Baldzuhn et al. 1989, 1991).

From the classical point of view there is some region in the experiment—first interferometer plate—where the particle must choose between two alternative routes though the interferometer. Therefore, it appears that by altering the apparatus at a later time one can affect that choice of routes the particle has made at an earlier time. This would imply hidden variables or a non-local interaction. According to the quantum-mechanical point of view the wave functions representing these alternatives do not allow the statement that the particle took one beam path or the other. Inequalities analogous to Bell's inequalities can be formulated and, therefore, such experiments can demarcate quantum theory from certain classes of so-called realistic theories (Greenberger et al. 1984). The neutron experiments clearly support the quantum mechanics predictions.

10.6 Neutron–Antineutron Oscillations

Neutron–antineutron oscillations are predicted due to the breaking of exact baryon-number conservation in unified theories of fundamental gauge interactions (Glashow 1980). The present experimental limit for this process is obtained from the nonobservation of antineutrons from an intense beam of reactor-produced neutrons which justifies a neutron–antineutron oscillation time longer than 10^7 s (Baldo-Ceolin 1984, Fidecaro et al. 1985, Baldo-Ceolin et al. 1990). This effect appears in neutron interferometry by causing a minute energy shift of the neutron coming from a baryon-number nonconservating term in the Hamiltonian (Casella 1984). Interferometers for ultra-cold neutrons, extremely long storage times, and a perfect compensation of the ordinary gravity and magnetic field effects would be required to realize such a measurement.

10.7 Non-Newtonian Gravity Effects

There are models that claim that besides the Newtonian contribution there exist additional terms to the gravitational interaction (Fischbach et al. 1986, Goldman et al. 1986, Dubbers and Schmidt 2011) of the form

$$V = -G_\infty \frac{m_1 m_2}{r}\left(1 + \eta e^{-r/\xi}\right). \tag{10.5}$$

Experiments with massive bodies on the Earth's surface gave no conclusive results concerning the existence or non-existence of such an additional term. Such a contribution would have consequences on baryon number and hypercharge coupling mechanisms and would also be a link for a unification of gravity, electro-weak, and strong interactions.

Bertolami (1986) calculated the additional phase shift due to the non-Newtonian term

$$\Delta\phi_{nN} = \frac{R\eta f\,(R/\xi)}{T \sin\theta_B}, \tag{10.6}$$

where R is the radius of the Earth and T is the distance between the interferometer plates. In the limit $R/\xi >> 1$ one gets $f(R/\xi) = 3\xi^2/2R$. Measurements with phase accuracies higher than 10^{-10} could at least put upper limits to the parameter of the non-Newtonian part of Eq. (10.5) and to the masses of hypothetical exchanged particles.

Comprehensive reviews on this topic have been given by Adelberger et al. (2003), Camacho and Camacho-Galvan (2007), and Antoniadis et al. (2011). They urged new investigations with neutrons for the gravitational force at small distances (1–10 μm). Neutrons provide the advantage that their Casimir or van der Waal forces are small or perhaps not existing. Experiments with ultra-cold neutrons set constraints on the possible magnitudes of η and ζ (Leeb and Schmiedmayer 1992; Abele et al. 2003; Nesvizhevsky et al. 2004, 2008). The optical potential of a material surface and the gravitational potential produce a triangular potential with energy eigenstates in the picoelectron volt range. Transitions between these states can be induced by oscillating surfaces and can be observed with ultra-cold neutrons. This has opened the field of picoelectron volt spectroscopy (Abele et al. 2010, Jenke et al. 2011, 2014). Neutron interference measurements may become even more sensitive to such additional gravity effects as stated by Greene and Gudkov (2007) and by Pokotilovski (2013). These authors propose two methods for measuring the additional non-Newtonian gravity effect. One method is based on the transmission of neutrons through narrow slits made from heavy materials which change the rectangular optical potential in a way to include the non-Newtonian term. The other one uses ultra-cold neutrons in the Lloyd mirror interferometer where a direct beam from a narrow slit and a totally reflected beam become superposed (Lloyd 1831, Cavey 2009, Pokotilovsky 2013).

This causes an additional term to be added to the usual Newtonian potential $V(z) = mgz$ above a surface of a material with a density ρ. One gets an additional term

$$V'(z) = 2\pi m\rho\eta\zeta^2 G e^{-z/\zeta}. \tag{10.7}$$

This term causes a change in the gravitational acceleration on the order of only $10^{-12}g$, where $g = 9.8$ m/s^2.

The second method uses the gravitational interaction between the neutron and atoms of a perfect single crystal as phase shifter in an interferometer in a mode indicated in Fig. 3.14. The phase shift depends in this case on the structure factor and changes strongly near to the exact Bragg direction as shown by Lemmel (2007, 2013). The form of the additional term (Eq. 10.5) causes a gravitational scattering length (in fm) of

$$b_G \approx -1.6 \times 10^{-6}(\eta\zeta^2), \tag{10.8}$$

where ζ is in meters. Thus very accurate scattering length measurements ($\Delta b/b \approx 10^{-6}$) are required to observe finite $\eta\zeta^2$ values. The feasibility and sensitivity of various neutronic methods have been discussed by Abele (2008), by Abele et al. (2010) and by Jenke et al. (2014).

Neutrons have the distinct advantage compared to atoms that they do not interact via van der Waals forces, which limits investigations with atoms as done by Perreault and Cronin (2005). Short-range forces may be spin-dependent as well, which gives a new horizon for such experiments (Antoniadis et al. 2011).

10.8 Spin–Rotation Coupling

The Sagnac effect (Section 8.2) may be regarded as a manifestation of the coupling of the orbital momentum of a particle to rotation ($\boldsymbol{\Omega} \cdot \boldsymbol{L}$, Eq. 8.6). This may be extended to the total angular momentum which contains the spin ($\mathcal{J} = \boldsymbol{L} + \boldsymbol{S}$). This yields an additional phase shift (Mashhoon 1988, 1999; Stedman 1997)

$$\Delta\phi_M = \frac{1}{\hbar}\,\boldsymbol{\Omega} \cdot \boldsymbol{S}\,\Delta t. \tag{10.9}$$

This coupling reveals the rotational inertia of an intrinsic spin.

A proposed neutron experiment uses longitudinally polarized neutrons passing through a rotating spin flipper (Mashhoon et al. 1998). An energy change $\Delta E = \hbar\Omega$ is predicted, which seems to be measurable with an arrangement similar to that used for the double-coil experiment described in Section 5.4 or that of the geometrical phase experiment described in Section 6.4.

10.9 Hanbury-Brown and Twiss Analog

The famous experiment of Hanbury-Brown and Twiss (1956) studied the correlation in the photo-current fluctuations from two detectors. These experiments showed an enhancement in the intensity correlation function at short time delays of thermal light due to the large intensity fluctuations in the thermal source. The effect can be understood in the context of classical optics (Loudon 1983) and in terms of quantum optics (Glauber 1965). Whereas for bosons a bunching effect at zero time delay between two detectors is expected, for fermions an anti-bunching effect is predicted which one understands in connection with the Pauli principle. This prediction has been verified for ultra-cold fermion (^{3}He) and boson gases (^{4}He) released from a magnetic trap and measured with a position-sensitive detector (Jeltes et al. 2007) and with Bose–Einstein condensates (Perrin et al. 2012). Anti-bunching effects have also been observed for electrons (e.g., Kiesel et al. 2002). Both atoms and electrons represent interacting objects, whereas neutrons can be considered as non-interacting particles, which makes a related neutron Hanbury-Brown and Twiss effect special.

In the neutron case the most direct measurement would be the determination of a zero-valued neutron–neutron triplet scattering length. This could be done by a crossed beam experiment with polarized neutrons, but the intensities at existing neutron sources are far too weak to start such an experiment today. Even the singlet scattering length ($a_{nn} = -18.6 \pm 0.3$ fm) has only been determined indirectly (Machleidt and Slaus 2001, Gardestig 2009). Thus, other methods must be exploited.

It has been mentioned in Section 1.3 that for any neutron beam the degeneracy parameter, i.e., the mean numbers of neutrons inside the phase space volume, is extremely small (10^{-15}–10^{-14}). Therefore, any higher order correlation function as defined in Eq. (4.93) becomes very difficult to observe. This is because the characteristic coherence lengths imposed on the beams are rather small, at least in some directions (e.g., Section 4.2). This means that for a measurement of the $n = 2$ self-correlation function, the sensitive area of the detector should match the transverse coherence lengths of the beam and the thickness should match the longitudinal coherence length to keep the detector response time in the order of the coherence time ($t_c = \Delta_x^c/v$). These constraints have not been overcome with the present-day instrumentation. The conditional probability of fermion arrivals is given by the self-intensity correlation

$$G_s^{(2)}(\mathbf{r}_1 0, \mathbf{r}_2 \Delta t) = < |\psi(\mathbf{r}_1, 0)|^2 \, |\psi(\mathbf{r}_2, \Delta t)|^2 >, \tag{10.10}$$

where ψ is the particle wave function expanded over both beam paths $\psi = \psi^{\mathrm{I}} + \psi^{\mathrm{II}}$ and r_1 and r_2 denote the detector positions in each beam path or for one placed behind the interferometer in the 0- and H-beams. This function shows for thermal light a characteristic bunching effect for equal arrival of photons at both detectors ($\Delta t = 0$), whereas for neutrons an anti-bunching is expected due to their fermion character (Fig. 10.2; Boffi and Cagliotti 1966, Silverman 1987). When the spatial and temporal coherence can be separated (Eq. 4.34) the coincidence count rate fluctuation in both detectors can be written as (e.g., Mandel and Wolf 1995, Fox 2006)

$$\overline{I_1 I_2} = \bar{I}_1 \bar{I}_2 \left(1 + \varepsilon \frac{t_c}{\Delta T} g_{12}(t)\right), \tag{10.11}$$

with

$$g_{12}(t) = \frac{1}{V^2} \iint |\gamma_{12}(\boldsymbol{r}_1, \boldsymbol{r}_2, t)|^2 \, d\boldsymbol{r}_1 \, d\boldsymbol{r}_2.$$

Here V is the sensitive volume of the detector, γ_{12} is the coherence function (Eq. 4.26), t_c is the coherence time, ΔT is the detector resolving time, and t is the delay time of the coincidence circuit. The parameter $\varepsilon = 1$ for bosons, and $\varepsilon = -1$ for fermions, whereas $\varepsilon = 0$ for classical particles. The widths of the bunching and anti-bunching phenomena are given by the coherence time which is on the order of nanoseconds in the neutron case (Section 4.2.6). Buffi and Cagliotti (1966, 1971) discussed the question for neutrons but a feasible experiment has been questioned for a long time (see also Klein et al. 1983). For the neutron case, a measurement time of 10,000 years has been estimated (Silverman 1988). Hanbury-Brown–Twiss effects have been observed for light sources with degeneracy parameters of 10^{-3}. Below these values random coincidences start to wash out the signal. The experimental difficulties for neutrons due to the low phase space density also persist in the case of two open entrance ports or other sophisticated proposals (Silverman 1988,

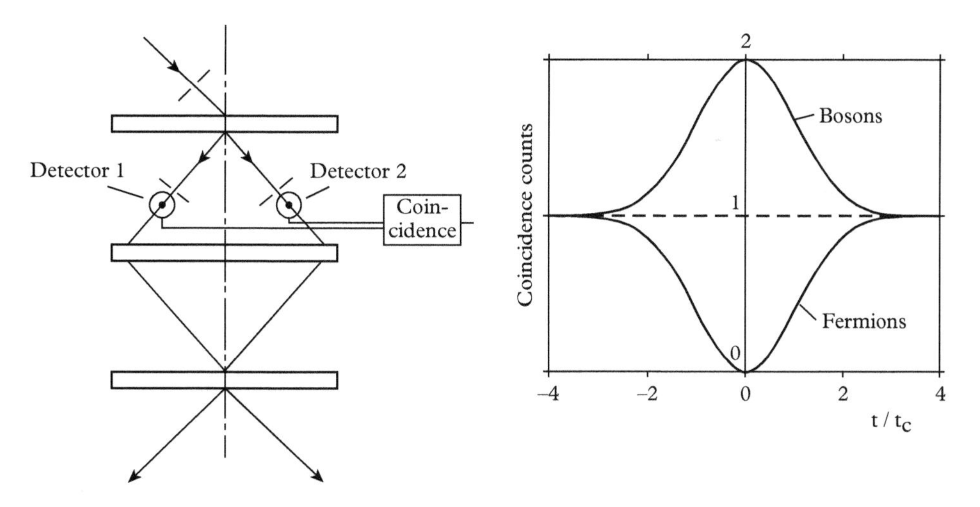

Figure 10.2 *Hanburry-Brown–Twiss arrangement and the expected outcome of a related experiment under ideal conditions*

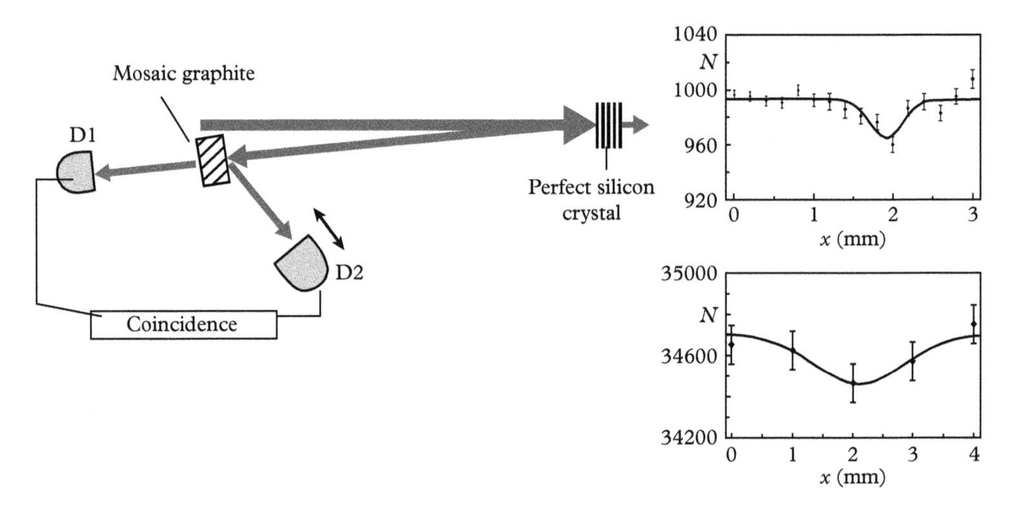

Figure 10.3 *Experimental arrangement (left) and results of the neutron Hanbury-Brown and Twiss experiment. Reprinted with permission from de Iannuzzi et al. 2006, copyright 2006 by the American Physical Society.*

Dobrynin and Lomorosov 1989). It should be mentioned that the phase space density does not enter Eq. (10.11), which may have been overlooked by these authors.

A rather clever arrangement has been used by Iannuzzi et al. (2006) to observe the free-fermion anti-bunching effect (Fig. 10.3). They produced a rather coherent beam by back reflection from a perfect silicon crystal (coherence time $\Delta t_c = \Delta_l^c/v \cong 17$ ns (see Eq. 4.73)) and used a mosaic graphite crystal as a beam splitter. They measured the coincidence count rate of two neutron detectors having a time resolution of 1.1 μs and whose distance from the beam splitter could be varied. They found a small dip when the detectors were equally distant from the beam splitter, which they interpreted as the fermion anti-bunching effect. In this lucky experiment the thickness of the graphite crystal had to provide a 50:50 overall beam splitting and the mosaic blocks must have had a dimension (≈2 μm) to split each wave packet coherently; and the time resolution of the detector must guarantee that neutrons arriving from all mosaic blocks are within the coincidence window. In this case the mosaic blocks act as very small sources (≈2 μm) producing very large coherence dimensions (several cm) at the position of the detectors. Since the orientation of the lattice planes of the different mosaic blocks are slightly different, they act incoherently on the beam, thereby creating these micro sources. A further experiment in this direction has been done with a position-sensitive detector and showed again the anti-bunching effect (Iannuzzi et al. 2011). Theoretical discussions of such measurements are given by Yuasa et al. (2008) and Varro (2008, 2011).

A more indirect but more feasible access to two-neutron correlation function exists by the double neutron emission process from nuclei which could, in principle, form the basis for two-particle neutron interferometry (Boal et al. 1990, Colonna et al. 1995). Neutron–neutron correlations are independent from the long-range Coulomb interaction which makes such investigations more sensitive to the space-time characteristics of the emitting nucleus. A typical example of such an idea is the deuteron desintegration by the reactions

$$n + D \rightarrow 2n + p \quad \text{and} \quad \pi^- + D \rightarrow 2n + \gamma. \tag{10.12}$$

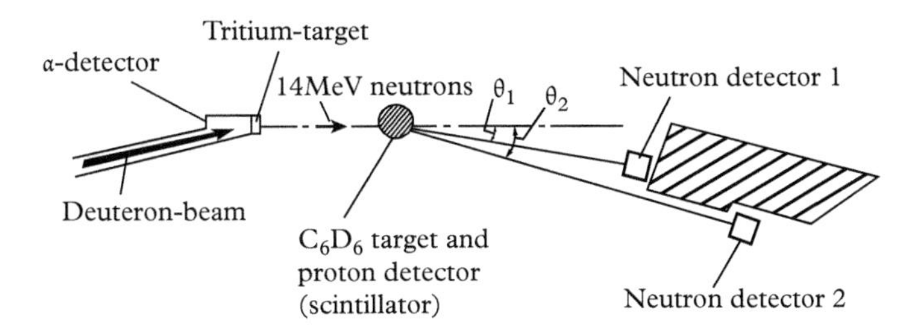

Figure 10.4 *Experimental arrangement for a two-particle neutron correlation experiment at high energies*

A conceptual arrangement is shown in Fig. 10.4. The proton or gamma quantum provides a start signal in a scintillator. Neutrons are emitted in the singlet state and have an energy of several megaelectron volts and have nearly the same direction and comparable energies. They are therefore candidates for such studies (Breunlich et al. 1974). The proper shielding of both detectors and fast electronic gates are important requirements to avoid cross-talk effects and to get reliable results (Cronqvist et al. 1992, Kun et al. 1992, De Yong et al. 1996). A possible anti-correlation has been observed for low relative and low total momentum pairs of neutrons.

The expected anti-correlation has also been observed in the double neutron decay of the ^{44}Ca compound nucleus (Dünnweber et al. 1990). The coincidence counting rate shows a marked dip for equal momenta of the emitted neutrons. The observed width can be associated with the decay width of 60 keV and can be explained by the overlapping of antisymmetrized wave packets of both neutrons.

In particle physics the Hanbury-Brown and Twiss phenomenon goes by the acronym GGLP effect after the authors of the first article in this field (Goldhaber, Goldhaber, Lee, Pais 1960). The different angular distributions for the like pairs ($\pi^+\pi^+$ or $\pi^-\pi^-$) and the unlike pairs ($\pi^+\pi^-$) gave rise to developing the time correlation for bosons and fermions (Koonin 1977). Many pair emission processes (pp, nn, np) in heavy ion collisions show related bunching and anti-bunching effects (Ghetti and Helgessoson 2005).

10.10 Search for Nuclear Entanglement

From deep inelastic neutron scattering experiments from liquid H_2O–D_2O mixtures an apparent observation of strong but short-lived entanglement of adjacent protons and deuterons has been claimed (Chatzidimitriu-Dreismann et al. 1997a). But from condensed matter physics much shorter relaxation times (coherence times) at room temperatures are expected and quasi-elastic neutron diffraction experiments confirmed this expectation (e.g., Springer 1972). Nevertheless, it has been suggested that such entanglement manifests itself as a deviation (up to 10%) from the scattering length density Nb_c, which determines the phase shift inside an interferometer (Chatzidimitriou-Dreismann 1997b). With different mixtures of H_2O–D_2O high precision phase shift measurement based on a non-dispersive sample arrangement has been performed as described in Section 3.1.2.1 (Fig. 3.6). No deviation greater than 0.4% from the conventional theory of the linearity of scattering length densities has been found (Ioffe et al. 1999). This is in agreement with the findings discussed in Section 10.3.

10.11 Confinement–Gravity Coupled Quantum Phase

The confinement-induced topological quantum phase has been described in Section 6.9. The transverse quantization of the wave function within narrow slits causes a slight delay of the transit time, as proposed by Levy-Leblond (1987) and Greenberger (1988). This effect appears as a phase shift as has been verified by the neutron interferometric method (Rauch et al. 2002). In this case a rectangular potential exists between the material walls. The phase shift of the basic mode is $\Delta\varphi = L\pi^2/ka^2$, where L denotes the length and a the width of the slit (Eq. 6.67). An additional effect is expected when the slits are in the horizontal direction and gravity causes a slightly tilted potential as shown in Fig. 10.5. In this case the quantization in the vertical direction becomes changed due to the presence of the gravitational field (Lushikov and Frank 1978). Interferometric measurements can compete with experiments done with ultra-cold neutrons which also demonstrate the quantization of neutron states above surfaces (Abele et al. 2003, 2010; Nesvizhevsky et al. 2004, 2008; Jenke et al. 2011).

The potential that the neutrons experiences in the vertical direction inside the channel becomes (Fig. 10.5)

$$V_y = mgy \tag{10.13}$$

and inside of the walls of the channel

$$V_y = \frac{2\pi\hbar^2}{m}\, b_c N + mgy. \tag{10.14}$$

The energy quantization above a flat surface and an infinitely high potential can be found from the roots of the Airy function (e.g., Flügge 1971)

$$A\left[-\left(\frac{2}{mg^2\hbar^2}\right)^{1/3} E\right] = 0. \tag{10.15}$$

The related energy levels are given in Table 10.1. The solutions become more complicated when the spatial limitation in the y-direction is taken into account. It shifts the energy levels slightly up, as shown in Table 10.1 as well. The related wave function for the lowest level is located around 10 μm above the surface and the higher order states show maxima and minima and spread from nearer distances to the surface to larger distances (Nesvizhevsky 1998). The neutron level population increases as $E^{3/2}$ and, therefore, the mixing of different quantum states must be considered.

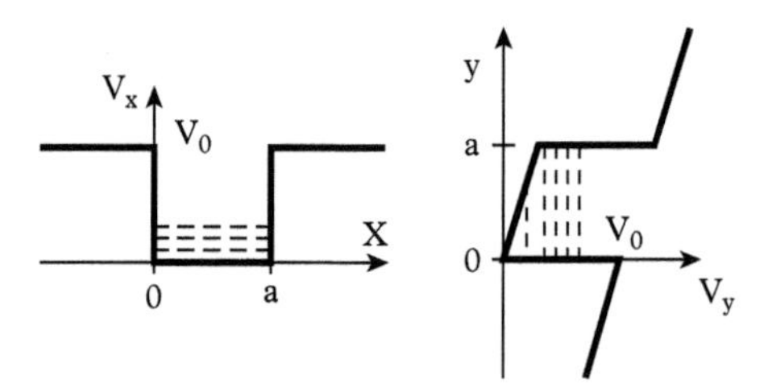

Figure 10.5 *Confinement potentials in the horizontal and vertical directions*

Table 10.1 *Neutron Energy Levels Inside Confinement Tubes with a Width of 20 μm (in 10^{-13} eV)*

Level no. (i)	1	2	3
Levels due to the geometry	5.113	20.45	46.01
Levels due to gravity at $a \to \infty$	14.4	25.3	34.2
Levels due to vertical geometry and gravity E_i	14.5	30.9	56.4

It causes a broadening of the density distribution function. Narrow channels ($a \sim E_1/mg$) increase the energy levels and act as a kind of filter, permitting the passage of neutrons with only low "vertical" energy. Pokotilovski (1998) calculated the corresponding energy levels and found for $a = 20\,\mu$m the values listed in Table 10.1.

For the basic mode ($n_x = 1$, $n_y = 1$), the related phase shift becomes

$$\Delta\varphi_{\mathrm{LG}} = \left(\frac{E_i m\, a^2}{\pi^2 \hbar^2} + \frac{1}{2}\right) \frac{L\pi^2}{ka^2}, \tag{10.16}$$

which is about 90% larger than those calculated without the effect of gravity (Eq. 6.65). Related experiments are in progress. The main difficulties lie in the high degree of collimation needed to excite only the low-lying levels. Otherwise, multimode excitation exists with the tendency of a quantum chaotic behavior. As mentioned in Section 6.4 the advantage of neutrons lies in the fact that they do not experience van der Waal and Casimir forces (Casimir and Polder 1948, Haroche and Raimond 1993, Bordag et al. 2001). The level structure may be influenced by speculative chameleon field as discussed in Section 10.17 and by Brax and Pignol (2011).

10.12 The Anandan Acceleration

In the course of analyzing the dynamics of a neutron's motion through a region of space containing both an electric field $\boldsymbol{E}$ and a magnetic field $\boldsymbol{B}$, Anandan (1989a, 1989b) noticed an interesting effect. Due to the neutron's precession in the magnetic field, its translational motion is coupled to its spin motion. Furthermore, its translational motion is coupled back to its spin motion via the spin–orbit coupling, which depends upon the electric field $\boldsymbol{E}$. He found that the classical mechanics describing this coupling to the EM field results in an acceleration given by the formula

$$a_{\mathrm{A}} = -(\gamma/mc)(\boldsymbol{\mu} \times \boldsymbol{B}) \times \boldsymbol{E}, \tag{10.17}$$

where γ is the neutron gyromagnetic ratio. This is a surprising effect since there appears to be an acceleration of the neutron which is dependent on the product of the electric and magnetic field strengths, and not dependent upon their gradients. That is, this acceleration persists even if the $\boldsymbol{E}$ and $\boldsymbol{B}$ fields are time independent and spatially uniform.

We begin by showing that this is, in fact, the case and then discuss whether such an effect is, at least in principle, observable by neutron interferometry. Consider a neutron moving through a region of space containing constant and uniform magnetic and electric fields, as shown in Fig. 10.6.

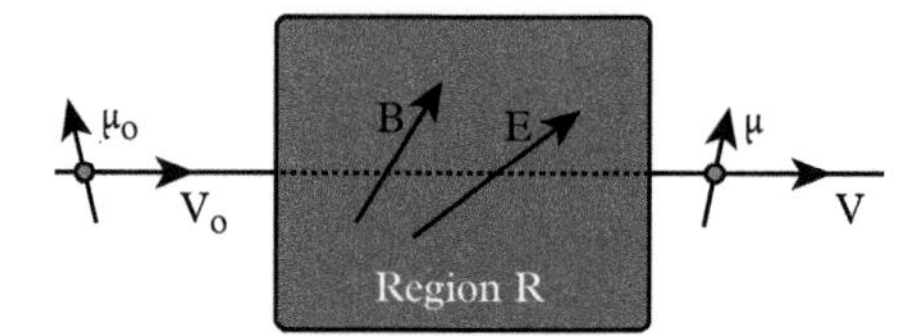

Figure 10.6 *Diagram showing a neutron of initial polarization $\boldsymbol{\mu}_0$ and velocity $\boldsymbol{v}_0$ traversing a region R containing both an electric field* **E** *and a magnetic field* **B**. *After traversing R its polarization is $\boldsymbol{\mu}$ and its velocity is $\boldsymbol{v}$*

The Hamiltonian for the neutron is therefore

$$\mathcal{H} = \frac{p^2}{2m} - \boldsymbol{\mu} \cdot \boldsymbol{B} - \boldsymbol{\mu} \times \boldsymbol{E} \cdot \boldsymbol{p}/mc, \tag{10.18}$$

where $\boldsymbol{p}$ is the neutron's canonical momentum

$$\boldsymbol{p} = m\boldsymbol{v} + \frac{1}{c}\boldsymbol{\mu} \times \boldsymbol{E}. \tag{10.19}$$

As in our description of the AC effect (see Section 6.1), this expression is obtained from the velocity $\boldsymbol{v} = \dot{\boldsymbol{r}}$, using Hamilton's equation. That is

$$\dot{\boldsymbol{r}} = \frac{\partial \mathcal{H}}{\partial \boldsymbol{p}} = \frac{\boldsymbol{p}}{m} - \frac{1}{mc}\,\boldsymbol{\mu} \times \boldsymbol{E}. \tag{10.20}$$

The second term in Eq. (10.19) is often called the electromagnetic or hidden momentum. Differentiating Eq. (10.20) with respect to time t, we obtain the acceleration $\boldsymbol{a}$, or the inertial force

$$\begin{aligned} m\boldsymbol{a} = m\ddot{\boldsymbol{r}} &= \dot{\boldsymbol{p}} - \frac{1}{c}\frac{\mathrm{d}}{\mathrm{d}t}(\boldsymbol{\mu} \times \boldsymbol{E}) \\ &= \dot{\boldsymbol{p}} - \frac{1}{c}\,\dot{\boldsymbol{\mu}} \times \boldsymbol{E} - \frac{1}{c}\,\boldsymbol{\mu} \times \dot{\boldsymbol{E}}. \end{aligned} \tag{10.21}$$

Since we are assuming that the electric field is static, the total time derivative of $\boldsymbol{E}$ is

$$\dot{\boldsymbol{E}} = (\boldsymbol{v} \cdot \nabla)\, \boldsymbol{E}. \tag{10.22}$$

The neutron's magnetic moment precesses in the effective magnetic field $\boldsymbol{B}' = \boldsymbol{B} - \frac{\boldsymbol{v}}{c} \times \boldsymbol{E}$. Thus, Bloch's equation for the precession of $\boldsymbol{\mu}(t)$ reads

$$\dot{\boldsymbol{\mu}} = \gamma\,\boldsymbol{\mu} \times \left(\boldsymbol{B} - \frac{\boldsymbol{v}}{c} \times \boldsymbol{E}\right). \tag{10.23}$$

The quantum mechanical correctness of this formula can be easily checked by evaluating the expectation value of the commutator $[\mathcal{H}, \boldsymbol{\mu}] = -i\hbar\,\dot{\boldsymbol{\mu}}$.

We now need to obtain an expression for $\dot{p}$, which is given by the second of Hamilton's equations, namely

$$\dot{p} = -\frac{\partial \mathcal{H}}{\partial r} = \frac{1}{mc} \nabla \left[(\mu \times E) \cdot p \right] + \nabla (\mu \cdot B). \tag{10.24}$$

To order $1/c$, we can replace p by mv on the right-hand side of this equation. Working out the gradient of the triple product, $(\mu \times E) \cdot v$, and requiring that $\nabla \cdot E = 0$ and that $\nabla \times E = 0$ in free space, we find that

$$\dot{p} = \frac{1}{c} \left[(v \cdot \nabla)(\mu \times E) - (\mu \cdot \nabla)(v \times E) \right] + \nabla (\mu \times B). \tag{10.25}$$

From Eq. (10.21), we see that the first term here is $(1/c)\dot{\mu} \times E$, which is identical, but of opposite sign, to the third term of Eq. (10.19). Thus, inserting the expression for $\dot{\mu}$ from Eq. (10.23), and this expression for $\dot{p}$ into Eq. (10.21) we finally obtain the equation of motion:

$$m\ddot{r} = \nabla (\mu \cdot B) - \frac{1}{c} (\mu \cdot \nabla)(v \times E) - \frac{\gamma}{c} \left[\mu \times \left(B - \frac{v}{c} \times E \right) \right] \times E. \tag{10.26}$$

The first term is the magnetic field force dependent only upon B, and is zero if B is uniform. The second term is the force on the neutron due to the electric field E. It is zero in the AC geometry (see Section 6.1), and also if E is uniform. The last term is due to the Anandan acceleration (Eq. 10.17). It is dependent upon the joint presence of both a magnetic field and an electric field. The mathematical analysis given here was first developed in this form by Casella and Werner (1992).

This result predicts that a neutron placed in a region of space containing time-independent, and spatially uniform electric and magnetic fields will begin to accelerate. This is surprising. But where does the energy come from to alter the neutron's kinetic energy? One should note also that $\dot{p} = 0$ in the region of uniform E and B, and furthermore the neutron's energy ε is conserved, since $\mathcal{H}$ is independent of time. That is, both the canonical momentum p and the energy ε are constants of the motion.

We will now explicitly solve the coupled equations of motion (Eqs. 10.23 and 10.26) for the spin and translational motions. Consider a neutron of initial velocity v_0 moving in the x-direction, and with polarization μ_0 in the y-direction. We take the electric and magnetic fields to be collinear, and pointing along the z-axis, as shown in Fig. 10.7. It is important to realize at the outset that the field $(v/c) \times E$ will, in general, be small in comparison to B. For example, for neutrons of wavelength $\lambda = 2.5$ Å, $v = 1.60 \times 10^5$ cm/s, and an electric field of $E = 250$ kV/cm $= 830$ statvolts/cm, $v_0 E/c = 0.046$ gauss. Furthermore, the Anandan force is weak, giving rise to a maximum acceleration

$$a_A = \frac{\gamma}{mc} \mu_n E\, B = 3.5 \times 10^{-6} E\,[\text{statvolts/cm}] \cdot B\,[\text{gauss}] = 0.15\text{cm/s}$$

for a magnetic field of 50 gauss. Consequently, the change in the neutron's velocity will be small for reasonable flight paths, say 10 cm, obtainable within perfect Si crystal interferometers. These facts allow us to solve Bloch's equation (10.23) independently from the equation of translational motion (Eq. 10.26). That is, we replace $v(t)$ by v_0 in Bloch's equation, and we see that the time rate of change of the three components of μ are

$$\dot{\mu}_x = \gamma \left(\mu_y B - \mu_z v_0 E/c \right),$$

$$\dot{\mu}_y = \gamma\, \mu_x B,$$

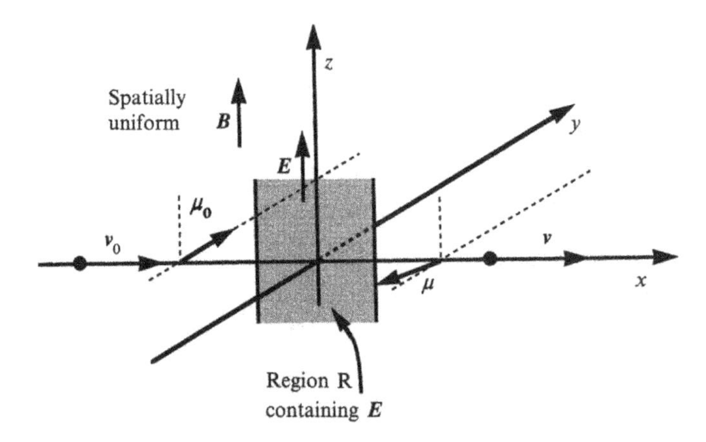

Figure 10.7 *Diagram showing a neutron initially polarized along the* y*-axis traversing a region R along the* x*-axis which contains an electric field* **E** *in the* z*-direction. There is a uniform magnetic field* **B** *in the* z*-direction over all the space*

and

$$\dot{\mu}_z = \gamma\ \mu_x v_0 E/c). \tag{10.27}$$

To first order in E, the solutions are

$$\mu_x(t) = \mu_n \sin\ \omega_L t,$$

$$\mu_y(t) = \mu_n \cos\ \omega_L t,$$

and

$$\mu_z(t) = (\mu_n v_0 E/cB)(1 - \cos\ \omega_L t), \tag{10.28}$$

where $\omega_L = \gamma B$ is the Larmor precession frequency.

We are now in a position to solve Eq. (10.26) for the translational motion. For the geometry under consideration, the magnetic and electric field forces act only when the neutron enters and traverses the region of (assumed) uniform $\boldsymbol{E}$ and $\boldsymbol{B}$. Thus, the equation of motion within the region R only involves the Anandan acceleration, namely

$$m\frac{d\boldsymbol{v}}{dt} = -\frac{\gamma}{c}(\boldsymbol{\mu} \times \boldsymbol{B}) \times \boldsymbol{E} = \frac{\gamma}{c}[(\boldsymbol{E}\cdot\boldsymbol{B})\boldsymbol{\mu} - (\boldsymbol{E}\cdot\boldsymbol{\mu})\boldsymbol{B}]. \tag{10.29}$$

The second term in the square brackets is of order E^2, and can therefore be neglected. The solution of this equation for the velocity is then given by

$$\boldsymbol{v}(t) = \boldsymbol{v}_0 + \frac{\mu_n E}{mc}\ [\hat{x}(1 - \cos\ \omega_L t) + \boldsymbol{y}\ \sin\ \omega_L t]. \tag{10.30}$$

Thus, the change in kinetic energy, to first order in E, is

$$\frac{1}{2}\ mv^2 - \frac{1}{2}\ mv_0{}^2 = \frac{\mu_n E v_0}{c}\ (1 - \cos\ \omega_L t). \tag{10.31}$$

The Zeeman energy is

$$-\mu_z B = -\frac{\mu_n E v_0}{c} (1 - \cos\, \omega_L t), \tag{10.32}$$

which is precisely equal, but of opposite sign to the change in the neutron's kinetic energy. The total energy

$$\varepsilon = <\mathcal{H}> = \frac{1}{2}\, mv^2 - \boldsymbol{\mu} \cdot \boldsymbol{B} = \frac{1}{2}\, m{v_0}^2, \tag{10.33}$$

is therefore conserved (Anandan and Rohrlich 2005). Consequently, we see that the energy necessary to accelerate the neutron via the Anandan force is extracted from the magnetic potential energy $(-\mu_z B)$.

As the neutron traverses the region of space containing the $\boldsymbol{E}$ and $\boldsymbol{B}$ fields, it precesses nearly in the xy-plane about the magnetic field $\boldsymbol{B} = B\hat{z}$. However, the motional field $-\frac{\boldsymbol{v}}{c} \times \boldsymbol{E} = (v_0 E/c)\hat{y}$ requires the plane of precession to be slightly inclined with respect to the xy-plane. This plane is perpendicular to the vector $B\hat{z} + (v_0 E/c)\hat{y}$, as we may have anticipated. The precise Larmor precession frequency is actually $\omega_L = \gamma B(1 + {v_0}^2 E^2/2c^2 B^2)$. For the values of E and B used above, the correction to ω_L is of order 10^{-8}.

According to Eq. (10.30), the neutron's trajectory is along a circle superposed upon a linear translation. How large is the radius R_0 of this circle? Using $\gamma = 2\mu_n/\hbar$, we find that this radius is

$$R_0 = \frac{1}{4\pi}\, \lambda_c \cdot E/B, \tag{10.34}$$

where $\lambda_c = h/mc = 1.32$ fm, which is the Compton wavelength for the neutron. Thus, for $B = 50$ gauss and $E = 830$ statvolts/cm, we find that $R_0 = 1.74$ fm; indeed, a very small perturbation on the trajectory. The center of the circle is at

$$\boldsymbol{R}(t) = \left(v_0 + \frac{\mu_n E}{mc}\right) t\hat{x} + R_0\hat{y}.$$

The corresponding maximum change in velocity is $\Delta v/v_0 = \mu_n E/mcv_0 = 10^{-12}$.

We now discuss the possibility of measuring the quantum-mechanical phase shift arising from the Anandan acceleration by neutron interferometry. Since both the canonical momentum $\boldsymbol{p}$ and the neutron's energy ε are constants of the motion within R_0, the change of phase is given by $d\phi = 1/\hbar(\boldsymbol{p} \cdot \delta \boldsymbol{x} - \varepsilon dt)$. However, when the neutron leaves the region R (Fig. 10.7) containing both an electric field $\boldsymbol{E}$ and a magnetic field $\boldsymbol{B}$ there is a jump in canonical momentum given by

$$\Delta p_x = \frac{1}{mc}\, \boldsymbol{\mu} \cdot \left(\boldsymbol{p} \times \frac{\Delta E}{\Delta x}\right) \Delta t = \mu_n E/c \tag{10.35}$$

at the exit boundary, which we take to be located at $x = D$, where $\varphi_L = \underline{\omega}_L t = \gamma BD/v_0 = \pi$. The neutron's velocity is conserved across this boundary and is given by

$$\boldsymbol{v} = (v_0 + \Delta v)\hat{x} = \left(v_0 + \frac{2\mu_n E}{mc}\right) \hat{x}, \tag{10.36}$$

according to Eq. (10.30). If the entire interferometer is placed in a uniform magnetic field $\boldsymbol{B}$, as shown in Fig. 10.8, the phase of the neutron wave packet traversing path II relative to path I is then

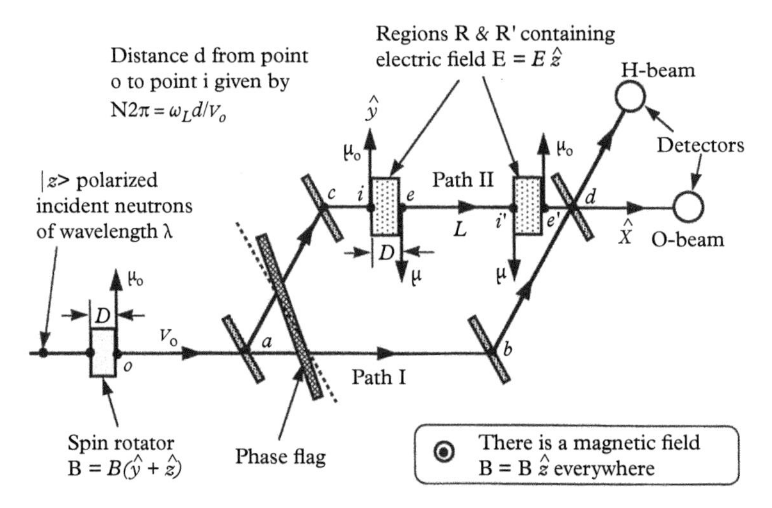

Figure 10.8 *This is a schematic diagram showing a proposed neutron interferometer experiment designed to measure the phase shift arising from the Anandan acceleration. The incident beam is polarized along the z-axis (perpendicular to the plane of the interferometer). There is a uniform magnetic field* $\mathbf{B} = B\hat{z}$ *permeating the entire region of the experiment. The initial* |z> *polarized state is rotated to a* |y> *polarized state by the spin rotator, which contains an additional magnetic field in the* y*-direction. The neutron wave packets will process about* $B\hat{z}$ *with the Larmor frequency* ω_L *on both paths, I and II, of the interferometer. The wave packet on path II traverses a region R containing an electric field* $\mathbf{E} = E\hat{z}$*. The distance from the point 0 to the point i along the trajectory is such that there are integral numbers of Larmor precessions. After traversing R the polarization is* $\boldsymbol{\mu} = -\boldsymbol{\mu}_0$ *and the velocity is increased to* $v_0 + \Delta v$*. At a distance* L *downstream from R the neutron traverses a second region R′ of electric field identical to R which decelerates the neutron back to its initial velocity* v_0*. The net effect of the electric field is to cause a phase shift* $\Delta\Phi_A$ *given by Eq. (10.37)*

$$\Delta\Phi_A = \frac{1}{\hbar}\, m\Delta v \cdot L = \frac{2\mu_n E}{\hbar c} \cdot L, \qquad (10.37)$$

where L is the distance from the exit point from the region R containing the electric field $\boldsymbol{E}$ to the entrant point of region R' containing the same E. This formula is identical to the result for the AC phase shift, except there, L was the distance traversed by the neutron within the electric field region. One should note that this phase shift is independent of the neutron wavelength. Taking $E = 830$ statvolts/cm again and $L = 10$ cm, we find that this phase shift is 5.08 mRad = 0.29°. For neutrons of wavelength $\lambda = 2.5$ Å and $B = 50$ gauss, the necessary length of the regions of electric field is $D = 0.55$ cm. The experimental procedure for observing this phase shift is to measure the phase difference between two interferograms obtained by rotating the phase flag, one with the electric field *on* and one with the electric field *off*. Such an experiment is currently in the planning stage at NIST.

The phase shift given by Eq. (10.37) could be increased by some integer n, simply by having a number of regions with electric field reversals as shown in the insert to Fig. 10.9, as first suggested by Anandan (1989b) and described in some detail by Wagh and Rakhecha (1997). As n becomes large the shift in velocity Δv approaches its maximum value

$$\Delta v = (\mu_n B/2\varepsilon_0)v_0, \tag{10.38}$$

where $\varepsilon_0 = \frac{1}{2}\, m v_0^2$. For $B = 50$ gauss, and $v_0 = 1.6 \times 10^5$ cm/s, $\Delta v/v_0 = 1.15 \times 10^{-8}$. The trajectory of the neutron's precession, $\boldsymbol{\mu}(t)$, is shown schematically in Fig. 10.9. In the first region of positive electric field, the neutron spin precesses from point a to point b on the circle in the plane perpendicular to $\boldsymbol{B}'$. In the second region having a negative electric field, the neutron precesses from point b to point c on a circle in the plane perpendicular to $\boldsymbol{B}''$. Then in the next region of positive electric field, the neutron precesses about $\boldsymbol{B}'$ again from point c to point d and so on. After traversing each region the z-component of the moment, μ_z, increases, finally reaching $\mu_z = \mu_0$ asymptotically. The decrease in Zeeman potential energy $-\mu_0 B$ goes into an increase of the neutron's kinetic energy and, thus, the increase of its velocity Δv given by Eq. (10.38).

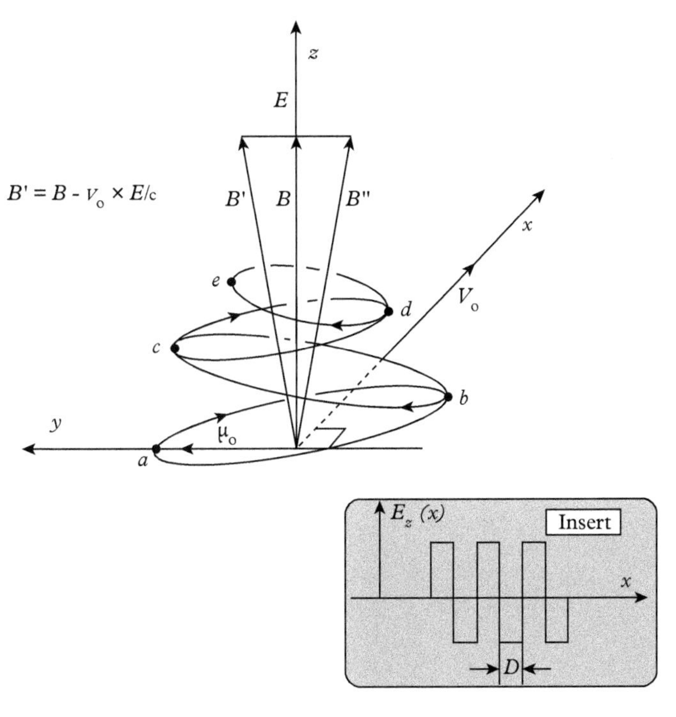

Figure 10.9 *The regions R and R′ of Fig. 10.8 could (in principle) be replaced by regions containing a staggered electric field as shown in the insert. The trajectory of the neutron's magnetic moment $\boldsymbol{\mu}(t)$ on the "spin sphere" is shown schematically here. This evolution of the neutron's polarization is explained in the text*

10.13 Search for Basic Dissipative Terms

In a rather general form the Born–Markov and the von Neumann equations (Eq. 4.139) can be written in the form

$$i\hbar \frac{\partial \rho}{\partial t} = H\rho(t) - \rho(t)H + iL(\rho(t)), \tag{10.39}$$

with

$$L(\rho(t)) = A\rho(t)A^* - A^*A\rho(t) + \rho(t)A^*A.$$

To assure an entropy increase A must be taken to be Hermitian. L produces dissipation and transitions from pure states to mixed states. This equation is independent of the microscopic mechanism responsible for the dissipative effects; it is the result of very basic physical assumptions, like probability conservation, entropy increase, and complete positivity (Spohn 1980, Scully and Zubairy 1997). This formulation has also been used for a loss of coherence induced by quantum gravity effects at the Planck's scale (Ellis et al. 1984; Hawking 1983, 1996). The dissipative term can be written by a symmetric 4 × 4 matrix acting on the column density vector $(\rho_0, \rho_1, \rho_2, \rho_3)$, which can be written in terms of the Pauli matrices ($\rho = \rho_i\sigma_i$, σ_0 being the identity):

$$L = -2\begin{pmatrix} 0 & 0 & 0 & 0 \\ 0 & a & b & c \\ 0 & b & \alpha & \beta \\ 0 & c & \beta & \gamma \end{pmatrix}. \tag{10.40}$$

For generic initial conditions describing the interferometer situation, the time dependence of the four components of the density can be calculated within first-order perturbative expansion (Benatti and Floreanini 1999). This gives an intensity behind the interferometer as

$$I_0 \propto \left[1 + e^{-A\Delta D/\hbar v} \cos\chi + \frac{|B|\,\Delta D}{\hbar\chi v} \sin\chi \cos\theta'\right], \tag{10.41}$$

where

$$A = \alpha + a$$
$$B = \alpha - a + 2ib = |B|\ e^{i\theta'},$$

and ΔD is the different path lengths within the phase shifter producing a phase shift χ. From measured interference patterns, the related parameters can be extracted. First attempts in this direction indicate an agreement with zero which imply $\gamma = \alpha$ and $b = c = \beta = 0$, and which is compatible with complete positivity. For α a value of $(0.71 \pm 0.21) \times 10^{-12}$ eV has been determined (Benatti and Floreanini 1999). This indicates a non-vanishing value for basic dissipative effects of the right order for a quantum gravity of "stringy" origin, but the results should be treated with care because many assumptions have been made and no dedicated experiment has yet been performed.

10.14 Proper Time Effects in Gravity Experiments

In Section 8.5.6 we derived a general phase shift for the gravitational measurements in the form (Eq. 8.108)

$$\phi \cong -\omega_c t + \frac{1}{\hbar}\int\left(\frac{mv^2}{2} - mU\right)dt, \tag{10.42}$$

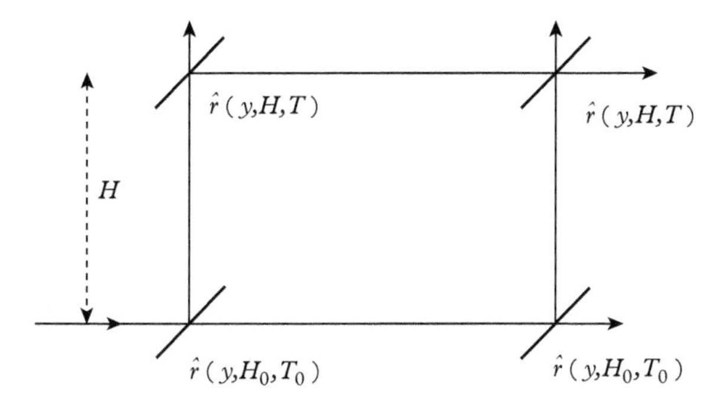

Figure 10.10 *Gravitation experiment with crystal plates at different heights in a gravity field and at different temperatures to keep the reflectivities unchanged*

where $\omega_c = mc^2/\hbar$ denotes the Compton frequency and $U = gH$ the gravitational potential along the beam paths at height H. In this connection it has been stated that the first term causes a general phase which cancels out over both beam paths. The second term represents energy conservation and is equivalent to the path integral method, which is used for all interferometer experiments (Sections 1.2 and 8.1).

The question arises whether the first term can be measured separately. In this context an experiment where the phase sensitivity of the crystal reflectivities at different heights are taken into account is proposed (Fig. 10.10). The phase sensitivity can be calculated from the dynamical diffraction theory and follows from the widths of the Pendellösung fringes as $y_H = 2.215/A$ with $A = \pi t/\Delta_0$ (see Eq. 11.51 and Fig. 10.10). For a symmetric (2, 2, 0) silicon interferometer and a neutron wavelength of $\lambda = 2$ Å this corresponds in momentum space to a sensitivity of $\delta k = 2.215/t$, where t denotes the thickness of the interferometer plates. The momentum shift due to gravity over a height H follows from energy conservation as $\delta k_g = m^2 gH/\hbar^2 k_0$, which gives a critical height for a phase-sensitive phase shift of $H_c = 2.215\hbar^2 k_0/m^2 gt \approx 5.6$ mm. The reflectivity of the crystal plates can be varied by temperature changes which alters the lattice constant according to the linear expansion coefficient $\alpha = 4.2 \times 10^{-6}/K$. For a $H =$ 5-cm-height interferometer with $t =$ 5-mm-thick crystal plates a temperature difference between the lower and upper crystal plates of 0.074 K would be adequate to compensate for the momentum change due to gravity. This can give a method to use the phase-sensitive crystal reflectivities to investigate different contributions to the overall phase of matter waves. For one spin state such a compensation and control of the crystal reflectivity can also be achieved by an appropriate magnetic field corresponding to $\delta k_m = \pm\mu Bm/\hbar^2 k$.

10.15 Neutron Fourier Spectroscopy

In neutron interferometry the phase of the neutron wave becomes an observable which is influenced by any momentum or/and energy change the beam experiences during its interaction with a sample. Any phase change is measured by a superposition with a coherent reference beam not being effected by the sample. In many cases a much higher sensitivity than in usual spectrometry can be achieved because now the usual constraints do not exist. Namely, that the momentum

Spectrocopy

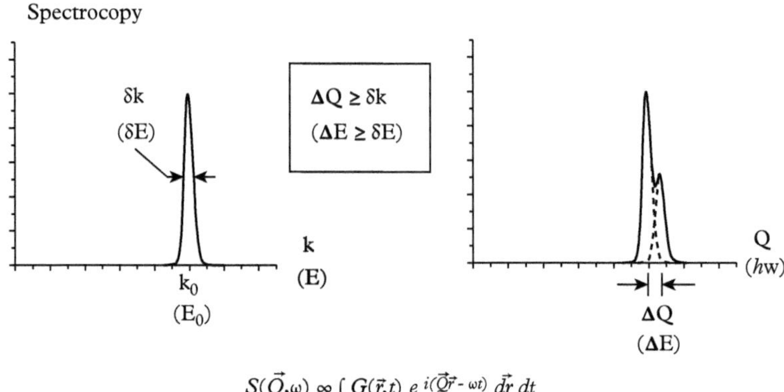

Fourier spectroscopy

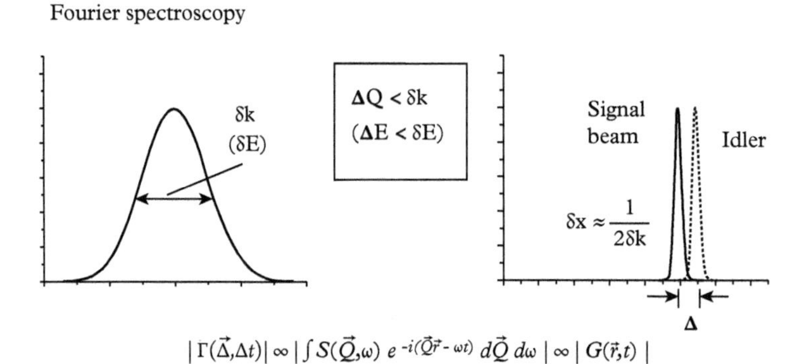

Figure 10.11 *Comparison of standard spectroscopy (above) and Fourier spectroscopy methods (below)*

(energy) change $\Delta Q(\Delta E)$ must be larger than the momentum (energy) width $\delta k(\delta E)$ of the beam. The new method shows many similarities to NMR and optical Fourier spectroscopy (e.g., Marshall and Verdun 1990). The Fourier spectrometric methods are mainly based on the van Cittert–Zernike theorem which gives the connection between the coherence function and the momentum distribution of the beam (see Eqs. 4.29 and 4.30). A schematic comparison of direct and Fourier spectroscopy is shown in Fig. 10.11.

In Section 4.2 it has been shown that the spatial and time-dependent coherence function is given by the Fourier transform of the momentum and energy distribution function (Eq. 4.28)

$$\Gamma^{(1)}(\mathbf{\Delta},\tau) \propto \int \rho(\mathbf{k},\omega)\, e^{i(\mathbf{k}\cdot\mathbf{\Delta}-\omega t)} d^3\mathbf{k}d\omega, \tag{10.43}$$

where $\left|\Gamma^{(1)}(\mathbf{\Delta},T)\right|$ denotes the contrast of the interference pattern (Eq. 4.30), and the intensity is

$$I(\mathbf{\Delta},\tau) = I_0\left[1+\left|\Gamma^{(1)}(\mathbf{\Delta},\tau)\right| \cos\left(\boldsymbol{k}\cdot\mathbf{\Delta}-\omega t\right)\right]. \tag{10.44}$$

This function can be measured by applying various spatial phase shifts $\mathbf{\Delta}$ or various temporal delays τ.

The analogies to the van Hove formalism of neutron scattering from condensed matter should be emphasized (van Hove 1954, Marshall and Lovesey 1971, Squires 1978). In this case, the space and time-dependent correlation function $G(\mathbf{r}, t)$ is obtained as a Fourier transform of the measured scattering functions $S(\mathbf{Q}, \omega)$

$$G(\boldsymbol{r}, t) \propto \int S(\boldsymbol{Q}, \omega)\, e^{i(\boldsymbol{Q}\cdot\boldsymbol{r}-\omega t)}\, \mathrm{d}^3\boldsymbol{Q}\, \mathrm{d}\omega. \tag{10.45}$$

For elastic scattering ($\omega = 0$) one gets

$$S(\boldsymbol{Q}) \propto \int e^{-\mathrm{i}\boldsymbol{Q}\cdot\boldsymbol{r}}\, G(\boldsymbol{r}, 0)\, \mathrm{d}^3 r, \tag{10.46}$$

where $G(\boldsymbol{r}, 0)$ denotes the radial pair-correlation function describing the probability of finding an atom (or a scattering length density element) at $\boldsymbol{r}$ when there is another one at $\boldsymbol{r} = 0$. For plane incident waves $S(\boldsymbol{Q})$ represents the momentum distribution after scattering. The momentum distributions of the incident beam ($g(\boldsymbol{k})$) can also be attributed to an effective correlation function $G_0(\mathbf{\Delta}, 0)$ where we get ($\boldsymbol{Q} = k - \boldsymbol{k}_0$)

$$g(\boldsymbol{Q}) \propto \int e^{-i\boldsymbol{Q}\cdot\boldsymbol{r}}\, G_0(\mathbf{\Delta}, 0)\, \mathrm{d}^3 r, \tag{10.47}$$

for a scattering vector $\boldsymbol{Q} = \boldsymbol{k} - \boldsymbol{k}_0$. A comparison with Eq. (10.44) shows the equivalence of the correlation and the coherence function

$$\Gamma_0(\mathbf{\Delta}) \hat{=} G_0(\mathbf{\Delta}, 0). \tag{10.48}$$

The momentum distribution of the beam after scattering from the sample is given by the convolution of the scattering function and the momentum distribution of the incident beam:

$$g_s(\boldsymbol{Q}) = S(\boldsymbol{Q}) * g(\boldsymbol{Q}). \tag{10.49}$$

This shows a close connection between the resolution function of a distinct experiment and the related beam properties described by their correlation (coherence) function (see Gähler et al. 1996, Rekveldt 1996, Gaehler et al. 1998).

When one considers the transmitted and scattered beam one gets the following momentum distribution behind the sample

$$g_{\mathrm{m}}(\boldsymbol{Q}) = e^{-\Sigma_t D} g(\boldsymbol{Q}) + \left(1 - e^{-\Sigma_s D}\right)\left(S(\boldsymbol{Q}) * g(\boldsymbol{Q})\right), \tag{10.50}$$

which yields after some algebraic calculations the coherence functions with and without sample

$$\frac{|\Gamma_{\mathrm{m}}(\mathbf{\Delta})|}{|\Gamma_0(\mathbf{\Delta})|} = e^{-\Sigma_t D} + \left(1 - e^{-\Sigma_s D}\right) G(\mathbf{\Delta}, 0). \tag{10.51}$$

The effective beam attenuation ($\exp(-\Sigma_t D)$) can be measured from the intensity ratio with and without the sample. Equation (10.51) shows that the correlation function can be obtained directly from the measurable coherence functions. The measurement scheme shown in Fig. 10.12 is a one-dimensional one. Therefore, the sample must be oriented in all three dimensions in order to obtain the whole correlation function (Rauch 1995b). A wide beam and a broad incident wavelength spectrum can be used. A very interesting application may also be a direct measurement of the surface profile in neutron reflectometry, but other small-angle scattering phenomena and structural investigations may also profit from this new technique. In this case the scattering function within the Rayleigh approximation is (Hayter et al. 1976, Lekner 1987)

$$I(Q_z) \propto S(Q_z) \propto \frac{1}{Q_z^{\,4}} \int e^{-iQ_z z} \frac{\mathrm{d}f}{\mathrm{d}z}(z)\mathrm{d}z, \tag{10.52}$$

where $\mathrm{d}f/\mathrm{d}z$ denotes the gradient of the scattering length density profile $f(z) = N(z)b_c(z)$ of the layered surface. Applying the van Cittert–Zernike theorem (Eq. 4.28) one obtains

$$\Gamma(\Delta_z) = \frac{1}{Q_z^{\,4}} \int \frac{\mathrm{d}f}{\mathrm{d}z}(z')\, F(z-z')\mathrm{d}z', \tag{10.53}$$

where

$$F(z-z') \propto \int \frac{1}{Q_z^{\,4}}\, e^{iQ_z(z-z')}\, \mathrm{d}Q_z, \tag{10.54}$$

for $Q_z > (4\pi Nb)^{1/2}$.

Equation (10.52) is equivalent to the form one gets by the Born approximation (or the Riemann integral)

$$I(Q_z) \propto \frac{1}{Q_z^2} \left| \int N(z)b_c(z)\, e^{iQ_z z}\, \mathrm{d}z \right|^2. \tag{10.55}$$

In both cases, this gives the well-known $1/Q_z^4$ *Fresnel reflectivity* of a flat surface of a compact material. Since the square of Fourier transforms is equivalent to the Fourier transform of

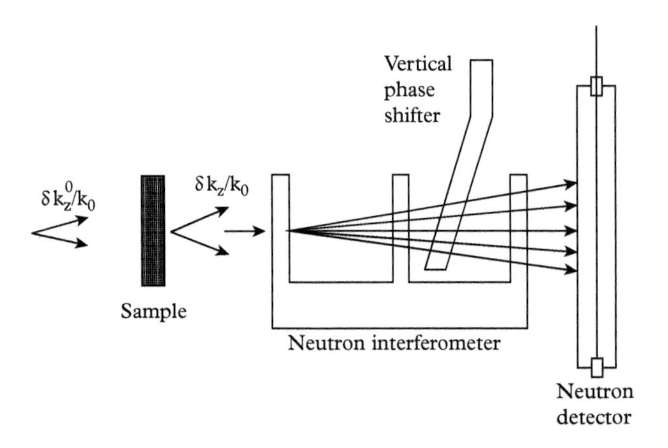

Figure 10.12 *Scheme of a spatial Fourier spectroscopy experiment*

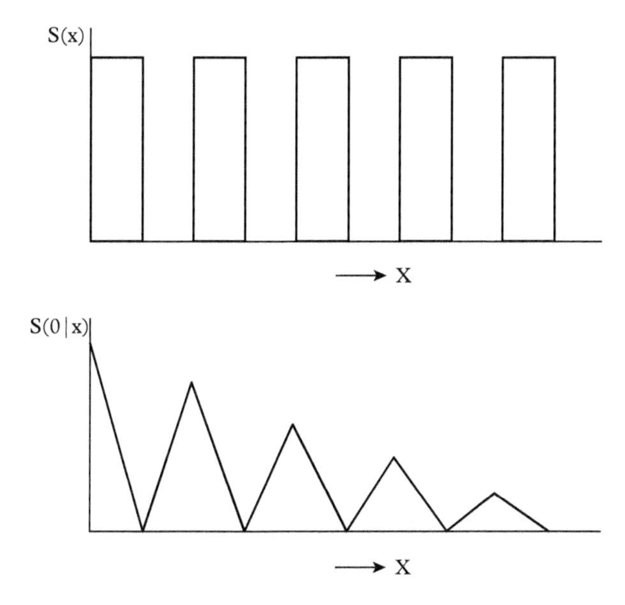

Figure 10.13 *Scattering length shape function (above) and its auto-correlation function (below) which relates to the interferometric coherence function*

the auto-correlation function of the scattering profile (Wiener–Khinchine theorem), one obtains the measurable correlation (coherence) function most easily from the auto-correlation function of the scattering profile. Figure 10.13 shows the expected reflection profile and the expected coherence function of a five-layer surface structure. Thus, a unique reconstruction of the surface structure should become feasible by this phase-sensitive interferometric method. Progress in exact phase determination in standard neutron reflectometry should be mentioned here (Lipperheide et al. 1995, Majkrzak and Berk 1995, Majkrzak et al. 1998). In this case the crystallographic phase problem has been solved by adding known reference layers onto the surface of the layered structure.

Using an incident beam described by stationary Schrödinger cat-like states may provide another method. Such states have been identified when the wave trains of the coherent beams inside an interferometer are shifted more than their coherence lengths (Jacobson et al. 1994; Section 4.5.2, Fig. 4.31). A similar situation exists in spin-echo spectroscopy (Fig. 4.32). In these cases, spatially separated coherent packets exit behind the interferometer which exhibit a marked modulation in momentum space.

In case of a Gaussian-shaped incident beam which equals a Gaussian resolution function, the neutrons "feel" a spatial region in the sample of about $\Delta^c \cong (2\delta k)^{-1}$ around the origin. On the other hand, a well-separated coherent double-peaked incident beam "feels" the physical situation in the sample mainly at the origin and at a distance of its separation Δ, which is adjustable by the phase shift applied inside the interferometer. In the limit of nearly δ-functions like double peaks an intensity measurement as a function of Δ directly gives the radial dependence of the van Hove correlation function

$$I(\Delta) \propto G(\Delta, 0). \tag{10.56}$$

It should be mentioned that this situation can more easily be approached when a rather polychromatic incident beam is used. This, of course, also increases the intensity considerably. No momentum scan would be necessary. Spin-echo systems (Section 2.4) have also been proposed for spatial Fourier spectroscopy for small-angle scattering and reflectometry applications (Rekveldt 1996) and realized by Bouman et al. (2008). This principle is based on the difference in Larmor precession angles in spin-echo coils when scattering from a sample between the two precession fields of a spin-echo spectrometer (Fig. 2.19). It can also be explained by the appearance of a spatially separated coherent beam (Fig. 2.20). When the measurements are performed as a function of the strength of the precession field, the real space correlation function can be obtained. All depolarization effects appear as dephasing effects. A comparison between scattering and spin-echo technique in the case of grating diffraction from a grooved silicon structure has been done by Trinker et al. (2007). Krouglow et al. (2008) gave a comprehensive description of the phase or polarization signal and the real-space structure of the sample.

Fourier spectroscopy can be extended to time-dependent phenomena which are associated with an energy change of the beam ($\omega \neq 0$). In this case, a purely energy-dependent phase shift must be applied inside the interferometer. This can be achieved by means of a neutron magnetic resonance energy transfer system. This idea has been tested in the past (Badurek et al. 1986; Section 5.3). A resonance spin flip inside a magnetic field B_0 is connected with an exchange of the Zeeman energy $\hbar\omega_r = 2\mu B_0$ (Alefeld et al. 1981b, Weinfurter et al. 1988; Fig. 5.8), where ω_r represents the frequency of the oscillating field (Larmor frequency). When a spin flip is applied to both coherent beams inside the interferometer, the wave function of the forward beam behind the interferometer is (see Section 5.4, Eqs. 5.14 and 5.19)

$$\psi \rightarrow e^{i(\omega-\omega_{r1})t} \,|\downarrow> + e^{i\chi}\, e^{i(\omega-\omega_{r2})t}\, |\downarrow>. \tag{10.57}$$

In most experimentally feasible situations, the net energy transfer is much smaller than the energy width δE of the beam $\hbar\Delta\omega_r = \hbar(\omega_{r1} - \omega_{r2}) = 2\mu(B_{01} - B_{02}) << \delta E$. If the energy transfer become comparable ($\hbar\Delta\omega_r \cong \delta E$) or when the temporal delay time $\Delta t = 2\mu\Delta B_0 t/E$ becomes comparable with the coherence time ($\Delta t^c \sim \hbar/2\delta E \cong \Delta_c/v$) a related damping factor appears

$$I \propto 1 + |\Gamma(\Delta t)|\cos(\Delta\omega_r.t), \tag{10.58}$$

with

$$|\Gamma(\Delta t)| = \exp\left[(-\Delta t.\delta E/\hbar)^2/2\right]. \tag{10.59}$$

This can be interpreted as the Fourier transform of the energy distribution of the beam (e.g., Lauterborn et al. 1995). When $\Delta\omega_r.t$ is replaced by $\Delta t.\omega$, where $\omega = E/\hbar$, the full analogy between spatial and temporal phase shifts according to Eq. (4.30) is recognized. Figure 10.14 shows a typical arrangement of how this technique can be used.

Analogies to the well-established spin-echo method (Mezei 1972, 1980; Felber et al. 1998) and to beam chopper Fourier spectroscopy should be mentioned (Colwell et al. 1968, Pöyry et al. 1975). Spin-echo time ($t_{se} = 2\mu BL/mv^3$) exactly describes this correlation time appearing in this kind of spectroscopy (see Eq. 2.37). The measured polarization of the beam behind a spin-echo arrangement (Fig. 2.19) is proportional to the Fourier transform of the energy transfer spectrum. Thus, the multiphoton exchange experiments within an oscillating field described in Section 5.5 are typical examples of such time-dependent interferometric Fourier methods. When

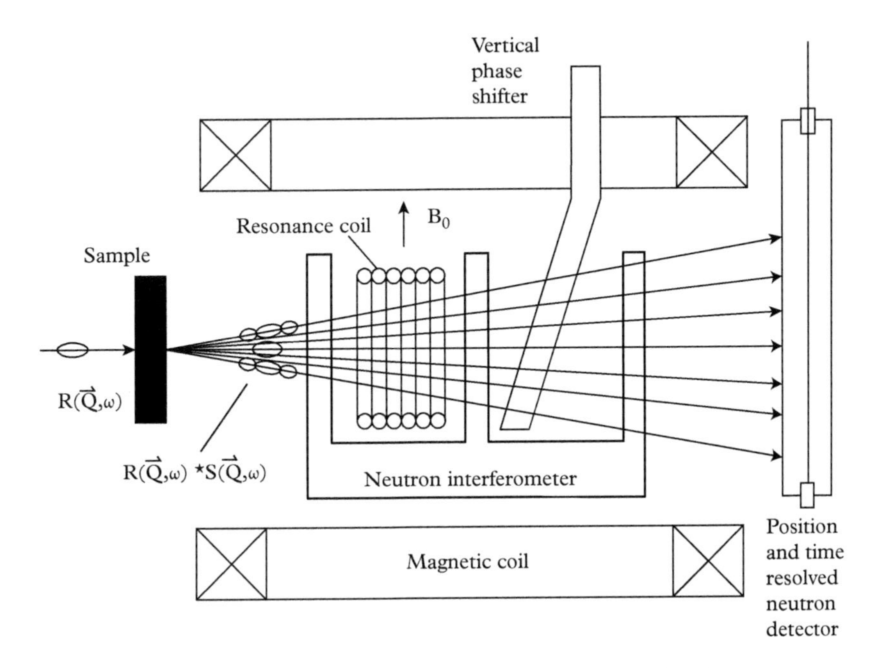

Figure 10.14 *Sketch of a feasible space-time interferometric Fourier spectroscopy experiment*

a scattering target is placed between the precession fields of a spin-echo arrangement the measured polarization of the scattered beam measures the Fourier transform of the scattering function $S(\mathbf{Q}, \omega)$ which is the time-dependent density–density correlation function of the scattering system (van Hove 1954, Gähler et al. 1996)

$$\begin{aligned} < \sigma_z >= & \int S(\mathbf{Q}, \omega) \cos \omega t_{se} \, d\omega \\ = I(\mathbf{Q}, t_{se}) = & \int G(\mathbf{r}, t_{se}) \, e^{iQ \cdot r} \, dr \,. \end{aligned} \tag{10.60}$$

Spatial and temporal Fourier spectroscopy can be done simultaneously when a proper neutron magnetic resonance system is added to the interferometer and the intensity is registered stereoscopically with the phase of the oscillating field. Thus, a measurement of the coherence functions provides direct access to the spatial and time dependence of the correlation functions describing the static and dynamic properties of condensed matter. The related experimental technique will be developed and tested for different simple substances where mainly large-scale structures and long-time fluctuations are of interest. A comprehensive theoretical treatment which describes the analogy between the neutron spectroscopic and the Fourier methods is still missing.

Rode and Jex (1999) performed inelastic X-ray interferometer experiments with vibrating central crystal plates. Ultra-sound waves of slightly different frequencies around 10 MHz were fed into the mirrors by quartz crystals attached on the sides of each crystal. Related beat effects have been observed (see Section 5.4) and an energy sensitivity of 1.56×10^{-20} eV has been achieved.

Similar experiments with neutrons are feasible and they can be adapted for high sensitivity thermal phonon and artificially induced phonon spectroscopy.

10.16 Time-Dependent Fizeau Phase Shift

In Chapter 8.5 we described the neutron Fizeau effect for a plane plate and got a phase shift of (Eq. 8.46; Fig. 8.18)

$$\Delta\Phi_{\text{Fizeau}} = -\frac{\alpha\beta\, k_x L}{2\,(1-\alpha)}, \tag{10.61}$$

where L is the thickness of the sample and $\alpha = w_x/v_x$ is the ratio of the velocities of the boundary in direction perpendicular to the surface and the neutron velocity in this direction. $ß = V_0/E_x$ denotes the ratio of the optical potential to the energy related to the x-component

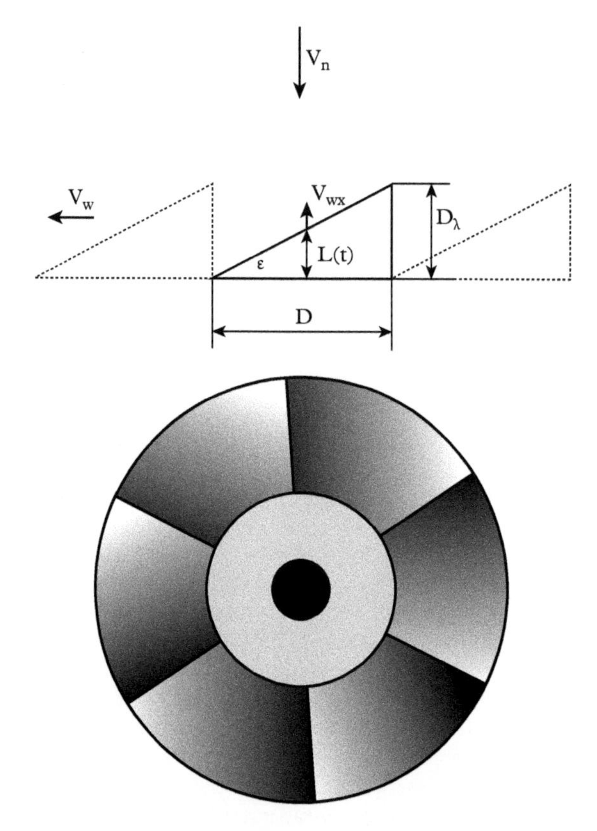

Figure 10.15 *Moving wedge shaped phase shifter to produce a time-dependent Fizeau effect*

of the neutron velocity. When one uses a moving wedge-shaped sample (Fig. 10.15) one can generalize this formula to include the variable thickness of the sample during motion ($x \equiv L(t) = D_\lambda v_w t/D = v_w t tg\varepsilon$; $D_\lambda = 2\pi/Nb_c\lambda$) giving the time-dependent phase shift

$$\Delta\Phi_{\text{Fizeau}}(t) = \frac{V_0 v_w^2}{\hbar(v_n - v_w tg\varepsilon)}.t. \tag{10.62}$$

The phase factor can be written as

$$e^{i\Delta\Phi(t)} = e^{i\Delta Et/\hbar}, \tag{10.63}$$

which gives an energy change of

$$\Delta E = \hbar\omega = \frac{V_0 v_w^2}{v_n(v_n - v_w tg\varepsilon)} \approx V_0 \frac{v_w^2}{v_n^2} \tag{10.64}$$

and a time-dependent interference pattern as

$$I_0 \propto [1 + \cos(\chi + \omega t)]. \tag{10.65}$$

For $V_0 = 9 \times 10^{-8}$ eV, $v_n = 1000$ m/s, $v_w = 10$ m/s one expects $E = 9 \times 10^{-12}$ eV = 9 peV. This may be small but measurable with interferometric methods (see Sections 4.6.2 and 5.4; Sulyok et al. 2012).

In an experiment a rotating disk can be used with a variable thickness and thickness steps larger than the lambda thickness D_λ. The situation is analogous to time-dependent phases caused by oscillating magnetic fields (Sections 5.3 and 4.6.2).

10.17 Search for Chameleon Fields

The accelerating expansion of the universe requires the existence of a long-range scalar interaction (Khoury et al. 2004). Slow neutron can feel such screening fields when they are near to massive surfaces or move within evacuated vessels within a neutron interferometer, where the chameleon produces a bubble-like profile (Brax et al. 2013, Yu and Oh 2014). The whole effect arises from a hypothetical coupling of matter to dark energy. Within the Ratra–Peeles model the chameleon field can be written as

$$V(\varphi) = \Lambda^4 f(\varphi/\Lambda) \approx \Lambda^4 + \frac{\Lambda^{4+n}}{\varphi^n}, \tag{10.66}$$

with $\Lambda^4 = 3\Omega_{A0} H_0^2 M_{\text{Pl}}^2 \approx 2.4 \times 10^{-12}$ GeV, where H_0 is the Hubble rate now and M_{Pl} is the reduced Planck mass, and $n > 0$. This potential yield a phase shift in an interferometer of

$$\delta\phi = \frac{\beta m^2}{k\hbar^2} \int \varphi(x)\text{d}x, \tag{10.67}$$

where the chameleon field fulfills within a medium with density ρ the equation

$$\frac{d^2\varphi}{dx^2} = V(\varphi) + \frac{\beta}{M_{\text{Pl}}}\rho. \tag{10.68}$$

This can be calculated for an empty and a gas-filled vessel and related measurements can estimate the interaction strength β of chameleons to matter.

Other proposals for measuring the scalar chameleon field by means of neutron interference experiments exist in the literature (e.g., Pokotilovski 2013).

10.18 Quantum Zeno Tomography

In Section 4.3.1 (Fig. 4.14) it has been shown that a surprisingly high-quality interference pattern can be observed when neutrons in one beam path of an interferometer are absorbed statistically. On that basis Facchi et al. (2002) proposed a multi-loop interferometer setup where this effect is multiplied and quantum Zeno tomography becomes feasible. Gain factors of about 100 can be anticipated.

10.19 Complementarity and Equivalence Tests with Unstable Particles

An unstable particle, like the neutron, used in a two-path interference experiment may decay inside the interferometer, leaving "which-path" information and showing still full interference features. This situation is comparable with the absorber experiments described in Section 4.3. Complementarity tests are usually based on the Greenberger–Englert relation (Eq. 4.76), where the path predictability P_D depends on the particle decay time $\tau = 1/\Gamma$. Related proposals have been put forward by Bonder et al. (2013) and Krause et al. (2014). The equivalence principle can be tested in a similar way and in both cases a proper time explanation is possible as well (see Section 8.5.6). Unstable particles can be characterized by a complex mass

$$m = m_0 - i\Gamma/2. \tag{10.69}$$

The general solution to the unperturbed time-independent wave equation in free space becomes (Greenberger and Overhauser 1979)

$$\psi(r) = A_0 \exp\left[ikr\left(1 + i\frac{m\Gamma}{2\hbar k^2}\right)\right]. \tag{10.70}$$

This adds in the COW experiment (Eq. 8.4) an imaginary phase shift

$$\Delta\Phi_{COW}^{imag} = i\left(\frac{g\Gamma A_0}{2v^3}\right)\sin\alpha \tag{10.71}$$

and an additional reduction of the contrast by

$$V_{COW} = V_{0,COW}\cosh^{-1}(\Delta\Phi_{COW}^{imag}) = V_{0,COW}\cosh^{-1}(\Delta/\Delta_d). \tag{10.72}$$

Δ denotes the shift of both wave packets (Eq. 4.44) and Δ_d the decay length (~2000 km; Table 1.1). This indicates that the effect would be very small and very difficult to measure. Short-lived atoms may be a better choice.

11 Perfect Crystal Neutron Optics

The classical Mach–Zehnder and Michelson interferometers operate on the principle of coherent division of the amplitude of an incident electromagnetic wave by partially reflecting, partially transmitting objects, called beam splitters. The perfect Si crystals in the neutron interferometer are the beam splitters and mirrors. Their operation depends upon the dynamical diffraction of neutrons having a de Broglie wavelength in the range of 1 to 5 Å. Although the results and basic physics of most of the experiments discussed in this book can be understood without a detailed knowledge of the wave fields inside the perfect crystal slabs of the interferometer, it has been found that the successful fabrication and utilization of this device requires a rather advanced analysis of the dynamical diffraction process occurring within each of the single-crystal elements. The main goal of this chapter is to provide the reader with a focused description of the physics of dynamical diffraction in perfect crystal media that is necessary to understand the interferometer in its essential details. In addition, there are a number of neutron physics experiments that do not use the interferometer, but are directly based upon the exquisite coherent wave fields accompanying the diffraction of neutrons in the perfectly periodic structure of silicon. The theoretical description follows the early development of X-ray dynamical diffraction theory (Darwin 1914; Ewald 1916, 1928; von Laue 1931), which has been adapted to the neutron case by Bonse and Graeff (1977) and by Rauch and Petrascheck (1978). Here we will also describe the first observation of the oscillation of the integrated reflectivity of Si crystals as a function of thickness by Sippel et al. (1965), and the observation of the Pendellösung fringe structure by Shull (1968, 1973). The related "effective inertial mass" experiments of Zeilinger et al. (1986) and Raum et al. (1995), which also use dynamical diffraction effects inside perfect crystals, have been discussed already in Section 8.2.3.

11.1 Transition from the Kinematical to Dynamical Diffraction

The first Born approximation assumes that the scattered wave is not rescattered within the crystal. This assumption leads to the kinematical theory of diffraction for X-rays and neutrons, and it predicts that the intensity of a Bragg reflected beam is proportional to the volume of the diffracting crystal. It is clear that if this were strictly true, the intensity of the diffracted beam could exceed the intensity of the incident beam, thus violating Liouville's theorem. The origin of this difficulty with the kinematical theory is that it ignores the obvious fact that the scattered waves will be substantially rescattered in a crystal whose size is larger than some characteristic attenuation length, ℓ_p, called

Neutron Interferometry. Second Edition. Helmut Rauch and Samuel A. Werner.

the primary extinction length. We will find that this length is typically 50 to 100 μm for thermal neutrons having wavelengths $\lambda = 1$ to 2 Å in perfect Si crystals under Bragg reflecting conditions. Thus, for crystals whose size is greater than ℓ_p the differential scattering cross-section $d\sigma/d\Omega$ will no longer be proportional to the square of the Fourier transform of the neutron–crystal interaction potential.

When a monochromatic incident neutron beam of wave vector $\boldsymbol{k}$ satisfies the Bragg condition for a reciprocal lattice vector $\boldsymbol{H}$, that is

$$\boldsymbol{k}' - \boldsymbol{k} = \boldsymbol{H}, \tag{11.1}$$

the diffracted beam of wave vector $\boldsymbol{k}'$ satisfies the Bragg condition for $-\boldsymbol{H}$, and is rescattered back into the incident beam. One would expect in general then that there will be a "dynamic" interchange of neutron intensity between the incident beam direction and the Bragg diffracted beam direction inside any crystal medium. This multiple scattering of the incident wave, first into the diffracted beam, then back again into the incident beam, and so on, represents the central physical issue in the dynamical theory of diffraction. Beginning with the historic work of Darwin (1914), von Laue (1931), Ewald (1916, 1928), Zachariasen (1967), and Kato (1974), this theory and its applications to the X-ray case is highly developed. Excellent reviews such as those by Batterman and Cole (1964), Jones (1963), and Pinsker (1977) have been written; and for the neutron case the reader is referred to the papers by Stassis and Oberteuffer (1974), Sears (1979), Rauch and Petrascheck (1978), and Lemmel (2013). There is a very large and interesting literature on the various theoretical and experimental aspects of X-ray, neutron, and electron diffraction by perfect crystals. We make no attempt here to review the entire theory or its experimental consequences. We will first focus our attention on the dynamical diffraction of neutrons by a perfect crystal in the symmetric Laue geometry shown in Fig. 11.1, which is the key optical element for our understanding of the perfect Si-crystal interferometer.

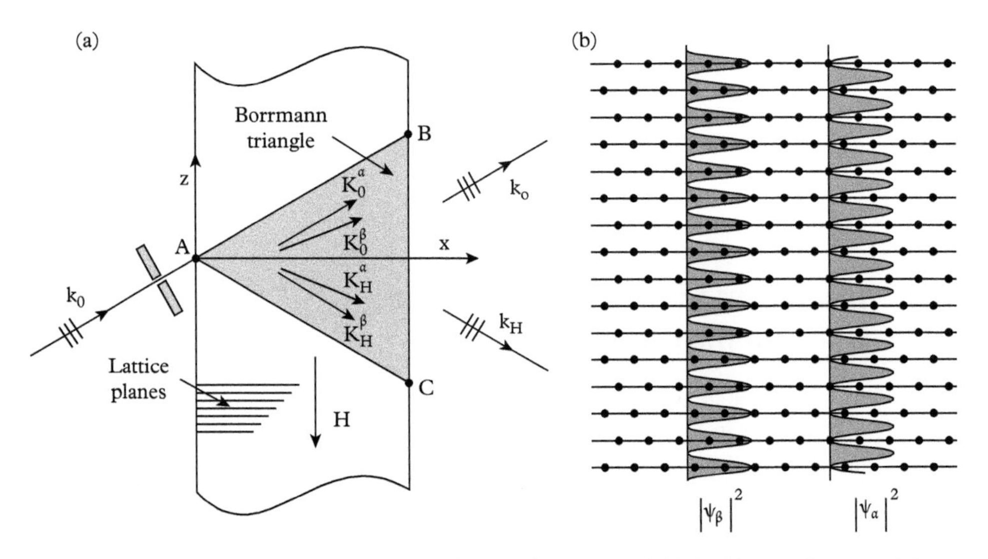

Figure 11.1 *Wave vectors (a) and wave amplitudes of both wave fields inside a perfect crystal (b)*

11.2 Dynamical Diffraction for the Symmetric Laue Case

Consider an incident monochromatic wave $A_0 e^{i\boldsymbol{k}_0 \cdot \boldsymbol{r}}$ with $\boldsymbol{k}_0$ oriented at or near the Bragg condition for the lattice planes shown in Fig. 11.1a. An incident wave vector that precisely satisfies the Bragg condition lies on the Brillouin zone (BZ) boundary corresponding to the reciprocal lattice vector $\boldsymbol{H}$, which is normal to the lattice planes spaced a distance $d = 2\pi/H$ apart. From the electron band theory of metals we know that the BZ boundaries are planes in $\boldsymbol{k}$-space representing the loci of energy gaps in the electron energy spectrum. The Fourier components V_H of the periodic crystal potential $V(\boldsymbol{r})$ mixes the neutron wave of wave vector $\boldsymbol{k}_0$ with the Bragg reflected wave of wave vector $\boldsymbol{k}_0 + \boldsymbol{H}$ forming a coherent state, which is a standing wave along $\boldsymbol{H}$. There are, in fact, two solutions, say ψ_+ and ψ_-, to the one-particle Schrödinger equation for a given $\boldsymbol{k}_0$. The one, ψ_- with an energy ε_- below the gap, has nodes between the atomic planes, whereas the other one, ψ_+ with energy ε_+ above the gap (see Fig. 11.2), corresponds to a standing wave with nodes at the atomic planes (as shown in Fig. 11.1b). Both wave fields propagate parallel to the reflecting planes with the velocity $v_{//} \cong v \cos\theta_B$. The extent of the admixture of the Bragg reflected wave into the incident waves diminishes as the orientation of $\boldsymbol{k}_0$ moves away from the exact Bragg condition.

When an incident monochromatic neutron beam enters a perfect crystal at the Bragg condition one would expect a similar splitting of the wave function into two components to occur. However, in this case the energy of the neutron given by

$$E_0 = \hbar^2 k_0^2 / 2m \tag{11.2}$$

is conserved. The tangential component (parallel to the crystal surface) of the wave vector $\boldsymbol{k}_0$ is also conserved as required by phase matching across the boundary. However, the component of the wave vector normal to the boundary is split into two parts. That is, the consequences of the energy gap at the BZ boundary in the electron case are manifested in dynamical neutron diffraction by a splitting of the allowed spectrum of wave vectors, giving rise to two solutions of the Schrödinger equation. These two solutions are standing waves labeled ψ_α and ψ_β in Fig. 11.1b, one in phase and one out of phase with the atomic planes.

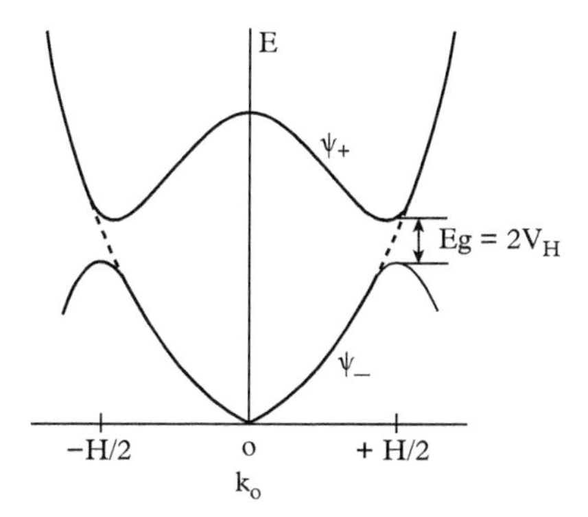

Figure 11.2 *Sketch of the dispersion surface of neutrons inside a perfect crystal*

The neutron wave field inside the crystal is a solution of the time-independent Schrödinger equation, which we write as

$$(\nabla^2 + k_0^2)\, \psi(\boldsymbol{r}) = v(\boldsymbol{r})\, \psi(\boldsymbol{r}), \tag{11.3}$$

where

$$v(\boldsymbol{r}) = \frac{2m}{\hbar^2}\, V(\boldsymbol{r}), \tag{11.4}$$

and k_0 is determined by the fixed incident energy ε_0 of the neutron. The scaled periodic potential $v(\boldsymbol{r})$ has dimensions of centimeters^{-2}. It has Fourier components $v_H{}'$ corresponding to every reciprocal lattice vector $\boldsymbol{H}'$. That is,

$$v(\mathbf{r}) = \sum_{H'} v_{H'}\, e^{i\boldsymbol{H}'\cdot\boldsymbol{r}}. \tag{11.5}$$

But it is only the Fourier components corresponding to the specific reciprocal lattice vectors $\pm\boldsymbol{H}$, normal to the Bragg reflecting planes of Fig. 11.1a, and the mean potential v_0 that have a significant influence on the reflected waves; thus, we write

$$v(\boldsymbol{r}) = v_0 + v_H e^{i\boldsymbol{H}\cdot\mathbf{r}} + v_{-H} e^{-i\boldsymbol{H}\cdot\mathbf{r}}. \tag{11.6}$$

In terms of the Fermi pseudo-potentials of the assembly of nuclei in the crystal

$$v(\boldsymbol{r}) = \sum_j b_j\, \delta(\boldsymbol{r} - \boldsymbol{r}_j). \tag{11.7}$$

Equating this expression for $v(\mathbf{r})$ to that in Eq. (11.6), multiplying by $e^{i\boldsymbol{H}\cdot\mathbf{r}}$ and integrating over a unit cell of the crystal, we find a relationship between v_H and the crystal structure factors F_H, namely

$$v_H = \frac{4\pi}{V_{\text{cell}}}\, F_H, \tag{11.8}$$

where V_{cell} is the volume of a unit cell, and the structure factor is the sum over the n atoms of a unit cell, namely

$$F_H = \sum_{j=1}^{n} b_j\, e^{i\boldsymbol{H}\cdot\mathbf{r}_j}. \tag{11.9}$$

The potential $v(\mathbf{r})$ in Eq. (11.6) couples the incident internal wave of wave vector $\boldsymbol{K}_0$ with the diffracted wave of wave vector

$$\boldsymbol{K}_H = \boldsymbol{K}_0 + \boldsymbol{H}, \tag{11.10}$$

such that wave field inside the crystal is of the form

$$\Psi(\boldsymbol{r}) = \psi_0\, e^{i\boldsymbol{K}_0\cdot\boldsymbol{r}} + \psi_H e^{i\boldsymbol{K}_H\cdot\boldsymbol{r}}. \tag{11.11}$$

The internal wave vector $\boldsymbol{K}_0$ differs slightly from the external incident wave vector $\boldsymbol{k}_0$ due to the index of refraction of the crystal medium. Substituting the expression (11.11) for $\Psi(\boldsymbol{r})$ into the Schrödinger equation (11.3), and equating coefficients of corresponding Fourier terms, we find that it is a solution only if the wave vector K_0 and the wave amplitude ψ_0 and ψ_H satisfy the following pair of coupled homogeneous equations:

$$(K^2 - K_0{}^2)\ \psi_0 - v_{-H}\psi_H = 0, \tag{11.12a}$$

and

$$-v_H\psi_0 + (K^2 - K_H{}^2)\ \psi_H = 0, \tag{11.12b}$$

where the wave vector K is an index of refraction modification of k_0, namely

$$K \equiv k_0(1 - V_0/\varepsilon_0)^{1/2}, \tag{11.13}$$

appropriate to non-Bragg reflecting conditions. Since Eqs. (11.12) are homogeneous, the determinant of the 2×2 matrix of the coefficients of ψ_0 and ψ_H must be zero, that is

$$\begin{vmatrix} K^2 - K_0{}^2 & -v_{-H} \\ -v_H & K^2 - K_H{}^2 \end{vmatrix} = 0 \tag{11.14}$$

or

$$(K^2 - K_0{}^2)\ (K^2 - K_H{}^2) = v_H v_{-H}. \tag{11.15}$$

This equation is called the *dispersion relation.* It should be regarded as an equation for the internal wave vector $\boldsymbol{K}_0$, which differs only slightly from $\mathbf{k}_0$. Since the internal wave vector for the diffracted wave, namely $\boldsymbol{K}_H$ is given by the Bragg condition, Eq. (11.10), we note that Eq. (11.15) is a quartic equation in $\boldsymbol{K}_0$. That is, strictly speaking, there are four allowed values for $\boldsymbol{K}_0$, each characterizing an eigenfunction of the Schrödinger equation, for a given $\boldsymbol{k}_0$. However, v_H is much smaller (by a factor 1 part in 10^5) than any one of the wave vectors $\boldsymbol{K}$, $\boldsymbol{K}_0$, or $\boldsymbol{K}_H$. Thus, we can write

$$K + K_0 \approx 2k_0 \quad \text{and} \quad K + K_H \approx 2k_0, \tag{11.16}$$

such that Eq. (11.15) can be written, to a very good approximation, as

$$(K - K_0)\ (K - K_H) = (v_H v_{-H})/4k_0{}^2. \tag{11.17}$$

This is a quadratic equation having two solutions for $\boldsymbol{K}_0$, which we will call $\boldsymbol{K}_0^{\alpha}$ and $\boldsymbol{K}_0^{\beta}$. Equation (11.17) can be understood graphically, as shown in Fig. 11.3, where the loci of allowed vectors $\boldsymbol{K}_0$ in $\boldsymbol{k}$-space are shown. The loci are two branches of a hyperbola, and they are called *dispersion surfaces.* The separation between the two branches of this dispersion surface is a characteristic inverse length

$$\kappa_0 = (v_H v_{-H})^{1/2}/k_0\ \cos\ \theta_B, \tag{11.18}$$

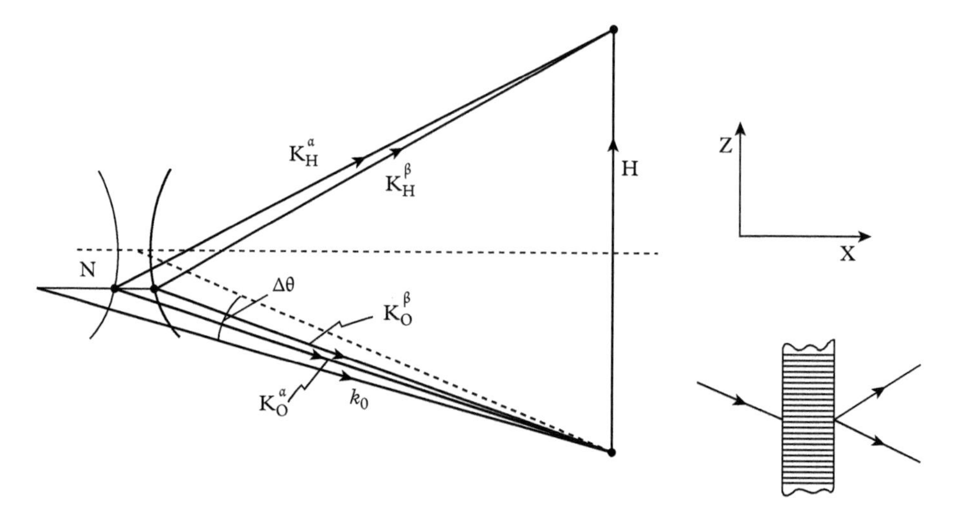

Figure 11.3 *Sketch of the dispersion surfaces and the excited wave vectors inside a perfect crystal. The part near to the dispersion hyperbola is enlarged by a factor of about 10^5 compared to the length of the wave vectors*

which is typically of order 100 cm^{-1}. It is very small compared to the wave vector $k_0 \sim 10^7$ cm^{-1}. Thus, the scale of the region of the hyperbola is greatly expanded in the drawing. Even though the splitting between the α- and β-branches of the dispersion surface is quite small, and the angle between the wave vectors, say $\boldsymbol{K}_0^\alpha$ and $\boldsymbol{K}_0^\beta$, is tiny compared to the nominal Bragg angle θ_B, this small region of **k**-space is where all the central features of the dynamical diffraction of neutrons take place.

From the above discussion we therefore conclude that a given external incident plane wave $A_0\, e^{i\mathbf{k}_0\cdot\mathbf{r}}$ generates two internal waves having wave vectors $\boldsymbol{K}_0^\alpha$ and $\boldsymbol{K}_0^\beta$, nearly collinear and equal to $\mathbf{k}_0$, but which differs from $\boldsymbol{k}_0$ by small components $\mathbf{N}_0^\alpha$ and $\mathbf{N}_0^\beta$ normal to the crystal surface. By Bragg's law, as given in vector form in Eq. (11.10), these two internal incident waves generate two diffracted waves having wave vectors $\boldsymbol{K}_H^\alpha$ and $\boldsymbol{K}_H^\beta$. Thus, we see that the total wave field inside the crystal consists of a coherent superposition of four plane waves, namely

$$\Psi(\boldsymbol{r}) = \psi_0^\alpha\, e^{i\boldsymbol{K}_0^\alpha\cdot\boldsymbol{r}} + \psi_0^\beta\, e^{i\boldsymbol{K}_0^\beta\cdot\boldsymbol{r}} + \psi_H^\alpha\, e^{i\boldsymbol{K}_H^\alpha\cdot\boldsymbol{r}} + \psi_H^\beta\, e^{i\boldsymbol{K}_H^\beta\cdot\boldsymbol{r}}. \tag{11.19}$$

Since the wave vectors $\boldsymbol{K}_0^\alpha$ and $\boldsymbol{K}_0^\beta$ belonging to the α- and β-branches of the dispersion surface differ by such small amounts, the neutron wave field will exhibit interesting interference "beats" on a macroscopic scale (∼ 100 μm), called Pendellösung interference fringes, which we will describe later. The two points, *a* and *b*, on the α- and β-branches of the dispersion surface, respectively, are picked out by the orientation of the external incident wave vector $\mathbf{k}_0$ with respect to the crystal lattice planes. These *tie points* move along the hyperbolas as the crystal is rotated through small angles Δθ away from the exact Bragg condition where the tie points are the diameter points of the hyperbolas. That is, a given incident $\mathbf{k}_0$ excites only two tie points. For a given orientation of $\mathbf{k}_0$

with respect to the reciprocal lattice vector $\boldsymbol{H}$, the ratio of the diffracted wave amplitudes ψ_H^α and ψ_H^β to the incident wave amplitudes ψ_0^α and ψ_0^β are fixed, and determined by Eqs. (11.12). The magnitudes of these amplitudes relative to the incident plane wave amplitude A_0 are adjusted to satisfy continuity of the total wave function across the entrant boundary. For the incident wave amplitudes we must have

$$\psi_0^\alpha + \psi_0^\beta = A_0. \tag{11.20a}$$

Since there is no external wave in the diffraction wave vector direction on the entrant surface, we must have

$$\psi_H^\alpha + \psi_H^\beta = 0. \tag{11.20b}$$

That is, the wave field corresponding to diffraction is zero very near the entrant surface, while taking a finite crystal depth for this part of $\Psi(\mathbf{r})$ to develop and become finite. It is convenient to define the small wave vector

$$v_H \equiv \frac{v_H}{2k_0}, \tag{11.21}$$

and write Eqs. (11.12) in matrix form

$$\begin{bmatrix} (K - K_0) & -\nu_{-H} \\ -\nu_H & (K - K_H) \end{bmatrix} \begin{pmatrix} \psi_0 \\ \psi_H \end{pmatrix} = 0. \tag{11.22}$$

For a crystal, like Si, that has essentially no absorption $\nu_H* = \nu_{-H}$. That is, the wave vectors are all real (as we have implicitly assumed above) and the ratios of the wave amplitudes are seen to be given by

$$\frac{\psi_H{}^{\alpha,\beta}}{\psi_0{}^{\alpha,\beta}} = \frac{K - K_0{}^{\alpha,\beta}}{\nu_{-H}} = \frac{\nu_H}{K - K_H{}^{\alpha,\beta}}. \tag{11.23}$$

The dispersion relation, with the definition of ν_H given above, is

$$(K - K_0{}^{\alpha,\beta})\,(K - K_H{}^{\alpha,\beta}) = |\nu_H|^2. \tag{11.24}$$

Once the tie points are selected by a given incident $\mathbf{k}_0$, the differences $(K–K_0{}^{\alpha,\beta})$ and $(K–K_H{}^{\alpha,\beta})$ are known. It is a straightforward matter to evaluate these differences from the geometry of Fig. 11.3. The results are

$$K - K_0^\alpha = |\nu_H|\left(-y - \sqrt{y^2 + 1}\right), \tag{11.25a}$$

$$K - K_0^\beta = |\nu_H|\left(-y + \sqrt{y^2 + 1}\right), \tag{11.25b}$$

$$K - K_H^\alpha = |\nu_H|\left(y - \sqrt{y^2 + 1}\right), \tag{11.25c}$$

and

$$K - K_H^\beta = |\nu_H|\left(y + \sqrt{y^2 + 1}\right), \tag{11.25d}$$

where we have defined the so-called "y-parameter," which is a scaled misset angle $\Delta\theta$, namely

$$y \equiv \Delta\theta\ (k_0 \sin 2\theta_B)/2\ |\nu_H|. \tag{11.26}$$

Using Eqs. (11.25) in the expression for the two amplitude ratios given in Eq. (11.23), and the two equations (11.20a) and (11.20b), satisfying the continuity of wave function boundary condition, we have four equations for the four unknown wave amplitudes $\psi_0{}^\alpha$, $\psi_0{}^\beta$, $\psi_H{}^\alpha$, and $\psi_H{}^\beta$. The solution is straightforward, and the results are

$$\psi_0^\alpha = \frac{1}{2}\left[1 - y(1+y^2)^{-1/2}\right] A_0, \tag{11.27a}$$

$$\psi_0^\beta = \frac{1}{2}\left[1 + y(1+y^2)^{-1/2}\right] A_0, \tag{11.27b}$$

$$\psi_H^\alpha = -\frac{1}{2}\ (1+y^2)^{-1/2}\left(\frac{\nu_H}{\nu_{-H}}\right)^{1/2} A_0, \tag{11.27c}$$

$$\psi_H^\beta = +\frac{1}{2}\ (1+y^2)^{-1/2}\left(\frac{\nu_H}{\nu_{-H}}\right)^{1/2} A_0. \tag{11.27d}$$

Furthermore, the wave vector $\mathbf{N}^{\alpha,\beta}$, giving the difference between $\mathbf{k}_0$ and $\mathbf{K}_0^{\alpha,\beta}$, namely

$$\mathbf{N}^{\alpha,\beta} = \boldsymbol{k}_0 - \mathbf{K}_0^{\alpha,\beta}, \tag{11.28}$$

can now also be written down explicitly in terms of the misset angle $\Delta\theta$, or more conveniently in terms of our dimensionless y-parameter. The result is

$$N^{\alpha,\beta} = \left\{\frac{\nu_0}{\cos\theta_B} + \frac{|\nu_H|}{\cos\theta_B}\left[-y \pm (1+y^2)^{1/2}\right]\right\}\hat{n}, \tag{11.29}$$

where $\hat{n}$ is the inward surface normal unit vector at the entrant surface. The negative sign applies to the α-branch and the positive sign applies to the β-branch of the dispersion surface, as shown in Fig. 11.3. The phase variation near a Bragg position has been calculated in more detail by Lemmel (2007) and is shown in his figures.

We now have a complete mathematical description of the wave fields inside the crystal. Our next task is to extend these solutions to the free space beyond the back face of the crystal. In free space, the wave vectors must again be of magnitudes $|\boldsymbol{k}_0|$ so that the energy is $E_0 = \hbar^2 k_0{}^2/2m$. The wave function leaving the crystal will consist of two parts, which we write as

$$\chi(\boldsymbol{r}) = \chi_0 e^{i\boldsymbol{k}_0\cdot\boldsymbol{r}} + \chi_H\, e^{i\boldsymbol{k}_H\cdot\boldsymbol{r}}, \tag{11.30}$$

where $|\boldsymbol{k}_H| = |\boldsymbol{k}_0|$. From the geometry of Fig. 11.3, we find that

$$\mathbf{k}_H = \mathbf{k}_0 + \boldsymbol{H} + \boldsymbol{\delta}, \tag{11.31}$$

where

$$\boldsymbol{\delta} = 2k_0\Delta\theta\ \sin\theta_B\,\hat{n}. \tag{11.32}$$

This additional small wave vector allows the neutron energy to be conserved for misset angles slightly off the exact Bragg condition. For a divergent beam it cancels out. The amplitudes χ_0 and χ_H of the outgoing waves are obtained by equating $\chi(\boldsymbol{r})$ as given by Eq. (11.30) to the internal wave field $\Psi(\boldsymbol{r})$. This is a relatively easy task to carry out explicitly. The results are

$$\chi_0(y) \equiv T(y) \cdot A_0 = \left[\cos\Phi - \mathrm{i}\, y(1+y^2)^{-1/2} \sin\Phi\right] e^{\mathrm{i}(\phi_1-\phi_0)} \cdot A_0 \tag{11.33}$$

and

$$\chi_H(y) \equiv R(y) \cdot A_0 = -\mathrm{i}\left(\frac{v_H}{v_{-H}}\right)^{-1/2} (1+y^2)^{-1/2} \sin\Phi\, e^{-\mathrm{i}(\phi_1+\phi_0)} \cdot A_0, \tag{11.34}$$

where the three phases φ_0, φ_1, and Φ are defined by

$$\phi_0 = v_0 D/\cos\theta_\mathrm{B} \tag{11.35a}$$

$$\phi_1 = y\,|v_H|\,D/\cos\theta_\mathrm{B} \tag{11.35b}$$

and

$$\Phi = |v_H|\ (1+y^2)^{1/2}\, D/\cos\theta_\mathrm{B}. \tag{11.35c}$$

Thus, we have explicit expressions for the transmission and reflection coefficients, $T(y) = \chi_0(y)/A_0$ and $R(y) = \chi_H(y)/A_0$, respectively, for a crystal of thickness D. The intensity of the diffracted beam is

$$I_H = |\chi_H|^2 = |A_0|^2 \sin^2\Phi/(1+y^2). \tag{11.36}$$

The intensity of the forward-diffracted beam is

$$I_0 = |\chi_0|^2 = |A_0|^2 \left(\cos^2\Phi + y^2(1+y^2)^{-1} \sin^2\Phi\right), \tag{11.37}$$

such that

$$|\chi_H|^2 + |\chi_0|^2 = |A_0|^2, \tag{11.38}$$

which is the necessary normalization required by conservation of neutrons passing through a non-absorbing medium.

The diffracted beam intensity I_H is a Lorentzian shaped peak, with rapid oscillations due to the $\sin^2\Phi$ factor. A plot of this peak on the y-scale along with the corresponding forward-diffracted intensity is shown in Fig. 11.4. The rapid oscillations are due to the interference between the α- and β-branch wave functions which have slightly different wave vectors, K_H^{α} and K_h^{β}. The oscillation frequency is determined by the parameter $v_H D/\cos\theta_\mathrm{B} = \pi D/\Delta_H$, where Δ_H is the Pendellösung period to be discussed in Section 11.4. Averaging over these Pendellösung interference fringes, one has

$$< I_H(y) > = (1/2)\,|A_0|^2 (1+y^2)^{-1}, \tag{11.39a}$$

and

$$< I_0(y) > = (1/2)|A_0|^2 \left[1 + y^2(1+y^2)^{-1}\right]. \tag{11.39b}$$

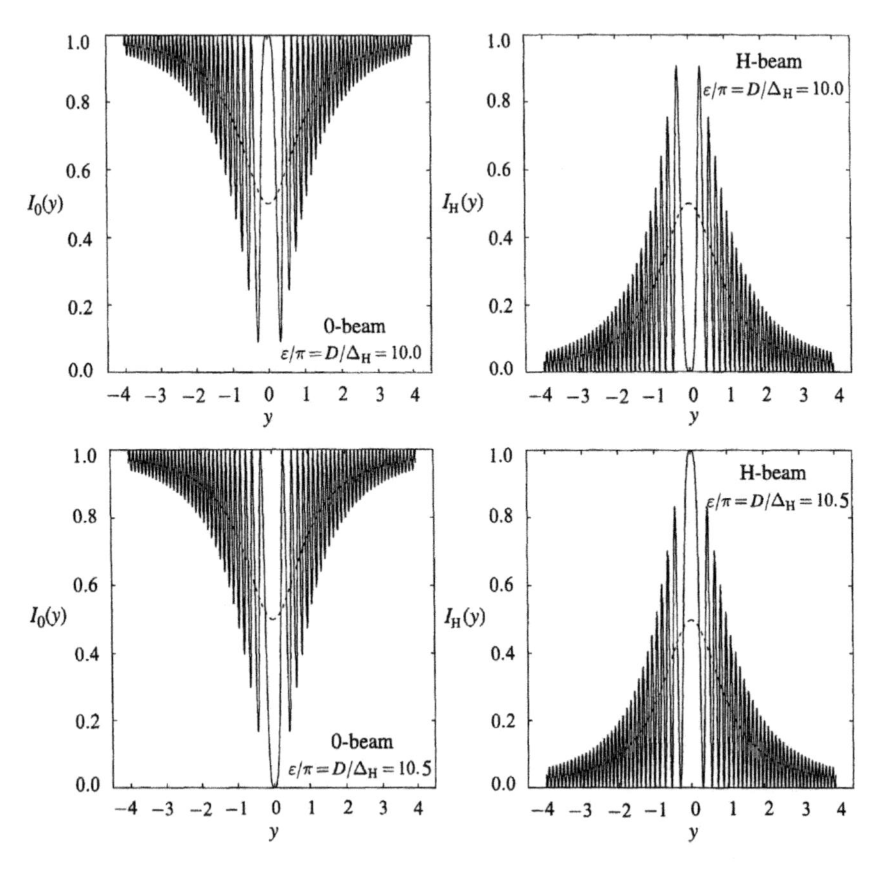

Figure 11.4 *Intensity profiles of the forward (0) and deflected (H) beam behind a perfect crystal slab for two different thicknesses. The dashed lines indicate the averaged intensity profiles*

The ratios of wave amplitudes χ_0 and χ_H to the incident plane wave amplitude A_0 given in Eqs. (11.33) and (11.34) are the transmission, T, and reflection, R, coefficients that we will need to describe the plane wave theory of the three-crystal LLL neutron interferometer. For absorbing crystals, the scattering amplitudes b are complex, as described in Chapters 1 and 3. Thus, the parameters ν_G characterizing the dynamical diffraction will also be complex. This creates no additional mathematical complication, and numerical evaluation of I_0 and I_H is straightforward.

11.3 Anomalous transmission, Angle Amplification, and High Collimation Effects

At the exact Bragg condition ($\Delta\theta = 0$, and the y-parameter $= 0$) the wave fields corresponding to the α- and β-branches are

$$\begin{aligned}\Psi_\alpha(\boldsymbol{r}) &= \psi_0^\alpha\, e^{i\boldsymbol{K}_0^\alpha\cdot\boldsymbol{r}} + \psi_H^\alpha\, e^{i\boldsymbol{K}_H{}^\alpha\cdot\boldsymbol{r}} \\ &= -A_0\, e^{i(K\cos\theta_B+\kappa_H/2)x}\sin(Hz/2)\end{aligned} \tag{11.40a}$$

and

$$\begin{aligned}\Psi_\beta(\boldsymbol{r}) &= \psi_0^\beta\, e^{\mathrm{i}\boldsymbol{K}_0^\beta\cdot\boldsymbol{r}} + \psi_H^\beta\, e^{\mathrm{i}\boldsymbol{K}_H^\beta\cdot\boldsymbol{r}} \\ &= -A_0\, e^{\mathrm{i}(K\cos\theta_B+\kappa_H/2)x}\cos(Hz/2).\end{aligned} \tag{11.40b}$$

These are travelling waves in the x-direction and standing waves in the z-direction. The density for the α-branch

$$|\Psi_\alpha|^2 = |A_0|^2 \sin^2\left(\frac{Hz}{2}\right) = |A_0|^2 \sin^2(\pi z/d) \tag{11.41a}$$

has nodes at the atomic planes, spaced a distance d apart, while the density for the β-branch

$$\left|\Psi_\beta\right|^2 = |A_0|^2 \cos^2\left(\frac{Hz}{2}\right) = |A_0|^2 \cos^2(\pi z/d) \tag{11.41b}$$

has nodes between the atomic planes as shown schematically in Fig. 11.1. (The spatial variables x and z here should not be confused with the y-parameter of the previous section.)

The phases of the wave functions (Eqs. 11.39 and 11.40) are implicitly been set equal to 0 for $z=0$, which corresponds to an atomic plane. This is the correct choice because it is only the (single) sinusoidal part of the periodic potential $V(\boldsymbol{r})$, namely

$$V_H\, e^{\mathrm{i}\boldsymbol{H}\cdot\boldsymbol{r}} + V_{-H}\, e^{-\mathrm{i}\boldsymbol{H}\cdot\boldsymbol{r}} = 2\,|V_H|\cos Hz, \tag{11.42}$$

that gives rise to the particular Bragg reflection on which we are concentrating our attention. This perfect registry of the neutron standing waves with the atomic planes of the lattice leads to the anomalous transmission, or Borrmann effect, that is familiar in X-ray dynamical diffraction. Since the α-branch neutron density is zero at the atomic planes ($z_n = nd$), the absorption of neutrons by the nuclei will be small for the neutron current carried by this part of the wave function, while absorption for the β-branch current will be enhanced. This effect was first observed by Knowles (1959) for the neutron case. He showed that the intensity of neutron-capture γ-rays varied in a particular way as a nearly perfect $CdSO_4$ crystal was rotated through the Bragg condition. Anomalous transmission effects have been observed in InSb crystals by Sippel et al. (1962) and Shilstein (1971). Of course, for Si the absorption cross-section, σ_a, is essentially zero, and anomalous transmission effects are not observable.

Using the four-component wave function of Eq. (11.19), we can calculate the current density $\mathcal{J}$ according to the quantum mechanical rule

$$\mathcal{J} = (\mathrm{i}\hbar/2m)\ [\Psi\nabla\Psi^* - \Psi^*\nabla\Psi]. \tag{11.43}$$

There are many interference terms when this is written out in detail. Some of these terms oscillate on the length scale of the unit cell, while others have a period of 10^5 to 10^6 unit cells. We are interested in these last terms which give rise to the Pendellösung interference effects, to be discussed later. We average the current density over the unit cell to eliminate the unobservable rapid oscillations. This averaged current density has three terms:

$$<\mathbf{J}> \;=\; <\mathbf{J}_\alpha> + <\mathbf{J}_\beta> + <\mathbf{J}_{\alpha\beta}>. \tag{11.44}$$

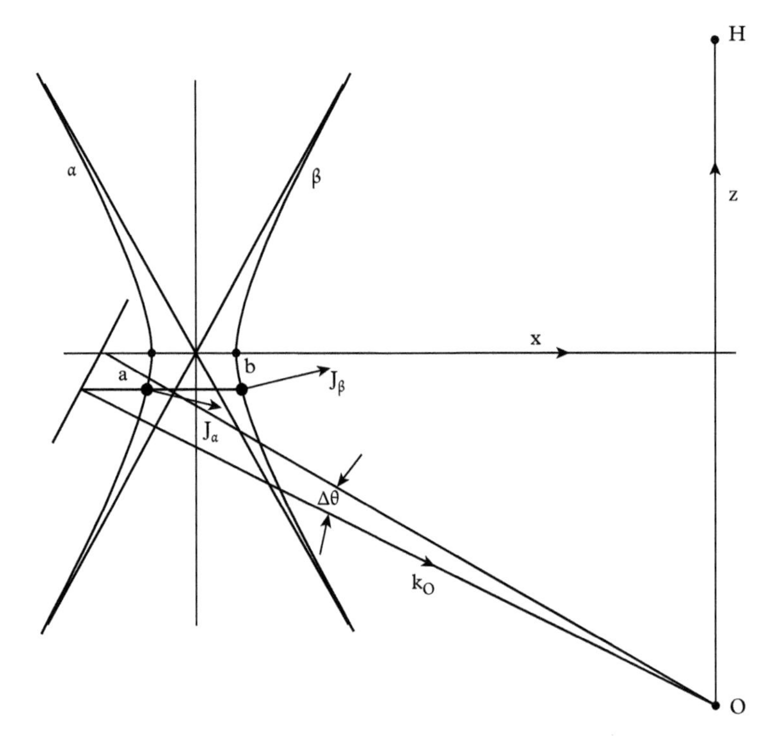

Figure 11.5 *Directions of the neutron current inside a perfect crystal*

The first term here is the current density carried out by the α-branch part of the wave function, namely

$$\Psi_\alpha(\mathbf{r}) = \psi_0^\alpha e^{i\mathbf{K}_0^\alpha \cdot \mathbf{r}} + \psi_H^\alpha e^{i\mathbf{K}_H^\alpha \cdot \mathbf{r}}. \tag{11.45}$$

The second term in Eq. (11.44) is the current density carried by the β-branch part of $\Psi(r)$, and the third term arises from interference effects between the α- and β-branch wave functions. It is an interesting and straightforward mathematical exercise to show that $<\mathbf{J}_\alpha>$ is normal to the α- branch dispersion surface and $<\mathbf{J}_\beta>$ is normal to the β-branch dispersion surface at the tie points *a* and *b* picked out by the orientation of the incident wave vector $\mathbf{k}_0$, as shown in Fig. 11.5. The last term $<\mathbf{J}_{\alpha\beta}>$ is directed along $\boldsymbol{H} = H\hat{z}$, being positive for certain depths x below the entrant surface and negative for other depths. The result is that the total current density oscillates between the incident beam direction and the diffracted beam direction as a function of the depth x in the crystal.

As the misset angle $\Delta\theta$ is changed from positive to negative values over a narrow range of only an arc second or so, the directions of the currents $<\mathbf{J}_\alpha>$ and $<\mathbf{J}_\beta>$ sweep over the angular range between $\hat{S}_0$ and $\hat{S}_H$, encompassing the entire *Borrmann triangle* ABC of Fig. 11.6. Consequently, it is seen that there is an enormous *angle amplification* effect. For a small misset angle $\Delta\theta$ of $\mathbf{k}_0$ away from the exact Bragg condition, the current $<\mathbf{J}_\alpha>$ will propagate across the crystal at an angle Ω with respect to the lattice planes while the current $<\mathbf{J}_\beta>$ will propagate across the crystal at an angle $-\Omega$. It is a matter of geometry to relate Ω to $\Delta\theta$, given the fact that $<J_\alpha>$ and $<J_\beta>$

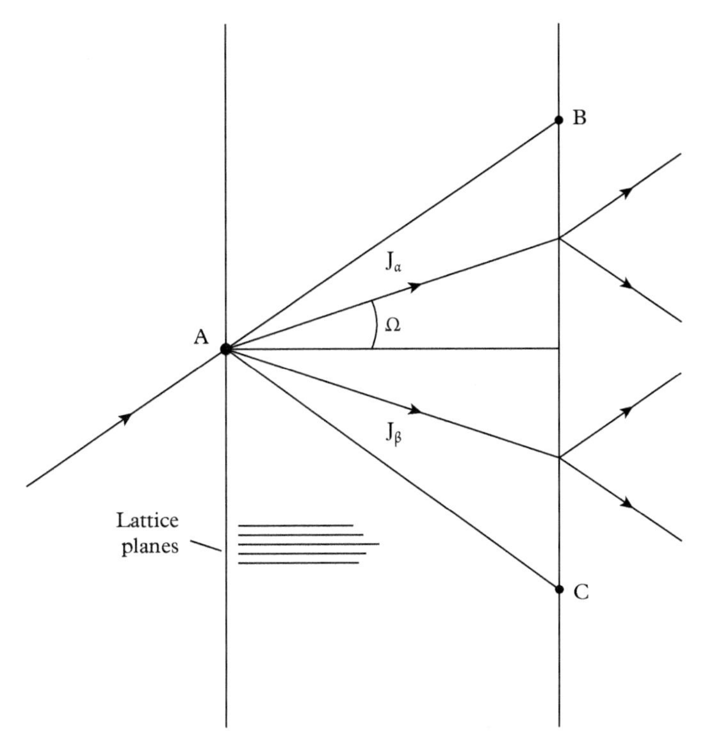

Figure 11.6 *Sketch of the Borrmann fan where waves become excited from point A when a slightly divergent beam (arc sec) enters the perfect crystal*

are normal to the dispersion surfaces at the tie points a and b. It is convenient to define

$$\Gamma \equiv \tan \Omega / \tan \theta_B \tag{11.46}$$

to characterize this effect. Using the geometry of Fig. 11.5 and the equation of the dispersion surface, Eq. (11.17), one finds the relationship between $\Delta\theta$ and Ω to be

$$\Delta\theta = \pm \frac{|V_H|}{E_0 \sin 2\theta_B} \left[\frac{\Gamma}{(1 - \Gamma^2)^{1/2}} \right], \tag{11.47}$$

where V_H is the Fourier component of the neutron–nuclear interaction potential energy corresponding to the reciprocal lattice vector, $\boldsymbol{H}$, and E_0 is the neutron's incident kinetic energy. Using Eq. (11.8) for $\boldsymbol{H} = (220)$ in Si, and taking $\varepsilon_0 = 20$ meV, we find that $|V_H/E_0| = 2.7 \times 10^{-7}$. This ratio sets the scale of the angle amplification. For small Ω, we can invert Eq. (11.48) to write Ω in terms of $\Delta\theta$, namely

$$\Omega = \left(\frac{2E_0 \sin^2 \theta_B}{|V_H|} \right) \Delta\theta. \tag{11.48}$$

For the (220) reflection in Si for $\lambda = 2$ Å ($E_0 = 20$ meV), $\theta_B \approx 31.6°$, which means that the angle amplification factor $(2E_0 \sin^2\theta_B / |V_H|)$ in this equation is 2×10^6.

Kikuta et al. (1975) exploited this angle amplification effect to measure the small directional changes of a neutron beam resulting from refraction by a prism, as shown in Fig. 11.7a. For thermal neutrons, angles of refraction are typically of order of 1 arc sec. Absorbing slits, one on the front face of a thick Si crystal and one on the back face, act as a *crystal collimator*, selecting only those neutrons which travel across the crystal parallel to the lattice planes. After leaving the first crystal, these neutrons enter a second thick crystal which is a part of a monolithic 2-crystal rigid structure. If it were not for the deflection of the wedge placed in the open space between the two crystals, the neutrons leaving the first crystal would travel straight across the second crystal parallel to the lattice planes. However, since the neutrons are deflected by the wedge, their paths split in the second crystal according to Eq. (11.48). The results of Kikuta's experiment are shown in Figs. 11.7b and 11.7c where the splitting in the second crystal is observed by scanning a slit across the back face of the second crystal.

This same technique was used by Zeilinger and Shull (1979) in a measurement of the longitudinal Zeeman effect. The related change of the neutron wavelength in the magnetic field is given by $\delta\lambda/\lambda = \pm\mu B/2E_0$. The results of this experiment are given in Fig. 11.8. The birefringence of a region of magnetic field for neutron waves is nicely displayed by these results. The fact that the energy shift due to the magnetic field was only 10^{-8} eV demonstrates the sensitivity of the angle amplification effect in perfect Si crystals. A comparison of Kikuta's result and that of Zeilinger and Shull also demonstrates the equivalence of an angular deflection and an energy change, because

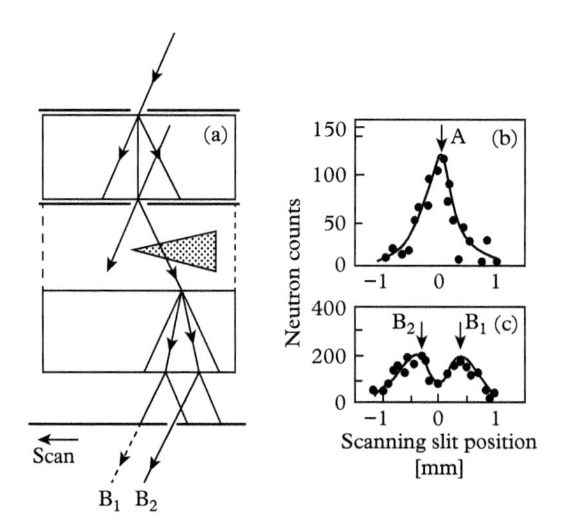

Figure 11.7 *Very small angle prism deflection between two successive Laue reflections. Reprinted with permission from Kikuta et al. 1975, copyright by Physical Society of Japan.*

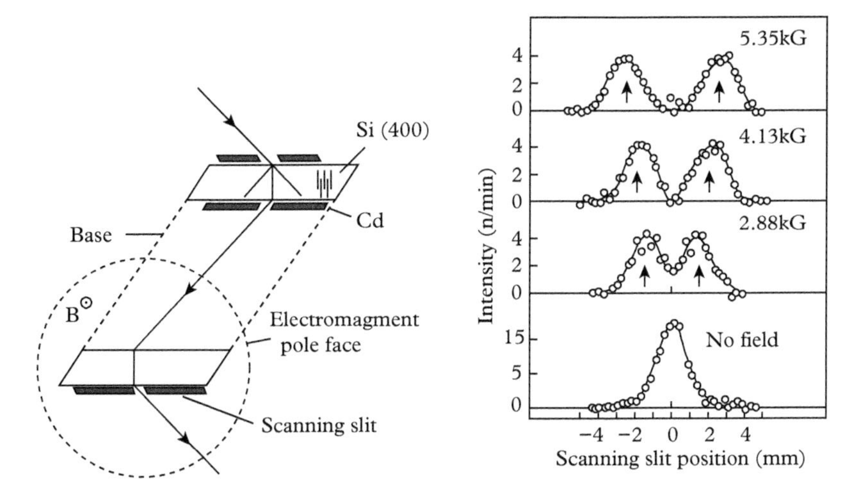

Figure 11.8 *Zeeman energy shift measurement by means of a double-Laue-reflection camera. Reprinted with permission from de Zeilinger and Shull 1979b, copyright 1979 by the American Physical Society.*

dynamical diffraction effects are sensitive to $\Delta\theta = \theta - \theta_B$ only. Deflection effects change θ, whereas an energy change varies θ_B. When inhomogeneous magnetic fields are applied, a force is acting on the neutron and a strong bending of the trajectories occur (Zeilinger et al. 1986, Raum et al. 1995). This can be described by an effective mass formalism as shown in Section 8.2.3.

Such systems have also been used for precise measurements of gravitational effects acting on the particle between and inside the crystal plates (Raum et al. 1995; Section 8.2.3, Fig. 8.16). According to a proposal of Summhammer (1996), a direct measurement of the multiphoton exchange in oscillating magnetic fields placed between the plates becomes feasible down to frequencies of $\nu_{hf} \sim 100$ kHz.

Ray optimal considerations and dynamical diffraction theory show various spatial and angular focusing effects for various perfect crystal arrangements (Indenbohm et al. 1974, 1976; Petrascheck 1976, 1986; Bonse and Graeff 1977; Zeilinger et al. 1979). In a spatial scan behind a double-crystal LL arrangement a marked peak in the center of the Borrmann fan is expected and has been observed experimentally (Zeilinger et al. 1979). The observation of spatial diffraction focusing effects requires a narrow entrance slit and a narrow scanning slit behind the interferometer (or a position-sensitive detector). This spatial focusing is analogous to the central peak phenomenon discussed in the previous section, where the narrow peak in the angular distribution has been used. In both cases, a rather broad "background" distribution exists also. The central peak follows from a coherent overlap of different wave components and inevitably a correct description needs a detailed plane (or better spherical) wave diffraction theory, which shows that the intensity in the focal area arises from neutrons that traveled in the α-branch wave field in the first crystal and in the β-branch wave field in the second one (or vice versa), which compensates for the overall phase shift. The diffraction focusing profile is obtained by the method of the stationary phase. In the Pendellösung-averaged and small absorption case (Petrascheck 1986) the forward (0-beam) intensity for crystals of thickness t is given by

$$I = \frac{A\pi}{8}\left[e^{-2A|\Gamma|} + \frac{1}{2\pi A}\sqrt{1-\frac{\Gamma^2}{4}}\,\Theta\left(1+\frac{|\Gamma|}{2}\right)\right], \tag{11.49}$$

where $\Gamma = x/t \tan\theta_B$ and Θ denotes the step function. The diffraction spot in the center of the Borrmann fan ($x = 0$) becomes very small, of order of a few microns but is usually broadened by the width of the entrance slit. Additional effects are expected for very narrow entrance slits that are on the order of the diffraction spot, as it is known from the X-ray case (Indenbohm et al. 1976, Aladzhadzhyan et al. 1977).

The neutrons contributing to the broad background can be partly (50%) focused to the focal spot by a properly shaped elliptical lens or two cylindrical lenses put into the beam behind the first and second crystal plates (Zeilinger and Horne 1986). However, in this case, the focusing condition for the central peak gets lost.

Neutrons crossing the crystal near to $\Omega \cong 0$ (Figs. 11.7 and 11.8) are transported through the crystal by multiple zigzag reflections between the reflecting planes. When the slits are wide open, this effect still persists but other beam paths contribute too. Thus, instead of measuring beam deflection in ordinary space, one can do that in momentum space by measuring multiple Laue-rocking curves. These are given by the convolution of Laue-reflection curves (Fig. 11.4) and show a very narrow central peak with a half-width given by the lattice constant and the thickness of the crystals, i.e., $\Delta\Theta \cong d_{hkl}/t$, which is on the order of 0.001 are sec (Bonse et al. 1977). This can be used for highly sensitive angular measurements.

The first measurements have been done using a monolithic two-plate arrangement, where the half-width of the narrow central peak was determined to be 0.007 arc sec (Bonse et al. 1979). In these experiments, the parallelism of the lattice planes is guaranteed by the monolithic design of the crystal and the high angular sensitivity is achieved by rotation of a wedge-shaped material around the beam axis. The beam deflection δ is controlled by the index of refraction n, the wedge angle β of the material, and the rotation angle α around the beam axis, namely

$$\delta = \left[2(1-n)\, tg\frac{\beta}{2}\right] \sin\alpha. \tag{11.50}$$

Figure 11.9 depicts the experimental setup and typical results for a triple Laue-diffraction arrangement, where the needle structure is even more pronounced than in the two-plate case. The figure shows the broadening of the central peak due to diffraction effects at a macroscopic slit having a width of $D = 2.5$ mm (Rauch et al. 1983). These multiple Laue-rocking curves have also been calculated analytically (Petrascheck and Rauch 1984b). The narrow central peak, which is to a first approximation independent from the wavelength spread and other uncertainties, reads for the two- and three-plate case as

$$\begin{aligned} I_{c.p}^{(2)} &\cong \frac{\pi}{8}\frac{\mathcal{J}_1(2Ay)}{2Ay} \\ I_{c.p}^{(3)} &\cong \frac{3\pi}{16}\left[2\frac{\mathcal{J}_2(4Ay)}{(2Ay)^2} + \frac{\mathcal{J}_2(4Ay)}{(4Ay)^2}\right] \end{aligned} \tag{11.51}$$

For a symmetric Laue reflection for Si, it has a full half-width of $y_H^{(2)} = 2.215/A$ and $y_H^{(3)} = 2.10/A$, respectively (where $A = Nb_c\lambda t/\cos\theta_B$). These quantities, on the angular scale, are $\delta\theta^{(2)} = 0.7 d_{hkl}/t$ and $\delta\theta^{(3)} = 0.67 d_{hkl}/t$, respectively. Since the half-width can be measured up to an accuracy of about 1%, one recognizes that this technique reaches an angular sensitivity up to about 2.5×10^{-5} arc sec, which may have applications for fundamental physics applications. The capability of this method for lowering the upper limit for an electric charge of the neutron deserves attention. In the X-ray domain very precise values for the structure factor of Si were

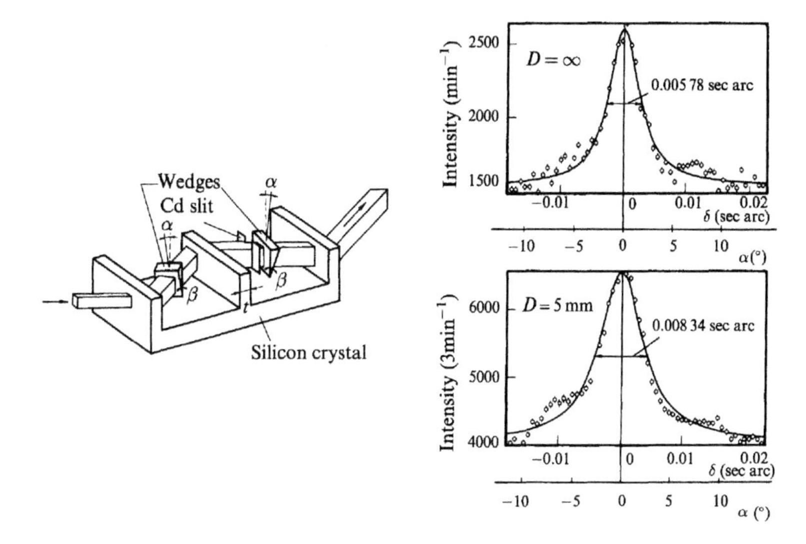

Figure 11.9 *Broadening of the central peak appearing in multiple Laue-diffractions due to single-slit diffraction of neutrons with a wavelength of 2 Å from a slit with a width of 5 mm (Rauch et al. 1983a)*

obtained with accuracies better than 10^{-3} from Laue-rocking curves of a polylithic double-crystal system (Teworte and Bonse 1984, Deutsch and Hart 1985).

The influence of the aperture and of different slit geometries on various diffraction focusing effect geometries was treated in detail by Teworte and Bonse (1987). They gave analytical expressions for situations where the slit is placed before or between the perfect crystal plates and they found good agreement with X-ray measurements where not only the broadening was observed but also a shift of the peak position. The slit diffraction effect also causes the contrast of the interference pattern to reduce when the slit width is reduced.

As mentioned earlier the neutrons crossing the crystal near a Bragg diffraction position experience *zigzag* reflections between the reflecting planes. This causes a time delay in comparison with a free evolution. The time spent within a crystal of thickness L becomes, according to Eq. (11.1a),

$$\tau_L = \frac{L}{v\cos\Theta_B} = \frac{d_{h,k,l}m}{\hbar\pi} L\tan\Theta_B. \tag{11.52}$$

This shows that the delay time increases strongly when the Bragg angle approaches 90°. Related measurements have been reported by Voronin et al. (2000). They used a time-of-flight technique and a glancing incident neutron beam and rotated a perfect crystal around a vertical axis (Fig. 11.10). The results show good agreement between theory and experiment (Fig. 11.11).

11.4 Pendellösung Interference Effects

Consider an experiment where the incident beam is limited by a slit which is narrow in comparison to the thickness D of the crystal. Then, for a given orientation $\Delta\theta$ of the incident wave vector $\boldsymbol{k}_0$, the wave fields $\Psi_\alpha(\mathbf{r})$ and $\Psi_\beta(\mathbf{r})$will only overlap in a small region of the crystal immediately

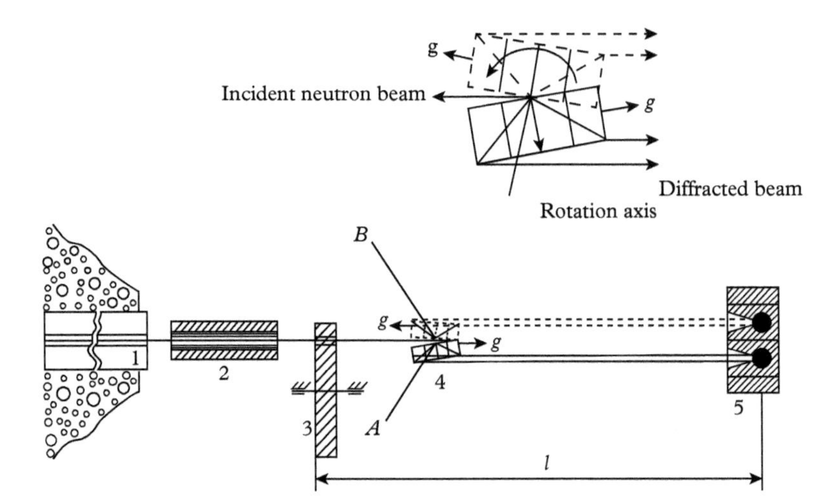

Figure 11.10 *Experimental arrangement to measure the delay time a neutron spends within a perfect quartz crystal near to a Bragg diffraction position. Voronin et al. 2000, with kind permission from Springer Science and Business Media.*

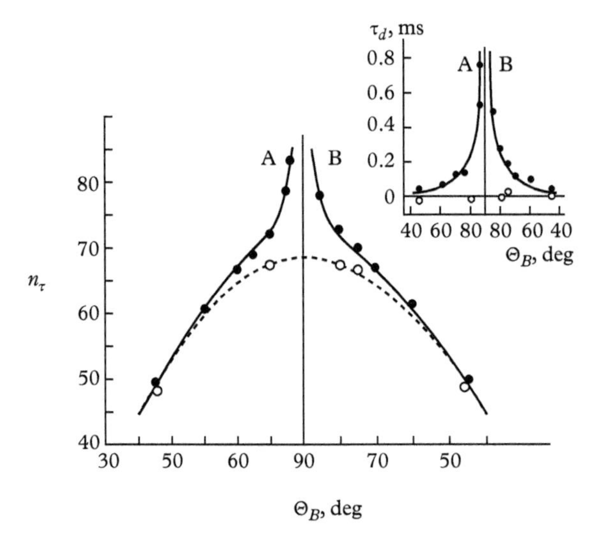

Figure 11.11 *Dependence of the time-of-flight of neutrons diffracted along the direct beam as a function of the Bragg angle. Voronin et al. 2000, with kind permission from Springer Science and Business Media.*

adjacent to the slit since the current densities $<J_\alpha>$ and $<J_\beta>$ propagate separately across the crystal at angles $\pm\Omega$ to the lattice planes. However, at the exact Bragg angle, $\Delta\theta = 0$, the current densities $<J_\alpha>$ and $<J_\beta>$ are both along the normal to the entrant surface of the crystal. Consequently, Ψ_α and Ψ_β overlap and interfere with each other across the entire thickness of the crystal. The total current density <J> will oscillate between the incident ($\hat{S}_0$) and diffracted ($\hat{S}_H$) wave directions with a period Δ_H corresponding to the difference in wave vectors

$$\left|K_0^\alpha - K_0^\beta\right| = \left|K_H^\alpha - \mathbf{K}_H^\beta\right| = \kappa_H \tag{11.53}$$

given by Eq. (11.18). Thus, the period of these Pendellösung oscillations is

$$\Delta_H = \frac{2\pi}{\kappa_H} = \frac{2\pi k_0 \cos\theta_B}{|v_H|} = \frac{\pi V_{\text{cell}} \cos\theta_B}{\lambda |F_H|}, \tag{11.54}$$

where we have used Eq. (11.8) to replace v_H with the structure factor F_H. For thermal neutrons in Si, Δ_H is typically 10^5 to 10^6 Å. At the exit face of the crystal, $x = D$, the neutron wave

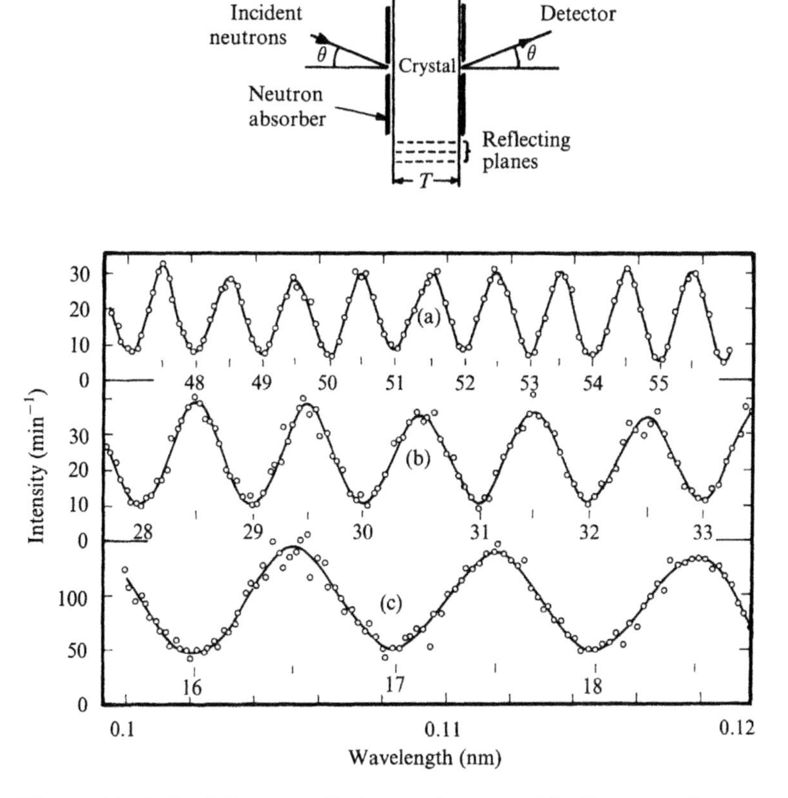

Figure 11.12 *Pendellösung oscillations at the center of the Borrmann fan as a function of the wavelength for three different crystal thicknesses: (a)* D = *1 cm, (b)* D = *0.5939 cm, and (c)* D = *0.3315 mm. Reprinted with permission from Shull 1968, copyright 1968 by the American Physical Society.*

function splits up into two waves, a single wave in the incident direction (the 0-beam) and a single diffracted wave (the H-beam). The relative intensities of these two beams depend upon the phase of the interference current densities $<J_{\alpha\beta}>$ at $x = D$. As the crystal thickness is varied, the neutron current leaving the crystal will first appear mainly in the 0-beam for very thin crystals, and then entirely in the H-beam at $x = \Delta_H/2$, and subsequently it will oscillate back and forth between these two beams with a period $\Delta D = \Delta_H$ as the crystal thickness is increased. Continuously varying the crystal thickness is not simple from an experimental point of view. Alternatively, since the period Δ_H depends upon the neutron wavelength λ, these Pendellösung oscillations can be observed experimentally by varying λ for a fixed crystal thickness. Such an experiment was carried out by Shull (1968, 1973), and the results are shown in Fig. 11.12. The selection of only those neutrons satisfying the exact Bragg condition, and therefore propagating directly across the crystal, was done by placing a slit on the exit face of the crystal. The diffracted current density oscillates according to the formula

$$I_H = I_0\left[1 - \cos\left(2\pi\frac{D}{\Delta_H}\right)\right] = I_0\left[1 - \cos\left(\frac{2DF_H\lambda}{V_{\text{cell}}\cos\theta_{\text{B}}}\right)\right]. \tag{11.55}$$

The period of the oscillations is predicted to be inversely proportional to the crystal thickness as observed experimentally and shown in Fig. 11.12. It is seen that these Pendellösung oscillations provide a very precise value for the structure factor F_H and thereby the scattering length b of Si. It should be noted that the Debye–Waller factor F_G enters here, and therefore the value of the scattering length b can be obtained by this method. The neutron interferometric technique measures the forward-scattering amplitudes for which the Debye–Waller factor is unity.

The description of the Pendellösung oscillations and fringes given here is somewhat oversimplified, and there are a number of important subtleties involved in properly accounting for these effects quantitatively. First of all, the incident beam is never sufficiently well collimated so as to generate separate plane-wave solutions inside the crystal. Indeed, the entire dispersion surface is generally simultaneous excited (illuminated) by the various divergent rays comprising the usual incident beam. These divergences need only be a few arc seconds to give rise to currents $<J_\alpha>$ and $<J_\beta>$ filling the entire Borrmann triangle. In order to describe these interference effects properly, it is necessary to specify the precise phase relationships between the various plane-wave components of the incident beam. With a narrow slit in the incident beam on the entrant face of the crystal, Shull (1973) carried out scans across the resulting diffracted beam, as shown in Fig. 11.13. The results are correctly described if one assumes that the incident beam is a monochromatic spherical wave, for which the amplitudes and phase relationships of the plane wave decomposition are known. For the X-ray case, spherical wave calculations were first carried out by Kato (1961, 1968). Subsequent applications of the spherical wave description to the neutron case and to the LLL interferometer were carried out by Petrascheck (1976), Petrascheck and Folk (1976), and Bauspiess et al. (1976). An alternative, perhaps more general, approach to this and other problems involving finite slits, based upon the Takagi–Taupin equations, is given in Section 11.6.

Pendellösung interference effects are also manifested in the integrated reflectivity of the Bragg reflected beam. That is, the integral of I_G in Eq. (11.36),

$$R_y = \int_{-\infty}^{\infty} I_H(y)\mathrm{d}y = |A_0|^2 \int_{-\infty}^{\infty} \frac{\sin^2\Phi(y)}{1+y^2}\,\mathrm{d}y, \tag{11.56}$$

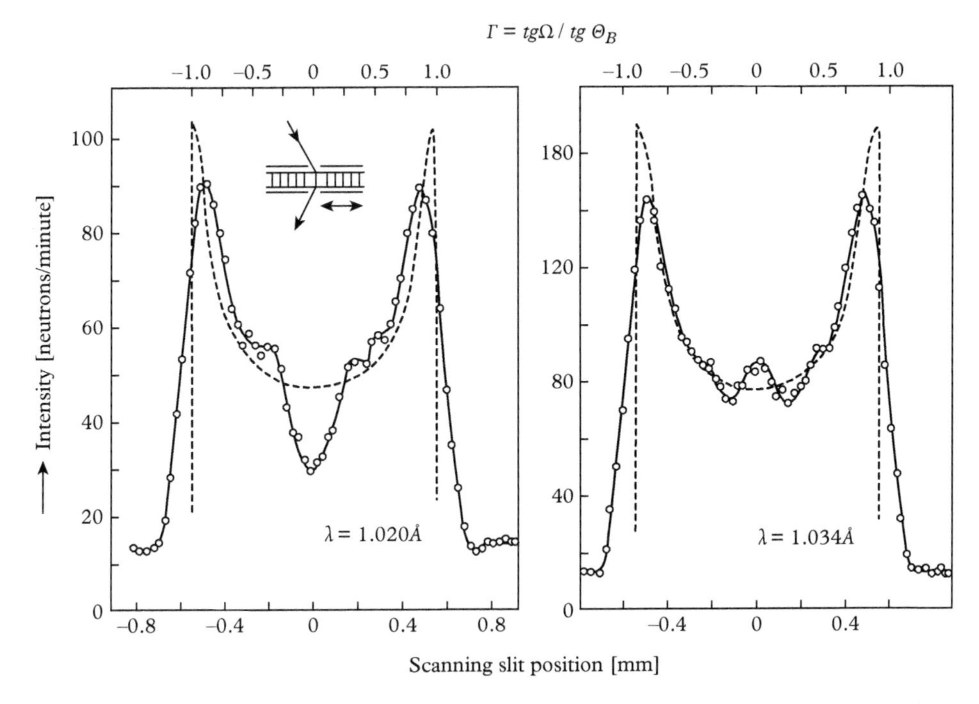

Figure 11.13 *Intensity profile within the Borrmann fan behind a Si-Laue reflection measured with a slit width of 0.13 mm. Reprinted with permission from Shull 1968, copyright 1968 by the American Physical Society.*

shows oscillations with a period equal to Δ_H as a function of the crystal thickness D. This is shown in Fig. 11.13. Difficult and time-consuming measurements of this effect using crystals of various thicknesses were made by Sippel et al. (1965). Their data are shown in Fig. 11.14. Note that the integrated reflectivity is linear in the crystal thickness D, i.e., proportional to the crystal volume illuminated, only up to about 30 μm. For larger D the effects of extinction and Pendellösung interference become significant. In the next section we give a brief discussion of the primary extinction length, ℓ_p.

11.5 Primary Extinction and the Width of a Bragg Reflection

The diffracted beam intensity, Eq. (11.36), is a function of the misset angle $\Delta\theta$, and is dependent upon two parameters, the crystal thickness D and the structure factor $|F_H|$. We explicitly display the dependence on $\Delta\theta$:

$$I_H(\Delta\theta) = \frac{|A_0|^2 \sin^2\left[a(1 + b^2(\Delta\theta)^2)^{1/2}\right]}{1 + b^2(\Delta\theta)^2}. \tag{11.57}$$

The two dimensionless parameters a and b are

$$a = |v_H|\, D/\cos\theta_B = \lambda\, |F_H|\, D/V_{cell} \cos\theta_B \tag{11.58}$$

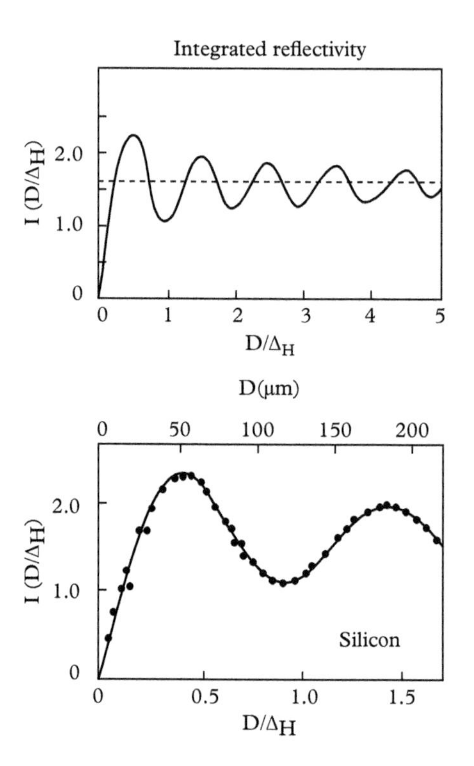

Figure 11.14 *Calculated (above) and measured (below) integrated reflectivity of a Laue reflection. Reprinted from Sippel et al. 1965, copyright 1965, with permission from Elsevier.*

and

$$b = k_0 \sin 2\theta_B / 2\,|\nu_H| = \pi \sin 2\theta_B\, V_{cell} \,/\, \lambda^2\ |F_H|. \tag{11.59}$$

This is a symmetric, rapidly oscillating function with a Lorentzian envelope that falls to ½ its maximum value when the y-parameter is unity, or equivalently when the misset angle is

$$\Delta\theta_{1/2} = \frac{1}{b} = \frac{\lambda^2\ |F_H|}{\pi V_{cell}\ \sin 2\theta_B}. \tag{11.60}$$

This width is typically less than 1 arc sec. It is sometimes called the *Darwin width*. From this formula we can estimate the primary extinction length, ℓ_p. The formula for $I_H(\Delta\theta)$ must go over to the kinematical diffraction results for sufficiently thin crystals. Just below the crystal surface, the internal wave field is dominated by $\Psi_0(\boldsymbol{r}) = \psi_0^\alpha\, e^{i\boldsymbol{K}_0^\alpha\cdot\boldsymbol{r}} + \psi_0^\beta\, e^{i\boldsymbol{K}_0^\beta\cdot\boldsymbol{r}}$, which diminishes as $\Psi_H(\boldsymbol{r}) = \psi_H^\alpha\, e^{i\boldsymbol{K}_H^\alpha\cdot\boldsymbol{r}} + \psi_H^\beta\, e^{i\boldsymbol{K}_H^\beta\cdot\boldsymbol{r}}$ builds up. At depths further below the entrant surface the neutron current oscillates back and forth between the incident beam direction $\hat{S}_0$ and the diffracted beam direction$\hat{S}_H$, giving rise to the Pendellösung phenomena discussed in the previous section.

Therefore, we should expect that the primary extinction length ℓ_p is intimately connected to the period of the Pendellösung oscillations, Δ_H. We can see this from the kinematical diffraction point of view, since the full effective Bragg reflection is established within a depth ℓ_p rather than D. We know from fundamental diffraction theory that reducing the size of the scattering system causes a broadening of the diffraction peak. In this case the effect is to cause the reciprocal lattice point $\boldsymbol{H}$ to appear broadened along an arc perpendicular to $\boldsymbol{H}$ in $\boldsymbol{k}$-space by an amount $2\pi/\ell_p$, that is 2π divided by the dimension of the scattering system. This is equivalent to a spread in crystal orientations $\Delta\theta$ of order $2\pi/\ell_p H$. Using Bragg's Law $H = 2k_0 \sin\theta_B$, this then implies that

$$\Delta\theta_{1/2} = \pi/\ell_p k_0 \sin\theta_B. \tag{11.61}$$

Equating this expression to the Darwin width given by Eq. (11.58) and solving for ℓ_p we obtain

$$\ell_p = \pi V_{cell} \cos\theta_B/\lambda\,|F_H|. \tag{11.62}$$

This is precisely the same expression as the equation for the Pendellösung oscillation period Δ_H given by Eq. (11.54). Thus, for crystals of thickness D less than say ⅓ of ℓ_p, the integrated Bragg reflection intensity will be proportional to the square of the structure factor and to the volume of the crystal irradiated. The ratio of the integrated intensity in Eq. (11.52) to the kinematical result is called the primary extinction factor ε_p, a number less than unity for all crystals of size greater than $\lambda_p \approx 100\ \mu$m.

11.6 The Takagi–Taupin Equations

In most experimental situations the incident neutron beam is shaped by a monochromator crystal placed some distance away from the neutron source (10–50 m and from the perfect single interferometer crystal (~ 2 m). The beam's spatial extent, angular divergence, and monochromicity are defined by the overall experimental layout, from the reactor source and a neutron guide to the final defining slits placed just in front of the first crystal of the interferometer. The beam is always divergent, not monochromatic and always of finite spatial width. Clearly, it cannot be fully described by a single plane wave, $A_0 e^{i\boldsymbol{k}_0\cdot\boldsymbol{r}}$. The geometry of the apparatus is important in the description of the coherence properties of the beam as discussed in some detail in Section 4.2.

So far our discussion of the dynamical theory of neutron diffraction has been based upon an incident monochromatic plane wave. There is an alternative approach to dynamical diffraction theory, originally developed by Takagi (1962, 1969) and Taupin (1961), to analyze the effects of strain on X-ray diffraction. Werner (1980) used this approach to calculate the effects of an external force (magnetic or gravitational) on the dynamical diffraction of neutrons. In this chapter we derive the Takagi–Taupin (T–T) equations and show how they can be solved in general, and in particular for an infinitely narrow incident beam defined by a slit.

The equation to be solved is the Schrödinger equation, Eq. (11.3). When the central wave vector $\boldsymbol{k}_0$ of the incident beam satisfies the Bragg condition for the reciprocal lattice vector $\boldsymbol{H}$, we know that the solution of the Schrödinger equation must be of the form

$$\Psi(\boldsymbol{r}) = \psi_0(\boldsymbol{r})\, e^{i\boldsymbol{K}_0\cdot\boldsymbol{r}} + \psi_H(\boldsymbol{r}) e^{i\boldsymbol{K}_H\cdot\boldsymbol{r}}, \tag{11.63}$$

where the wave vector $\boldsymbol{K}_H \equiv \boldsymbol{K}_0 + \boldsymbol{H}$, as defined in the plane wave treatment of dynamical diffraction. However, here we allow the amplitudes $\psi_0(\boldsymbol{r})$ and $\psi_H(\boldsymbol{r})$ to be functions of position to be determined by the Schrödinger equation and the appropriate boundary conditions. We are free to fix the internal wave vector $\boldsymbol{K}_0$ to be equal to $\boldsymbol{k}_0$, and thereby to incorporate all effects of the index of refraction in the functions $\psi_0(\boldsymbol{r})$ and $\psi_H(\boldsymbol{r})$. The magnitude of $\boldsymbol{K}_H$ is then seen to depend upon the crystal misset angle $\Delta\theta$ (see Fig. 11.15); it is easy to see that

$$K_H{}^2 \approx k_0{}^2(1 - 2\Delta\theta \sin 2\theta_B). \tag{11.64}$$

Substituting Eq. (11.63) into the Schrödinger equation, and equating the coefficients of $e^{i\boldsymbol{K}_0\cdot\boldsymbol{r}}$ and $e^{i\boldsymbol{K}_H\cdot\boldsymbol{r}}$, we obtain a pair of coupled differential equations for the amplitude functions

$$(k_0{}^2 - v_0 - K_0{}^2)\, \psi_0(\boldsymbol{r}) + 2i\, \boldsymbol{K}_0 \cdot \nabla\psi_0(\boldsymbol{r}) - v_{-H}\psi_H(\boldsymbol{r}) = 0, \tag{11.65a}$$

and

$$-v_H\psi_0(\boldsymbol{r}) + 2i\, \boldsymbol{K}_H \cdot \nabla\psi_H(\boldsymbol{r}) + (k_0{}^2 - v_0 - K_H{}^2)\, \psi_H(\boldsymbol{r}) = 0. \tag{11.65b}$$

Terms involving $\nabla^2\psi_0(\boldsymbol{r})$ and $\nabla^2\psi_H(\boldsymbol{r})$ have been neglected since they are small compared to $\boldsymbol{K}_H \cdot \nabla\psi_H(\boldsymbol{r})$ and $\boldsymbol{K}_0 \cdot \nabla\psi_0(\boldsymbol{r})$. As discussed in several ways already, a strongly diffracted beam

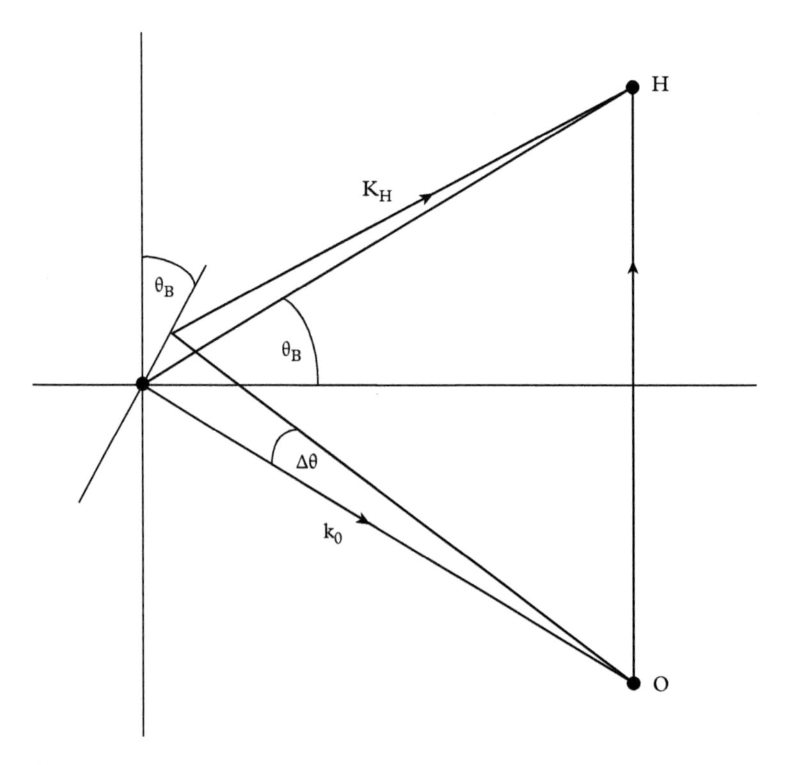

Figure 11.15 *Wave vectors used in the Takagi–Taupin theory*

occurs only over a very narrow range of misset angles $\Delta\theta$, which means that the direction of $\boldsymbol{K}_H$ is essentially fixed along $\hat{S}_H$, and $\boldsymbol{K}_0$ is essentially fixed along $\hat{S}_0$. Thus, we can write the second terms in Eqs. (11.65a) and (11.65b) as

$$\boldsymbol{K}_0 \cdot \nabla\psi_0 \approx k_0 \frac{\partial \psi_0}{\partial S_0} \quad \text{and} \quad \boldsymbol{K}_H \cdot \nabla\psi_H \approx k_0 \frac{\partial \psi_H}{\partial S_H}, \tag{11.66}$$

where the non-orthogonal coordinate frame (S_0, S_H) is the natural one. Consequently, the amplitudes Ψ_0 and Ψ_H are most conveniently expressed as functions of S_0 and S_H, which means that Eqs. (11.65a) and (11.65b) take the form

$$-\nu_0\psi_0 + i\frac{\partial \psi_0}{\partial S_0} - \nu_{-H}\psi_H = 0, \tag{11.67a}$$

and

$$-\nu_H\psi_0 + i\frac{\partial \psi_H}{\partial S_H} - (\beta - \nu_0)\psi_H = 0, \tag{11.67b}$$

where the definition of $v_H$is given by Eq. (11.21), and β is given by the misset angle $\Delta\theta$ as

$$\beta \equiv k_0 \Delta\theta \sin 2\theta_B. \tag{11.68}$$

The above pair of coupled differential equations can be further simplified by defining

$$\psi_0(S_0, S_H) \equiv \exp\left[-\mathrm{i}\nu_0(S_0 + S_H) + \mathrm{i}\beta S_H\right]\ U_0(S_0, S_H) \tag{11.69a}$$

and

$$\psi_H(S_0, S_H) \equiv \exp\left[-\mathrm{i}\nu_0(S_0 + S_H) + \mathrm{i}\beta S_H\right] U_H(S_0, S_H). \tag{11.69b}$$

Using these expressions in Eqs. (11.67a) and (b) gives the remarkably simple and symmetric pair of coupled equations for the phase-modified amplitude functions U_0 and U_H:

$$\frac{\partial U_0}{\partial S_0} + \mathrm{i}\nu_{-H} U_H = 0, \tag{11.70a}$$

and

$$\frac{\partial U_H}{\partial S_H} + \mathrm{i}\nu_H U_0 = 0. \tag{11.70b}$$

These equations are the neutron version of the X-ray *Takagi–Taupin equations*.

From a physical point of view, the origin of these equations is clear. The first equation shows that the rate of change of the incident wave amplitude U_0 is proportional to the diffracted wave amplitude U_H; that is, the diffracted wave is being rescattered back into the incident beam direction. Similarly, the second equation shows that the rate of change of the diffracted wave amplitude U_H is proportional to the incident wave amplitude U_0. This is a mathematical embodiment of the underlying multiple scattering processes that couples U_0 to U_H as schematically illustrated in Fig. 11.16. The characteristic length scale of this multiple scattering process is v^{-1}, which is typically 50 to 100 mm.

Substituting the second of these equations into the first one, yields a second-order partial differential equation for the diffracted wave amplitude $U_H(S_O, S_H)$, namely

$$\frac{\partial^2 U_H}{\partial S_0 \partial S_H} + \nu^2 U_H = 0, \tag{11.71}$$

where

$$\nu^2 \equiv \nu_H \nu_{-H} = \frac{\lambda^2 |F_H|^2}{V_{\text{cell}}^2}. \tag{11.72}$$

An identical equation applies to the incident wave amplitude, $U_0(S_0, S_H)$. This equation must be solved subject to the continuity of the wave function and the current density as required by quantum mechanics. However, we can envisage certain aspects of the solution by inspection from the drawing of a typical multiple-reflection path. In order for the wave to be a part of the diffracted beam at the exit face of the crystal, it must have made an odd number $(2n + 1)$ of reflections. The one shown in Fig. 11.16 is a 5-reflection path. The total diffracted wave is the sum over all possible $(2n + 1)$ reflection paths. Since $v \cdot \Delta S$ is the probability for reflection in a distance ΔS, the mathematical solution for U_H must involve v^{2n+1} for any $(2n + 1)$-reflection path. We will see that this physical idea is borne out by the explicit calculation of U_H. Furthermore, it is clear that $U_H(S_0,S_H)$ can only be finite within the Borrmann triangle ABC, since no multiple-reflection path allows the neutron wave entering the crystal at point A to get into a region outside the Borrmann triangle.

The general solution to the elliptical differential equation for $U_H(S_0, S_H)$ is (Werner et al. 1986),

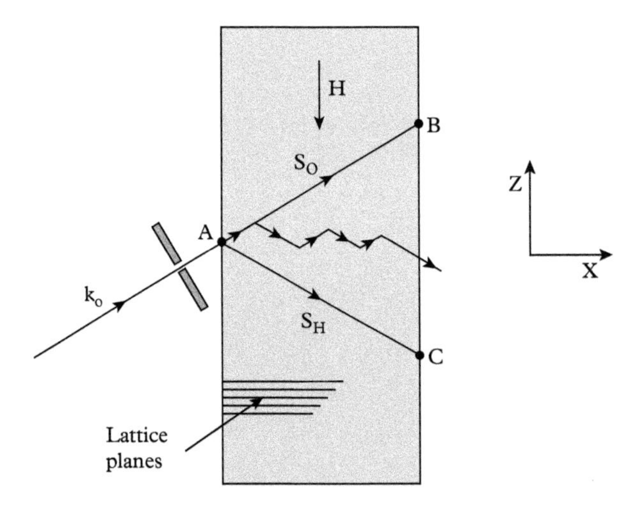

Figure 11.16 *Flight paths through the perfect crystal when local changes of the lattice parameters exist*

$$U_H(S_0, S_H) = \sum_{n=-\infty}^{\infty} a_n \left(\frac{S_0}{S_H}\right)^{n/2} \mathcal{J}_n\left(2\nu\sqrt{S_0 S_H}\right), \tag{11.73}$$

where $\mathcal{J}_n(Z)$ are the ordinary Bessel functions of order n, and the a_n are coefficients to be determined from the boundary conditions. Suppose now that the incident beam is confined by a very narrow slit as shown in Fig. 11.16. This incident "pencil" beam can be described by the wave function

$$\phi_0(\boldsymbol{r}) = A_0\, \delta(S_H)\, e^{i\boldsymbol{k}_0\cdot\boldsymbol{r}}, \tag{11.74}$$

where $\delta(S_H)$ is a Dirac delta function. Along the lines AB and AC, the diffracted wave amplitude remains constant. This is best seen from considering the line AC first, and then using symmetry arguments for the line AB. This requires all $a_n = 0$ to be zero except for a_0. We find therefore that

$$U_H(S_0, S_H) = -i\nu_H \mathcal{J}_0\left(2\nu\sqrt{S_0 S_H}\right). \tag{11.75}$$

The incident wave amplitude $U_0(S_0, S_H)$ is found from Eq. (11.70b) by differentiating $U_H(S_0, S_H)$ with respect to S_H, that is

$$\begin{aligned} U_0(S_0, S_H) &= i\nu_H{}^{-1} \frac{\partial U_H(S_0, S_H)}{\partial S_H} \\ &= \nu\sqrt{\frac{S_0}{S_H}} \mathcal{J}_1\left(2\nu\sqrt{S_0 S_H}\right). \end{aligned} \tag{11.76}$$

Equations (11.72) and (11.73) provide us with a complete description of the wave fields within and on the boundaries of the Borrmann triangle ABC. It is of interest to calculate the intensities I_0 and I_H on the back face of the crystal, that is, along the line BC. The coordinates S_0 and S_H can be written in terms of the orthogonal (x, z) coordinates:

$$S_0 = \frac{1}{2}\left(\frac{x}{\cos\theta_B} + \frac{z}{\sin\theta_B}\right) \tag{11.77a}$$

and

$$S_H = \frac{1}{2}\left(\frac{x}{\cos\theta_B} - \frac{z}{\sin\theta_B}\right). \tag{11.77b}$$

Using these coordinate transformation equations, we could plot I_0 and I_H as a function of z at $x = D$, the thickness of the crystal. However, it is more convenient to write the argument of the Bessel functions in terms of the angle parameter $\Gamma = \tan\Omega/\tan\Theta_B$ used in Section 11.3 (Eq. 11.46), namely

$$2\nu\sqrt{S_0 S_H} = \frac{\nu D}{\cos\theta_B}\sqrt{1-\Gamma^2} = \pi\frac{D}{\Delta_H}\sqrt{1-\Gamma^2}, \tag{11.78}$$

where Δ_H is the Pendellösung period, given by Eq. (11.54). Thus, the profile of the diffracted current density across the back face of the crystal from point B to point C is

$$I_H(\Gamma) = |U_G|^2 = \nu^2 |A_0|^2 \mathcal{J}_0^{\,2}\left(\pi \frac{D}{\Delta_H}\sqrt{1-\Gamma^2}\right). \tag{11.79}$$

Similarly, the profile of the beam in the incident direction (the 0-beam) leaving the crystal is given by

$$I_0(\Gamma) = \nu^2 |A_0|^2 \left(\frac{1-\Gamma}{1+\Gamma}\right) \mathcal{J}_1^{\,2}\left(\pi \frac{D}{\Delta_H}\sqrt{1-\Gamma^2}\right). \tag{11.80}$$

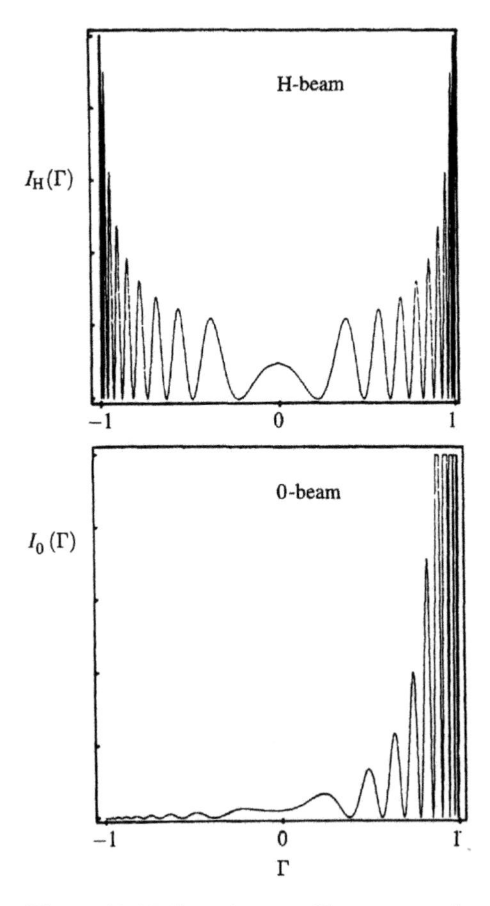

Figure 11.17 *Intensity profile across the Borrmann fan when the incident beam has a small misset angle. The curves are calculated for a crystal thickness* D/D$_0$ = *10 (Bonse and Graeff 1977)*

Γ ranges from +1 at point C to −1 at point B. The amplitude A_0 of the incident wave is assumed to be uniform in the misset angle $\Delta\theta$. When plotting the exit beam profiles versus Γ, we must take the Jacobian of transforming $\Delta\theta$ or $y(\Delta\theta$ to $\Gamma = \Gamma(\Delta\theta)$ into account. This Jacobian requires replacing $|A_0|^2$ by $|A_0|^2/(1-\Gamma^2)^{3/2}$. This will be discussed in Section 11.7.2. A plot of I_H and I_0 is given in Fig. 11.17 for $D/\Delta_H = 10$. These interesting profiles are the result of Pendellösung interference between the α- and β-branch wave functions, when summed over the angular divergence of the incident (spherical) wave. We note that I_H is symmetric in Γ and peaks at the two edges of the Borrmann fan. This can be understood from the fact that $<J_\alpha>$ and $<J_\beta>$ are directed normal to the dispersion surfaces, and there are many points on these surfaces having normals approaching $\hat{S}_0$ and $\hat{S}_H$. At one edge of the Borrmann fan (near point B) the current leading to a strong diffracted beam is carried by the α-branch, and at the other edge of the Borrmann fan (near point C) the current is carried by the β-branch. The measured profiles in Shull's (1973) experiment discussed in the previous section can be understood as the convolution of the finite slit resolution with the dynamical diffraction profiles calculated here. The profile of the beam leaving the crystal in the incident direction (the 0-beam) peaks near point B. The difference in the profiles of $I_H(\Gamma)$ and $I_0(\Gamma)$ are important in understanding the profiles of the beams traversing the neutron interferometer, which will be discussed in Section 11.8.

It was suggested at the beginning of this section that the contribution of all $(2n+1)$-times reflected waves to U_H should involve a coefficient v^{2n+1}. Although we did not derive U_H here from the point of view of summing over all possible paths leading to the $(2n+1)$-multiple reflection term, we can identify this term in the result by looking at the series expansion of the Bessel function $\mathcal{J}_0$, namely

$$U_H = -i\nu_H \sum_{n=0}^{\infty} (-1)^n \left(\nu\sqrt{S_0 S_H}\right)^{2n} / (n!)^2. \tag{11.81}$$

For example, the 5-times reflected term involves $v_H v_{-H} v_H v_{-H} v_H$ or $v_H v^4$. That is, the incident wave is first reflected, say to the right by v_H, then to the left by v_{-H} (back into the incident direction), then again to the right by v_H, and so on.

11.7 Theory of the Perfect Silicon Crystal Neutron Interferometer

A full and complete analysis of the perfect Si-crystal neutron interferometer requires a detailed description of the coherent wave fields which propagate through the device. The starting point for this analysis is the plane wave dynamical diffraction theory for a symmetric Laue-geometry crystal slab, as developed in the previous section. Repeated, sequential application of the transmission and reflection coefficients for each of the crystals of the interferometer leads to the formulas describing the 0- and H-beam wave fields and the intensities leaving the interferometer, as shown in Fig. 11.18. We first pursue this problem in Section 11.7.1 along the lines initially developed by Rauch and Suda (1974), Petrascheck (1976), and Werner (1976). It is essential that both the α- and β-branch wave fields for each crystal, and the consequent Pendellösung interferences, are retained throughout the calculation. It is for this reason that the detailed structure of the wave fields is considerably more complex in the neutron case for which absorption is zero than in the X-ray case where absorption is significant (Bonse and Hart 1965).

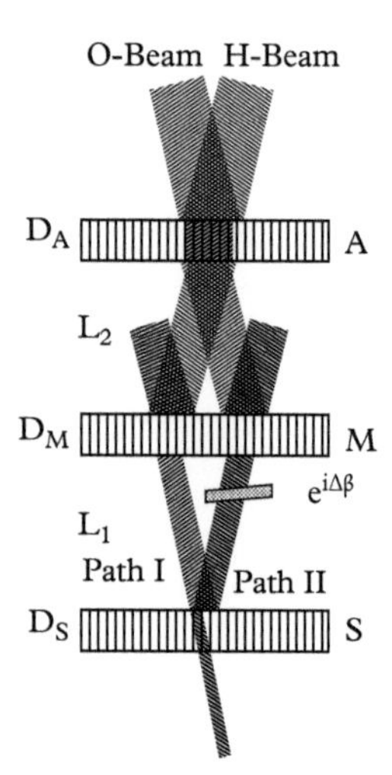

Figure 11.18 *Scheme of the wave fields inside a perfect crystal interferometer*

For practical reasons, the beam incident upon the interferometer is always restricted in its lateral extent by apertures and slits. If the entrant slit is sufficiently narrow, a spherical wave description of the resulting wave fields is necessary. Detailed spherical wave calculations have been carried out by Petrascheck (1976), Petrascheck and Folk (1976), and Bauspiess et al. (1976). These calculations are rather complex, but are necessary to analyze defocusing effects due to non-ideal geometry, such as errors in the dimensionality of the interferometer resulting from its original fabrication. In Section 11.7.2 we will give a brief discussion of the results of these spherical wave calculations, with special emphasis on the loss of contrast due to defocusing (Kischko and Bonse 1985). A numerical study of the symmetric LLL X-ray interferometer based upon the Takagi–Taupin equations has been published by (Accotto et al. 1994). Applications of this method to the zero absorption neutron case would appear to be a natural extension. This has not yet been done.

11.7.1 Plane Wave Theory of the LLL Interferometer

In this section we use the dynamical diffraction results of Section 11.2 to develop a theory of the symmetric LLL neutron interferometer. We assume that the interferometer is illuminated by a plane wave as shown in Fig. 11.18. We denote the wave functions leaving the interferometer by $U_0 = u_0 e^{i\boldsymbol{k}_0\cdot\boldsymbol{r}}$ and $U_H = u_H e^{i\boldsymbol{k}_H\cdot\boldsymbol{r}}$, corresponding to the 0-beam and the H-beam, respectively.

In deriving the formulas for the transmission coefficient $T(y)$ and the reflection coefficient $R(y)$ in Section 11.2 we implicitly assumed that the origin of coordinates was in the plane corresponding to the entrant surface of the symmetric Laue-geometry crystal. In order to use these results to calculate the diffracted and transmitted waves leaving the middle *mirror* crystal, labeled M in Fig. 11.18, and subsequently the waves leaving the *analyzer* crystal, labeled A, we must take into account the displacement of the plane containing the origin of coordinates, first by $L_1 + D_S$ for the mirror crystals, and the by $L_1 + L_2 + D_S + D_M$ for the analyzer crystal. It is easy to show that this displacement of the origin of coordinates by a distance $\boldsymbol{l}$ has the effect of adding a phase factor $e^{i\boldsymbol{l}\cdot\boldsymbol{\delta}}$ to the reflection coefficient $R(y)$, but has no effect on the transmission coefficient $T(y)$. Here $\boldsymbol{\delta}$ is defined by Eq. (11.32), and can be rewritten in terms of the Pendellösung length Δ_H given by Eq. (11.54), and the y-parameter as (Eq. 11.26)

$$\boldsymbol{\delta} = \frac{2\pi y}{\Delta_H}\hat{n}. \tag{11.82}$$

To obtain the wave functions $U_0(\boldsymbol{r})$ and $U_H(\boldsymbol{r})$ leaving the interferometer consider paths I and II separately. For path I we obtain for the amplitude

$$u_0^{\mathrm{I}} = A_0 T_H^{\mathrm{S}}(y) R_H^{\mathrm{MI}}(y) R_{-H}^{\mathrm{A}}(-y) e^{-i\delta(D_{\mathrm{M}}^{\mathrm{I}}+L_2^{\mathrm{I}})} \tag{11.83a}$$

and

$$u_H^{\mathrm{I}} = A_0 T_H^{\mathrm{S}}(y) R_H^{\mathrm{MI}}(y) T_H(-y) e^{i\delta(D_{\mathrm{S}}+L_1^{\mathrm{I}})}. \tag{11.83b}$$

Similarly, for path II the amplitudes of the waves leaving the interferometer are

$$u_0^{\mathrm{II}} = A_0 R_H^{\mathrm{S}}(y) R_{-H}^{\mathrm{MII}}(-y) T_H^{\mathrm{A}}(y) e^{-i\delta(D_{\mathrm{M}}^{\mathrm{II}}+L_1^{\mathrm{II}})} e^{i\Delta\beta} \tag{11.84a}$$

and

$$u_H^{\mathrm{II}} = A_0 R_H^{\mathrm{S}}(y) R_{-H}^{\mathrm{MII}}(-y) R_H^{\mathrm{A}}(y) e^{i\delta(D_{\mathrm{M}}^{\mathrm{II}}+L_2^{\mathrm{II}})} e^{i\Delta\beta}. \tag{11.84b}$$

Here we have included the externally adjustable phase $\Delta\beta$ in beam path II. In the above equations the superscripts S, MI, MII, and A on the transmission and reflection coefficients indicate the appropriate crystal elements along each of the paths I or II. The subscripts H or $-H$ on the reflection coefficients give the sign of the appropriate reciprocal lattice vector giving rise to the Bragg reflection. We note that for an incoming plane wave with $\boldsymbol{k}_0$ oriented at a positive angle $\Delta\theta$ (positive y) with respect to the nominal (central) $\boldsymbol{k}_0$, the angle of the diffracted wave vector $\boldsymbol{k}_H$ when the wave arrives at the mirror crystal will be $-\Delta\theta$ (negative y). Since $R(y)$ and $T(y)$ depend upon the appropriate crystal thickness D, the superscripts S, MI, MII, and A are necessary. The distances between the crystal slabs are denoted by $L_1{}^{\mathrm{I}}$ and $L_2{}^{\mathrm{I}}$ on path I and by $L_1{}^{\mathrm{II}}$ and $L_2{}^{\mathrm{II}}$ on path II.

The intensity in the 0-beam is calculated by evaluating

$$I_0(y) = \left|u_0^{\mathrm{I}}(y) + u_0^{\mathrm{II}}(y)\right|^2, \tag{11.85}$$

and the H-beam intensity is

$$I_H(y) = \left|u_H^{\mathrm{I}}(y) + u_H^{\mathrm{II}}(y)\right|^2. \tag{11.86}$$

For an ideally perfect interferometer we take

$$D_{\mathrm{S}} = D_{\mathrm{M}}^{\mathrm{I}} + D_{\mathrm{M}}^{\mathrm{II}} = D, \tag{11.87a}$$

and

$$L_1^{\mathrm{II}} = L_1^{\mathrm{II}} = L_2^{\mathrm{I}} = L_2^{\mathrm{II}} = L. \tag{11.87b}$$

Under these conditions, we obtain the following expressions for the intensities:

$$I_0(y) = |A_0|^2 A(y,\varepsilon)\,[1 + \cos(\Delta\beta)], \tag{11.88a}$$

and

$$I_H(y) = |A_0|^2\,[B(y,\varepsilon) - A(y,\varepsilon)\cos(\Delta\beta)], \tag{11.88b}$$

where the functions $A(y, \varepsilon)$ and $B(y, \varepsilon)$ are

$$A(y,\varepsilon) = \frac{2\sin^4(\Phi)\left[y^2 + \cos^2(\Phi)\right]}{\left[1+y^2\right]^3}, \tag{11.89a}$$

and

$$B(y,\varepsilon) = \frac{\sin^2(\Phi)}{(1+y^2)^3}\left[(\cos^2(\Phi)+y^2)^2 + \sin^4(\Phi)\right]. \tag{11.89b}$$

These functions are shown in Fig. 11.19 for selected values of ε. Averaging over the Pendellösung oscillations, that is over the angle Φ, one obtains

$$< A(y,\varepsilon) > = (1/8)(6y^2 + 1)(1 + y^2)^{-3}, \tag{11.90a}$$

and

$$< B(y,\varepsilon) > = (1/8)(4y^4 + 2y^2 + 3)(1 + y^2)^{-3}. \tag{11.90b}$$

The angle Φ is defined by Eq. (11.35c), which we write as

$$\Phi = \varepsilon(1 + y^2)^{1/2}. \tag{11.91}$$

The parameter ε is conveniently written in terms of the Pendellösung length Δ_H as

$$\varepsilon = \frac{|v_H|\,D}{\cos(\theta_B)} = \pi\frac{D}{\Delta_H}. \tag{11.92}$$

Similarly, the y-parameter can be written as

$$y = \mathbf{k}_0\,\Delta\theta\,\sin(2\theta_B)/2\,|v_H| = \Delta\theta\,\frac{\Delta_H}{d_{hkl}}, \tag{11.93}$$

where d_{hkl} is the lattice plane spacing for the Bragg reflection $\boldsymbol{H}_{hkl}$ having Miller indices (hkl).

Since the incident beam always has a divergence much greater than the inherent width of the reflectivity (the Darwin width), we must integrate the expressions (11.89a) and (11.89b) over $\Delta\theta$, or equivalently over y, to obtain the experimentally measured integrated intensities which we will call $I_0(\Delta\beta)$ and $I_H(\Delta\beta)$. That is,

$$I_0(\Delta\beta) = N_0\left[a(\varepsilon) + a(\varepsilon)\cos(\Delta\beta)\right], \tag{11.94a}$$

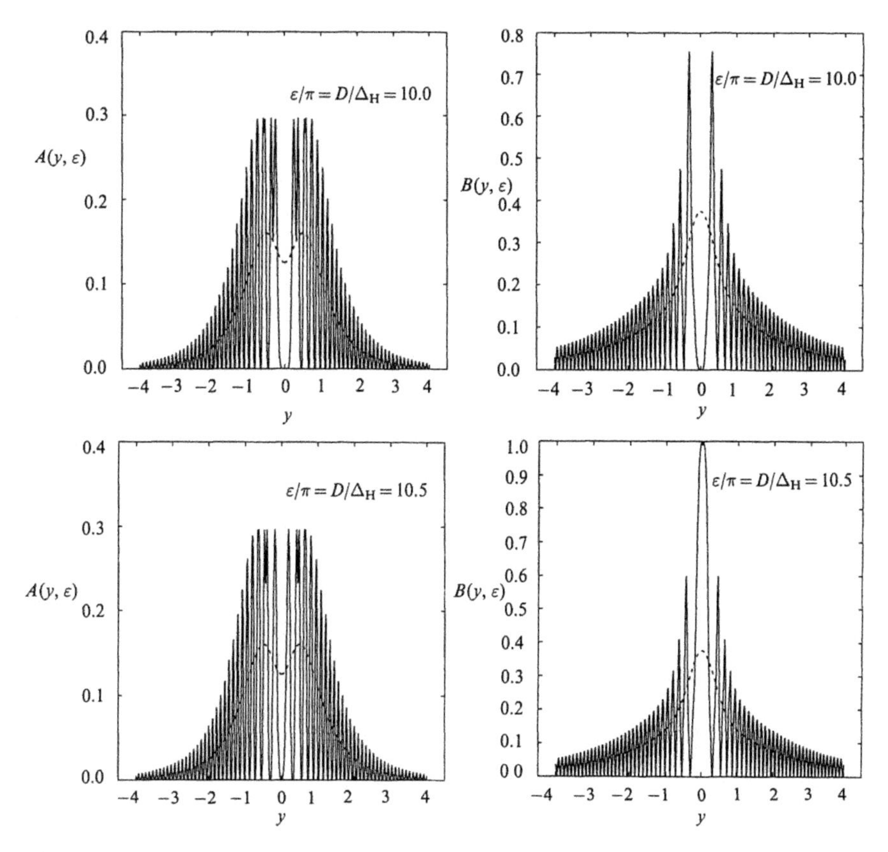

Figure 11.19 *Intensity profiles behind a symmetric LLL-interferometer for two crystal thicknesses*

and

$$I_H(\Delta\beta) = N_0\left[b(\varepsilon) - a(\varepsilon)\cos(\Delta\beta)\right], \tag{11.94b}$$

where $N_0 = |A_0|^2$ is the effective incident beam intensity. The coefficients a and b depend upon the parameter ε and are given by the integrals:

$$a(\varepsilon) = \int_{-\infty}^{\infty} A(y,\varepsilon)\mathrm{d}y, \tag{11.95a}$$

and

$$b(\varepsilon) = \int_{-\infty}^{\infty} B(y,\varepsilon)\mathrm{d}y. \tag{11.95b}$$

The result of a numerical evaluation of these integrals is shown in Fig. 11.20. For the (220) reflection in Si, the Pendellösung length $\Delta_H = 64\ \mu$m, for $\lambda = 2$-Å neutrons. Thus, for an interferometer

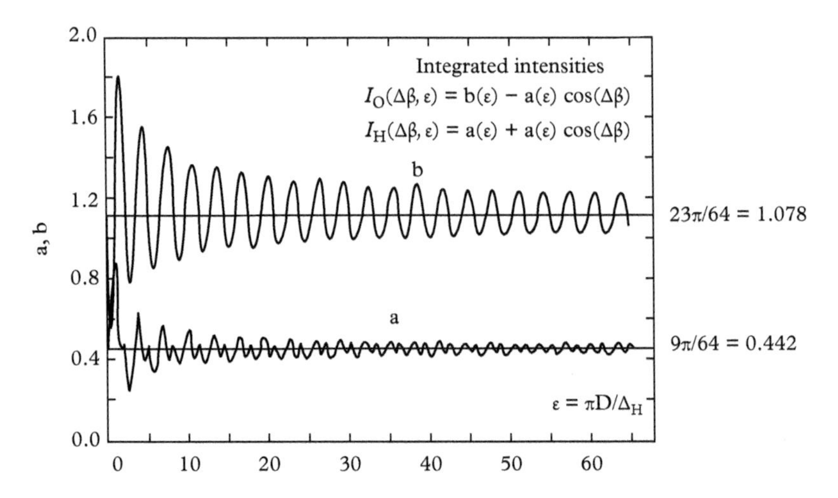

Figure 11.20 *Integral reflectivities of a symmetric LLL-interferometer arrangement*

with crystal plate thicknesses $D = 1$ mm, the parameter $\varepsilon = 49$. Asymptotically, for large ε, $a(\varepsilon) \to 9\pi/64$ and $b(\varepsilon) \to 23\pi/64$, so that the ratio $b/a = 23/9 = 2.55$. The oscillations in $a(\varepsilon)$ and $b(\varepsilon)$ are a direct consequence of the Pendellösung interferences occurring in each of the three crystal plates of the interferometer.

We see that the 0-beam interferogram given by Eq. (11.94) is predicted to exhibit 100% contrast. The amplitudes of the oscillations of the H-beam are 180° out of phase with the 0-beam, and the contrast is much less. The reason for this is that the wave function $U_H^I(\boldsymbol{r})$ involves two transmission coefficients and one reflection coefficient, i.e., T^2R, while the wave function $U_H^{II}(\boldsymbol{r})$ involves three reflections, i.e., R^3. As the phase shift $\Delta\beta$ within the interferometer is varied, the intensity is swapped back and forth between the 0-beam and the H-beam, thus conserving total neutron intensity, that is

$$I_0(\Delta\beta) + I_H(\Delta\beta) = N_0[a(\varepsilon) + b(\varepsilon)]. \tag{11.96}$$

Even though the predicted 100% contrast for an ideal interferometer is never realized in practice, this neutron conserving relation is always valid.

11.7.2 Beam Profiles and Ray Tracing through the Interferometer

In the previous section we developed a theory of the LLL interferometer based upon an incident plane wave of essentially infinite lateral extent. We now describe the operation of the interferometer from a somewhat different point of view. Consider an interferometer which is illuminated through an entrant slit which is narrow compared to the thickness D of each of the three crystal plates as shown in Fig. 11.21. The incident beam can be viewed as a set of divergent rays corresponding to the angles $\Delta\theta$ differing from the exact Bragg angle. The phase relationship between these rays depends upon the coherence properties of the incident beam. If the nominally monochromatic incident beam is produced by a perfect crystal monochromator, and the entrant slit is

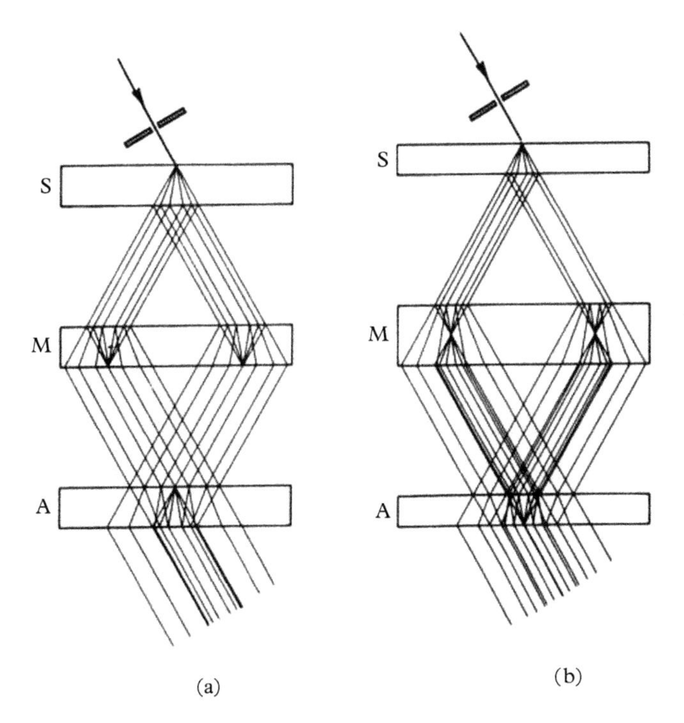

Figure 11.21 *Rays through LLL-interferometers of plane waves incident with slightly different angles of incidence*

very narrow, the phase relationships between the various rays is appropriately given by the Fourier decomposition of a spherical wave. However, if the monochromator is a mosaic crystal and the incident slit is not so narrow as to cause appreciable diffraction (say greater than 10^5 Å), then we may view the incident beam as an incoherent superposition of plane waves, each propagating separately through the interferometer. That is, interference between rays of differing $\Delta\theta$ (or equivalently, differing values of the y-parameter) is washed out and is not observable. This is the point of view that is inherent in the construction of the multiple-ray diagram shown in Fig. 11.21. We will now use this idea to calculate the beam profiles within the interferometer, and more importantly, the spatial profiles of the beams leaving the third plate of the interferometer. An extension of the methods developed here have recently been used by Littrell et al. (1998b) to include the effects of the Earth's gravitational potential.

The fate of a given incident ray, oriented at a misset angle $+\Delta\theta$, as it propagates through a perfect symmetric LLL interferometer is shown in Fig. 11.22. These misset angles are again specified by the y-parameter (Eq. 11.26). The current density $<J_\alpha>$ corresponding to the α-branch of the dispersion surface propagates across the first crystal at an angle $+\Omega(\Delta\theta)$ as shown in Fig. 11.6, which is determined by the misset angle $\Delta\theta$. Using the definition of $\Gamma(\Delta\theta$ given by Eqs. (11.46) and (11.47), and the definition of $y(\Delta\theta)$, we have

$$\Gamma = \frac{\tan\Omega}{\tan\theta_B} = \frac{y}{(1+y^2)^{1/2}}. \tag{11.97}$$

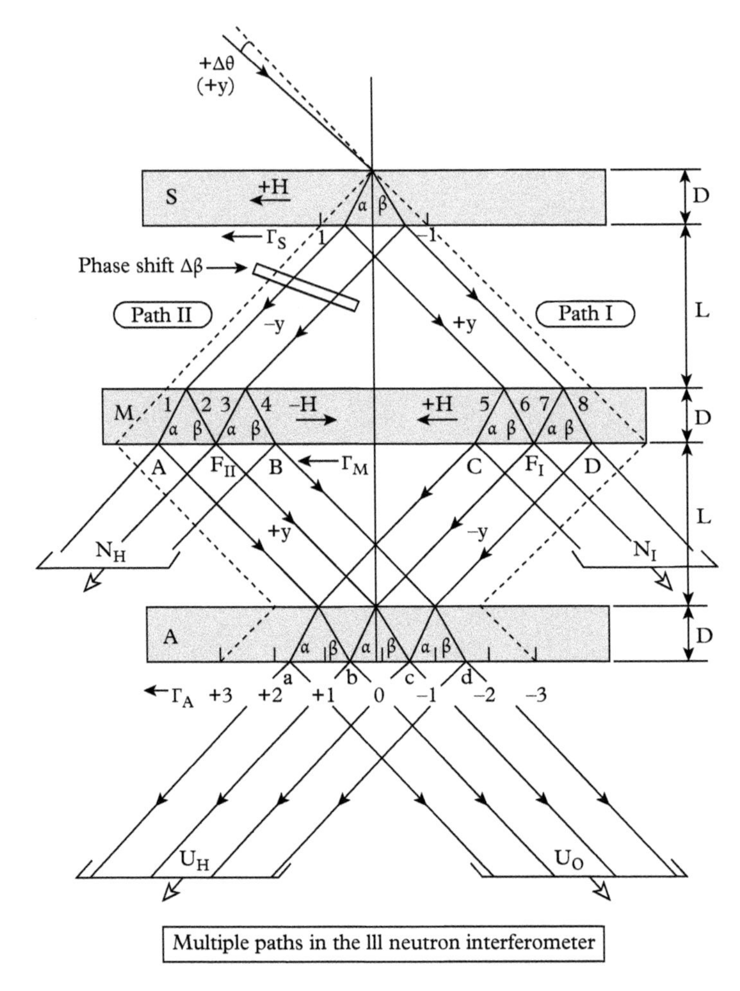

Figure 11.22 *Rays excited inside a LLL interferometer from a given incident ray with a misset angle* $\Delta\Theta$

The inverse relation between y and Γ is

$$y = \frac{\Gamma}{(1-\Gamma^2)^{1/2}}, \tag{11.98}$$

where the notation we use will generally suppress the dependence of Ω, Γ, and y on $\Delta\theta$. The current density $< J_\beta >$ corresponding to the β-branch of the dispersion relation propagates across the first crystal at an angle $-\Omega(\Delta\theta)$. Thus, for a given incident ray, characterized by $y(\Delta\theta)$, we see that the four exit rays from the first "splitter" crystal, labeled S, give rise to eight rays traversing the second "mirror" crystal, labeled M. It is interesting and important to note that the dynamical diffraction process gives rise to a focusing effect at the points F_I and F_{II} on the exit face of the middle crystal. Rays 2 and 3 come together at F_{II} and rays 6 and 7 come together at F_I for all entrant values of the y-parameter. Of the 12 rays leaving the middle crystal, 6 rays combine and interfere in the third *analyzer* crystal, labeled A. The other 6 rays miss the analyzer crystal, and represent non-interfering beams, labeled N_I and N_{II} in Fig. 11.22. There are 4 points, labeled *a,b,c* and *d* on the exit face of the third crystal, which give rise to 4 rays contributing to the 0-beam wave function $U_0(\boldsymbol{r})$, and 4 rays contributing to the H-beam wave function $U_H(\boldsymbol{r})$. As the y-parameter is varied, the points *a* and *d* move across the exit face of the analyzer crystal in the range (–3, 3) for values of $\Gamma_A = \pm 3\Gamma$; whereas, the points *b* and *c* move across the exit face in the range (–1, 1) for values of $\Gamma_A = \pm\Gamma$. Following Horne (1986), we will call the paths leading to the exit points *b* and *c* the *primary* paths, and the paths leading to the exit points *a* and *d* the *maverick* paths.

With this introduction to the multiple trajectories through the interferometer, it is a straightforward bookkeeping matter to calculate the spatial profiles of the beams at various points within and exiting the interferometer, using the wave amplitudes calculated in Section 11.2 (Eqs. 11.27). Here we express these amplitudes in terms of $\Gamma = \Gamma(y)$, namely

$$\psi_0^\alpha(y) = \frac{1}{2}(1 - \Gamma), \tag{11.99a}$$

$$\psi_0^\beta(y) = \frac{1}{2}(1 + \Gamma), \tag{11.99b}$$

$$\psi_H^\alpha(y) = -\frac{1}{2}\,(1 - \Gamma^2)^{1/2}(\nu_H/\nu_{-H})^{1/2}, \tag{11.99c}$$

$$\psi_H^\beta(y) = \frac{1}{2}\,(1 - \Gamma^2)^{1/2}(\nu_H/\nu_{-H})^{1/2}. \tag{11.99d}$$

Since the spatial points where the α- and β-branch currents (for a given y) leave the first crystal do not coincide (except for $y = 0$), the continuity condition at the exit face of the first crystal give exit wave amplitudes separately for the α and β waves. The corresponding amplitudes are given by the branch-specific transmission and reflection coefficients, namely

$$T_\alpha(y) = \psi_0^\alpha(y)e^{-iN_\alpha(y)D}, \tag{11.100a}$$

$$T_\beta(y) = \psi_0^\beta(y)e^{-iN_\beta(y)D}, \tag{11.100b}$$

$$R_\alpha(y) = \psi_H^\alpha(y)e^{-iN_\alpha(y)D-i\delta(y)D}, \tag{11.100c}$$

$$R_\beta(y) = \psi_H^\beta(y)e^{-iN_\beta(y)D-i\delta(y)D}. \tag{11.100d}$$

Here $N_\alpha(y)$ and $N_\beta(y)$ are the normal vectors given by Eq. (11.29), and $\delta(y)$ is given by Eq. (11.32). It is easy to show that they can be expressed in terms of $\Gamma(y)$ as

$$N_\alpha(y) = \frac{\pi}{\Delta_0} - \frac{\pi}{\Delta_H}\left(\frac{\Gamma + 1}{(1 - \Gamma^2)^{1/2}}\right), \tag{11.101a}$$

$$N_\beta(y) = \frac{\pi}{\Delta_0} - \frac{\pi}{\Delta_H}\left(\frac{\Gamma - 1}{(1 - \Gamma^2)^{1/2}}\right), \tag{11.101b}$$

and

$$\delta(y) = \frac{2\pi y}{\Delta_H} = \frac{2\pi}{\Delta_H}\left(\frac{\Gamma}{(1-\Gamma^2)^{1/2}}\right), \tag{11.102}$$

where Δ_0 and Δ_H are the Pendellösung lengths defined by Eq. (11.54)

$$\Delta_0 = \frac{\pi \cos\theta_B}{\nu_0} \quad \text{and} \quad \Delta_H = \frac{\pi \cos\theta_B}{\nu_G}. \tag{11.103}$$

It will be noted in the following derivations that the Pendellösung phases in Eqs. (11.100) will not enter the expressions for the intensities when the magnitudes of the wave functions are squared.

The contribution of the α-branch current density to the spatial profile of the diffracted beam leaving the first crystal along path II is given by

$$I_{\alpha,r}(\Gamma) = |R_\alpha(y)|^2 \cdot \mathcal{J} = \frac{1}{4}(1-\Gamma^2)\mathcal{J}, \tag{11.104}$$

where the Jacobian is obtained from Eq. (11.98), namely

$$\mathcal{J} = \frac{\mathrm{d}y}{\mathrm{d}\Gamma} = (1-\Gamma^2)^{-3/2}. \tag{11.105}$$

It converts the distribution in misset angle $\Delta\theta$ (or y) to a distribution in Γ which specifies the spatial positions of the exit rays. For the α-branch current, the exit position on the first crystal is $\Gamma_S = \Gamma$. For the β-branch current, $\Gamma_S = -\Gamma$, and its contribution to the diffracted beam profile is

$$I_{\beta,r}(\Gamma) = \left|R_\beta(y)\right|^2 \cdot \mathcal{J} = \frac{1}{4}(1-\Gamma^2)^{-1/2}. \tag{11.106}$$

Thus, the total diffracted (reflected) beam profile on path II after the first crystal is

$$I_r(\Gamma_S) = I_{\alpha,r}(\Gamma = \Gamma_S) + I_{\beta,r}(\Gamma = -\Gamma_S) = \frac{1}{2}(1-\Gamma_S{}^2)^{-1/2}, \tag{11.107}$$

which diverges at both edges of the reflected (r) beam, but has an integrated intensity

$$\int_{-1}^{+1} I_r(\Gamma_S)\mathrm{d}\Gamma_S = \frac{\pi}{2}. \tag{11.108}$$

The spatial profile of the forward-diffracted (transmitted) beam leaving the first crystal is given by

$$I_{\alpha,t}(\Gamma) = \left|T_\alpha(y)\right|^2 \cdot \mathcal{J} = \frac{1}{4}(1-\Gamma)^2(1-\Gamma^2)^{-3/2}, \tag{11.109}$$

and

$$I_{\beta,t}(\Gamma) = \left|T_\beta(y)\right|^2 \cdot \mathcal{J} = \frac{1}{4}(1+\Gamma)^2(1-\Gamma^2)^{-3/2}, \tag{11.110}$$

where we must substitute $\Gamma = \Gamma_S$ in the expression for $I_{\alpha,t}(\Gamma)$ for the exit point of the α-branch current, and $\Gamma = -\Gamma_S$ in the expression for $I_{\beta,t}(\Gamma)$ to obtain the spatial profile of the transmitted (t) beam leaving the first crystal. That is,

$$I_t(\Gamma_S) = I_{\alpha,t}(\Gamma = \Gamma_S) + I_{\beta,t}(\Gamma = -\Gamma_S) = \frac{1}{2}(1-\Gamma_S)^2\left(1-\Gamma_S^2\right)^{-3/2}, \tag{11.111}$$

which diverges at the right-hand edge of the beam ($\Gamma_S = -1$). The integrated transmitted intensity also diverges. The profiles $I_r(\Gamma_S)$ and $I_t(\Gamma_S)$ shown in Fig. 11.23 are the averages over the Pendellösung oscillations given by Eqs. (11.79) and (11.80) and shown in Fig. 11.18. The intensities $I_r(\Gamma_S)$ and $I_r(\Gamma_S)$ are identical to the expressions (11.39a) for $<I_H(y)>$ and $<I_0(y)>$ obtained earlier if we substitute $y = {}_S(1-\Gamma_S^2)^{-1/2}$ and multiply by the Jacobian $\mathcal{J} = (1-\Gamma_S^2)^{-3/2}$.

We now derive expressions for the beam profiles leaving the middle crystal. The reflected beam on path II leaves the middle crystal along three rays emanating from the points A, B, and F_{II}. Similarly, the reflected beam on path I leaves the middle crystal along three rays emanating from the points C, D, and F_I. We note that the focal points F_I and F_{II} remain fixed as $y(\Delta\theta)$ varies, while the other four points (A, B, C, D) move in the range (–2, 2) of Γ_M, where the midpoint of the range is at F_I for beam path I and at F_{II} for beam path II. The locations of the A and C are given by $\Gamma_M = +2\Gamma$, while the locations of B and D are given by $\Gamma_M = -2\Gamma$.

The reflected (r) beam intensity leaving the middle crystal from point A is

$$I_{A,r}(\Gamma) = \left|R_\alpha(y)R_\alpha(-y)\right|^2 \cdot \mathcal{J}. \tag{11.112}$$

In writing this equation we have noted that the wave vector of the reflected beam from the first crystal is given by $\boldsymbol{k}_H = \boldsymbol{k}_0 + \boldsymbol{H} + \boldsymbol{\delta}$ (see Eq. 11.31), where the addition of $\boldsymbol{\delta}$ means that the rays approach the reflecting planes in the middle crystal at a misset angle, $-\Delta\theta$, corresponding to a

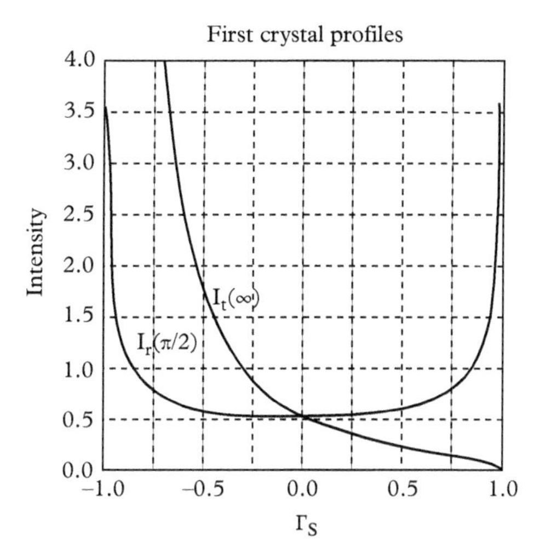

Figure 11.23 *Averaged beam profiles behind the first interferometer plate (r: reflected beam, t: transmitted beam)*

negative y-parameter as indicated in Fig. 11.22. When plotted as a function of Γ_M, this reflected intensity is

$$I_{A,r}(\Gamma_M) = \frac{1}{32}\left(1 - (\Gamma_M/2)^2\right)^{1/2}. \tag{11.113}$$

This equation involves an extra factor of ½ coming from $d\Gamma/d\Gamma_M$. Similarly, the reflected intensity leaving the point B is given by

$$I_{B,r}(\Gamma) = \left|R_\beta(y)R_\beta(-y)\right|^2 \cdot \mathcal{J}, \tag{11.114}$$

so that

$$I_{B,r}(\Gamma_M) = \frac{1}{32}\left(1 - (\Gamma_M/2)^2\right)^{1/2}. \tag{11.115}$$

Neutrons can arrive at the focal point F_{II} via two routes, paths 2 and 3 in the middle crystal. The reflected beam intensity leaving the point F_{II} is therefore

$$I_{F_{II},r}(\Gamma) = \left|R_\alpha(y)R_\beta(-y) + R_\beta(y)R_\alpha(-y)\right|^2 \cdot \mathcal{J} = \frac{1}{4}\left(1 - \Gamma^2\right)^{1/2}. \tag{11.116}$$

Since this intensity all leaves the middle crystal at the fixed focal point F_{II}, its distribution in the variable Γ_M is a delta function, namely

$$I_{F_{II},r}(\Gamma) = (\pi/8)\delta(\Gamma_M), \tag{11.117}$$

where the normalization is given by the integrated intensity

$$\int_{-1}^{+1} I_{F_{II},r}(\Gamma)d\Gamma = \pi/8. \tag{11.118a}$$

The integrated intensity of the reflected beam leaving the points A is

$$\int_{-1}^{+1} I_{A,r}(\Gamma)d\Gamma = \int_{-2}^{+2} I_{A,r}(\Gamma_M)d\Gamma_M = \pi/32, \tag{11.118b}$$

and similarly for points B we have

$$\int_{-1}^{+1} I_{B,r}(\Gamma)d\Gamma = \int_{-2}^{+2} I_{B.r}(\Gamma_M)d\Gamma_M = \pi/32. \tag{11.118c}$$

Thus, ⅔ of the integrated reflected intensity on path II leaving the middle crystal comes from the focal point F_{II}. A plot of the spatial profile of this interfering (reflected) beam on path II is shown in Fig. 11.24. The numbers in parentheses are the integrated intensities.

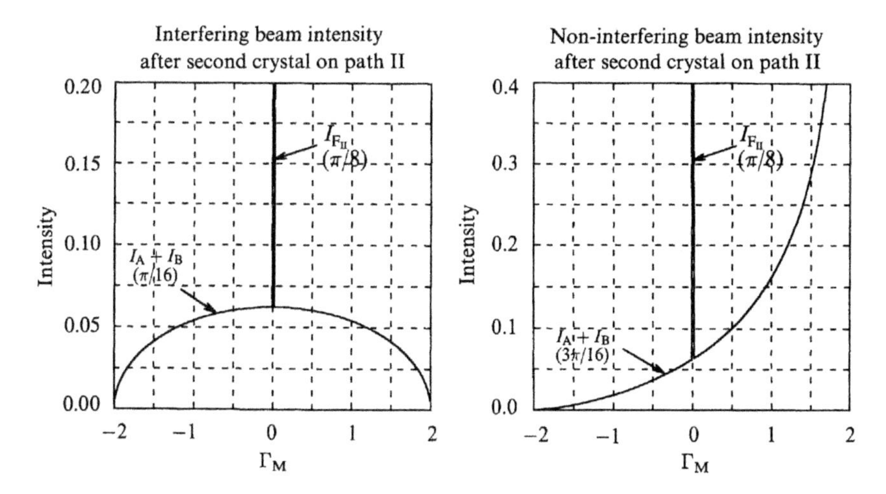

Figure 11.24 *Averaged interfering (left) and non-interfering (right) intensities of beam path II behind the second interferometer plate*

The distribution of the intensity of the non-interfering (transmitted) beam, labeled N_{II} in Fig. 11.22, is

$$I_{A,t}(\Gamma_M) = I_{B,t}(\Gamma_M) = \frac{1}{32}\,[1 + (\Gamma_M/2)]^2\left[1 - (\Gamma_M/2)^2\right]^{-1.2}, \tag{11.119}$$

and

$$I_{F_{II},t}(\Gamma_M) = \frac{\pi}{8}\,\delta\,(\Gamma_M)\,. \tag{11.120}$$

The normalization of the delta function here is given by the integral

$$\int_{-1}^{+1} I_{F_{II},t}(\Gamma)\,d\Gamma = \int_{-1}^{+1} \frac{1}{4}\,\Gamma^2\left(1 - \Gamma^2\right)^{-1/2} d\Gamma = \frac{\pi}{8}. \tag{11.121}$$

The integrated non-interfering (transmitted) intensities leaving points A and B are each $3\pi/32$. Thus, the total integrated intensity in the non-interfering (transmitted) beam is $\pi/8 + 3\pi/32 + 3\pi/32 = 5\pi/16$, while the total integrated intensity in the interfering (reflected) beam on path II is $\pi/8 + \pi/32 + \pi/32 = 3\pi/16$ (sum of intensities from points F_{II}, A and B). That is, ⅜ of the incoming intensity to the middle crystal on path II contributes to the reflected interfering beam which is directed toward the third crystal to combine and interfere with the beam traversing path I, while ⅝ of the incoming intensity is lost to the transmitted non-interfering beam, labeled N_{II}. The spatial distribution of this non-interfering beam is also shown in Fig. 11.24.

We now derive expressions for the beam intensities leaving the middle crystal from points C, D, and F_I on path I. The reflected beam intensity from point C is

$$I_{C,r}(\Gamma) = \left|T_\alpha(y)R_\alpha(y)\right|^2 \cdot \mathcal{J} = \frac{1}{16}(1 - \Gamma)^2\left(1 - \Gamma^2\right)^{-1/2}, \tag{11.122}$$

and when plotted versus $\Gamma_M = 2\Gamma$, we have

$$I_{C,r}(\Gamma_M) = \frac{1}{32}(1 - \Gamma_M/2)^2\left(1 - (\Gamma_M/2)^2\right)^{-1/2}. \tag{11.123}$$

For point D, the reflected intensity is

$$I_{D,r}(\Gamma) = \left|T_\beta(y)R_\beta(y)\right|^2 \cdot \mathcal{J} = \frac{1}{16}(1+\Gamma)^2\left(1-\Gamma^2\right)^{-1/2}. \tag{11.124}$$

However, at point D we have $\Gamma_M = -2\Gamma$, such that

$$I_{D,r}(\Gamma_M) = \frac{1}{32}(1 - \Gamma_M/2)^2\left(1 - (\Gamma_M/2)^2\right)^{-1/2}, \tag{11.125}$$

which is identical to $I_{C,r}(\Gamma_M)$. The integrated intensities of $I_{C,r}(\Gamma_M)$ and $I_{D,r}(\Gamma_M)$ are each equal to $3\pi/32$. There are two paths, 6 and 7, which contribute to the reflected beam leaving the focal point F_I. Therefore, we see that the reflected beam intensity from this point is

$$I_{F_I,r}(\Gamma) = \left|T_\alpha(y)R_\beta(y) + T_\beta(y)R_\alpha(y)\right|^2 \cdot \mathcal{J} = \frac{1}{4}\Gamma^2(1-\Gamma^2)^{-1/2}. \tag{11.126}$$

It has an integrated intensity

$$\int_{-1}^{+1} I_{F_I,r}(\Gamma)\mathrm{d}\Gamma = \frac{\pi}{8}, \tag{11.127}$$

which gives the normalization when plotted versus Γ_M, namely

$$I_{F_I,r}(\Gamma_M) = \frac{\pi}{8}\delta(\Gamma_M). \tag{11.128}$$

Thus, we see that 2/5 of the intensity in the reflected beam on path I leaving the middle crystal comes from the focal point F_I. The profile of the interfering beam leaving the middle crystal on path I is shown in Fig. 11.25.

To complete the calculation of the beam intensities leaving the middle crystal we now evaluate the intensity of the transmitted non-interfering beam, labeled N_I in Fig. 11.22. For point C we have

$$I_{C,t}(\Gamma) = \left|T_\alpha(y)T_\alpha(y)\right|^2 \cdot \mathcal{J} = \frac{1}{16}(1-\Gamma)^4\left(1-\Gamma^2\right)^{-3/2}. \tag{11.129}$$

At point C, $\Gamma_M = 2\Gamma$, so that when plotted versus Γ_M we have

$$I_{C,t}(\Gamma_M) = \frac{1}{32}(1 - \Gamma_M/2)^4\left(1 - (\Gamma_M/2)^2\right)^{-3/2}. \tag{11.130}$$

For point D, where $\Gamma_M = -2\Gamma$, an analogous calculation shows that

$$I_{D,t}(\Gamma_M) = I_{C,t}(\Gamma_M). \tag{11.131}$$

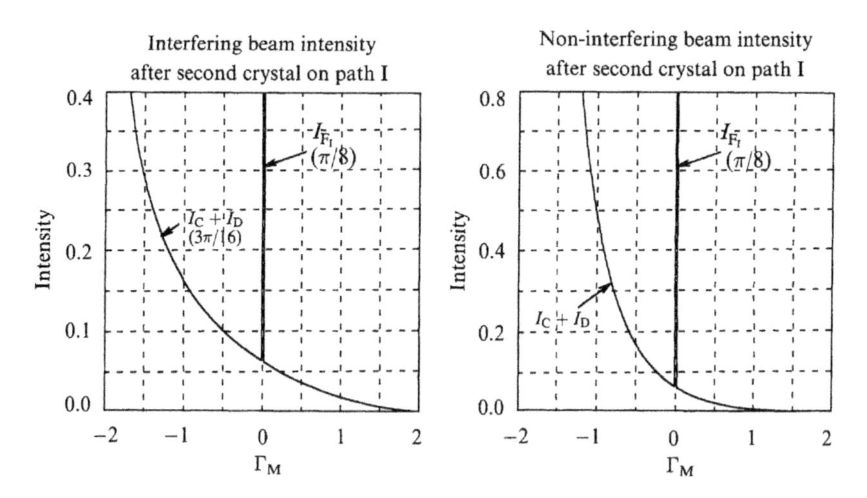

Figure 11.25 *Averaged interfering (left) and non-interfering (right) intensities of beam path I behind the second interferometer plate*

The integrated intensities of both $I_{C,t}$ and $I_{D,t}$ diverge due to the divergence of their distributions at the right-hand edge of the beam profiles at $\Gamma_M = -2$. The non-interfering transmitted beam, labeled N_I in Fig. 11.22, leaving the focal point F_I is

$$I_{F_I,t}(\Gamma) = \left|T_\alpha(y)T_\beta(y) + T_\beta(y)T_\alpha(y)\right|^2 \cdot \mathcal{J} = \frac{1}{4}\left(1-\Gamma^2\right)^{1/2}. \tag{11.132}$$

Its integrated intensity is $\pi/8$, such that its distribution in Γ_M is a delta function given by

$$I_{F_I,t}(\Gamma_M) = \frac{\pi}{8}\delta(\Gamma_M). \tag{11.133}$$

The distribution of the non-interfering beam N_I leaving the middle crystal is shown in Fig. 11.25.

It is a straightforward matter to show that the sum of the transmitted (non-interfering) beam and the reflected (interfering) beam leaving the middle crystal from points A, B, and F_{II} is equal to the beam incident upon the middle crystal on path II, namely $(1/2)(1-\Gamma^2)^{-1/2}$, for every value of $\Gamma(\Delta\theta)$. A similar calculation for path I gives the necessary conservation condition, where the incident intensity equals the sum of the reflected and transmitted intensities, namely $(1/2)(1+\Gamma^2)(1-\Gamma^2)^{-3/2}$, coming from the points C, D, and F_I.

We are now prepared to calculate the spatial profiles of the interfering beams leaving the third crystal, labeled A in Fig. 11.22. For a given incident wave vector $\boldsymbol{k}_0$ oriented at the misset angle $\Delta\theta$, the wave function leaving the interferometer is

$$\Psi(\boldsymbol{r}) = u_0 e^{i\boldsymbol{k}_0\cdot\boldsymbol{r}} + u_H e^{i\boldsymbol{k}_H\cdot\boldsymbol{r}}, \tag{11.134}$$

where the intensity emanates from the exit face of the analyzer crystal at the points *a*, *b*, *c*, and *d*. As $\Delta\theta$ is varied, the points *a* and *d* move on the exit face in the range (−3, 3) of the parameter Γ_A. Similarly, the points *b* and *c* move in the range (−1, 1) of the parameter Γ_A. We begin with a calculation of the intensity corresponding to the *maverick* paths leading to the exit points *a* and *d*. For point *a* we have for its contribution to the 0-beam

$$I_0^a(\Gamma) = \left|u_0^a\right|^2 \cdot \mathcal{J} = \left|R_\alpha(y)R_\alpha(-y)T_\alpha(y)e^{i\Delta\beta} + T_\alpha(y)R_\alpha(y)R_\alpha(-y)\right|^2 \cdot \mathcal{J}$$
$$= \frac{1}{32}(1-\Gamma)^2(1-\Gamma^2)^{1/2}(1+\cos\Delta\beta), \tag{11.135}$$

or, when plotted versus $\Gamma_A = 3\Gamma$ we have

$$I_0^a(\Gamma_A) = \frac{1}{96}(1-\Gamma_A/3)^2(1-\Gamma_A/3)^2)^{1/2}(1+\cos\Delta\beta). \tag{11.136}$$

The additional factor of ⅓ comes from the Jacobian $d\Gamma_A/d\Gamma$. The phase difference between all rays on path II and all rays on path I is $\Delta\beta$, provided by a phase-shifting flag on path II. The contribution of point *a* to the H-beam intensity is given by

$$I_H^a(\Gamma) = \left|u_H^a\right|^2 \cdot \mathcal{J} = \left|R_\alpha(y)R_\alpha(-y)R_\alpha(y)e^{i\Delta\beta} + T_\alpha(y)R_\alpha(y)T_\alpha(-y)\right|^2 \cdot \mathcal{J}$$
$$= \frac{1}{32}(1-\Gamma^2)^{3/2}(1+\cos\Delta\beta). \tag{11.137}$$

When this part of the H-beam intensity is plotted versus $\Gamma_A = 3\Gamma$, we have

$$I_H^a(\Gamma_A) = \frac{1}{96}(1-(\Gamma_A/3)^2)^{3/2}(1+\cos\Delta\beta). \tag{11.138}$$

We note that the contribution of the point *a* to the interferograms $I_0(\Gamma_A, \Delta\beta)$ and $I_H(\Gamma_A, \Delta\beta)$ are *in-phase*, going up and down together as a function of the phase shift $\Delta\beta$, but with different amplitudes of oscillation. The reason for this is that only α-branch currents in each of the three crystals contribute to the intensity leaving point *a* on the exit face of the analyzer crystal.

For the second *maverick* path leading to the exit point *d*, we note that only the β-branch currents in the three crystals contribute. The resulting intensity profiles are identical to those coming from point *a*, namely

$$I_0^d(\Gamma_A) = I_0^a(\Gamma_A) \qquad \text{and} \qquad I_H^d(\Gamma_A) = I_H^a(\Gamma_A), \tag{11.139}$$

where we have used the fact that at point *d*, $\Gamma_A = -3\Gamma$.

Calculation of the beam intensities leaving points *b* and *c* is somewhat more complicated. The total wave function for the *primary* paths on paths I and II involves three terms each. For the 0-beam intensity emanating from point *b* we have

$$I_0^b(\Gamma) = \left|u_0^b\right|^2 \cdot \mathcal{J} = \left|\left(R_\alpha(y)R_\alpha(-y)T_\beta(y) + R_\alpha(y)R_\beta(-y)T_\alpha(y) + R_\beta(y)R_\alpha(-y)T_\alpha(y)\right)e^{i\Delta\beta}\right.$$
$$\left. + \left(T_\alpha(y)R_\alpha(y)R_\beta(-y) + T_\alpha(y)R_\beta(y)R_\alpha(-y) + T_\beta(y)R_\alpha(y)R_\alpha(-y)\right)\right|^2 \cdot \mathcal{J}$$
$$= \frac{1}{32}\left(1-\Gamma^2\right)^{1/2}(3\Gamma-1)^2(1+\cos\Delta\beta). \tag{11.140}$$

At point *b*, we note the $\Gamma_A = \Gamma$ since each term in this equation involves two α-branch currents and one β-branch current. The contribution of the point *b* to the H-beam is

$$I_H^b(\Gamma) = \left|u_H^b\right|^2 \cdot \mathcal{J} = \left|\left(R_\alpha(y)R_\alpha(-y)R_\beta(y) + R_\alpha(y)R_\beta(-y)R_\alpha(y) + R_\beta(y)R_\alpha(-y)R_\alpha(y)\right)e^{i\Delta\beta}\right.$$
$$\left. + \left(T_\alpha(y)R_\alpha(y)T_\beta(-y) + T_\alpha(y)R_\beta(y)T_\alpha(-y) + T_\beta(y)R_\alpha(y)T_\alpha(-y)\right)\right|^2 \cdot \mathcal{J}$$
$$= \frac{1}{64}\left[9\left(1-\Gamma^2\right)^{3/2} + (3\Gamma^2+1)^2(1-\Gamma^2)^{-1/2} - 6(3\Gamma^2+1)(1-\Gamma^2)^{1/2}\cos\Delta\beta\right]. \tag{11.141}$$

The "primary" paths leading to point c involve two β-branch currents and one α-branch current. A calculation analogous to the one above for point b gives for the contribution of point c to the 0-beam intensity

$$I_0^c(\Gamma) = \frac{1}{32}(1-\Gamma^2)^{1/2}(3\Gamma+1)^2(1+\cos\Delta\beta), \tag{11.142}$$

and the H-beam intensity is symmetric in Γ and identical to that for point b, that is

$$I_H^c(\Gamma) = I_H^b(\Gamma). \tag{11.143}$$

For point c, $\Gamma_A = -\Gamma$; thus, when plotted versus Γ_A we see that the 0-beam intensities from points b and c are also equal.

We now have a complete mathematical description of the spatial profiles of the 0-beam and H-beam leaving the third crystal of the interferometer. The spatial profile of the 0-beam is shown in Fig. 11.26. We note that it is asymmetric about $\Gamma_A = 0$, with two peaks occurring at $\Gamma_A = 0.874$ and -0.763, coming from the *primary* paths (points b and c). The spatial profile of the H-beam, shown in Fig. 11.26, is symmetric about $\Gamma_A = 0$, showing singular peaks at $\Gamma_A = \pm 1$. The *maverick* path contributions to both the 0-beam and the H-beam spatial distributions are smooth functions, going to 0 at $\Gamma_A = \pm 3$. The integrated intensities are the areas under these profiles, and are shown in the parentheses in Fig. 11.27. We note that for the 0-beam 13/18 of the total intensity is due to the primary paths and falls in the range (−1,1) of Γ_A; while for the H-beam 43/46 of the total integrated intensity comes from the *primary* paths and falls in the range (−1,1) of Γ_A. The phases of these beams are treated in more detail by Lemmel (2013).

The Γ-dependent interferograms for both the 0-beam and the H-beam are each composed of a "maverick" and a "primary" component. The interferograms are of the general forms

$$I_0(\Gamma, \Delta\beta) = a_0(\Gamma) + b_0(\Gamma)\cos\Delta\beta, \tag{11.144}$$

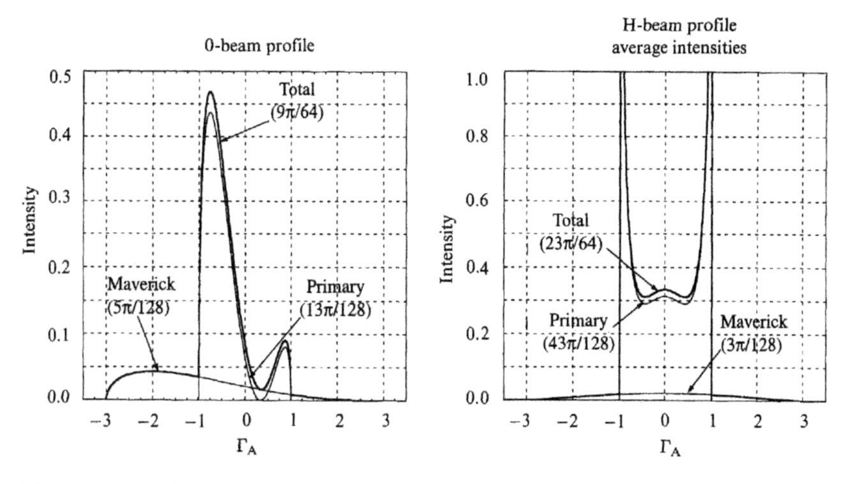

Figure 11.26 *Averaged spatial intensity profiles behind the interferometer for the 0- and H-beams. The primary part indicates the interfering and the maverick part the non-interfering contributions*

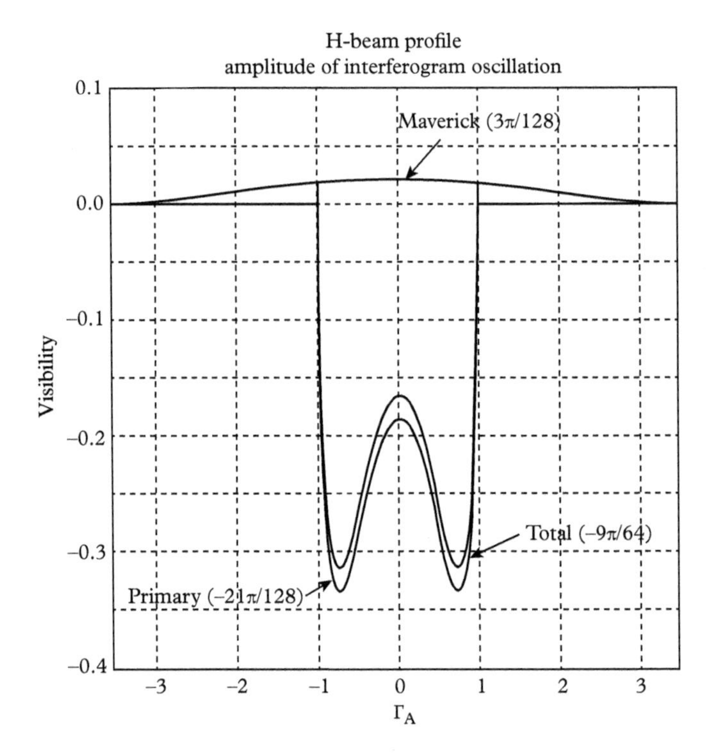

Figure 11.27 *Averaged visibility of the H-beam behind the interferometer (– sign indicates opposite phase to 0-beam)*

and

$$I_{\mathrm{H}}(\Gamma,\Delta\beta)=a_{\mathrm{H}}(\Gamma)+b_{\mathrm{H}}(\Gamma)\cos\Delta\beta. \tag{11.145}$$

To conserve neutron flux as the phase $\Delta\beta$ is varied, $b_{\mathrm{H}}(\Gamma)$ must be equal to $b_0(\Gamma)$, but of opposite sign; and the mean intensities given by $a_{\mathrm{H}}(\Gamma)$ and $a_0(\Gamma)$ must sum to give the total intensity of the two beams on paths I and II incident upon the third crystal for every value of $\Gamma(\Delta\theta)$. We now explicitly establish that these requirements are satisfied. From Eqs. (11.136)–(11.140), we can write down expressions for the mean intensities $a_{\mathrm{H}}(\Gamma)$ and $a_0(\Gamma)$, and for the amplitudes of oscillation $b_{\mathrm{H}}(\Gamma)$ and $b_0(\Gamma)$ of the interferograms. The mean intensity of the 0-beam is given by

$$\begin{aligned}a_0(\Gamma)&=a_0^{\mathrm{m}}(\Gamma)+a_0^{\mathrm{p}}(p)\\&=\frac{1}{32}\left(1-\Gamma^2\right)^{1/2}\left[(1-\Gamma)^2+(1+\Gamma)^2+(3\Gamma-1)^2+(3\Gamma+1)^2\right]\\&=\frac{1}{8}(1-\Gamma^2)^{1/2}(1+5\Gamma^2),\end{aligned} \tag{11.146}$$

where the first two terms in the middle equation are for the maverick m paths (points *a* and *d*), and the second two terms are for the primary p paths (points *b* and *c*). The mean intensity of the H-beam is given by

$$
\begin{aligned}
a_H(\Gamma) &= a_H^m(\Gamma) + a_H^p(p) \\
&= \frac{1}{32}(1-\Gamma^2)^{1/2}[2(1-\Gamma^2)^2 + 9(1-\Gamma^2)^2 + (3\Gamma^2+1)^2] \\
&= \frac{1}{8}(1-\Gamma^2)^{-1/2}(5\Gamma^4 - 4\Gamma^2 + 3),
\end{aligned} \tag{11.147}
$$

where the first term in the middle equation is for the maverick m paths, and the second and third terms come from the primary p paths. The sum of these mean intensities of the 0-beam and H-beam is therefore seen to be

$$
a_0(\Gamma) + a_H(\Gamma) = \frac{1}{2}(1-\Gamma^2)^{-1/2}. \tag{11.148}
$$

In a similar manner, using Eqs. (11.113)–(11.132), we can calculate the total intensity (for a given Γ) on paths I and II entering the third crystal. After a small amount of algebra, it is found to be given by

$$
[I_{F_I}(\Gamma) + I_C(\Gamma) + I_D(\Gamma)] + [I_{F_{II}}(\Gamma) + I_A(\Gamma) + I_B(\Gamma)] = \frac{1}{2}(1-\Gamma^2)^{-1/2}, \tag{11.149}
$$

which agrees with the expression (11.148) for the mean intensity leaving the third crystal in the 0-beam plus the H-beam.

The total amplitude of the 0-beam interferogram is the same as its mean value given by Eq. (11.146). That is,

$$
\begin{aligned}
b_0(\Gamma) &= b_0^m(\Gamma) + b_0^p(p) \\
&= \frac{1}{8}(1-\Gamma^2)^{1/2}(1+5\Gamma^2)
\end{aligned} \tag{11.150}
$$

The amplitude of the oscillations of the H-beam interferogram is

$$
\begin{aligned}
b_H(\Gamma) &= b_H^m(\Gamma) + b_H^p(\Gamma) \\
&= \frac{1}{32}[2(1-\Gamma^2)^{3/2} - 6(3\Gamma^2+1)(1-\Gamma^2)^{1/2}] \\
&= -\frac{1}{8}(1-\Gamma^2)^{1/2}(1+5\Gamma^2).
\end{aligned} \tag{11.151}
$$

Thus, $b_H(\Gamma) = -b_0(\Gamma)$, as required for conservation of neutron intensity, independent of the phase shift $\Delta\beta$. It should be noted, however, that when plotted versus Γ_A, the profile of the oscillation amplitude b_H of the H-beam (Fig. 11.27) is quite different from the profile of the 0-beam oscillation amplitude, b_0.

In most neutron interferometry experiments, the detectors are sufficiently wide open so as to integrate over the spatial profiles of the beams leaving the interferometer. We are therefore interested in the integrated intensities

$$
\mathrm{I}_0(\Delta\beta) \equiv \int_{-1}^{+1} I_0(\Gamma, \Delta\beta)\,d\Gamma = A_0 + B_0\cos\Delta\beta, \tag{11.152a}
$$

and

$$\mathrm{I_H}(\Delta\beta) \equiv \int_{-1}^{+1} I_\mathrm{H}(\Gamma, \Delta\beta)\mathrm{d}\Gamma = A_\mathrm{H} + B_\mathrm{H}\cos\Delta\beta, \tag{11.152b}$$

where the mean integrated intensities are

$$A_0 = \int_{-1}^{+1} a_0(\Gamma)\mathrm{d}\Gamma = \frac{1}{8}\int_{-1}^{+1} (1-\Gamma^2)^{1/2}(1+5\Gamma^2)\mathrm{d}\Gamma = \frac{9\pi}{64}, \tag{11.153a}$$

and

$$A_\mathrm{H} = \int_{-1}^{+1} a_\mathrm{H}(\Gamma)\mathrm{d}\Gamma = \frac{1}{8}\int_{-1}^{+1} (1-\Gamma^2)^{-1/2}(5\Gamma^4 - 4\Gamma^2 + 3)\mathrm{d}\Gamma = \frac{23\pi}{64}. \tag{11.153b}$$

Thus, we see that the ratio of the average counting rates in the H-beam detector to that in the 0-beam detector is predicted to be

$$\frac{<\mathrm{I_H}(\Delta\beta)>}{<\mathrm{I_0}(\Delta\beta)>} = \frac{A_\mathrm{H}}{A_0} = \frac{23}{9} = 2.555, \tag{11.154}$$

a prediction that is borne out by experiment. The integrated 0-beam amplitude of oscillation of the interferogram is equal to its mean, i.e., $B_0 = A_0$. For the H-beam, the amplitude of oscillation of the interferogram is given by

$$B_\mathrm{H} = \int_{-1}^{+1} b_\mathrm{H}(\Gamma)\mathrm{d}\Gamma = -\frac{9\pi}{64}, \tag{11.155}$$

which is equal to B_0, but of opposite sign. Therefore, we see that the predicted contrast of the 0-beam is 100%, but the contrast of the H-beam is

$$\frac{\mathrm{I_H(max)} - \mathrm{I_H(min)}}{\mathrm{I_H(max)} + \mathrm{I_H(min)}} = \frac{B_\mathrm{H}}{A_\mathrm{H}} = \frac{9}{23} = 0.391, \tag{11.156}$$

or 39.1%.

If the detectors are not sufficiently wide open to integrate over the entire widths of the exit beams, the amplitudes of oscillation of the 0- and H-beam interferograms will not be equal in magnitude. The assumption that the detectors are integrating over the entire beam profiles is not always easily met, primarily because the maverick paths extend the profiles out to $\Gamma_\mathrm{A} = \pm 3$. The profiles shown in Figs. 11.26 and 11.27 are drawn for an incident beam slit of zero width. For an incident beam defined by a slit of width W, the exit beam profiles are all broadened by W. That is, experimentally measured profiles should be compared with distributions obtained after convoluting the calculated profiles above with a slit transmission function. Furthermore, the incident beam is always divergence in angle, thus further broadening the spatial profiles of the exit beams calculated above. Typically, the incident beam divergence is of order ½°. Thus, for an interferometer that is

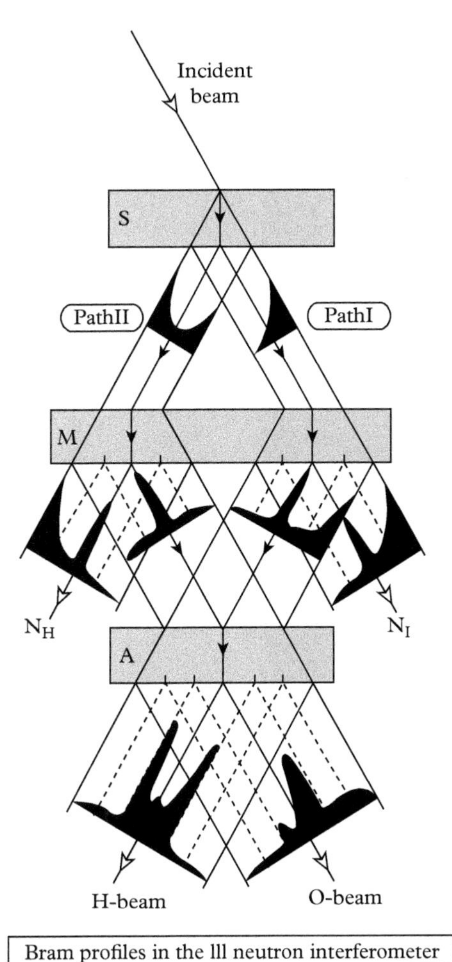

Figure 11.28 *Schematic view of the averaged intensity profiles within and behind a symmetric LLL interferometer*

10 cm in size, the beam divergence will contribute a spatial broadening of about 1 mm to the exit beam profiles. In Fig. 11.28 we schematically summarize the beam profiles within the interferometer, and the profiles of the average intensities leaving the third crystal in the 0- and the H-beams.

Various experimental attempts have been made to measure the intensity profiles at least those behind the interferometer. Figure 4.28 shows such an example where the profile of the 0-beam has been measured by means of a scanning slit (Bauspiess et al. 1978). The qualitative agreement with the theoretical prediction is visible (Fig. 11.24). A more comprehensive study was performed by Kischko (1983) who used a photographic film and a Gd-converter to obtain the profiles behind the interferometer for various thicknesses of the interferometer plates (Fig. 11.29).

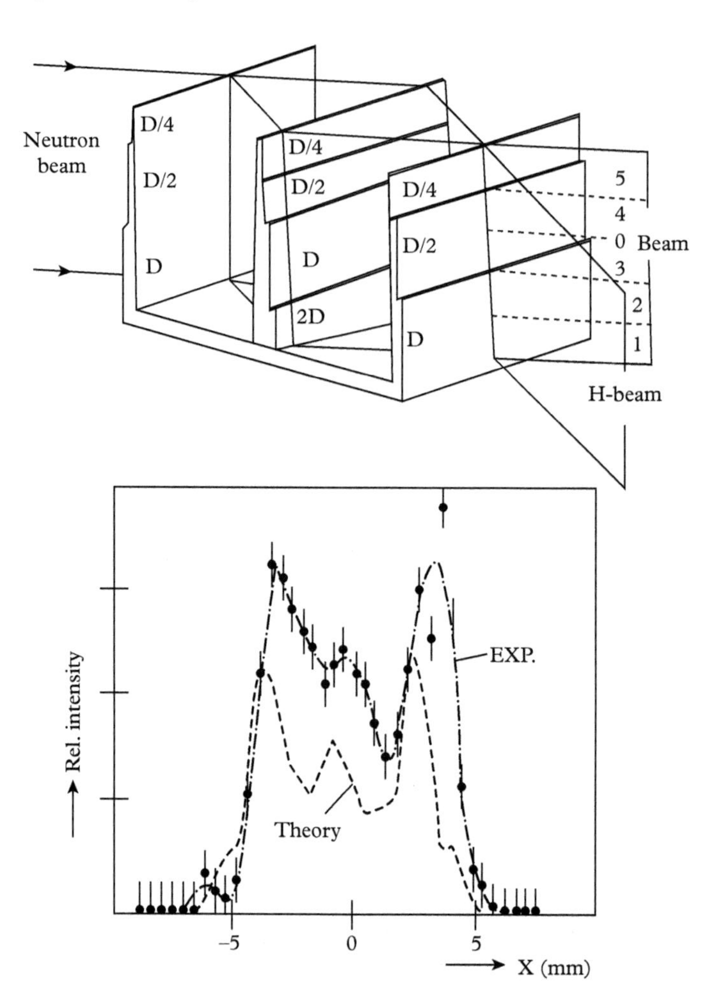

Figure 11.29 *Sketch of the step interferometer crystal and a typical result for the H-beam in the case of equal crystal thicknesses corresponding to the vertical segment labeled 2 above (Kischko 1983)*

The uncertainties of the photographic method, the rather poor geometrical accuracies of the interferometer crystal, and the deconvolution with the slit width of the incident beam make a quantitative comparison rather difficult. A position-sensitive detector was used by Ioffe et al. (1995) to measure the interferometer output beams but the resolution was not sufficient to observe details of the profile. A more detailed study of the intensity and contrast profile has been done by Lemmel (2010) who used an advanced position sensitive detectorand used these measurements for adjusting high contrast (see Fig. 2.9).

A futher comment should be made concerning the loss beams at the second crystal plate and the non-perfect contrst in the H-beam. In this cases the reflection (transmission) function reaches 1 or 0 which means these neutrons exhibit beam path information and, therefore, they do not contribute to the contrast.

In the previous chapters we have assumed that the geometry of the interferometer is perfect; that is, we have assumed that the thicknesses of the crystal blades are all equal, and that the distance L_1 between the first two blades is equal to the distance L_2 between the middle blade and the third blade. These conditions of perfection are never met in practice. The lack of perfection of the machining of the crystal leads to defocusing effects which reduce the observed interferogram contrast. These defocusing effects are in addition to the mechanisms of contrast reduction discussed in detail in Chapter 4. In the next section we will briefly describe these geometrical defocusing effects.

11.7.3 The Defocused Interferometer

In this case the distances of the crystal plates or the thickness of the plates may be different (Fig. 11.30). The wave functions can be calculated in the same manner as before but the geometrical factors entering Eqs. (11.100) change. For both beam paths one gets another Γ-factor

$$\Gamma_{\mathrm{I,II}} = \frac{y}{D \tan \Theta_{\mathrm{B}}} \mp \frac{\Delta t}{D}, \tag{11.157}$$

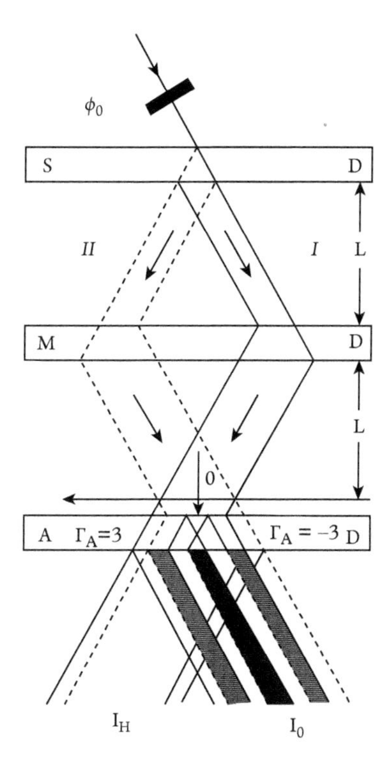

Figure 11.30 *Beam path inside a defocused interferometer. The hatched regions indicate where the Borrmann fan from both beam paths overlaps*

where Δt denotes the defocus. In Section 4.2.2 it has been shown that defocusing can also occur due to a non-dispersively arranged phase shifter. If $\Delta t << D$ and thick crystals $A >> \pi$ one gets after some lengthy calculations (Bonse and de Kaat 1971, Bauspiess et al. 1976, Petrascheck 1976, Petrascheck and Folk 1976)

$$\begin{aligned} I_0(\Gamma) &\propto (1-3\Gamma)^2(1-\Gamma^2)^{1/2}\cos^2\left(\frac{A\Gamma\,\Delta t/D}{(1-\Gamma^2)^{1/2}}\right) \\ &+\frac{1}{3}\left(1-\frac{\Gamma}{3}\right)^2 1-\Gamma^2/9)^{1/2}\cos^2\left(\frac{A\Gamma\,\Delta t/D}{3(1-\Gamma^2/9)^{1/2}}\right) \text{ for } \Gamma_{\rm I}, \Gamma_{\rm II}, < 1 \\ &\propto \frac{1}{3}(1-\Gamma/3)^2(1-\Gamma^2/9)^{1/2}\cos^2\left(\frac{A\Gamma\,\Delta t/D}{3(1-\Gamma^2/9)^{1/2}}\right) \\ &\text{for } 1 \le |\Gamma_{\rm I}|, |\Gamma_{\rm II}| < 3. \end{aligned} \tag{11.158}$$

This formula gives typical and nearly equidistant defocusing fringes at (here given on the y-scale)

$$y_n \cong \frac{3D\Delta_0 \tan\Theta_{\rm B}}{\Delta t}\cdot n. \tag{11.159}$$

This feature can be used for precise registration of small motions and they are essential for the adjustment of split perfect crystal interferometers which exists for X-rays (Deslattes and Henins 1973, Becker et al. 1981). The auto-correlation function of Eqs. (11.158) gives, after some further simplifications, the transverse coherence function as discussed in Section 4.2.2 (Eq. 4.54; Holy 1980, Petrascheck 1988).

Defocusing may also occur due to distorted crystals. In this case the Takagi–Taupin theory can be used as well. Applications for high-sensitive internal strain and dislocation studies are known. An overview has been given by Gronkowski (1991).

12

Interpretational Questions and Conclusions

We have shown many successful applications of quantum mechanics in this book and it would be a major problem not having the fantastic quantum theory available. Despite its successes, however, the basic conceptual framework and its interpretation have been considered by many scientists to be unsatisfactory. We will outline the standard Copenhagen–Göttingen interpretation and some of the alternative approaches in this chapter. Quantum mechanics must be seen as a major part of more general theories of physics. Inevitably many of the questions that arise are matters of opinion rather than facts and, therefore, many physicists consider that they belong more properly to philosophy than to physics. However, the conceptual basis of quantum mechanics is so fundamental to our understanding of nature that it should surely be important for physicists and philosophers as well. The conception of *determinism* differs substantially from classical physics because quantum theory predicts only the relative probabilities of different outcomes, of different events. Thus, Einstein's dream that a complete theory should describe the outcome of individual observational events is still lacking. The feature of *non-locality* is another new concept of most interpretations which makes space-like separated events instantaneously entangled. The well-known Einstein–Bohr debate about the *completeness* of quantum theory, the question whether the universe behaves *deterministically* or *non-deterministically*, and the meaning of *free will* are brought into focus in this chapter (e.g., Schilpp 1949, Bell 1964, Withaker 1995, Groeblacher et al. 2007, Conway and Kochen 2009, Englert 2013). We must come back to the underlying fundamental motivation of all thinking people to find the *causes* for *effects*, and that *causes* should always precede *effects*.

The various kinds of particle optics experiments with electrons, neutrons, atoms, molecules, and clusters have attracted strong interest on interpretational questions in quantum mechanics. This is mainly because they demonstrate that massive and composite particles can be described by wave fields, engendering a unification of massless and massive entities, that is, of light and matter. In neutron interferometry widely separated beams in ordinary and/or momentum space can be produced. These situations are well described by quantum mechanics but they elucidate the wave–particle dualism in a new manner. There are a number of achievements of neutron interferometry which may contribute new insights to the discussion of the interpretation of quantum mechanics.

(a) Spatially split-beam interferometry with massive and composite particles is feasible and high-contrast interference patterns can be obtained. Thus, each neutron having evolved with a given "split" history joins behind the interferometer in either the 0-beam or H-beam, depending on the relative phase shift it has experienced in the separated beam

Neutron Interferometry. Second Edition. Helmut Rauch and Samuel A. Werner.

paths. Thus, it seems that the neutron after superposition carries merger information about the physical situation in both widely separated beam paths, which have evolved simultaneously over time (Chapter 2, Fig. 2.1).

(b) All measured results belong to self-interference phenomena where only one neutron exists at a time within the region of the interferometer (Section 1.3). Experiments with pulsed beams had a mean occupation number per neutron burst far below unity and the interference pattern exists there as well (Section 4.5.5).

(c) Post-selection experiments have shown that an interference pattern can be restored even when at high interference order, where the interference contrast essentially disappears due to the dispersive action of the phase shifter (Section 4.5.2). At the same time an interference modulation of the momentum distribution appears. This can be observed by proper post-selection methods. Various retrieval methods of interference phenomena are found to be applicable (Section 4.5).

(d) The superposition states behind the interferometer have highly non-classical features and they are very sensitive to any kind of decoherencing and dephasing effects (Section 4.5.2). They are *Schrödinger cat-like* states since the neutron occupies two different regions in space.

(e) Intrinsic features of the neutron spin-½ system leave their fingerprint on the interference pattern. In this connection the 4π-symmetry of spinor wave functions and the quantum spin superposition law have been verified (Sections 5.1 and 5.2). Here again, the self-interference phenomenon becomes very apparent.

(f) Phase shifts can not only be produced by conservative potentials (nuclear, magnetic, gravitational), but also by potentials, which do not result in a classical force acting on the neutron. The actions of an electric field (Aharonov–Casher effect), of a purely time-dependent magnetic field (scalar Aharonov–Bohm effect), and of a geometrically formed magnetic field interaction (Berry phase) have been observed and quantitatively measured (Chapter 6).

(g) Energy sensitivities on the order of 10^{-19}eV have been achieved, which is more than 10 orders of magnitude below the energy width of the beams (~ 0.2 meV). This indicates how extremely tight energy transfers in both beam paths affect each neutron wave function (Section 5.4).

(h) Delayed choice experiments with a Jamin multilayer interferometer have shown that the action of the first beam splitter is not dependent upon whether an interference or a beam path detection experiment follows (Section 10.5, Kawai et al. 1998b).

(i) Split-beam interference experiments where beams with different directions of their momenta and spin-echo experiments where beams with different momenta along the same beam path interfere are thought to be equivalent. This shows that interference effects in ordinary and momentum space should be seen as a common interference phenomenon in phase space. This can be visualized by Wigner functions (Section 4.5.3).

(j) In Chapter 7 it has been shown that different degrees of freedom of a single neutron can be entangled and can be used as a basis for Bell-like inequalities measurements. These measurements verified quantum contextuality as an additional basic feature of nature.

In all kinds of discussions regarding the epistemological impact of quantum theory, it should be kept in mind that solutions of the Schrödinger equation only exist when boundary and/or initial conditions are defined; and that measurement results are obtained only when a number of events arising from a similarly prepared ensemble are taken, or when a single system is measured numerous times. In any case, our knowledge is limited to the knowledge of the wave function which follows from the Schrödinger equation (Eq. 1.2) when a quantum state Ψ is exposed to a certain interaction described by a Hamiltonian H and prepared according to certain boundary conditions. This should be taken as seriously as the Schrödinger equation itself,

$$\psi(\boldsymbol{r} = \boldsymbol{R},\ t = t_0) = \psi_{\mathrm{B}}(\boldsymbol{R}, t_0). \tag{12.1}$$

Therefore, the linear superposition principle which follows from the structure of the Schrödinger equation is a basic requisite of quantum mechanics. A measurement consists, in general, of a preparation stage, which can be described also by a Hamiltonian, and the irreversible collapse stage within a detector where the most evident coupling of the quantum and classical worlds happens; and where a Hamiltonian description cannot be given. Here we focus on interpretational questions related to quantum mechanics. In a broader view one should be aware of the fact that this theory is one pillar only of general quantum field theories, bridging general relativity and quantum theory. Such field theories are often related to superstring theories describing the fundamental components of matter (electrons, neutrons, protons, quarks, etc.) as strings, which vibrate in space-time at a fundamental frequency and its harmonics (Schwarz 1982, Green and Schwarz 1985, Polchinski 1998). Typical dimensions of strings are 10^{-35}m and it has been found that there is one, and only one, form of symmetry (SO(32)) in a 32-dimensional space which is free of anomalies and infinities. In this theory Fermions vibrate in ten dimensions, whereas bosons need 26 dimensions for an adequate description. Various compaction procedures which yield a $SU_3 \times SU_2 \times SU_1$ symmetrization are feasible, where SU_3 describes the symmetry groups associated with the standard model of quarks and gluons while $SU_2 \times SU_1$ represents the symmetry group for the electroweak interaction. In this respect, the discussions of quantum mechanics and especially on matter wave interferometry may help the discussion of the grand unification perspective as well.

What might be the open questions in quantum mechanics to be further discussed? The predictions of quantum mechanics derived from wave function consist of probability amplitudes. The operational significance of the resulting relative frequencies of occurrence has been tested up to an incredible accuracy. Nevertheless, epistemological questions remain which should be discussed in order to avoid unnecessary confusion and dispute:

1. What is the meaning of the wave function?
2. How is the measurement process described?
3. How can a classical world appear out of quantum mechanics?
4. How can non-locality and contextuality be explained?
5. Is quantum evolution reversible?
6. Is there a measurement problem?

Since it has always been the goal of physics to give an objective realistic description of the world, it might seem that this goal is most easily achieved by interpreting any observed quantum state as an element of reality. Such ideas are very common in the literature, or they appear as implicit assumptions. However, the assumption that a quantum state is a property of an individual physical system and indicates local realism leads to contradictions and must be abandoned. The lack of separability of the quantum system from the measuring apparatus and even between two widely separated and noninteracting quantum systems puts the question forward as to whether the quantum-mechanical description of physical reality can be considered to be complete (Einstein et al. 1935). In that context Einstein (1949) strongly advocated an underlying ensemble interpretation which may be alleviated by a slightly more general definition of a quantum state that includes the state preparation procedure. Separability in ordinary space would be a physically more relevant objective criterion, particularly because all known interactions decrease rapidly with distance in ordinary space. But one may argue that separability in ordinary space is not sufficient to ensure separability in configuration space where the wave function is defined. We will follow this idea throughout this chapter.

The predictions of quantum mechanics derived from a wave function consist of probabilities and the operational significance of a probability is the relative frequency of events (i.e., $|\psi(\boldsymbol{r},t)|^2$). In Chapter 4 it has been shown that the auto-correlation function (i.e., $\Gamma(\boldsymbol{\Delta},t) = <\psi(\boldsymbol{r}',t')\ \psi(\boldsymbol{r}'',t'')>$, $\boldsymbol{\Delta} = \boldsymbol{r}' - \boldsymbol{r}''$, $t = t' - t''$), which defines the coherence function, or the Wigner function, which describes a quasi-distribution function, can be used to connect predictions of the theory with measurable quantities. Thus, we are bound to invoke an ensemble of similar systems, regardless of how we originally interpreted the wave function.

Neutrons which are moving in free space represent a pure state $|\psi>$. It is generally is a coherent superposition of plane wave solutions (Eq. 1.28). In the ordinary space representation one gets

$$|\psi> \; = \; <\boldsymbol{r}|\psi> \; = (2\pi)^{-3/2} \int a(\boldsymbol{k}-\boldsymbol{k}_0)\, e^{i(\boldsymbol{k}\cdot\boldsymbol{r}-\omega_k t)}\, \mathrm{d}^3 k, \tag{12.2}$$

where $|a(\boldsymbol{k}-\boldsymbol{k}_0)|^2 \propto g(\boldsymbol{k}-\boldsymbol{k}_0)$ denotes the momentum distribution of the state with a continuous set of energy eigenvalues $E_k = \hbar\omega_k = \hbar^2 k^2/2m$. Such a pure state can be a coherent state as well when the related spatial and momentum distribution functions have Gaussian forms and a Poissonian particle distribution function. In this case the minimum uncertainty relation ($\Delta x \Delta k = ½$) is fulfilled and the variance of the particle number is $\Delta N = \sqrt{N}$. The quantum state $|\psi>$ in the momentum representation is given by

$$|\psi> \; = \; <\boldsymbol{k}|\psi> \; = (2\pi)^{-3/2} \int e^{-i\boldsymbol{k}\cdot\boldsymbol{r}} <\boldsymbol{r}|\psi>\, \mathrm{d}^3 \boldsymbol{r}. \tag{12.3}$$

The question whether $|\psi>$ provides a complete and exhaustive description of an individual system (variant A), or it describes the statistical properties of an ensemble of similarly prepared systems (variant B) must be elucidated in more detail. Interpretation A may be the more common one (e.g., Bohr 1958, Hartle 1968, Aharonov et al. 1993) but interpretation B, originally used by Einstein et al. (1935), has been consistently adapted over the past decades (e.g., Ballentine 1990). Interpretation A can be split into two variants. Either each neutron is emitted in a single energy (plane wave) but the particular energy varies from one neutron to the next (A_1) or each neutron is emitted as a wave packet that has an energy (momentum) spread equal to the energy (momentum) spread of the beam (A_2). In the first case (A_1) the state operator which is a solution of the von Neumann (or quantum Liouville) equation

$$i\hbar\frac{\partial\rho}{\partial t} = [\mathcal{H}, \rho], \tag{12.4}$$

which reads for one coordinate

$$\rho(x,x') = <x|\rho\,|x'> = \int \psi_k(x,t)\ \psi_k^*(x',t)\ g(k-k_{0x})\mathrm{d}\,k$$
$$= \int e^{ik(x-x')}\ g(k-k_0)\mathrm{d}\,k. \tag{12.5}$$

This is time-independent and obtained by an averaging procedure. In the second case (A_2) one does not speculate about the individual wave functions ψ_k but uses the steady-state condition $\partial\rho/\partial t = 0$ (or $[\mathcal{H}, \rho] = 0$), where $\mathcal{H} = \hbar^2 k^2/2m$ poses a complete set of eigenvectors $<x\,|\,\psi>= \exp(ikx)$ and one obtains

$$\rho(x,x') = \int <x\,|\psi><\psi\,|\,x'>g(k-k_0)\mathrm{d}\,k = \int e^{ik(x-x')}g(k-k_0)\mathrm{d}\,k. \tag{12.6}$$

This is equivalent to Eq. (12.5) and indicates that all observable quantities, including the interference pattern, are insensitive to the variants A_1 or A_2.

Although quantum theory is an extremely successful theory with broad consequences in our understanding of nature, in modern technology and epistemology there is still a lack of unique interpretation. Jauch et al. (1967) concluded that as long as one accepts the validity of the superposition principle and the linearity of the Schrödinger equation there is no escape from epistemological dilemmas in interpreting quantum mechanics. The wave-particle duality was criticized by Popper (1967) as a non-realistic element in the theory and Feynman et al. (1965) concluded that it is the only mystery of quantum mechanics. The debate about interpretational questions has been intensified since experiments with single ions and atoms (Wineland et al. 1984, Meschede et al. 1985, Walther 1998), with entangled photon states (e.g., Aspect et al. 1982), with molecules and clusters (Chapman et al. 1995, Schoellkopf and Toennies 1996, Arndt et al. 1999, Cronin et al. 2009), and last but not least with single neutrons became feasible. Quantum mechanics challenges our conceptions of reality, objectivity, and separability. Mermin (1987) gave a rather compelling argument that the predictions of quantum theory are very mysterious if one tries to deny the existence of superluminal information transfer. Most supporters of the non-locality issue (e.g., Shimony 1987) tie their analysis to the conclusion that any local hidden-variable reality conflicts with the concept of quantum theory. Hence, it is not clear whether the "locality" assumption, rather than the "reality" assumption fails. An experimental test of non-local realism has been published, thus apparently rendering local realistic theories untenable (Groeblacher et al. 2007). The formalism of quantum theory has the power to predict only the probabilities that events which are localized in space and time occur (Haag 1990, 2013; Englert 2013).

The measurement problem, i.e., the transition from the quantum to a classical world and the non-locality feature of the theory, is still a challenging question. Here neutron interference experiments can at least shed some new light to deepen our understanding of nature. A key issue may be seen on several state superposition experiments where probability amplitudes arising from separated beam paths become superposed. Indeed even the initial states must be seen as superposition states. That is, the wave packet in a beam experiment may describe the neutron state entering the interferometer (see Eq. 12.3 and the discussion there)

$$|\psi> = \sum_{n=0}^{N} a_n\,|\psi_n> \xrightarrow[N\to\infty]{} \int_0^\infty a(k)|\psi(k)>\mathrm{d}\,k. \tag{12.7}$$

Due to the finite distance (L) between the source and the detector, the limit $N \to \infty$ is somewhat unphysical because the momentum space cannot be divided into elements narrower than $\Delta k = 2\pi/L$. Nevertheless, for most practical purposes the integral form of Eq. (12.7) is adequate.

In the quantum measurement process a superposition state between the quantum object $|\psi_0>$ and the apparatus $|\mathcal{A}_0>$ is produced, which results in an entanglement between the quantum object and the environment (Zeh 1970). Since quantum mechanics is supposed to be universal, the evolution of the joint system $|\psi>|\mathcal{A}>$ is governed by the linear Schrödinger equation under the action of a unitary operator U

$$|\psi_0>|\mathcal{A}_0> \xrightarrow{U} \sum_{n=0}^{N} a_n|\psi_n>|\mathcal{A}_n>. \qquad (12.8)$$

This is a state where the measuring apparatus is in a superposition of distinct states corresponding to all the possible eigenvalues. However, this is not in accordance with experience where we always find the pointer of the apparatus in one of the possible states, e.g., $n = m$

$$|\psi_0>|\mathcal{A}_0> \xrightarrow{?} a_m|\psi_m>|\mathcal{A}_m>. \qquad (12.9)$$

This is the measurement problem of quantum mechanics. It indicates that a measurement does not reveal a pre-existing value of the measured property. On the contrary, the outcome of a measurement is brought into being by the act of measurement itself as a joint manifestation of the state of the probed system and the probing apparatus (Mermin 1993). In Heisenberg's terminology a measurement is a transition process from a possible to an actual event. The statistical distribution of many such encounters is a proper matter for scientific inquiry. Thus, the basic question is whether properties of individual systems possess values prior to the measurement that reveals them. Efforts to construct deeper levels of description where individual systems do have pre-existing values are known as hidden-variable programs and they will be discussed later on in this chapter.

When calculating the probability density of finding a particle and an apparatus in a certain state, all interference terms disappear and a classical probability distribution remains. The collapse of the wave field remains the most puzzling and counterintuitive aspect of the interpretation of quantum mechanics. Possible mechanisms and the boundary between the preparation and the irreversible collapsed state are debated in the literature. A way out of this dilemma was proposed by von Neumann (1932), who postulated that whenever a measurement occurs the unitary evolution is replaced by a projection onto the eigenstate associated with the measured value. This assertion has been widely criticized and the main problem is that nowhere in the formalism is it found what uniquely characterizes a measurement (see, e.g., Wheeler and Zurek 1983, Bell 1987, Mermin 1993, Namiki et al. 1997). The most well-known criticism was put forward by Schrödinger (1935) when he proposed his famous "cat paradox," where the question is raised as to when and where the instantaneous wave function reduction takes place and what is the role of the observer. In this connection he introduced so-called *entangled states* with the property that a quantum system consisting of more than one particle cannot be separated (factorized) except in a measurement where the phase relationship between them are destroyed. The simplest case is a two-particle spin-$\frac{1}{2}$ system in the singlet state where the entangled state can be written as

$$\psi_{\alpha,\beta}^{\sin glet} \propto \left[|+>_\alpha |->_\beta - |->_\alpha |+>_\beta \right] \psi_\alpha \psi_\beta, \qquad (12.10)$$

where α and β denote the two spin-½ particles, and $|+>$ and $|->$ are the spin eigenvectors. The spatial wave functions ψ_α and ψ_β are often considered to be separated when the wave packets no longer overlap. When thinking about neutron post-selection experiments as described in Section 4.5, this may be a questionable assumption (see also Rauch 1993). Nevertheless, many theoretical analyses accept the spatial separation and proceed with the spin part of Eq. (12.10) alone. Here the warning is expressed to neither neglect the intrinsic coupling (entanglement) of the quantum system to the apparatus nor the coupling of the quantum features in phase space. Single particles can exhibit entanglement as well, because general entanglement is defined as the relation between different degrees of freedom (Englert 1999, 2013). This leads to the phenomenon of quantum contextuality discussed in Chapter 7. In a neutron interferometer the spatial and the spin parts of the wave function are, in general, entangled.

The question whether quantum mechanics is complete or some hidden variables may exist was initiated by Einstein, Podolsky, and Rosen (1935), who at least believed that such a theory is possible. An "impossibility proof" for such hidden-variable theories was given by von Neumann (1932), but later on it was shown (e.g., Jammer 1974) that the proof excludes local hidden-variable theories but keeps a loophole for non-local hidden parameter theories (Bohm 1952a, 1952b; Bell 1966). Experimental tests of various hidden-variable theories became feasible. Bell (1965) formulated inequalities between measurable count rates which demarcate between quantum theory and a broad class of local hidden-variable theories. The experiments are mainly related to the spin-part of Eq. (12.10) and register counting rates for different settings of spin analyzers placed at opposite and widely separated sides of an emitting source. Most experiments have been performed with correlated photons (Aspect et al. 1981, 1982; Perrie et al. 1985). Reviews of such experiments are given by Ballantine (1987), Selleri (1990), and Home and Selleri (1991). All results confirm the quantum-mechanical predictions and eliminate local hidden-variable theories. Greenberger, Horne, and Zeilinger (1989) formulated an interesting new theorem that considers a system of three mutually well-separated but correlated particle systems. In this case an even stronger refutation of local realist theories can be given. Non-locality of quantum mechanics provides the basis for quantum teleportation. It may also open new horizons for advanced technologies (Bennett et al. 1993, Bouwmeester et al. 1997). Nevertheless, some critical comments about non-locality still exist in the literature (Englert 2013).

Typical Bell-type experiments require more than one particle to become and behave entangled. In the meantime, it has been recognized that behind this phenomenon there is a more general principle which is called the Kochen–Specker (1967) phenomenon. In this case different degrees of freedom of single particles become entangled and show the phenomenon of quantum contextuality. This topic has been addressed for neutrons in Chapter 7. It shows that there are no hidden variables feasible which determine the outcome of a measurement before the measurement has been done. Vice versa, the outcome of an experiment may depend upon what was the outcome of a previously or simultaneously performed experiment on a compatible variable. That is, it depends upon which context the measurement is done. Thus the question whether local reality exists is attacked (Groeblacher et al. 2007). In the neutron case an entangled state between spin and beam path has been produced and the measured observables fulfill a Bell-like inequality, thus signifying quantum contextuality as an independent feature valid at least for entangled states (Hasegawa et al. 2003, Bartosik et al. 2009). The debate continues whether this conclusion can be extended to all quantum states (e.g., Nieuwenhuizen et al. 2007). It seems that, as always, quantum mechanics wants to have the last word: it stubbornly refuses to admit hidden variables even under seemingly innocent conditions. It turns out that neutrons can be prepared in such a way that spin and momentum measurements, although nominally still independent, are so strongly correlated that non-contextual hidden variables cannot explain this strong correlation (Weihs 2007). Thus, unless one allows the existence of contextual hidden variables with very strange mutual influences, one must abandon them—and, by extension, "realism" in quantum physics—altogether. By analyzing the concept of contextuality in terms of pre- and post-selection methods (Section 4.5) it is possible to assign definite values to observables which by "weak measurements" do not disturb pre- and post-selection (Aharonov et al. 1988, Tollaksen 2007). In more general terms it can be stated that just non-locality and contextuality cause stronger correlations than those given by classical theory. Or as Asher Peres (1993) claims: *quantum phenomena are more disciplined than classical ones.*

Operational quantum mechanics tries to combine quantum mechanics with general relativity and explore how *action* enters the world (Piron 1964, Aerts and Aerts 2004). In this sense the observer becomes a crucial part of the game and even single observable measurements must be seen in context with other observables (Khennikov 2009, Allahverdyan et al. 2013).

Before continuing with a general discussion of various interpretations of quantum mechanics a comment regarding the spatial part of Eq. (12.10) should be made. When a wave packet formalism is used for the wave functions ψ_α and ψ_β, the spatial separation occurs for the packets, but not necessarily for the components of the packets (see Eq. 12.2). By means of proper post-selection procedures (Section 4.5) additional information about features of the physical system can be obtained. For example, when one describes a gamma cascade transition between long-lived nuclear states via a metastable level the width of the packets is given by the lifetime τ of the metastable level $\delta E \propto \delta k \propto \tau^{-1}$ (Fig. 12.1). When the energy conservation for individual pairs is taken into account

$$k_\alpha + k_\beta = k_{0\alpha} + k_{0\beta} = \text{constant}, \tag{12.11}$$

one obtains for each photon pair a characteristic spatial modulation due to the interference of the beams (Rauch 1993)

$$I(k_\alpha, k_\beta, r) \quad \propto \quad 1 + \cos\left[2(k_\alpha - k_\beta)r\right]. \tag{12.12}$$

These modulations exist far beyond the dimensions of the packets ($r > (2\delta k)^{-1}$) and their observation depends on the momentum resolution of the apparatus. A measurement on one side of the source of one of the two photons instantaneously reduces the whole wave function to that of the second photon. This indicates the phase space coupling of both spin and ordinary space variables, which seems to be an important feature of quantum mechanics.

Within standard quantum physics locality is not the only concept that must be abandoned. In a more general sense, classical realism also cannot exist and microscopic objects are simply not the kind of thing that can possess properties independently of the macroscopic apparatus used to observe them. However, as pointed out by Schrödinger in 1935 there is no good reason to accept this division of the world into a microscopic regime where quantum mechanics reigns and a macroscopic one governed by classical physics. Decoherence cannot constitute a resolution to that problem.

Before briefly describing various alternative interpretations of quantum mechanics it should be mentioned that the overwhelming majority of physicists today seem to believe that quantum theory is a complete theory and local realism cannot be associated with it. This can also be understood as a consequence of the general nonseparability feature of any physical system (d'Espagnat 1976, Legett 2008). Nevertheless, a mind-independent realism may be maintained, guaranteeing the existence of nature even without human consciousness (e.g., d'Espagnat 2011). The whole mystery of quantum physics and the various efforts to resolve them started with the so-called Einstein–Bohr debates. This starting point has been elucidated recently by Whitaker (1996, 2012) and nicely summarized by Greenberger et al. (2009).

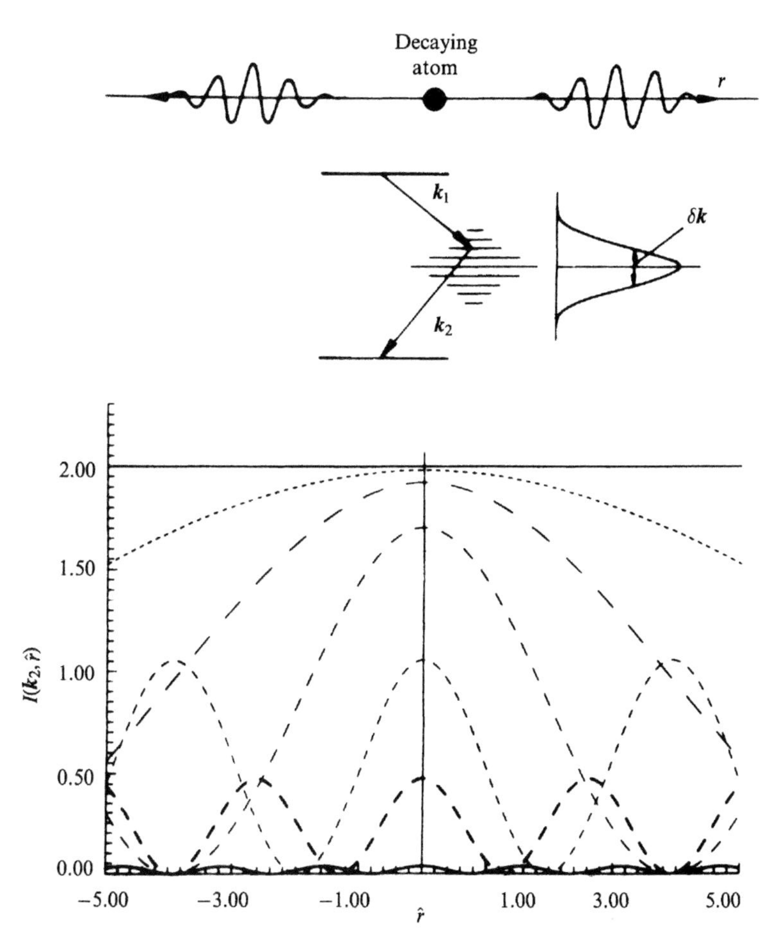

Figure 12.1 *Sketch of a correlated atomic photon decay and the expected intensity modulation for distinct pairs of photons with different momentum differences* $|k_1 - k_2| = n\delta k$ *with* $n = 0, 0.1, 0.2, 1, 2, 3$ *(from above to below). The distance from the source is in units of the packet length* $1/\delta k$

12.1 Interpretations and Approaches

12.1.1 Copenhagen–Göttingen interpretation

This is the standard and most pragmatic interpretation of quantum mechanics as it can be found in most textbooks and review articles (e.g., Jammer 1966, 1974; Stapp 1972; Baggot 1992; Sakurai 1994). It describes facts in the form of the results of physical experiments. It does not ask how nature is, but how it acts. Basic features are the Heisenberg (1927) uncertainty relation, the statistical interpretation of Born (1926), the complementarity principle of Bohr (1928), and the identification of the wave function with our knowledge of the system (Hartle 1968). The wave function represents a tendency or potentiality for various events to occur with different probabilities. A measurement does not reveal pre-existing values of the measured system (Omnès 1994). In Popper's terminology this means that only "propensities" exist (Popper 1973, 1994). Thus, nothing can be said about what happens between two observations.

When the complementarity principle is accepted, the quantum system can appear as a particle or as a wave, depending upon the experimental conditions imposed on the system (Leggett 1986). This is the basis for wave–particle duality where one can describe an object either as a wave *or* as a particle but never both simultaneously in the same experimental setup. There is no ontological duality in the object itself, its behavior being determined by its inseparable interaction with the apparatus (Bohr 1934). A particle entering a beam splitter, for example, has a 50% chance of veering either right or left. If it is a classical particle it would unambiguously take one of the two paths. A quantum particle, however, is placed in a superposition of both paths until a measurement is made.

Inside the interferometer the neutron must be considered as a wave and there exists a simple way to explain the behavior of that wave during the interaction with a crystal lattice (Fig. 12.2). The action of different phase shifts causes a different position of the wave oscillations peaks and valleys compared to the crystal lattice which causes different output neutron waves. Whether the

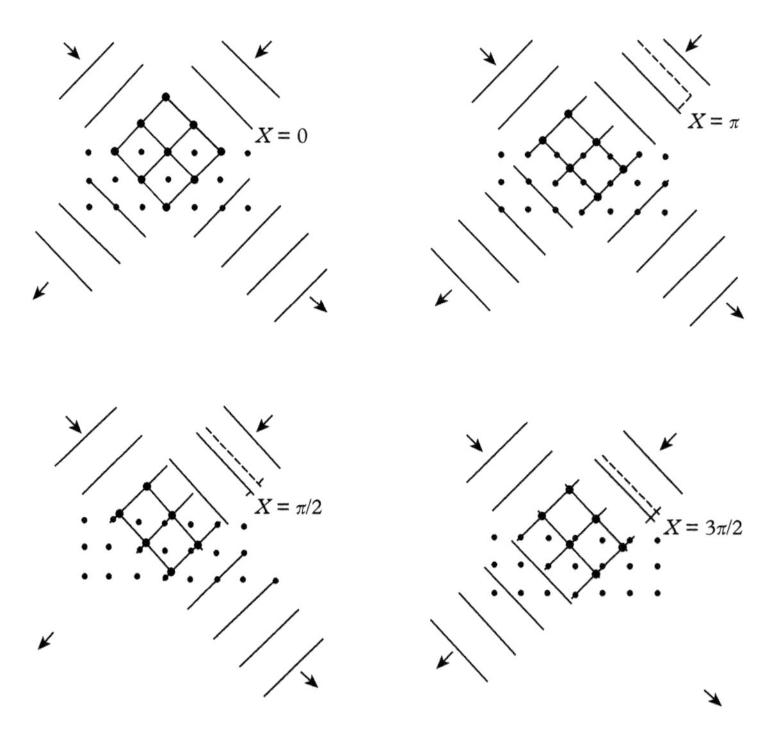

Figure 12.2 *Interaction of classical waves with a periodic lattice. Notice that the nodes of the wave field are at different positions in relation to the crystal lattice, which creates different outgoing waves*

complementarity principle can be considered as a consequence of the Heisenberg uncertainty relation or as an additional axiom is still an open question (e.g., Scully et al. 1991, Storey et al. 1994, Englert et al. 1995, Wiseman and Harrison 1995, Wiseman et al. 1995). After reading the Einstein–Bohr debates (see Schilpp 1949, Whitaker 1996) concerning the recoil double-slit gedanken experiment with electrons, most physicists believe in an intrinsic connection. Some of the neutron interference experiments which cause a labeling of neutrons along the beam paths inside the interferometer by spin rotation (Section 5.2) or energy transfer (Section 5.3) can give the impression that the uncertainty principle is not needed for an appropriate explanation. But a more detailed analysis shows that following erasure and post-selection the experiment is not really finished after superposition. Experiments with labeled atoms are also used to push this interpretational question into a direction relying more on the complementarity principle than on the uncertainty relation (Duerr et al. 1998, Dürr and Teufel 2009).

The Copenhagen interpretation imposes the collapse of the wave function during the measurement process but does not address the borderline region between the quantum and the classical worlds which is intrinsically needed to describe the measurement apparatus, always seen as a classical object (Griffiths 1984). This cut between quantum and classical world is mandatory within this view and therefore a wave function for the whole universe cannot exist. Nevertheless, nowadays it becomes respectable and even fashionable to describe the whole universe quantum mechanically (Weinberg 1978, Hawking 1990). The question of the border line has been discussed in more detail by Omnès (1994). He also analyzed the magnetic Josephson effect analog as measured by neutron interferometry and as described in Section 5.4.

Machida and Namiki (1980) and independently Araki (1980) developed a measurement theory in which they maintain that quantum mechanics can describe the whole process of a quantum-mechanical measurement when the measuring apparatus and its interaction with the quantum-mechanical object are properly formulated. This is an improved approach compared to the von Neumann approach. It is based on the many-Hilbert-space theory and supposes distinct detector models and introduces so-called decoherence parameters (Joos and Zeh 1984, Namiki and Pascazio 1993). This gives a direct connection to the decoherence approach to be discussed later. Stochastic quantization in configuration space is also such an approach making use of hypothetical processes acting in a fictitious time (Namiki 1992). These hypothetical processes are described by a random force added to the basic classical equations of motion. They consider the Planck constant as a sort of diffusion constant giving the laws of quantum mechanics in the thermal equilibrium limit. In the very extreme case Wigner (1970) postulated that the real collapse of the wave function happens when the measured information enters our mind through our senses and ends up in our consciousness. This is not a physics matter that is subject to physical investigation, at least at the present time.

12.1.2 The Ensemble or Statistical Interpretation

This is a rather easily defendable interpretation. It starts with a large number of identically prepared systems (the ensemble) for which the wave function provides a deterministic description (Born 1926; Einstein 1936; Ballentine 1970, 1990). Identically prepared systems mean that the system has been prepared and limited by some distinct boundary conditions. The measurement problem is a "non-problem" because we should not ask about the outcome of a single-event experiment, and the state vector is not itself an element of quantum reality, but is only a means to calculate the probability distributions for various observables (Penrose 2004). Nevertheless, several dynamical models for describing the outcome of a single run have been developed

(Jammer1989, Allahverdyan et al. 2013). Standard quantum statistical mechanics alone appears sufficient to explain the occurrence of a unique answer in each run.

12.1.3 The Spontaneous Localization Approach

This approach tackles the question of how a superposition state between the object and the apparatus can suppress all possible terms except one for the macroscopic system needs some modification of the Schrödinger equation, thus avoiding macroscopic superposition. This line of thought is called the spontaneous localization program or the Ghirardi–Rimini–Weber concept (Ghirardi et al. 1986), where the modified Schrödinger equation contains a term for random and spontaneous wave function collapses. Due to the stochastic character of the additional term the collapse occurs at randomly distributed times. The parameters for that term should be determined by experiment but they are not yet found in a reliable form.

A slight modification of this model is the continuous spontaneous localization model in which a Brownian motion noise term coupled non-linearly to a local mass density is added to the Schrödinger equation (Ghirardi et al. 1990). The noise is responsible for the spontaneous collapse of the wave function. Feasible parameters for the strength and the spatial correlation length of the noise field have been discussed by Adler and Bassi (2009). Most estimates show that the effects are completely camouflaged by environmental decoherence (Tegmark 1993).

12.1.4 The Decoherence Approach

This approach was introduced by Zeh (1970). It has many appealing aspects and has been addressed several times in this book. It starts with the observation that quantum-mechanical predictions are different for closed and open systems. There are always influences from the environment that destroy the phase relation between the superposed substates of a real quantum state (Eq. 12.8). Therefore, this dephasing washes out the interference terms and results in a statistical mixture which describes the probabilities for the specific outcomes of an experiment (Zeh 1970; Zurek 1981, 1991, 1998a; Omnès 1994; Giulini et al. 1996; Haroche and Raimond 2008). The decoherence approach is seen as a rather pragmatic one and it cannot exactly discriminate microsystems and macrosystems. It deals with the quantum system and the apparatus on an equal footing, both being described by the same quantum laws. The irreversibility and non-unitary behavior emerge naturally from the Schrödinger equation, when looking at the evolution of a small subsystem entangled with a large reservoir, where the emerging irreversibility results from the cumulative effect of perfectly reversible microscopic events. The observer decides to forget about the environment and renounce keeping track of its correlation, which transfers a quantum to a classical situation. Zeh (1970) and Zurek (1993, 1998b) showed that special *einselected* pointer states are rather robust in their interaction with the environment. That provides the basis for the so-called *existential interpretation*, which defines just these einselected states as relatively objective existing features. Einselection means a kind of environment-induced superselection of states, which creates properties similar to classical states. In the interaction between the quantum system and the environment most of the entangled states decohere very rapidly, but some states (einselected pointer states) become very stable. Such pointer states of an apparatus communicate intensively with the environment, indicating a repeated measurement process which stabilizes the quantum system.

It may become possible in the future to observe more and more interference effects among macroscopically distinguishable states of the apparatus and to suppress decoherence effects by means of advanced measuring methods. Therefore, several authors claim that this approach cannot solve the fundamental quantum measurement problem (Leggett 1986, 1994; Bell 1990; d'Espagnat 1995; Home 1997; Namiki et al. 1997). Various calculations show that the influence of fluctuation is minimized for Gaussian (coherent) states, but strongly increases for non-classical states, like Schrödinger cat-like states (Walls and Milburn 1985, Glauber 1986, Schleich et al. 1991, Rauch and Suda 1998). It gives a realistic interpretation of the measurement process, it provides a theory of dissipation, and it introduces an arrow of time with no full reversibility of events at least in cases where a macroscopic number of particles are involved. Here it has some common grounds with the event-driven view of quantum physics (Haag 1990, 2013; Englert 2013). In the course of discussions with the opponents a revised kind of realism which contains non-separability has been introduced. Epistemological aspects of this view are discussed by Omnès (1999). Thus, the decoherence approach describes the measuring process in the framework of standard quantum theory and its unitary dynamics, which may be summarized with the statement as *collapse without a collapse*. It should be mentioned that decoherence effects are also expected to appear due to quantum gravity effects and the corresponding quantization of space-time, which guarantees the appearance of a classical world during the evolution of the universe (Joos 1986, Penrose 1986, Kiefer 2000).

12.1.5 The Consistent History Approach

This scheme is basically a decoherence approach but it also makes an attempt to realistically interpret the Copenhagen view (Griffiths 1984, Omnès 1988, Gell-Mann and Hartle 1993). It can be applied for the whole universe, because it does not need the environment. Here one discusses state sequences ("histories") rather than states at a single instant. Measured quantities are always correlated to decohering histories, whereas all cross-correlations between micro- and macrophysical variables remain unobservably small for all times. Therefore, in recent years it attracted in a lot of attention since it seemed to yield a solution to the conceptual and interpretational problems of standard quantum mechanics. It supports a realistic interpretation of quantum mechanics where a measurement reveals what is actually there (Griffiths 1999). Nevertheless, there remain concerns that all decoherent history theories also do not meet the requirements of a "realistic" description of the physical world, that is, the very reason for which they have been developed (Bassi and Ghirardi 1999).

The further development of this approach is called *consistent stochastic approach* (Griffiths 2010), which distinguishes non-locality and instantaneous non-local influences since the last term violates special relativity (see e.g., Albert and Galchen 2009) and abandons *local realism* and *free will*. In this approach one avoids the use of classical concepts in a quantum context and states that quantum mechanics is fundamentally stochastic since the world is seen as stochastic as well. Consistent history formulations imply definite properties to a quantum system even before measurement. The theory (and the quantum state) is still non-local but non-local influences are absent. There are correlations but no causes. In this approach the de Broglie–Bohm pilot wave (de Broglie 1927) and the Ghirardi–Rimini–Weber approach (Ghirardi et al. 1986) become compatible with special relativity.

12.1.6 The Transactional Interpretation

This approach provides a description of the wave function (state vector) as an actual wave physically present in real space and provides a mechanism for the occurrence of non-local correlations due to the use of advanced waves having negative energy and traveling in the negative time direction. It is a time-symmetric formulation and considers source and absorber on an equal footing, sending "offer waves" and "confirmation waves," respectively. Confirmation waves are determined by the complex conjugate of the wave function. The wave collapse occurs as the formation of a transaction, which occurs by the interaction of offer and confirmation waves (Cramer 1986). In its heart it treats the reality of possibilities (Kastner 2012). Advanced or confirmation waves bring with them a number of problems related to causality requiring a hierarchy of transactions (Boisvert and Marchildon 2013).

12.1.7 The Guide or Pilot Wave Interpretation

This is the fundamental idea of de Broglie (1926, 1960). It can be taken as a kind of preliminary version of the transactional interpretation, but still rests on local realism by an underlying mechanism for the interplay of waves and particles. The waves guide the particle but cannot carry energy or momentum. Schrödinger (1927) also argued for a close analogy between matter waves and classical waves and claimed a particle character only when it interacts with a target or an apparatus. This also leads to conceptual problems mainly due to its local character. It got some renaissance due to the recent discussion on Compton frequency effects (Mueller et al. 2010).

Recent model experiments with liquid droplets bouncing on the surface of an oscillating liquid may couple with the surface waves it generates and thus start to propagate. The resulting "walker" is a macroscopic object that associates the droplet and its wave, which results in surprising "quantum-like" effects when such a system passes through single and double slits (Couder et al. 2005, Eddi et al. 2009, Couder and Fort 2012, Brandy and Anderson 2014). In this description the Planck constant does not show up; the wave is emitted by the particle and propagates at a fixed velocity on a material medium. Surface waves which guide the droplets are Lorentz covariant with a characteristic speed. Attraction and repulsion between bouncing pairs of droplets depend on the relative phase between the droplets. All these macroscopic phenomena can be correlated to the behavior of quantum-mechanical particles.

12.1.8 The Information Interpretation

This approach interprets the quantum state as a mathematical tool that encodes subjective information about potential results of experiments. It sees the collapse as a sudden improvement of knowledge of the experimenter (Fuchs and Peres 2000, Ferrero 2003). In this respect it also contains elements of subjectivity and does not constitute a reality without observers (Groeblacher et al. 2007). This interpretation takes into account that our description of the physical world is represented by propositions, i.e., information gained by classical measuring results. It assumes the principle of quantization of information. An elementary system can be represented by a single proposition, i.e., a single bit of information with "true" being identified with the bit "1" and "false" with the bit "0" (Zeilinger 1999a,b; Paterek et al. 2010). When one uses "spin up" of a particle as such a single bit, any measurement other than the spin-up direction must contain an element of randomness and cannot be reduced to unknown hidden properties, since they would then represent more than a single bit of information. In the case of two spin particles one has two bits of information available. In this case one can specify the truth of the following two propositions:

- The two spins are different along z;
- The two spins are the same along z.

The corresponding quantum state is uniquely defined; it is an entangled state as given by Eq. (12.10). Entanglement is therefore a consequence of the fact that the total information is used to define joint and not individual properties of the composite system. The individual properties remain completely undefined, i.e., random. There is a book that tries to popularize this view (Vedral 2011).

12.1.9 Many-World Interpretation

This is the original idea of Everett (1957). It sticks to pure Schrödinger evolution and denies that collapse ever happens. Here all superposition states between the system and the apparatus are ascribed on an equal status and the outcome of the measurement takes different values in different worlds and even the observer is split into these different worlds (Everett III 1957, De Witt 1973, Squires 1985). This interpretation is intrinsically connected to hidden variables which determine which results are observed in which world. It restores individual determinism and solves the measurement problem by denying the collapse, but uses very abstract arguments. In the many-world interpretation the wave function provide a complete description of the state of the system but adds the conscious awareness of an observer at random with postulated weight factors specified by the standard quantum theory. Thus, it connects quantum theory and consciousness. Through the interplay of consciousness with the wave function definite outcomes are supposed to emerge. In the so-called extended Everett concept one derives the main features of consciousness (intuition and truth) from quantum mechanics (Mensky 2010). It may turn out that we do not know enough about consciousness and its relation to the physical world to solve the quantum mystery (Squires 1990). An obvious criticism is that the problem is simply moved from pure physics to the more speculative area of the theory of mind, for which we do not have as yet a sound formulation. The question how strong the interaction must be to split worlds remain open.

In the following we describe examples of hidden-variable theories using pilot waves (de Broglie 1927) and quantum potentials will be presented (Bohm 1952a, 1952b). According to various "no hidden-variable proofs" only non-local hidden-variable theories seem to be worthy of further discussion (von Neumann 1932, Kochen and Specker 1967, Greenberger et al. 1989, Hardy 1993). There are still scientists who believe locality and, therefore, no action at the distance focus all problems to the randomness and irreversibility of events (Englert 2013, Vaidman 2014).

12.1.10 Bohm's Quantum Potential

This approach is strictly connected to the Schrödinger quantum mechanics formalism (Bohm 1952a, 1952b). It includes the de Broglie (1927) pilot wave theory to a great extent and it gives a complete deterministic view of quantum phenomena due to adding a (non-local) quantum potential to the Schrödinger equation. The de Broglie–Bohm theory treats wave and particle on an equal footing and therefore it is a kind of a causal interpretation (Selleri 1993, Holland 1999, Dürr and Teufel 2009). Since this quantum potential is constructed out of the Schrödinger equation the predictions are equivalent to the solution of the Schrödinger equation, but they nicely show how an interference pattern is built up by deterministically determined individual trajectories (e.g., Philippidis et al. 1979, Bohm and Hiley 1993, Holland 1993). The arrow-of-time is explicitly recognized and local realism is preserved. The price to pay is that the guiding quantum potential becomes non-local. Extensions toward special relativity effects failed since the quantum potential does not disappear for increasing distances from objects.

The time-dependent Schrödinger equation (Eq. 1.2) can be rewritten by means of the Madelung (1928) transformation as

$$\psi(\boldsymbol{r}, t) = R(\boldsymbol{r}, t) e^{\mathrm{i}S(\boldsymbol{r},t)/\hbar}, \tag{12.13}$$

which gives two differential equations for the real functions $R(\boldsymbol{r}, t)$ and $S(\boldsymbol{r}, t)$ as the basis for the Bohm formulation of quantum physics:

$$\frac{\partial R^2}{\partial t} + \nabla\left(R^2 \frac{\nabla S}{m}\right) = 0, \tag{12.14a}$$

and

$$\frac{\partial S}{\partial t} + \frac{1}{2m}(\nabla S)^2 + V - \frac{\hbar^2}{2m}\frac{\nabla^2 R}{R} = 0. \tag{12.14b}$$

These equations have known analogs in classical physics and fluid dynamics, aside from the term describing the quantum potential, namely

$$Q = -\frac{\hbar^2}{2m}\frac{\nabla^2 R}{R}. \tag{12.15}$$

Since the density $\rho(\boldsymbol{r}, t) = R^2(\boldsymbol{r}, t) = |\psi(\boldsymbol{r}, t)|^2$, Eq. (12.14a) describes the continuity equation of a particle ensemble traveling with a momentum

$$m\boldsymbol{v}(\boldsymbol{r}, t) = \nabla S(\boldsymbol{r}, t). \tag{12.16}$$

Equation (12.14b) has the same structure as the classical Hamilton–Jacobi equation, but here with the additional quantum potential. Whenever the ψ-function is known, the quantum potential and then the individual trajectories can be calculated. Dewdney (1985) calculated the quantum potential and the individual trajectories for a Mach–Zehnder interferometer with square potentials describing the beam splitter and the beam combiner and a Gaussian incident wave function (Fig. 12.3). The non-crossing feature of the trajectories is a characteristic feature of these quantum trajectories.

The non-local behavior of the quantum potential for different phase shifts is visible. Similar calculations are known for the double-slit situation (Philippides et al. 1979, Sanz and Miret-Artés 2002) and for the spin-superposition case (Dewdney et al. 1986). The non-local character of this causal theory has been elucidated in detail by Dewdney et al. (1989), Brown et al. (1995), and Dürr and Teufel (2009). A direct verification of the quantum potential seems to be impossible because any measuring probe would change this potential substantially. Analogies between the Wignerian and Bohmian interpretation of the two-slit experiment have been addressed by Wiseman (1998). Nevertheless, a satisfactory Lorentz invariant Bohmian theory still needs to be found. *Weak* measurements may overcome these restrictions (Aharonov et al. 1988). This has been shown by Kocsis et al. (2011), who performed a weak measurement of the particle momentum in a double-slit experiment. The particle momentum was postulated according to the result of a *strong* measurement of the position of the particle. Similar conclusions have been drawn in a delayed choice experiment by Jasques et al. (2007). A detailed study of the present status and a proposal for further work can be found by Davidovic and Sanz (2013) and by Braverman and Simon (2013).

Trajectories can also be obtained on the basis of the Feynman path integral method as done by Gondran and Gondran (2005). They calculated the individual paths of cold atoms from an atom

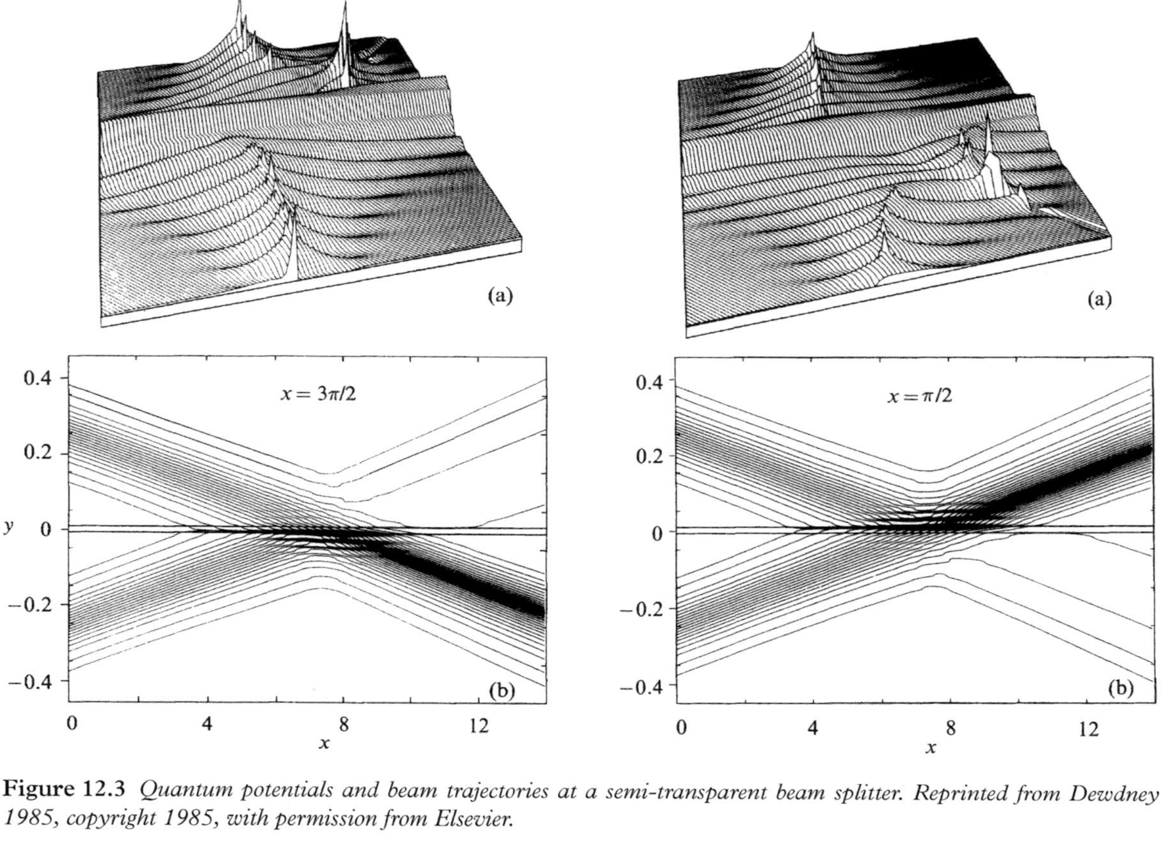

Figure 12.3 *Quantum potentials and beam trajectories at a semi-transparent beam splitter. Reprinted from Dewdney 1985, copyright 1985, with permission from Elsevier.*

cloud through a double-slit arrangement and found good agreement with experimental results (Shimizu et al. 1992). In this case the particles are not only represented by their wave function but also by the motion of their center of mass; and the trajectories result just from this motion (Holland 1999, Davidovic et al. 2013). The coupling of liquid droplet motions onto a liquid coupled by means of the waves that they generated themselves is another example of visualizing the simultaneous trajectory and wave behavior (see Section 12.1.7 and Couder et al. 2005, Eddi et al. 2009, Couder and Fort 2012).

If two particles are in an entangled (correlated) state, then, because of the quantum field, guiding the trajectory of the second particle depends upon the trajectory of the first. If now a field is suddenly turned on in a region where the first particle happens to be, the subsequent motion of the second particle can be drastically altered in a manner that does not diminish with the distance between the two particles (Fig. 12.2). Connections with the principle of quantum contextuality, as discussed in Chapter 7 become visible.

A maximally realistic causal quantum theory avoiding the asymmetrical treatment of position and momentum as it happens in the de Broglie–Bohm theory has been formulated by Roy and Singh (1999). In this case a positive definite phase space density reproduces the correct quantum probabilities as marginal. This theory also defines quantum potentials which guide the particles in ordinary and momentum space.

12.1.11 Event-Based Models

It should be mentioned that there exist even more causal and "realistic" pictures of quantum mechanics. Strictly speaking the event-based model is not an interpretation of quantum physics but it is a new approach to understanding quantum-like phenomena in a new light. Particle-only models try to describe quantum effects without using the Schrödinger equation. They use computer algorithms to describe individual particles and detection events (De Raedt et al. 2005, Zhao et al. 2008, Jin et al. 2010, De Raedt et al. 2012, De Raedt and Michielsen 2014). Two beam interferences of a Fresnel biprism, double-slit diffraction, and interference experiments have been investigated in detail. The individual particles do not have any direct interaction between them but indirectly through the common interaction with the source, the detector, the polarizers, the beam splitters, the collimators, etc. An event-based model for the detectors is assumed where the detection process involves some kind of memory. The single particles are seen as messengers between the source and the detector, which carry specific messages, e.g., the time-of-flight between the specific place of emission at the source and the specific place of arrival at the detector. The time-of-flight and the spin of a neutron can be encoded in a two-dimensional complex valued vector

$$y = \begin{pmatrix} e^{i\psi^{(1)}} \cos(\theta/2) \\ e^{i\psi^{(2)}} \sin(\theta/2) \end{pmatrix}. \tag{12.17}$$

As the messenger moves for a certain time T the message changes as

$$y' \rightarrow e^{ivT} y, \tag{12.18}$$

where v denotes a characteristic frequency determined by the intensity of the beam. T may be different in both beam paths when a phase shifter is applied. The message itself is influenced by the source, the beam splitter, and the phase shifter. The value of the previous messenger can be stored within the system and an internal vector x, which depends on random variables, is added, and can be interpreted as a kind of learning parameter. More details can be found in de Raedt et al. (2012).

Many interference experiments with various particles have been simulated and good agreement between simulation and the standard quantum calculation has been achieved. A simulation of the neutron interference absorber experiments is discussed in Section 4.3.1, where both cases, the statistical and the deterministic one, are described correctly. This model does not rely on any wave equation and satisfies Einstein's criterion of local causality. It is a "cause-and-effect" description in terms of discrete-event, particle-like processes. Here the question remains how an event-based detector with memory function can be justified.

A particle-only view is also promoted by Utsuro and Ignatovich (2010), who argue that entangled states can be described by simple product states which means that non-locality is rejected as well. The momentum and position of a particle are not considered as eigenvalues of their operators but as expectation values of these operators. They suggest that related EPR experiments should be re-analyzed including the background and a non-linear system of classical equations should replace the linear Schrödinger equation.

12.1.12 No-Problem Approach

This view is based upon events and on the probabilities for events. These events are well localized in space and time (Haag 1990b). Such events are irreversible and leave a mark behind, a definite trace, and they are randomly realized. This principle of random realization ensures that the events do happen in accordance with their probabilities of occurrence (Haag 1996, 2013). The randomness may be explained by the ignorance of the past of a quantum system. In the neutron case we do not know from which fission process the neutron comes, in which direction and with which energy it has been emitted, which moderation process with many collisions it has experienced, where it has left the moderator, etc. We also do not know the time and position when and where all these processes happened. We construct the wave function from the little knowledge we have at the beginning of an experiment. From these known and unknown features one can construct equations of motion which are intrinsically irreversible, although the related Schrödinger equation remains time-reversal symmetric (Englert 2013). In the abstract the author declares "Quantum theory is a well-defined local theory with a clear interpretation. No 'measurement problem' or any other fundamental matters are waiting to be settled." The description uses the statistical operator, which tells us our knowledge about the preparation of the system and it does not change until we acquire additional information. Thus, the statistical operator is not a physical object or a property of a physical object; it describes the object by encoding what we know about it. The fact that quantum processes are fundamentally probabilistic is sufficient to explain Bell's theorem and preserving locality to the quantum system itself. Thus, we have an interplay of local phenomena caused by local interactions and a mathematical formalism boosted by an epithemological interpretation.

12.2 Conclusions

The numerous experiments performed by neutron interferometry agree well with the predictions of quantum mechanics. It can be stated that new instruments, like the neutron interferometer, provide a deeper insight into the curious features of quantum mechanics (see, for example, Bromberg 2008, Snow 2013, Klepp et al. 2014). From an epistemological point of view they show the non-local features of that theory and they show how new insights into the nature of quantum systems depend on the quality and sensitivity of a measurement. Thus, the border between the micro-world and macro-world seems to be shifting according to the exquisite standards of measurement technique. Decoherence and dephasing effects play an important role in this transition process. From the experimental point of view the results are well described by the formalism of quantum mechanics, which gives a sound basis for accepting phenomena which are inherently in a domain outside of our daily experience. Hidden variables are not needed for the interpretation. However, various experiments have shown that properties of the wave function can be measured even when they are virtually masked by destructive interactions and fluctuations. In this respect more extended wave packets than the original one have been observed by various post-selection experiments (see Section 4.5). This shows that a wave packet can be affected by its surroundings over distances considerably larger than its spatial extent, thereby indicating a general non-separability feature of nature. Resonance enhancement effects are feasible in such situations. A unique distinction between various interpretations of quantum mechanics which are actually based on the same formalism seems to be impossible, even in principle. Nevertheless, the intellectual challenge remains to put quantum phenomena more and more onto a macroscopic scale and thereby make such phenomena more understandable according to the experience of ordinary people. This probably would mitigate the apparent magic and mystery of quantum mechanics. No theory can be so robust as not to require some further modification resulting from new experiments. This challenge remains on the horizon as a challenge to the next generation of experimentalists. Be guided by theory, but stick with the data.

References

Abele H. (2008) *Prog.Part.Nucl.Phys.* **60**, 1
Abele H., Baeßler S., Westphal A. (2003) *Lect.Notes Phys.* **631**, 355
Abele H., Jenke T., Leeb H., Schmiedmayer J. (2010) *Phys.Rev.* **D81**, 065019
Abernathy D.I., Gruebel G., Brauer S., McNulty I., Stephenson G.B., Mochrie S.G.J., Sandy A.R., Mulders N., Sutton M. (1998) *J. Synchrotron Rad.* **3**, 37
Abragam A. (1972) *2nd General Conf. EPS*, Wiesbaden 1972, 177
Abragam A., Bacchella G.L., Glaettli H,. Meriel P., Pinot M., Piesvaux J. (1973) *Phys. Rev. Lett.* **31**, 776
Accotto A., Vittore E., Zosi G. (1994) *Z. Physik* **B95**, 151
Achiwa H., Hino M., Tasaki S., Ebisawa T., Akiyoshi T., Kawai T. (1996) *J. Phys. Soc. Japan* **65**, 183
Adam P., Jansky J., Vinogradov A.V. (1991) *Phys. Lett.* **A160**, 506
Adams C.S., Sigel M., Mlynek J. (1994) *Phys. Rep.* **240**, 143
Adelberger E.G., Heckel B.R., Nelson A.E. (2003) *Annu. Rev. Nucl. Part. Sci.* **53**, 77
Adler S.L. (1995) *Quaternionic quantum mechanics and quantum fields.* Oxford University Press, Oxford
Adler S.L., Bassi A. (2009) *Science* **325**, 275
Adli D.G., Summerfield G.C. (1984) *Phys. Rev.* **A30**, 119
Aerts D. (1986) *J. Math. Phys.* **27**, 203
Aerts D., Aerts S. (2004) *Towards a general operational and relativistic framework for quantum mechanics and relativity theory.* Springer Science, Berlin
Aerts D., Reignier J. (1991) *Helv. Phys. Acta* **64**, 527
Afriat A., Selleri F. (1999) *The Einstein, Podolsky and Rosen Paradox in atomic, nuclear and particle physics.* Plenum Press, New York
Afshar S.S., Flores E., McDonald K.F., Knoesel E. (2007) *Found. of Physics* **37**, 295
Agamalyan M.M., Drabkin G.M., Sbitnev V.I. (1988) *Phys. Rep.* **168**, 265
Agarwal G.S., James D.F.V (1993) *J. Mod. Optics* **40**, 1431
Agarwal G.S., Singh R.P. (1996) *Phys. Lett.* **A217**, 215
Aharonov Y. (1984) *Proc. Int. Symp. Found. Quantum Mechanics*, Tokyo, Phys. Soc. Japan, p. 10
Aharonov Y., Albert D.Z., Vaidman L. (1988) *Science* **60**, 1351 and *Phys. Rev. Lett.* (1988) **60**, 447
Aharonov Y., Anandan J. (1987) *Phys. Rev. Lett.* **58**, 1593
Aharonov Y., Anandan J., Vaidman L. (1993) *Phys. Rev.* **A47**, 4616
Aharonov Y., Bohm D. (1957) *Phys. Rev.* **108**, 1070
Aharonov Y., Bohm D. (1959) *Phys. Rev.* **115**, 485
Aharonov Y., Casher A. (1984) *Phys. Rev. Lett.* **53**, 319
Aharonov Y., Kaufherr T., Popescu S., Reznik B. (1998) *Phys. Rev. Lett.* **80**, 2023
Aharonov Y., Pearle P., Vaidman L. (1988) *Phys. Rev.* **A37**, 4052
Aharonov Y., Popescu S., Rohrlich D., Skrzypszyk P. (2013) *New J. Phys.* **15**, 113018
Aharonov Y., Rohrlich D. (2005) *Quantum Paradoxes.* Wiley-VCH, Weinheim
Aharonov Y., Susskind L. (1967) *Phys. Rev.* **158**, 1237
Aharonov Y., Vaidman L. (1990) *Phys. Rev.* **A41**, 11
Aharonov Y., Vaidman L. (1991) *J. Phys. A. Math. Gen.* **24**, 2315
Ahluwalia D.V., Burgard V. (1996) *Gen. Relativ. Gravity* **28**, 1161
Aitchison I.J.R., Wanelik K. (1992) *Proc. R. Soc.* **A439**, 25
Aladzhadzhyan G.M., Bezirganyan P.A., Semerdzhyan O.S., Vardanyan D.M. (1977) *Phys. Stat. Sol.* **(a) 43**, 399
Alefeld B., Badurek G., Rauch H. (1981a) *Phys. Lett.* **83A**, 32

Alefeld B., Badurek G., Rauch H. (1981b) *Z. Physik* **B41**, 231
Alefeld G., Voelkl J. (eds.) (1978) *Hydrogen in metals I*, Top. Appl. Phys. 28. Springer, Berlin
Aleksejeva A., Barkanova S., Krasta T., Prokofjevs P., Tambergs J., Waschkowski W., Samosvat G.S. (1997) *Phys. Scripta* **56**, 20
Alexandrov Y.A. (1983) *Sov. J. Nucl. Phys.* **37**, 149
Alexandrov Y.A., Vavra J., Vrana M., Kulda I., Machekhina T.A., Mikula P., Michalec R., Nazarov V.M., Okorokov A.I., Peresedov V.V., Runov L.N., Sedlakova L.N., Chapula B. (1985) *Sov. Phys. JETP* **62**, 19
Alexandrov Yu.A. (1992) *Fundamental properties of the neutron.* Clarendon Press, Oxford
Alexandrov Yu.A. (1994) *Phys. Rev.* **C49**, 2297
Alexandrov Yu.A., Kulda J., Lukas P., Sedlakova L., Vrana M. (1989) *Z. Physik* **A334**, 359
Alexeev V.L., Federov V.V., Lapin E.G., Laushkin E.K., Rurniantzev V.L., Sumbaev O.I., Voronin V.V. (1989) *Nucl. Instr. Meth.* **A284**, 181
Allman B.E., Cimmino A., Griffin S.L., Klein A.G. (1999) *Found. Physics* **29**, 325
Allman B.E., Cimmino A., Klein A.G., Opat G.I., Kaiser H., Werner S.A. (1992) *Phys. Rev. Lett.* **68**, 2409
Allman B.E., Jacobson D.L., Lee W.-T., Littrell K.C., Werner S.A. (1998) *Nucl. Instr. Meth.* **A412**, 392
Allman B.E., Kaiser H., Werner S.A., Wagh A.G., Rakhecha V.C., Summhammer J. (1997) *Phys. Rev.* **A56**, 4420
Allman B.E., Lee W.T., Motrunich O.I., Werner S.A. (1999) *Phys. Rev.* **A** (in press)
Allman B.E., Littrell K., Klein A.G., Cimmino A, Opat G., Werner S.A. (1996) *J. Phys. Soc. Japan* **65**, 102
Allman B.E., McMahon P. J., Nuget K.A., Paganin D., Jacobson D.L. Arif M., Werner S.A. (2000) *Nature* **408**, 158
Allman B.E., Nugent K.A. (2006) *Physica* **B385-386**, 1395
Allahverdyan A.E., Balian R., Nieuwenhuizen T.M. (2013) *Phys. Rep.* **525**, 1
Altarev I.S., Borisov Y.V., Borovikova N.V., Ivanov S.N., Kolomensky E.A., Lasakov M.S., Lobashev V.A., Nazarenko V.A., Piroshkov A.P., Serebrov A.P., Sobolev Y.V., Shulgina E.V., Yegorov A.I. (1992) *Phys. Lett.* **B276**, 242
Alvarez L.W., Bloch F. (1940) *Phys. Rev.* **57**, 111
Amidor I. (1999) *The Theory of Moiré Phenomena*, Kluwer, Dordrecht
Amselem E., Radmark M., Bourennane M., Cabello A. (2009) *Phys. Rev. Lett.* **103**, 160405
Amselem E., Bourennane M., Budroni C., Cabello A., Gühne O., Kleinmann M., Larson J.-A., Wiesniak M. (2013) *Phys. Rev. Lett.* **110**, 078901
Anandan J. (1977) *Phys. Rev.* **D15**, 1448
Anandan J. (1979), *Il Nuovo Cim.* **53A**, 221
Anandan J. (1982) *Phys. Rev. Lett.* **48**, 1660
Anandan J. (1989a) *Proc. Found. Quantum Mechanics*, Tokyo, 98
Anandan J. (1989b) *Phys. Lett.* **A138**, 347
Anandan J. (1992) *Nature* **360**, 307
Anandan J., Christian J., Nanelik K. (1997) *Am. J. Phys.* **65**, 180
Anandan J., Hagen C.R. (1994) *Phys. Rev.* **A50**, 2860
Anandan J., Stodolsky L. (1987) *Phys. Rev.* **D35**, 2597
Anderson M.H., Ensher J.R., Matthews M.R., Wieman C.E., Cornell E.A. (1995) *Science* **269**, 198
Ando M., Hosoya S. (1972) *Phys. Rev. Lett.* **29**, 281
Antonidis I., Baessler S., Büchner M., Federov V.V., Hoedl S., Lasmbrecht A., Nesvizhevsky V.V., Pignol G., Protasov K.V., Sobolev Yu. (2011) *C. R. Phys.* **12**, 755
Appel A., Bonse U. (1991) *Phys. Rev. Lett.* **67**, 1673
Araki H. (1980) *Prog. Theor. Phys.* **64**, 719
Arif M., Jacobson D.L. (1997) NIST, Gaithersburg, private communication
Arif M., Brown D.E., Greene G.L., Clothier R., Littrell K. (1994) *SPIE* **2264**, 20

Arif M., Dewey M.S., Greene G.L., Jacobson D.L., Werner S.A. (1994) *Phys. Lett.* **A184**, 154
Arif M., Kaiser H., Clothier R., Werner S.A., Berliner R., Hamilton W.A., Cimmino A., Klein A.G. (1988), Physica **B151**, 63
Arif M., Kaiser H., Clothier R., Werner S.A., Hamilton W.A., Cimmino A., Klein A.G. (1989) *Phys. Rev.* **A39**, 931
Arif M., Kaiser H., Werner S.A., Cimmino A., Hamilton W.A., Klein A. G., Opat G.L. (1985) *Phys. Rev.* **A31**, 1203
Arif M., Kaiser H., Werner S.A., Willis J.O. (1987) *Phys. Rev.* **A35**, 2810
Arndt M., Nairz O., Voss-Andreae J., Keller C., Van der Zouw G., Zeilinger A. (1999) *Nature* **401**, 680
Arnold L.G. (1973) *Phys. Lett.* **44B**, 401
Arthur J., Shull C.G., Zeilinger A. (1985) *Phys. Rev.* **B32**, 5753
Arzumanov S.S., Masalovich S.V., Strepetov A.N., Frank A.I. (1984) *JETP-Lett.* **39**, 590
Ashkar R., Schaich W.L., de Haan V.O., van Well A.A., Dalgliesh R. (2011) *J. Appl. Phys.* **110**, 102201
Ashkar R., Stonaha P., Washington A.L., Shah V.R., Fitzsimon M.R., Maranville B., Majkrzak, Lee W.T., Schaich W.L., Pynn R (2010) *J. Appl. Cryst.* **43**, 455
Aspect A., Grangier P., Roger G. (1981) *Phys. Rev. Lett.* **47**, 460
Aspect A., Grangier P., Roger G. (1982) *Phys. Rev. Lett.* **49**, 1804
Atwood D.K., Horne M.A., Shull C.G., Arthur J. (1984) *Phys. Lett.* **52**, 1673
Audretsch J. (1981) *Phys. Rev.* **D24**, 1470
Audretsch J., Lammerzahl C. (1983) *J. Phys.* **A16**, 2457
Awaya K., Tomita M. (1997) *Phys. Rev.* **A56**, 4106
Bacon G.E. (1975) *Neutron diffraction*, 3rd edn. Clarendon Press, Oxford
Bacon G.E. (ed.) (1986) *Fifty years of neutron diffraction.* Adam Hilger, Bristol
Badurek G., Buchelt R.-J., Englert B.G., Rauch H. (2000) *Nucl. Instr. Meth.* **A440**, 252
Badurek G., Gösselberger Ch., Jericha E. (2011) *Physica* **B406**, 2458
Badurek G., Rauch H., Suda M., Weinfurter H. (2000) *Opt. Comm.* **176**, 13
Badurek G., Rauch H., Summhammer J. (1983b) *Phys. Rev. Lett.* **51**, 1015
Badurek G., Rauch H., Summhammer J. (1988a) *Physica* **B151**, 82
Badurek G., Rauch H., Summhammer J., Kischko U., Zeilinger A. (1983a) *J. Phys.* **A16**, 1133
Badurek G., Rauch H., Tuppinger D. (1986) *Phys. Rev.* **A34**, 2600
Badurek G., Rauch H., Wilfing A., Bonse U., Graeff W. (1979) *J. Appl. Cryst.* **12**, 186
Badurek G., Rauch H., Zeilinger A. (1980a) in *Neutron Spin Echo* (ed. F. Mezei), Lecture Notes Phys. **128**, p. 136. Springer, Berlin
Badurek G., Rauch H., Zeilinger A. (1980b) *Z. Phys.* **B38**, 303
Badurek G., Rauch H., Zeilinger A. (eds.) (1988b) *Matter wave interferometry.* North Holland, Amsterdam
Badurek G., Rauch H., Zeilinger A., Bauspiess W., Bonse U. (1976) *Phys. Rev.* **D14**, 1177
Badurek G., Weinfurter H., Gaehler R., Kollmar A., Wehinger S., Zeilinger A. (1993) *Phys. Rev. Lett.* **71**, 307
Badurek G., Westphal G.P., Ziegler P. (1974) *Nucl. Instr. Meth.* **120**, 351
Baggot J. (1992) *The meaning of quantum theory.* Oxford University Press, Oxford
Baldo-Ceolin M. (1984) *J. de Phys.* **45**, C3–173
Baldo-Ceolin M., Benetti P., Bitter T., Bobisut F., Calligarich E., Dolfini R., Dubber D., Eisert F., El-Muzeini P., Genoni M., Gibin D., Gigli-Berzolari A., Gobrecht K., Gugliemi A., Kessler M., Kinkel U., Klemt E., Ladever M., Lippert W., Mattioli F., Mauri F., Mezzetto M., Piazzoli A., Puglieri G., Rappoli A., Raselli G.L. Scannicchio D., Sconza A., Vascon M., Visentin L., Werner R. (1990) *Phys. Lett.* **B236**, 95
Baldzuhn J., Martienssen W. (1991) *Z. Physik* **B82**, 309
Baldzuhn J., Mohler E., Martienssen W. (1989) *Z. Physik* **B77**, 347
Ballantine L.E. (1970) *Rev. Mod. Phys.* **42**, 358
Ballantine L.E. (1987) *Am. J. Phys.* **55**, 785
Ballantine L.E. (1990) *Quantum mechanics.* Prentice Hall, Englewood Cliffs, NJ

Baron M. (2005) doctoral thesis, Techn. Univ., Vienna (in German)
Baron M., Rauch H. (2011) *AIP Conf. Proc.* **1327**, 89
Bartell L.S. (1980) *Phys. Rev.* **D21**, 1698
Bartosik H., Klepp J., Schmitzer C., Sponar S., Cabello A., Rauch H., Hasegawa Y. (2009) *Phys. Rev. Lett.* **103**, 040403
Bartscher M., Bonse U. (1998) *Cryst. Res. Tech.* **33**, 535
Baruchel J. (1992) *Neutron News* **3**, 20
Barut A.O. (1988) *Found. Phys. Lett.* **18**, 95
Barut A.O., Bozic M. (1990) *Phys. Lett.* **149**, 431
Barut A.O., Bozic M., Maric Z., Rauch H. (1987) *Z. Physik* **A328**, 1
Baryshevski J., Martienssen W. (1991) *Z. Phys.* **B82**, 309
Baryshevski V.G., Cherepitsa S.V., Frank A.I. (1991) *Phys. Lett.* **A153**, 299
Baryshevski V.G., Podgoretskii M.I. (1965) *Sov. Phys. JETP* **20**, 704
Baryshevski V.G., Zaitseva A.M. (1990) *Phys. Stat. Sol.* **(b)157**, 129
Bassi A., Ghirardi G.C. (1999) *Phys. Lett.* **A257**, 24
Basu S., Bandhyopadhyay S., Kar G., Home D. (2001) *Phys.Lett.* **A279**, 281
Batterman B.W., Cole H. (1964) *Rev. Mod. Phys.* **36**, 681
Baumann J., Gähler R., Kalus J., Mampe W., Alefeld B. (1988) *Physica* **A151**, 130
Bauspiess W. (1977) thesis, University of Dortmund
Bauspiess W., Bonse U., Graeff W. (1976a) *J. Appl. Cryst.* **9**, 68
Bauspiess W., Bonse U., Graeff W., Rauch H. (1977) *J. Appl. Cryst.* **10**, 338
Bauspiess W., Bonse U., Rauch H. (1976b) *Proc. Conf. Neutron Scatt.* Gatlinburg/TN, Vol. II, p. 1094
Bauspiess W., Bonse U., Rauch H. (1978) *Nucl. Instr. Meth.* **157**, 495
Bauspiess W., Bonse U., Rauch H., Treimer W. (1974) *Z. Phys.* **271**, 177
Bayh W. (1962) *Z. Phys.* **169**, 492
Beck M., Smithey D.T., Raymer M.G. (1993) *Phys. Rev.* **A48**, R890
Becker P., Dorenwendt K., Ebeling G., Launer R., Lucas W., Probst R., Rademacher H.-J., Reim G., Seyfried P., Siegert H. (1981) *Phys. Rev. Lett.* **23**, 1540
Becker P., Seyfried P., Siegert R. (1982) *Z. Phys.* **48**, 17
Beckmann F. (1998) thesis, University of Dortmund
Beckmann F., Bonse U., Busch O., Grünnewig O. (1997) *J. Comp. Assist. Tomogr.* **21**, 539
Bell J. (1964) *Physics* **1**, 195
Bell J.S. (1966) *Rev. Mod. Phys.* **38**, 447
Bell J.S. (1987) *Speakable and unspeakable in quantum mechanics.* Cambridge University Press, Cambridge
Bell J.S. (1990) *Phys. World* 38, 33
Benatti F., Floreanini R. (1999) *Phys. Lett.* **B451**, 422
Benenson R.E., Rimawi K., Sexton E.H., Center B. (1973) *Nucl. Phys.* **A212**, 147
Bennett C.H., Brassard G., Crepeau C., Josza R., Peres A., Wooters W.K. (1993) *Phys. Rev. Lett.* **70**, 1895
Bergamin A., Cavagnero G., Mana G. (1989) *Z. Phys.* **76**, 25
Berman P.R. (ed.) (1997) *Atom interferometry.* Academic Press, Oxford
Bernabeu J., Ericson T.E.O. (1972) *Phys. Lett.* **42B**, 93
Bernhoeft N., Hiess A., Langridge S., Stunault A., Wermeille D., Vettier C., Lander G.H., Huth M., Jourdan M., Adrian H. (1998) *Phys. Rev. Lett.* **81**, 3419
Bernstein H.J. (1967) *Phys. Rev. Lett.* **18**, 1102
Bernstein H.J. (1979) in *Neutron Interferometry* (eds. U. Bonse and H. Rauch), p. 231. Clarendon Press, Oxford
Bernstein H.J. (1985) *Nature* **315**, 42
Bernstein H.J., Phillips A.V. (1981) *Sci. Am.* **245**, 123
Bernstein H.J., Zeilinger A. (1980) *Phys. Lett.* **75A**, 169

Berry M.V. (1984) *Proc. Roy. Soc. London* **A392**, 415
Berry M.V. (1988) *Sci. Am.*, Dec.
Bertlmann R.A., Durstberger K., Hasegawa Y. (2006) *Phys. Rev.* **A73**, 022111
Bertlmann R.A., Zeilinger A. (eds.) (2002) *Quantum [un]speakables.* Springer, Berlin
Bertolami O. (1986) *Mod. Phys. Lett.* **A1**, 383
Beyer H., Nitsch J. (1983), Phys. Lett.**127B**, 336
Bhandari R. (1991) *Physica* **B175**, 111
Bhandari R. (1993) *Phys. Lett.* **A180**, 15
Bhandari R. (1997) *Phys. Rep.* **281**, 1
Bhandari R. (1999) *Phys. Rev. Lett.* **83**, 2089
Bhandari R., Samuel J. (1988) *Phys. Rev. Lett.* **60**, 1211
Bialynicki-Birula I., Mycielski J. (1976) *Ann. Phys. (N.Y.)*, **100**, 62
Bialynicki-Birula I., Mycielski J. (1979) *Phys. Ser.* **20**, 539
Bitter T., Dubbers D. (1987) *Phys. Rev. Lett.* **59**, 251
Björk G., Söderholm J., Trifonov A., Tsegaye T., Karlsson A. (1999) *Phys. Rev.* **A60**, 1874
Black T.C., Huffman P.R., Jacobson D.L., Snow W.M., Schoen K., Arif M., Kaiser H., Lamoreaux S.K., Werner S.A. (2003) *Phys. Rev. Lett.* **90**, 192502
Blanchard P., Jadczyk A. (1993) *Phys. Lett.* **A175**, 157
Blatt J.M., Weisskopf V.F. (1952) *Theoretical nuclear physics.* Wiley, New York
Bloch F., Siegert A. (1940) *Phys. Rev.* **57**, 522
Boal D.H., Gelbke C.K., Jennings B.K. (1990) *Rev. Mod. Phys.* **62**, 553
Boersch H., Hamisch H., Grohmann K., Wohlleben D. (1961) *Z. Phys.* **65**, 79
Boersch H., Lischke B. (1970) *Z. Phys.* **237**, 449
Boeuf A., Bonse U., Caciuffo R., Fournier J.M., Manes L., Kischko U., Rustichelli F., Wroblewski T. (1985) *Acta Cryst.* **B41**, 81
Boeuf A., Caciuffo R., RebonatoR., Rustichelli F., Founier J.M., Kischko U., Manes L. (1982) *Phys. Rev. Lett.* **49**, 1086
Boffi S., Caglioti G. (1966) *Il Nouvo Cim.* **41B**, 247
Boffi S., Caglioti G. (1971) *Il Nuovo Cim.* **3B**, 262
Bohm D. (1952a) *Phys. Rev.* **85**, 166
Bohm D. (1952b) *Phys. Rev.* **89**, 180
Bohm D., Hiley B.J. (1993) *The undivided universe: an ontological interpretation of quantum theory.* Routledge, London
Bohm D., Vigier J.P. (1984) *Phys. Rev.* **96**, 208
Bohm D.J., Dewdney C., Hiley B.J. (1985) *Nature* **315**, 294
Bohr N. (1928) in *Atti del Congresso Int. Di Fisici Como* Vol. **2**, p. 565. Zanchelli, Bologna
Bohr N. (1934) *Atomic theory and the description of nature.* Cambridge University Press, Cambridge
Bohr N. (1958) *Atomic physics and human knowledge.* Wiley, New York
Boisvert J.-S., Marchildon L. (2013) *Found. Phys.* **43**, 294
Bonder Y., Fischbach E., Hernandez-Colonada H., Krause D.E., Rohrbach Z., Sudarsky D. (2013) *Phys. Rev.* **D** 87, 125021
Bonse U., Beckmann F., Bartscher M., Biermann T., Busch F., Günnewig O. (1997) *SPIE* **3149**, 108
Bonse U., Graeff W. (1977) *Top. Appl. Phys.* **22**, 93
Bonse U., Graeff W., Rauch H. (1979) *Phys. Lett.* **69A**, 420
Bonse U., Graeff W., Teworte R., Rauch H. (1977) *Phys. Stat. Sol.* (**a**) **43**, 487
Bonse U., Hart M. (1965a) *Appl. Phys. Lett.* **6**, 155
Bonse U., Hart M. (1965b) *Appl. Phys. Lett.* 7, 238
Bonse U., Johnson Q., Nichols M., Nusshardt R., Krasnicki S., Kinney J. (1986) *Nucl. Instr. Meth.* **A246**, 644
Bonse U., Kischko U. (1982) *Z. Phys.* **A305**, 171
Bonse U., Rauch H. (eds.) (1979) *Neutron interferometry.* Clarendon Press, Oxford

Bonse U., Rumpf A. (1984) *Acta Cryst.* **A40**, C-359
Bonse U., Rumpf A. (1986) *Phys. Rev. Lett.* **56**, 2441
Bonse U., Rumpf A. (1988) *Phys. Rev.* **A37**, 1059
Bonse U., Te Kaat E. (1971) *Z. Phys.* **243**, 14
Bonse U., Teworte R. (1980) *J. Appl. Cryst.* **13**, 410
Bonse U., Uebbing R, Bartscher M., Nusshardt M. (1994) *Metrologia* **31**, 195
Bonse U., Wroblewski T. (1983) *Phys. Rev. Lett.* **51**, 1401
Bonse U., Wroblewski T. (1984) *Phys. Rev.* **D30**, 1214
Bordag M., Mokideen U., Mostepanenko V.M. (2001) *Phys. Rep.* **353**, 1
Borde Ch.J. (1989) *Phys. Lett.* **A140**, 10
Born A., Wolf E. (1975) *Principles of optics*. Pergamon Press, Oxford
Born M. (1926) *Z. Physik* **38**, 803
Boucherle J.X., Kischko U., Schweizer J. (1985) *Acta Cryst.* **A41**, 589
Bouwman W.G., Plomp J., de Haan V.O., Kraan W.H., van Well A.A., Habicht K., Keller T., Rekveldt M.T. (2008) *Nucl. Instr. Meth.* **A586**, 9
Bouwman W.P., Krouglov T.V., Plomp J., Grigoriev S.V., Kraan W.H., Rekveldt M.T. (2004) *Physica* **B350**, 140
Bouwmeester D., Pan J.-W., Mattle K., Eibl M., Weinfurter H., Zeilinger A. (1997) *Nature* **390**, 575
Boyer T.H. (1987a) *Phys. Rev.* **A36**, 5083
Boyer T.H. (1987b) *Nuovo Cim.* **B100**, 685
Bozic M., Maric Z. (1991) *Phys. Lett.* **A158**, 33
Bozic M., Vukovic L., Davidovic M., Sanz A.S. (2011) *Phys. Scripta* **T143**, 014007
Brandy R., Anderson R. (2014) arXiv:1401.4356v [quant-ph]
Braunstein S.L., McLachlan R.I. (1987) *Phys. Rev.* **A35**, 1659
Braverman B., Simon C. (2013) *Phys. Rev. Lett.* **110**, 060406
Brax P., Pignol G. (2011) *Phys. Rev. Lett.* **107**, 111301
Brax P., Pignol G., Roulier D. (2013) arXiv:1306.6536v1 [quant.ph]
Brax P., van de Bruck C., Davis A.-C., Khoury J., Weltman A. (2004) *Phys. Rev.* **D70**, 123518
Breit G., Wigner E.P. (1936) *Phys. Rev.* **49**, 519
Breunlich W.H., Tagesen S., Bertl W., Chalupka A. (1974) *Nucl. Phys.* **A221**, 269
Bromberg J.J. (2008) *Hist. Stud. Nat. Sci.* **38**, 325
Brown H.R. (1996) *Perspectives on quantum reality* (ed. R. Clifton), p. 183. Kluwer, Netherlands
Brown H.R., Dewdney C., Horton G., (1995) *Found. Phys.* **25**, 329
Brown H.R., Summhammer J., Callaghan R.E., Kaloyerou P. (1992) *Phys. Lett.* **A163**, 21
Brukner C., Zeilinger A. (1997) *Phys. Rev.* **A56**, 3804
Brune M., Hagley E., Dreyer E., Maitre X., Maali A., Wunderlich C., Raimond J.M., Haroche S. (1996) *Phys. Rev. Lett.* **77**, 4887
Bunakov V.E., Gudkov V.P. (1981) *Z. Phys.* **A303**, 285
Bunatian G.G., Nikolenko V.G., Popov A.B., Samosvat G.S., Tretyakova T.Y. (1997) *Z. Phys.* **A359**, 337
Buonomano V. (1980) *Il Nuovo Cim.* **B57**, 146
Buonomano V. (1988) *Physica* **B151**, 349
Buonomano V. (1989) *Found. Phys. Lett.* **2**, 565
Buras B., Kjems J.K. (1973) *Nucl. Instr. Meth.* **106**, 461
Busch P. (1987) *Found. Phys.* **17**, 905
Busch P., Grabowski M., Lahti P.J. (1995) *Operational quantum physics*. Springer, Berlin
Busch P., Shilladay C. (2006) *Phys. Rep.* **435**, 1
Butt N.M., Bashir J., Nasir Khan M.J. (1993) *Mat. Sci.* **28**, 1595
Butt N.M., Bashir J., Willis B.T.M., Heger G. (1988) *Acta Cryst.* **A44**, 396
Büttiker M. (1983) *Phys. Rev.* **B27**, 6178
Buzek V., Keitel C.H., Knight P.L. (1995) *Phys. Rev.* **A51**, 2594

Buzek V., Knight P.L. (1991) *Opt. Comm.* **81**, 331
Buzek V., Knight P.L. (1995) *Progr. Opt.* **34**, 1
Byrne J. (1978) Nature 275, p. 188
Byrne J. (1994a) *Neutrons, nuclei and matter*. Institute of Physics Publications, Bristol
Byrne J. (1994b) *Neutron News* **5**, 15
Cabello A., Filipp S., Rauch H., Hasegawa Y. (2008) *Phys. Rev. Lett.* **100**, 130404
Caldeira A.O., Leggett A.J. (1983) *Ann. Phys. (N.Y.)* **149**, 374
Caldeira A.O., Leggett A.J. (1985) *Phys. Rev.* **A31**, 1095
Camacho A., Camacho-Galvan A. (2007) *Rep. Prog. Phys.* **70**, 1937
Cappelletti R.L. (2012) *Phys. Lett* **A 376**, 2096
Caprez A., Barwick B., Batelaan H. (2007) *Phys. Rev. Lett.* **99**, 210401
Carlson J., Schiavilla R. (1998) *Rev. Mod. Phys.* **70**, 743
Carnal O., Mlynek J. (1991) *Phys. Rev. Lett.* **66**, 2689
Carruthers P., Nieto M.N. (1968) *Rev. Mod. Phys.* **40**, 411
Casella R.C. (1984) *Phys. Rev. Lett.* **53**, 1033
Casella R.C., Werner S.A. (1992) *Phys. Rev. Lett.* **69**, 1625
Casimir H.B.G., Polder D. (1948) *Phys. Rev.* **73**, 360
Cavey W.M. (2009) *Acustics Today*, April, 14
Chadwick J. (1932) *Nature* **129**, 312; and *Proc. R. Soc. London* **A136**, 692
Chambers R.G. (1960) *Phys. Rev. Lett.* **5**, 3
Champenois C., Jacque S., Leportre S., Büchner M., Terenec G., Vigue J. (2008) *Phys. Rev.* **A77**, 013621
Chapman M.S., Ekstrom C.R., Hammond T.D., Rubenstein R.A., Schmiedmayer J., Wehinger S., Pritschard D.E. (1995) *Phys. Rev. Lett.* **75**, 3783
Chatzidimitriou-Dreismann C.A. (1997) *Adv. Chem. Phys.* **99**, 393
Chatzidimitriou-Dreismann C.A., Abdul-Redah T., Streffer R.M.F., Mayers J. (1997) *Phys. Rev. Lett.* **79**, 2839
Chen C.R., Payne G.L., Friar J.L., Gibson B.F. (1991) *Phys.Rev.***C44**, 50
Cheng Ta-Pei (2012) *Relativity, Gravitation, and Cosmology*, pp. 124–5. Oxford University Press, Oxford
Chiao R.Y., Antaramian A., Ganga K.M., Jiao H., Wilkinson S.R. (1988) *Phys. Rev. Lett.* **60**, 1214
Chiao R.Y., Wu Y.S. (1986), *Phys. Rev. Lett.* **57**, 933
Chiu C., Stodolsky L. (1980), Phys. Rev. **D22**, 1337
Christ J., Springer T. (1962) *Nukleonika* **4**, 23
Christensen J., Wilcox W., Lee F.X., Zhou L. (2005) *Phys. Rev.* **D72**, 034503
Cimmino A., Opat G. I., Klein A.G., Kaiser H., Werner S.A., Arif M., Clothier R. (1989) *Phys. Rev. Lett.* **68**, 380
Cinelli C., Barbieri M., Perris R., Matolini P., De Martini F. (2005) *Phys.Rev.Lett.* **95**, 240405
Cini M., Serva M. (1992) *Phys. Lett.* **A167**, 319
Clauser J.F., Dowling J.P. (1996) *Phys.Rev.* **A53**, 4587
Clauser J.F., Horne M.A., Shimony A., Holt R.A. (1969) *Phys. Rev. Lett.* **23**, 880
Clauser J.F., Shimony A. (1978) *Rep. Progr. Phys.* **41**, 1982
Clothier R. (1991), PhD thesis, University of Missouri, Columbia
Clothier R., Kaiser H., Werner S.A., Rauch H., Woelwitsch H. (1991) *Phys. Rev.* **A44**, 5357
Cohen J.M., Mashoon B. (1993) *Phys. Lett.* **A181**, 353
Cohen-Tannoudji C., Dupont-Roc J., Grynberg G. (1992) *Atom–photon interaction*, p. 460. Wiley, New York
Cohen-Tannoudji C., Haroche S. (1969) *J. Phys. (Paris)* **30**, 125, 153
Colella R., Overhauser A.W., Werner S.A. (1975) *Phys. Rev. Lett.* **34**, 1472
Collonna N., Bowman D. R., Celano L., Erasmo G.D., Fiore E.W., Pantelea A., Vaticchio V., Tagliente G., Pratt S. (1995) *Phys. Rev. Lett.* **75**, 4190

Colwell J.H., Mueller P.H., Whitemore W.L. (1968) *Proc. Inel. Scatt. Neutrons*, Vol. II, p. 429. IAEA, Vienna
Comay E. (1998) *Phys. Lett.* **A250**, 12
Comay E. (2000) *Phys.Rev.* **A62**, 042102
Conway J., Kochen S. (2006) *Found. Phys.* **36**, 1441
Cook R.J., Milonni P.W. (1987) *Phys. Rev.* **A35**, 5081
Coselia R.C., Werner S.A. (1992), *Phys. Rev. Lett.* **69**, 1625
Couder Y., Fort E. (2012) *J. Phys.* **361**, 012001
Couder Y., Protiere S., Fort E., Boudaoud A. (2005) *Nature* **437**, 208
Cowley J. M. (1981) *Diffraction physics*. North-Holland, Amsterdam
Cramer J.G. (1986) *Rev. Mod. Phys.* **58**, 647
Cribier D., Jacrot B., Rao L.M., Farnoux B. (1964) *Phys. Lett.* **9**, 106
Croca J.R., Guraccio A., Selleri F. (1988) *Found. Phys. Lett.* **1**, 101
Cronin A.D., Schmiedmayer J., Pritchard D.E. (2009) *Rev. Mod. Phys.* **81**, 1051
Cronqvist M., Jonson B., Nilsson T., Nyman G., Riisager K., Rott H.A., Skeppstedt O., Tengblad O., Wilhelmsen K. (1992) *Nucl. Instr. Meth.* **A317**, 273
Cser L., Farago B., Krexner G., Sharkov I., Török G. (2004) *Physica* **B350**, 113
Cser L., Krexner G., Török G. (2001) *Europhys.Lett.* **54**, 747
Cser L., Török G., Krexner G., Sharkov I., Farago B. (2002) *Phys. Rev. Lett.* **89**, 175504–1
Cusatis C., Hart M. (1975), in *Anomalous Scattering* (eds. S. Ramaseshan and S.C. Abrahams). Munksgaard, Copenhagen
d'Espagnat B. (2011) *Found. Phys.* **41**, 1703
Dabbs J.W.T., Harvey J.A., Paya D., Horstmann H. (1965) *Phys. Rev.* **139**, 765
D'Ariano G., Paris M.G.A. (1997) *Phys. Rev.* **A55**, 2267
Darwin C.G. (1914) *Phil. Mag.* **27**, 315, 675
Dattoli G., Matrone G., Prosperi D. (1977) *Lett. Nuovo Cim.* **19**, 601
Davidovic M.D., Snaz A.S. (2013) *Europhys. News* **44**, 33
Davidovic M., Sanz A.S., Bozic M., Arsenovic D. (2013) *Phys. Scripta* **T153**, 014015
De Chiara G., Palma G.M. (2003) *Phys. Rev. Lett.* **91**, 090404
De Haan V.-O., Plomp J., Rekveldt T.M., Kraan W.H., van Well Ad.A. (2010) *Phys. Rev. Lett.* **104**, 010401
De Polavieja G.G. (1997) *Phys. Lett.* **A232**, 1
De Broglie L. (1923a) *Comptes Redus* **177**, 507
De Broglie L. (1923b) *Nature* **112**, 540
De Broglie L. (1925) *Ann. de Phys.* 10e ser. **3**, 22
De Broglie L. (1926) *C. R. Acad. Sci.* **183**, 447
De Broglie L. (1927) *J. Phys. (France)* **8**, 225
De Broglie L. (1960) *Nonlinear wave mechanics, a causal description*. Elsevier, Amsterdam
Debye P. (1913) *Verh. Dt. Phys. Ges.* **15**, 678
Denkmayr T., Geppert H., Sponar S., Lemmel H., Matzkin A., Tollaksen J., Hasegawa Y. (2014) arXiv: 1312.3775v1 [quant-ph]
De Raedt H., De Raedt K., Michielsen K. (2005) *Europhys. Lett.* **69**, 861
De Raedt H., Jin F., Michielson K. (2012) *Quantum Matter* **1**, 20
De Raed H., Michielsen K. (2014) *Front. Phys.* **2**, 14/2
DeMartini F., DeDomenicis L., Clossocolanti V., Milani G. (1992) *Phys. Rev.* **A45**, 5144
Denkmayr T., Geppert H., Sponar S., Lemmel H., Matzkin A., Tollaksen J., Hasegawa Y. (2014) *Nature Communications/5.4492/DOI: 10.1038/ncomms5492/*
Descatis R.P. (1637) *Discurse de la Methode* (ed. Ian Maire). Dioptrique, Leiden
Deslattes R.D., Henins A. (1973) *Phys. Rev. Lett.* **31**, 972
D'Espargnat B. (1976) *Conceptual foundations of quantum mechanics*. Perseus Books, Reading, MA
D'Espargnat B. (1979) *Sci. Am.* **9**, 218

D'Espargnat B. (1995) *Veiled reality—an analysis of present-day quantum mechanical concepts.* Addison-Wesley, Reading, MA
Deutsch I.H. (1991) *Am. J. Phys.* **59**, 834
Deutsch M., Hart R (1985) *Acta Cryst.* **A41**, 48
Dewdney C. (1985) *Phys. Lett.* **A109**, 377
Dewdney C., Guerret P., Kyprianidis A., Vigier J.P. (1984) *Phys. Lett.* **102A**, 291
Dewdney C., Holland P.R., Kyprianidis A. (1986) *Phys. Rev.* **A119**, 259
Dewdney C., Holland P.R., Kyprianidis A., Vigier J.P. (1988) *Nature* **336**, 536
DeWitt B.S., Graham R.D. (1973) *The many-world interpretation of quantum mechanics.* Princeton University Press, Princeton
Dexhage K.H. (1984), in *Progress in Optics* **12** (ed. E. Wolf), p. 165. North-Holland, Amsterdam
DeYoung P.A., Bennink R., Butler T., Chung W., Dykstra C., Gilfoyle G., Hinnefeld J., Kaplan M., Kolota J.J., Kryger R.A., Kugi J., Mader C., Nimchek M., Santi P., Snyder A. (1996) *Nucl. Phys.* **A597**, 127
Dietze H.D., Nowak E. (1981) *Z. Phys.* **B44**, 245
Dirac P.A.M. (1927) *Proc. R. Soc. London* **A114**, 243
Dirac P.A.M. (1930) *The principles of quantum mechanics.* Clarendon Press, Oxford
Dirac P.A.M. (1945) *Rev. Mod. Phys.* **17**, 195
Dobrynin Yu.L., Lomonosov V.V. (1988), *Sov. Phys. JETP* **68**, 1122
Dodonov V.V., Kalmykov S.Yu., Man'ko V. I. (1995) *Phys. Lett.* **A199**, 123
Dodonov V.V., Man'ko O.V., Man'ko V.I. (1994) *Phys. Rev.* **A49**, 2993
Drabkin G., Ioffe A., Kirsanov S., Mezei F., Zabijakin V. (1994) *Nucl. Instr. Meth.* **A348**, 198
Drabkin G.M., Trunov V.A., Runov V.V. (1968) *Sov. Phys. JETP* **27**, 194
Drabkin G.M., Zhitnikov R.A. (1960) *Sov. Phys. JETP* **11**, 729
Dresden M., Yang C.N. (1979) *Phys. Rev.* **D20**, 1846
Dubbers D. (1976) *Z. Phys.* **A276**, 245
Dubbers D. (1989) *Nucl. Instr. Meth.* **A284**, 22
Dubbers D., El-Muzeini P., Kessler M., Last J. (1989) *Nucl. Instr. Meth.* **A275**, 294
Dubbers D., Schmidt M.G. (2011) *Rev. Mod. Phys.* **83**, 1111
Dubbers D., Stöckmann H.-J. (2013) *Quantum physics: the bottom-up approach.* Springer Berlin-Heidelberg
Dubus F., Bonse U., Zawisky M., Baron M., Loidl R. (2005) *IEEE Trans. Nucl. Sci.* **52**, 364
Dunningham J., Vedral V. (2011) *Introductory quantum physics and relativity.* Imperial College Press, London
Dünnweber W., Lippich W., Otten D., Assmann W., Hartmann K., Hering W., Konnerth D., Trombik W. (1990) *Phys. Rev. Lett.* **65**, 297
Dürr S., Nonn G., Rempe G. (1998) *Nature* **395**, 33
Dürr S., Rempe G. (2000) *Am. J. Phys.* **68**, 1021
Dürr D., Teufel S. (2009) *Bohmian Mechanics.* Springer, Berlin, Heidelberg
Durt T. (1999) *Int. J. Theor. Phys.* **38**, 457
Dyumin A.N., Korenblimi Y., Ruban V.A., Takerev B.B. (1980) *JETP Lett.* **311**, 384
Ebisawa T., Funahashi H., Tasaki S., Otake Y., Kawai T., Hino M., Achiwa N. (1996) *J. Phys. Soc. Japan*, Suppl. A **65**, 66
Ebisawa T., Tasaki S., Kawai T., Akiyoshi T., Utsuro M., Otake Y., Funahashi H., Achiwa N. (1994) *Nucl. Instr. Meth.* **A344**, 597
Ebisawa T., Tasaki S., Kawai T., Hino M., Achiwa N., Otake Y., Funahashi H., Yamazaki D., Akiyoshi T. (1998b) *Phys. Rev.* **A57**, 4720
Ebisawa T., Yamazaki D., Tasaki S., Hino M., Kawai T., Iwata Y., Achiwa N., Kanaya T., Soyama K. (1999) *Phys. Lett.* **A259**, 20
Ebisawa T., Yamazaki D., Tasaki S., Kawai T., Hino M., Akiyoshi T., Achiwa N., Otake Y. (1998a) *J. phys. Soc. Japan* **67**, 1569

Eddi A., Fort E., Moisy F., Couder Y. (2009) *Phys.Rev.Lett.* **102**, 240401
Eder G., Zeilinger A. (1976) *Il Nuovo Cim.* **34B**, 76
Eder K., Gruber M., Zeilinger A., Gähler R., Mampe W. (1991) *Physica* **B172**, 329
Edwards S.F. (1958) *Phil. Mag.* **3**, 1020
Ehrenberg W., Siday R.W. (1948) *Proc. R. Soc. London* **B62**, 8
Einstein A. (1905) *Ann. Phys. (Leipzig)* **17**, 891
Einstein A. (1949), in *Albert Einstein: Philosopher-Scientist* (ed. P.A. Schilpp). Harper & Row, New York
Einstein A. (1936) *J. Franklin Institute* 221, 349
Einstein A., Podolski B., Rosen N. (1935) *Phys. Rev.* **47**, 777
Ekstein H. (1953) *Phys. Rev.* **89**, 490
Ekstrom Ch. R., Keith D.W., Pritschard D.E. (1992) *Appl. Phys.* **B54**, 369
Ellis J., Hagelin J.S., Nanopoulos D.V., Srednicki M. (1984) *Nucl. Phys.* **B241**, 381
Elsasser W.M. (1936) *C. R. Acad. Sci. Paris* **202**, 1029
Englert B.-G. (1996) *Phys. Rev. Lett.* 77, 2154
Englert B.-G. (2013) *Eur. Phys. J.* **67**, 238
Englert B.-G., Bergou J.A. (2000) *Opt. Comm.* **179**, 337
Englert B.-G., Schwinger J., Scully M.O. (1988) *Found. Phys.* **18**, 1045
Englert B.-G., Scully M.O., Walther H. (1995) *Nature* **375**, 367
Erdösi D., Huber M., Hiesmayr B.C. Hasegawa Y. (2011) *New J. Phys.***15**, 023033
Erhart J., Sponar S., Sulyok G., Badurek G., Ozawa M., Hasegawa Y. (2012) *Nat. Phys.* **8**, 185
Everett III, H. (1957) *Rev. Mod. Phys.* **29**, 454
Ewald P.D. (1916) *Ann. Phys. (Leipzig)* **49**, 1, 117; **54**, 519
Ewald P.D. (1928) *Ann. Phys. (Leipzig)* **87**, 55
Fabri E., Picasso L.E. (1986) *Phys. Lett.* **A119**, 268
Fabry C., Perot A. (1899) *Ann. Chim. Phys.* **16**, 116
Facchi P., Hradil Z., Krenn G., Pascazio S., Rehacek J. (2002) *Phys. Rev.* **A66**, 012110
Facchi P., Mariano A., Pascazio S. (2001) *Phys. Rev.* **A63**, 052108
Faktis D., Morris G.M. (1988) *Opt. Lett.* **13**, 4
Fally M., Klepp J., Tomita Y., Nakamura T., Pruner C., Ellabban M.A., Rupp R.A., Bichler M., Drevensek Olenik I., Kohlbrecher J., Eckerlebe H., Lemmel H., Rauch H. (2010) *Phys. Rev. Lett.* **105**, 123904
Federov V.V., Volonin V.V, Lapin E.G. (1992) *J. Phys.* **G18**, 1133
Fedorov V.V., Voronin V.V., Lapin E.G., Sumbaev O. I. (1995) *Pis'ma ZhTF* **21**, 50
Felber J., Gähler R, Rausch C., Golub R. (1996) *Phys. Rev.* **A53**, 316
Felber J., Gähler R., Golub R., Hank P., Ignatovich V., Keller T., Rauch U. (1999) *Found. Phys.* **29**, 381
Felber J., Gähler R., Golub R., Prechtel K. (1998) *Physica* **B252**, 34
Felcher G.P., Peterson S.W. (1975) *Acta Cryst.* **A31**, 76
Fermi E. (1936) *Ric. Sci.* **1**, 13
Fermi E., Marshall L. (1947) *Phys. Rev.* **71**, 666
Fermi E., Zinn W.H. (1946) *Phys. Rev.* **70**, 103
Fernandez D.J., Rosas-Ortic O. (1997) *Phys. Lett.* **A236**, 75
Ferraro M. (2003) *Physics* **33**, 665
Ferry D.K., Grubin H., Jacobini C., Jauho A.-J. (eds.) (1995) *Quantum transport in ultrasmall devices*. Plenum, New York
Feshbach H., Porter C.E., Weisskopf V.F. (1954) *Phys. Rev.* **96**, 448
Feynman R.P. (1948) *Rev. Mod. Phys.* **20**, 367
Feynman R.P. (1961) *Quantum electrodynamics*. Benjamin, New York
Feynman R.P., Leighton R.B., Sands K. (1965) *The Feynman lectures on physics*, Vol. III. Addison-Wesley, Reading, MA
Feynman R.P., Vernon F. (1963) *Ann. Phys. (N.Y.)* **24**, 118
Feynman R.P., Vernon F.I., Heilwarth R.W. (1957) *J. Appl. Phys.* **28**, 49

Fidecaro G., Fidecaro M., Lancere L., Marchioro A., Mampe W., Baldo-Ceolin M., Mattioli F., Puglierin G., Batty C.J., Green K., Prosper H.B., Sharman P., Pendlebury J.M., Smith K.F. (1985) *Phys. Lett.* **B156**, 122
Filipp S., Hasegawa Y., Loidl R., Rauch H. (2005) *Phys. Rev.* **A72**, 021602
Filipp S., Klepp J., Hasegawa Y., Plonka-Spehr Ch., Schmidt U., Geltenbort P., Rauch H. (2009) *Phys. Rev. Lett.* **102**, 030404
Finkelstein K.D., Jauch J.M., Schiminovich S., Speiser D. (1962) *J. Math. Phys.* **3**, 207
Finkelstein K.D., Shull C.G., Zeilinger A. (1986) *Physica* **B136**, 131
Fischbach E., Sudarsky D., Szafer A., Talmatje C., Aaronson S.H. (1986) *Phys. Rev. Lett.* **56**, 3
Fischer H.E., Neuefeind J., Simonson J.M., Loidl R., Rauch H. (2008) *J. Phys. Condens. Matter* **20**, 045221
Fizeau H.L. (1851) *C. R. Hebd. Seances Acad. Sci.* **33**, 349
Fizeau H.L. (1859) *Ann. Chim. Phys.* **57**, 385
Fluegge S. (1971) *Practical quantum mechanics*, problem 40. Springer, Berlin
Foldy L.L. (1951) *Phys. Rev.* **83**, 688
Foldy L.L. (1958) *Rev. Mod. Phys.* **30**, 471
Forte M. (1982) *Lett. Nuovo Cim.* **34**, 296
Forte M., Henkel B.R., Ramsay N.F., Green K., Greene G.I., Byrne J., Pendlebury J.M. (1980) *Phys. Rev. Lett.* **45**, 2088
Forte M., Zeyen C.M.E. (1989) *Nucl. Instr. Meth.* **A284**, 147
Fox M. (2006) *Quantum optics*. Oxford University Press, Oxford
Francon K. (1979) *Optical image formation and processing*. Academic Press, New York
Frank A.I. (1989) *Nucl. Instr. Meth.* **A284**, 161
Frank A.I. (2013) *J. Phys.* **528**, 012029
Frank A.I., Bondarenko I.V., Kozlov A.V. Hoghoj P., Ehlers G. (2001) *Physica* **B 297**, 307
Frauenfelder H., Henley E.M. (1974) *Subatomic physics*. Prentice Hall, New York
Freund A.K., Kischko U., Bonse U., Wroblewski T. (1985) *Nucl. Instr. Meth.* **A234**, 495
Freyberger M., Henri M., Schleich W.P. (1995) *Quantum Semiclass. Opt.* 7, 187
Freyberger M., Kienle S.H., Yakovlev V.P. (1997) *Phys. Rev.* **A56**, 195
Friedrich H., Heintz W. (1978) *Z. Phys.* **B31**, 423
Fuchs C.A., Peres A. (2000) *Phys. Today* **53**, 70
Funahashi H., Ebisawa T., Haseyama T., Hino M., Masaike A., Otake Y., Tsuguchika T., Tasaki S. (1996) *Phys. Rev.* **A54**, 649
Gabor E. (1956) *Rev. Mod. Phys.* **28**, 260
Gaehler R., Felber J., Mezei F., Golub R. (1998) *Phys. Rev.* **A58**, 280
Gaehler R., Golub R. (1984) *Z. Phys.* **B56**, 5
Gaehler R., Golub R. (1987) *Z. Phys.* **B65**, 269
Gaehler R., Golub R. (1988) *J. Phys., France* **49**, 1195
Gaehler R., Golub R., Habicht K., Keller T., Felber J. (1996) *Physica* **B229**, 1
Gaehler R., Kalus J., Mampe K. (1980) *J. Phys.* **E13**, 546
Gaehler R., Kalus J., Mampe K. (1982) *Phys. Rev.* **D25**, 2887
Gaehler R., Klein A.G., Zeilinger A. (1981) *Phys. Rev.* **A23**, 1611
Gaehler R., Zeilinger A. (1991) *Am. J. Phys.* **59**, 316
Gähler R., Golub R., Keller T. (1992) *Physica* **180–1**, 899
Galapon E.A. (2009) *Phys. Rev.* **A80**, 030102
Gardestig A. (2009) *J. Phys. G: Nucl. Phys.* **36**, 053001
Gea-Banacloche J. (1990) *Phys. Rev. Lett.* **65**, 3385
Gell-Mann M., Hartle J.B. (1993) *Phys. Rev.* **D47**, 3345
Gerhardt H., Buechler U., Litfin G. (1974) *Phys. Lett.* **49A**, 119
Gerhardt H., Welling H., Frohlich D. (1973) *Appl. Phys.* **2**, 91
Gericke M.T., Bowman J.D., Johnson M.B. (2008) *Phys.Rev.* **C78**, 044003

Geszti T. (1998) *Phys. Rev.* **A58**, 4206
Ghetti R., Colonna N., Helgesson J. (1999) *Nucl. Instr. Meth.* **A421**, 542
Ghetti R., Helgesson J. (2005) *Nucl. Phys.* **A 752**, 480
Ghirardi G.C. (1999) *Phys. Lett.* **A262**, 1
Ghirardi G.C., Pearle P., Rimini A. (1990) *Phys.Rev.* **A42, 78**
Ghirardi G.C., Rimini A., Weber T. (1986) *Phys. Rev.* **D34**, 470
Ghose P. (1999) *Testing quantum mechanics on new grounds.* Cambridge University Press, Cambridge
Ghose P. (2009) *NeuroQuantology* 7, 623
Gibbin J. (1998) *The search for superstrings, symmetry an the theory of everything.* Little, Brown, Boston
Gilder L. (2008) *The age of entanglement: when quantum physics was reborn.* Knopf, New York
Gillies G.T. (1997), Rep. Progr. Phys. **60**, 151
Gillot J., Lepoutre S., Gauguet A., Vigué, Büchner M. (2014) *Eur. Phys. Lett.* (in print)
Giulini D., Joos E., Kiefer C., Kupsch J., Stamatescu I.-O., Zeh H.D. (1996) *Decoherence and the appearance of a classical world in quantum theory.* Springer, Berlin
Glashow S.L. (1980), in *Quarks and leptons* (ed. M. Levy). Plenum, New York
Glättli H., Bacchella G.L., Fourmond M., Malinovski A., Meriel P., Pinot M., Roubeau R., Abragam A. (1979) *J. de Physique* 7, 629
Glättli H., Goldman M. (1987) *Meth. Exp. Phys.*, Vol. 23, part C, Chap. 21, p. 241. Academic Press, New York
Glauber R.J. (1963) *Phys. Rev.* **130**, 2529; **131**, 2766
Glauber R.J. (1965) in *Quantum Optics and Electronics* (eds. C. DeWitt, A. Blandini, C. Cohen-Tannoudji). Gordon & Beach, New York
Glauber R.J. (1968) in *Fundamental Problems in Statistical Mechanics* (ed. E.G.D. Cohen), p. 140. North Holland, Amsterdam
Glauber R.J. (1986) in *New Techniques and Ideas in Quantum Mechanics* (ed. D.M. Greenberger), p. 336. Academic Science, New York
Gösselsberger C., Bacak M., Gerstmayr T., Gumpenberger S., Hawlik A., Hinterleitner B., Jericha E., Nowak S., Welzl A., Badurek G. (2013) *Phys. Proc.* **42**, 106
Goldberger M.L., Seitz F. (1947) *Phys. Rev.* **71**, 294
Goldhaber A.S. (1989) *Phys. Rev. Lett.* **62**, 482
Goldhaber G., Goldhaber S., Lee W., Pais A. (1960) *Phys.Rev.* **120**, 300
Goldman T., Hughes R.J., Nieto M.M. (1986) *Phys. Lett.* **B171**, 217
Goldstein H. (1980) *Classical mechanics*, 2nd.edn. Addison-Wesley, Reading, MA
Golub R., Gähler R., Keller T. (1994) *Am. J. Phys.* **62**, 779
Golub R., Lamoreaux K. (1994) *Phys. Rep.* **237**, 1
Golub R., Pendlebury J.M. (1972) *Contemp. Phys.* **13**, 519
Golub R., Pendlebury J.M. (1979) *Rep. Progr. Phys.* **42**, 439
Gondran M., Gondran A. (2005) *Am. J. Phys.* **73**, 507
Goos F., Hänchen H. (1947) *Ann. Phys. (Leipzig)* **436**, 333
Goos F., Hänchen H. (1949) *Ann. Phys. (Leipzig)* **440**, 251
Gottfried K. (1966) *Quantum mechanics*, p. 188. Benjamin, New York
Gottfried K., Weisskopf V.F. (1984) *Concepts in particle physics.* Clarendon Press, Oxford
Goy P., Raimond J.M., Gross M., Haroche S. (1983) *Phys. Rev. Lett.* **50**, 1903
Graeff W., Bauspiess W., Bonse U., Rauch H. (1978) *Acta Cryst.* **A34**, S238
Graf A., Rauch H., Stern T. (1979) *Atomkernenergie* **33**, 298
Grangier P., Aspect A. (1985) *Phys. Rev. Lett.* **54**, 418
Grangier P., Roger G., Aspect A. (1986) *Europhys. Lett.* **1**, 173
Green G.M. (1978) *Phys. Rev.* **A18**, 1057
Green M.B., Schwarz J.H. (1985) *Phys. Lett.* **B151**, 21
Greenberger D.M. (1983) *Rev. Mod. Phys.* **55**, 875
Greenberger D.M. (1988) *Physica* **B151**, 374

Greenberger D.M., Atwood D.K., Arthur J., Shull C.G., Schlenker M. (1981) *Phys. Rev. Lett.* **47**, 751

Greenberger D.M., Hentschel K., Weinert F. (2009) *Compendium of quantum physics*. Springer, Berlin

Greenberger D.M., Horne M., Shull C.G., Zeilinger A. (1984) *Proc. Int. Symp. Found. Quantum Mechanics*, Tokyo, Phys. Soc. Japan, p. 294

Greenberger D.M., Horne M., Zeilinger A. (1989) in *Bell's Theorem, Quantum Theory, and the Concepts of Universe* (ed. M. Kafatos), p. 73. Kluwer, Dordrecht

Greenberger D.M., Overhauser A.W. (1979) *Rev. Mod. Phys.* **51**, 43

Greenberger D.M., Schleich W.P., Rasel E.M. (2012) *Phys. Rev.* **A86**, 063622

Greenberger D.M., Yasin A. (1988) *Phys. Lett.* **128**, 391

Greene G.L., Gudkov V., (2007) *Phys. Rev.* **C75**, 015501

Greene G.L., Ramsey N.F., Mampe W., Pendlebury J.M., Smith K., Dress W.B., Miller P.D., Perrin P. (1979) *Phys. Rev.* **D20**, 2139

Griesshammer H.W. (2007) in *Proc. 5th Int. Workshop on Chiral Dynamics, Theory and Experiment* (ed. M. Ahmed, H. Gao, H.R. Weller, and B. Holstein). World Scientific, Singapore

Griffin S., Cimmino A., Klein A.G., Opat G., Allman B., Werner S.A. (1996) *J. Phys. Soc. Japan* **65**, 71

Griffiths R.B. (1984) *J. Stat. Phys.* **36**, 219

Griffiths R.B. (1999a) *Phys. Rev.* **A60**, R5

Griffiths R.B. (1999b) *Phys. Lett.* **A261**, 227

Griffiths R.B. (2011) *Found. Phys.* **41**, 705

Grigoriev S.V., Chetverikov Yu.O., Metelev S.V., Kraan W.H. (2006) *Phys. Rev.* **A74**, 043605

Grigoriev S.V., Chetverikov Yu.O., Syromyatnikov A.V., Kraan W.H., Rekveldt M.Th. (2003) *Phys. Rev.* **A68**, 033603

Grigoriev S.V., Kraan W.H., Mulder F.M., Rekfeldt M.Th. (2000) *Phys. Rev.* **A62**, 63601

Gröblacher S., Paterek T., Kaltenbaek R., Brukner C., Zukovski M., Aspelmeyer M., Zeilinger A. (2007) *Nature* **446**, 871

Gronkowski J. (1991) *Phys. Rep.* **206**, 1

Gruber M., Eder K., Zeilinger A., Gähler R., Mampe W. (1989) *Phys. Lett.* **140**, 363

Gühne O., Kleinmann M., Cabello A., Larsson J.A., Kirchmair G., Zähringer F., Gerritsma R., Roos C.F. (2010) *Phys. Rev.* **A 81**, 022121

Gureyev T.E., Roberts A., Nugent K.A. (1995) *J. Opt. Soc. Am.* **A12**, 1942

Gustavson T.L., Bouyer P., Kasevich M.A. (1997) *Phys. Rev. Lett.* 78, 2046

Gustavson T.L., Landragin A., Kasevich M.A. (2000) *Class. Quantum Grav.* 78, 2046

Haag R. (1990a) *Nucl. Phys.* **B18**, 135

Haag R. (1990b) *Comm. Math. Phys.* **132**, 245

Haag R. (1996) *Local quantum physics*. Springer, Heidelberg

Haag R. (2013) *Found. Phys.* **43**, 1295

Haavig D.L., Reifenberger R. (1982) *Phys. Rev.* **B26**, 6408

Hackermüller L., Hornberger K., Brezger B., Zeilinger A., Arndt M. (2004) *Nature* **427**, 711

Hackermüller L., Uttenthaler S., Hornberger K., Reiger G., Bretzger B., Zeilinger A., Arndt M. (2003) *Phys. Rev. Lett.* **91**, 090408

Hafele J.C., Keating R.E. (1962) *Science* **177**,, 166

Hafner M., Summhammer J. (1997) *Phys. Lett.* **A235**, 563

Hagen C. (1990) *Phys. Rev. Lett.* **64**, 2347

Halban H.V., Preiswerk P. (1936) *C. R. Acad. Sci. Paris* **203**, 73

Halpern O., Hamermesh H., Johnson M.H. (1941) *Phys. Rev.* **59**, 981

Halpern O., Holstein T. (1941) *Phys. Rev.* **59**, 960

Halpern O., Johnson M.H. (1939) *Phys. Rev.* **55**, 898

Hamilton W.A., Klein A.G., Opat G.I. (1983) *Phys. Rev.* **A28**, 3149

Hammerschmied S., Rauch H., Clerc H., Kischko U. (1981) *Z. Phys.* **A302**, 163

Hanbury Brown R., Twiss R.Q. (1956) *Nature* **171**, 27; **178**, 1046

Hanney J.H. (1985) *J. Phys.* **A18**, 221

Hardy L. (1993) *Phys. Rev. Lett.* **71**, 1665
Haroche S., Kleppner D. (1989) *Phys. Today* **42**, 24
Haroche S., Raimond J.M. (1993) *Sci. Am.* **268**, 26
Haroche S., Raimond J.-M. (2006) *Exploiting the quantum.* Oxford University Press, Oxford
Harris P.G., Baker C.A., Iaydjiev P., Ivanov S., May D.J.R., Pendebury J.M., Shiers D., Smith K.F., van der Grinten M., Geltenbort P. (1999) *Phys. Rev. Lett.* **82**, 904
Hart M. (1975) *Proc. R. Soc. London* **A346**, 1
Hart M., Bonse U. (1970) *Phys. Today*, **23**, 26
Hart M., Siddons D.P. (1978) *Nature* **275**, 45
Hartle J.B. (1968) *Am. J. Phys.* **36**, 704
Hasegawa Y., Badurek G. (1999) *Phys. Rev.* **A59**, 4614
Hasegawa Y., Erdösi D. (2011) *AIP Conf. Proc.* **1384**, 213
Hasegawa Y., Kikuta S. (1991) *Jap. J. Appl. Phys.* **30**, L316
Hasegawa Y., Kikuta S. (1994) *Phys. Lett.* **A195**, 43
Hasegawa Y., Loidl R., Badurek G., Baron M., Rauch H (2006) *Phys. Rev. Lett.* **97**, 23041
Hasegawa Y., Loidl R., Badurek G., Baron M., Rauch H. (2003) *Nature* **425**, 45
Hasegawa Y., Loidl R., Badurek G., Durstberger-Rennhofer K., Sponar S., Rauch H. (2010) *Phys. Rev.* **A81**, 032121
Hasegawa Y., Loidl R., Baron M., Badurek G., Rauch H. (2006) *Physica* **B385-6**, 1377
Hasegawa Y., Menhart S., Meixner R., Badurek G. (1997) *Phys. Lett.* **A234**, 322
Hasegawa Y., Zawisky M., Rauch H., Ioffe A.I. (1996) *Phys. Rev.* **A53**, 2486
Hasselbach F. (1995) in *Fundamental Problems in Quantum Physics* (eds. M. Ferrera and A.v.d. Merwe), p. 123 Kluwer, Dordrecht
Hasselbach F. (2010) *Rep. Prog. Phys.* **73**, 016101
Hasselbach F., Nicklaus M. (1993) *Phys. Rev.* **A48**, 143
Hauge E.H., Stovneng J.A. (1999) *Rev. Mod. Phys.* **61**, 917
Hawking S. (1983) *Math. Phys.* **87**, 395
Hawking S. (1996) *Phys. Rev.* **D37**, 3099
Hawking S.W. (1990) *A brief history of time.* Guild, London
Hayter J.B., Penfold J. (1979) *Z. Phys.* **B35**, 199
Hayter J.B., Penfold J., Williams W.G. (1976) *Nature* **262**, 569
Heckel B. (1989) *Nucl. Instr. Meth.* **A284**, 66
Heckel B., Forte M., Schaerpf O., Green K., Greene G.L., Ramsey N.F., Byrne J., Pendlebury J.M. (1984) *Phys. Rev.* **C29**, 2389
Heil W., Anderson K., Hofmann D., Humblot H., Kulda J., Lelievre-Berna E., Schärpf O., Tasset F. (1998) *Physica* **241–3**, 56
Heineger F., Herden A., Tschudi T. (1983) *Opt. Comm.* **48**, 237
Heinrich M., Petrascheck D., Rauch H. (1988) *Z. Phys.* **B72**, 357
Heinrich M., Wölwitsch H., Rauch H. (1989) *Physica* **B156 & 157**, 588
Heisenberg W. (1927) *Z. Phys.* **43**, 172
Henley E.M. (1966) in *Isobaric Spin in Nuclear Physics* (eds. J.D. Fox and D. Robson), p. 3. Academic Press, New York
Henny M., Oberholzer S., Strunk C., Heinzel T., Ensshin K., Holland M., Schoenenberger C. (1999) *Science* **284**, 296
Herman G.T. (1980) *Image reconstruction from projections.* Academic Press, New York
Herrmann P., Steinhauser K.A., Gähler R., Steyerl A., Mampe W. (1985) *Phys. Rev. Lett.* **54**, 1969
Herzog T.J., Kwiat P.G., Weinfurter H., Zeilinger A. (1995) *Phys. Rev. Lett.* **75**, 3034
Hilbert S.A., Caprez A., Batelaan H. (2011) *New J. Phys.* **13**, 093025
Hillery M., O'Connell R.F., Scully M.O., Wigner E.P. (1984) *Phys. Rev.* **106**, 121
Hils Th., Felber J., Gähler R., Glaeser W., Golub R., Habicht K., Wille P. (1998) *Phys. Rev.* **A58**, 4784
Hino M., Achiwa N., Tasaki S., Ebisawa T., Akiyoshi T. (1995) *Physica* **B213 & 214**, 842

Hino M., Achiwa N., Tasaki S., Ebisawa T., Kawai T., Akiyoshi T., Yamazaki D. (1998) *Phys. Rev.* **A57**, 4720
Hino M., Achiwa N., Tasaki S., Ebisawa T., Kawai T., Akiyoshi T., Yamazaki D. (1999) *Phys. Rev.* **A59**, 2261
Hittmair O. (1972) *Lehrbuch der Quantentheorie.*Verlag K. Thiemig, Munich
Ho B., Morgan M.J. (1994) *Austr. J. Phys.* **47**, 245
Hofmann H.M., Hale G.M. (2003) *Phys. Rev.* **C68**, 021002
Hohensee M.A., Chu S., Peters A., Müller H. (2011) *Phys. Rev. Lett.* **106**, 151102
Holladay W.G. (1998) *Am. J. Phys.* **66**, 27
Holland P. (1993) *The quantum theory of motion.* Cambridge University Press, Cambridge
Holland P.R. (1999) *Phys. Rev.* **A60**, 4326
Holy V. (1980) *Phys. Stat. Sol. (b)* **101**, 575
Home D. (1997) *Conceptual foundations of quantum physics.* Plenum Press, New York
Home D., Selleri F. (1991) *Riv. Nuova Cim.* **14**, 1
Home D., Pan A.K., Banerjee A. (2013) *Eur. J. Phys.* **D67**, 72
Hornberger K., Uttenthaler S., Brezger B., Hackermuller L., Arndt M., Zeilinger A. (2003) *Phys. Rev. Lett.* **90**, 160401
Horne M.A. (1986) *Physica* **137B**, 260
Horne M.A., Jex I., Zeilinger A. (1999) *Phys. Rev.* **A59**, 2190
Horne M.A., Zeilinger A. (1979) in *Neutron Interferometry* (eds. U. Bonse and H. Rauch), p. 350. Clarendon Press, Oxford
Horne M.A., Zeilinger A., Klein A.G., Opat G.I. (1983) *Phys. Rev.* **A28**, l
Hoskinson E., Sato Y., Packard R.E. (2006) *Phys.Rev.* **B74**, 100509R
Hradil Z., Myska R., Perina J., Zawisky M., Hasegawa Y., Rauch H. (1996) *Phys. Rev. Lett.* **76**, 4295
Huang Y.-F., Li C.-F., Zhang Y.-S., Pan J.-W., Guo G.-C. (2003) *Phys. Rev. Lett.* **90**, 250401
Huber M.G., Arif M., Black T.C., Chen W.C., Gentile T.R., Hussey D.S., Pushin D.A., Wietfeldt F.E., Yang L. (2009a) *Phys. Rev. Lett.* **102**, 200401
Huber M.G., Arif M., Black T.C., Chen W.C., Gentile T.R., Hussey D.S., Pushin D.A., Wietfeldt F.E., Yang L. (2009b) *Phys. Rev. Lett.* **103**, 179903
Huber M.G., Arif M., Chen W.C., Gentile T.R., Hussey D.A., Black T.C., Pushin D.A., Shahi C.B., Wietfield F.E., Yang L. (2014) *Phys. Rev. C* (in print)
Huesmann R., Balzer C., Courteille P., Neuhauser W., Toschek P.E. (1999) *Phys. Rev. Lett.* **82**, 1611
Huffman P.R., Jacobson D.L., Schoen K., Arif M., Black T.C., Snow W.M., Werner S.A. (2004) *Phys. Rev.* **C70**, 014004
Hughes R.J. (1993) *Contemp. Phys.* **34**, 171
Hulet R.G., Hilfer E.S., Kleppner D. (1985) *Phys. Rev. Lett.* **50**, 2137
Hussain H., Imoto N., Loudon R. (1992) *Phys. Rev.* **A45**, 1987
Iaconis C., Walmsley I.A. (1996) *Opt. Lett.* **21**, 1783
Iannuzzi M., Orecchini A., Sacchetti F., Facchi P., Pascazio S. (2006) *Phys. Rev. Lett.* **96**, 080402
Iannuzzi M., Messi R., Moricciani D., Orecchini A., Saccetti F., Facchi P., Pascazio S. (2011) *Phys. Rev.***A84**, 015601
Imoto N. (1996) in *Quantum Physics, Chaos Theory and Cosmology* (eds. M. Namiki et al.). American Institute of Physics, Woodbury, NY
Indenbohm V.L., Slobodetkii I.S., Truni KG. (1974) *Phys. Stat. Sol.* **B73**, K9
Indenbohm V.L., Suvorov E.V., Slobodetskii I.I. (1976) *Sov. Phys. JETP* **44**, 187
Ioffe A. (1984) *Nucl. Instr. Meth.* **228**, 141
Ioffe A. (1986), Nucl. Instr. Meth. **A268**, 169
Ioffe A. (1997) *Physica* **B234-6**, 1180
Ioffe A., Arif M., Jacobson D.L., Mezei F. (1999) *Phys. Rev. Lett.* **82**, 2322
Ioffe A., Ermakov O., Karpikhin I., Krupchitsky P., Mikula P., Lukas P., Vrana M. (2000) *Eur. Phys. J.* **A7**, 197

Ioffe A., Jacobson D.L., Arif M., Vrana M., Warner S.A., Fischer P., Greene G., Mezei F. (1998a) *Phys. Rev.* **A58**, 1475
Ioffe A., Jacobson D.L., Arif M., Vrana M., Warner S.A., Fischer P., Greene G., Mezei F. (1998b) *Physica* **141-143**, 130
Ioffe A., Lukas P., Mikula P., Vrana M., Alefeld B. (1995) *Physica* **B213-214**, 833
Ioffe A., Lukas P., Mikula P., Vrana M., Zabijukin V. (1994) *Z. Phys.* **A348**, 243
Ioffe A., Mezei F. (2001) *Physica* **B297**, 303
Ioffe A., Neov S. (1997) *Physica* **B234-6**, 1183
Ioffe A., Turkevich Yu.G., Drabkin G.M. (1981) *JETP Lett.* **33**, 374
Ioffe A., Vrana M. (1997) *Phys. Lett.* **A231**, 319
Ioffe A., Vrana M., Zabiyakin V. (1996) *J. Phys. Soc. Japan* **65**, 82
Ioffe A., Zabiyakin V.L., Drabkin G.M. (1985) *Phys. Lett.* **111**, 373
Isgur N. (1999) *Phys. Rev. Lett.* **83**, 272
Isgur N., Karl G. (1977) *Phys. Lett.* **B72**, 109
Isgur N., Karl G. (1978) *Phys. Rev.* **D18**, 4187
Ishikawa T. (1988) *Acta Cryst.* **A44**, 496
Izyumov Y.A., Ozerov R.P. (1970) *Magnetic neutron diffraction.* Plenum Press, New York
Jacobson D.L., Allman B.E., Zawisky M., Werner S.A., Rauch H. (1996) *J. Jap. Phys. Soc.* **A65**, 94
Jacobson D.L., Werner S.A., Rauch H. (1994) *Phys. Rev.* **A49**, 3196
Jacques V., Wu E., Grosshans F., Treussart F., Grangier P., Aspect A., Roch J.-F. (2007) *Science* **315**, 966
Jacquey M., Miffre A., Büchner M., Trenec G., Vigue J. (2006) *Europhys. Lett.* **75**, 688
Jaeger G., Shimony A., Vaidman L. (1995) *Phys. Rev.* **A51**, 54
James D.F.V., Wolf E. (1991) *Phys. Lett.* **A157**, 6
James R.W. (1950) *The optical principles of the diffraction of X-rays.* Bell, London
James R.W. (1963) *Solid State Phys.* **15**
Jammer M. (1966) *The conceptual development of quantum mechanics.* McGraw Hill, New York
Jammer M. (1974) *The philosophy of quantum mechanics*, p. 121. Wiley, New York
Janski J., Vinogradov A.V. (1990) *Phys. Rev. Lett.* **64**, 2771
Jauch J.M., Wigner E.P., Yanase M.M. (1967) *Il Nuovo Cim.* **B48**, 144
Jaynes E.T., Cummings F.W. (1963) *Proc. IEEE* **51**, 89
Jeltes T., McNamara J.M., Hogervorst W., Vassen W., Krachmalnicoff V., Schellekens M., Perrin A., Chang H., Boiron D., Aspect A., Westbrook C.I. (2007) *Nat. Lett.* **445**, 402
Jenke T., Cronenberg G., Burgdörfer J., Chizhova L.A., Geltenbort P., Ivanov A.N., Lauer T., Lins T., Rotter S., Saul H., Schmidt U., Abele H. (2014) *Phys. Rev. Lett.* **112**, 15105
Jenke T., Geltenbort P., Lemmel H., Abele H. (2011) *Nat. Phys.* **7**, 468
Jericha E., Carlile C.J., Rauch H. (1996) *Nucl. Instr. Meth.* **A379**, 330
Jin F., Yuan S., De Raedt H., Michielsen K., Miyashita S. (2010) *J. Phys. Soc. Japan* **79**, 074401
Joos E. (1986) *Phys. Lett.* **A116**, 6
Joos E. (1996) in *Decoherence and the Appearance of a Classical World in Quantum Theory* (eds. D. Giulini, E. Joos, C. Kiefer, J. Kupsch, I.-O. Stamatescu, H.D. Zeh), p. 35. Springer, Berlin
Joos E. (2006) in *Quantum Decoherence* (eds. B. Duplantier, J.M. Raimond, V. Rivasseau), *Prog. Math. Phys.* **48**, 177
Joos E., Zeh H.I. (1985) *Z. Phys.* **B59**, 223
Jordan T.F. (1983) *Phys. Rev.* **96A**, 457
Josephson B.D. (1974) *Rev. Mod. Phys.* **46**, 251
Just W., Schneider C.S., Ciscewski R., Shull C.G. (1973) *Phys. Rev.* **B7**, 4142
Kaan W.H., Tricht v.J.B., Rekveldt M.Th. (1988) *Nucl. Instr. Meth.* **A276**, 521
Kaiser H., Arif M., Berliner R., Clothier R., Werner S.A., Cimmino A., Klein A.G., Opat G.I. (1988) *Physica* **151B**, 68
Kaiser H., Arif M., Werner S.A., Willis J.O. (1986) *Physica* **136B**, 134

Kaiser H., Armstrong N.L., Wietfeldt F.E., Huber M., Black T.C., Arif M., Jacobson D.L., Werner S.A. (2007) *Physica* **B385–6**, 1384
Kaiser H., Clothier R., Werner S.A., Rauch H., Woelwitsch H. (1992) *Phys. Rev.* **A45**, 31
Kaiser H., George E.A., Werner S.A. (1984) *Phys. Rev.* **A29**, 2276
Kaiser H., Rauch H. (1999) in *Bergmann-Schaefer*, Vol. III, *Optics* (ed. H. Niedrig), p. 1043. W. de Gruyter, Berlin
Kaiser H., Rauch H., Badurek G., Bauspiess W., Bonse U. (1979) *Z. Phys.* **A291**, 231
Kaiser H., Rauch H., Bauspiess W., Bonse U. (1977) *Phys. Lett.* **71B**, 321
Kaiser H., Werner S.A., Arif M., Klein A.G., Opat G.I., Cimmino A. (1991) *ICAP XII*, Ann Arbor, AIP Conf. Proc. (ed. J.C. Zorn and R.R. Lewis), p. 233
Kaiser H., Werner S.A., George E.A. (1983) *Phys. Rev. Lett.* **50**, 560
Kaloyerou P.N., Brown H.R. (1992) *Physica* **B176**, 78
Kamesberger J., Zeilinger A. (1988) *Physica* **B151**, 193
Kaneno T. (1960) *Progr. Theor. Phys.* **23**, 17
Karlsson A., Björk., Forsberg E. (1998) *Phys. Rev. Lett.* **80**, 1198
Kasevich J.R., Skagerstam B.S. (1985) *Coherent States*. World Scientific, Singapore
Kasevich M., Chu S. (1991) *Phys. Rev. Lett.* **67**, 181
Kasevich M., Chu S. (1992) *Appl. Phys.* **B54**, 321
Kastner R.E. (1993) *Am. J. Phys.* **61**, 852
Kastner R.E. (2012) *The transactional interpretation of quantum mechanics*. Cambridge University Press, Cambridge
Kato N. (1961) *Acta Cryst.* **14**, 627
Kato N. (1968) *J. Appl. Phys.* **39**, 2231
Kato N. (1974) in *X-Ray Diffraction* (ed. L.V. Azaroff). McGraw-Hill, New York
Kato N. (1991) *Acta Cryst.* **14**, 526 and 627
Kawai T., Ebisawa T., Tasaki S., Hino M., Yamazaki D., Kakata H., Akiyoshi T., Matsumoto Y., Achiwa N., Otake Y. (1998a) *Physica* **B241–3**, 133
Kawai T., Ebisawa T., Tasaki S., Hino M., Yamazaki D., Akiyoshi T., Matsumoto Y., Achiwa N., Otake Y. (1998b) *Nucl. Instr. Meth.* **A410**, 259
Kawano S., Kawai T., Kawaguchi A. (eds.) (1996) *J. Phys. Soc. Japan* **65**, Suppl.A
Kearney P.D., Klein A.G., Opat G.I., Gaishler R. (1980) *Nature* **287**, 313
Keith D.W., Ekstrom C.R., Turchette Q.A., Pritchard D.E. (1991) *Phys. Rev. Lett.* **66**, 2693
Keith D.W., Schattenburg M.L., Smith H.I., Pritchard D.E. (1988) *Phys. Rev. Lett.* **61**, 1580
Keller C., Schmiedmayer J., Zeilinger A., Nonn T., Dürr S., Rempe G. (1999) *Appl. Phys.* **B69**, 303
Keller M. (1961) *Z. Phys.* **164**, 292
Keller T., Golub R., Mezei F., Gähler R. (1998) *Physica* **B241-243**, 101
Kelly J.J. (2002) *Phys. Rev.* **C66**, 065203
Kendrick H., King J.S., Werner S.A., Arrot A. (1970) *Nucl. Instr. Meth.* **79**, 82
Ketter W., Heil W., Badurek G., Baron M., Jericha E., Loidl R., Rauch H. (2006) *Eur. Phys. J.* **A27**, 243
Khoury J., Weltman A. (2004) *Phys. Rev.* **D69**, 044026
Khrennikov A. (2009) *Contextual approach to quantum formalism*. Springer Science, Berlin
Kiefer C. (2000) in *Decoherence: Theoretical, Exüperimental, and Conceptional Problems* (eds. Ph. Blanchard, D. Guilini, E. Joos, C. Kiefer, I.-O. Stamatescu). Springer, Berlin
Kievsky A. (1997) *Nucl. Phys.* **A624**, 125
Kikuta S., Ishikawa I., Kohra K., Hioshino S. (1975) *J. Phys. Soc. Japan* **39**, 471
Kim J.I., Fronseca-Romero K.M., Horiguti A.M., Davidovich L., Nemes M.C., De Toledo Piza A.F.R. (1999) *Phys. Rev. Lett.* **82**, 4737
Kim J., Lee K.H., Lim C.H., Kim T., Ahn C.W., Cho G., Lee S.W. (2013) *Rev. Sci. Instrum.* **84**, 063705
Kinney J.H., Nicols M.C. (1992) *Ann. Rev. Mat. Sci.* **22**, 121
Kirchmair G., Zähringer F., Gerritsma R., Kleinmann M., Gühne O., Cabello A., Blatt R., Roos C.F. (2009) *Nature* **460**, 494

Kischko U. (1983) thesis, University of Dortmund
Kischko U., Bonse U. (1985) *J. Appl. Cryst.* **18**, 326
Kischko U., Schweizer J., Tasset F. (1982) *Z. Phys.* **A307**, 163
Kitaguchi M., Funahashi H., Nakura T., Hino M., Shimizu H.M. (2003) *Phys. Rev.* **B 67**, 033609
Klauder J.R. (1960) *Ann. Phys.* **11**, 123
Klauder J.R., Skagerstam S. (eds.) (1985) *Coherent states.* World Scientific, Singapore
Klein A.G. (1986) *Physica* **137B**, 230
Klein A.G. (1988) *Physica* **B151**, 44
Klein A.G., Kearney P.D., Opat G.I., Cimmino A., Gähler R. (1981c) *Phys. Rev. Lett.* **46**, 959
Klein A.G., Kearney P.D., Opat G.I., Gähler R. (1981b) *Phys. Lett.* **83A**, 711
Klein A.G., Opat G. L. (1976) *Phys. Rev. Lett.* **37**, 238
Klein A.G., Opat G.I., Cimminio A., Zeilinger A., Treimer W., Gähler R. (1981a) *Phys. Rev. Lett.* **46**, 1551
Klein A.G., Opat G.I., Hamilton W.A. (1983) *Phys. Rev. Lett.* **50**, 536
Klein A.G., Werner S.A. (1983) *Rep. Progr. Phys.* **46**, 259
Klein A.G., Werner S.A. (1991) *Neutron News* **2**, 19
Klempt E. (1976) *Phys. Rev.* **D13**, 3125
Klepp J., Sponar S., Hasegawa Y. (2014) *Prog. Theor. Exp. Phys.* **2014**, 082A01
Klepp J., Tomita Y., Pruner C., Kohlbrecher J., Fally M. (2012) *Appl. Phys. Lett.* **101**, 154104
Klink W.H. (1998) *Ann. Phys.* **260**, 27
Knowles J.W. (1956) *Acta Cryst.* **9**, 61
Kochen S., Specker E. (1967) *J. Math. Mech.* **17**, 59
Kocsis S., Braverman B., Ravets S., Stevens S.J., Mirin R.P., Shalm L.K., Steinberg A.M. (2011) *Science* **332**, 1170
Kodama T., Osakabe N., Endo J., Tonomura A., Ohbayashi K., Urakami T., Ohsuka S., Tsuchiya H., Tsuchiya Y., Uchikawa Y. (1998) *Phys. Rev.* **A57**, 2781
Koester L. (1 977) *Springer Tr. Mod. Phys.* **80**, 1
Koester L. (1965) *Z. Phys.* **182**, 328
Koester L. (1967) *Z. Phys.* **198**, 187
Koester L. (1976) *Phys. Rev.* **D14**, 907
Koester L., Knopf K. (1971) *Z. Naturforsch.* **26a**, 391
Koester L., Nistler W. (1975) *Z. Phys.* **A272**, 189
Koester L., Rauch H., Seymann E. (1991) *At. Data Nucl. Data Tables* **49**, 65
Koester L., Waschkovski W., Kluever A. (1982) *Physica* **137B**, 137
Koester L., Waschkowski W., Kluever A. (1986) *Physica* **137B**, 282
Koester L., Waschkovski W., Mitsyna L.V., Samosvat G. S., Prokofjevs P., Tambergs J. (1995) *Phys. Rev.* **C51**, 3363
Kolomenski E.A., Lobashev V.M., Perozhkov A.N., Smotritsky L.M., Titov N.A., Vesna V.A. (1981) *Phys. Lett.* **107B**, 272
Kono N., Machida K., Namiki M., Pascazio S. (1996) *Phys. Rev.* **A54**, 1064
Koonin S.E. (1977) *Phys. Lett.* **703**, 43
Kopecky S., Harvey J.A., Hill N.W., Krenn M., Pernicka M., Riehs P., Steiner S. (1997) *Phys. Rev.* **C56**, 2229
Kopecky S., Riehs P., Harvey J.A., Hill N.W. (1994) *Proc. Int. Conf. Nuclear Data for Science and Techn.* (ed. J.K. Dickens), Gatlinburg, p. 233
Kopecky S., Riehs P., Harvey J.A., Hill N.W. (1995) *Phys. Rev. Lett.* **74**, 2427
Kostorz G. (1979) in *Treatise on Condensed Science and Technology*, Vol. 13 (ed. G. Kostorz), p. 227. Academic Press, New York
Kozuma M., Deng L., Hagley E.W., Wen J., Lutwak R., Helmerson K., Rolston S.L., Phillips W.D. (1999) *Phys. Rev. Lett.* **82**, 871
Kraan W.H., Grigoriev S.V., Rekveldt M.Th. (2004) *Europhys. Lett.* **66**, 164

Kraan W.H., Grigoriev S.V., Rekveldt M.Th. (2010) *Phys. Rev.* **A 82**, 013619
Kraan W.H., van Trich J.B., Rekfeldt M.Th. (1989) *Nucl. Instr. Meth.* **A276**, 521
Kraus K. (1987) *Phys. Rev.* **D35**, 3070
Krause D.E., Fischbach E., Rohrbach Z. (2014) *Phys. Lett.* **A** (in print)
Krouglov T., de Schlepper I.M., Bouwman W.G., Rekveldt M.T. (2003) *J. Appl. Cryst.* **36**, 117
Kroupa G., Bruckner G., Bolik O., Zawisky M., Hainbuchner M., Badurek G., Buchelt R.J., Schricker A., Rauch H. (2000) *Nucl. Instr. Meth.* **A440**, 604
Krüger E. (1980) *Nukeonika* **25**, 889
Krupchitsky P.A. (1987) *Fundamental Research with Polarized Neutrons.* Springer, Berlin
Kübler O., Zeh H.D. (1973) *Ann. Phys. (Paris)* **76**, 405
Kun S.Y., Gentner R., Larsen L. (1992) *Z. Phys.* **342**, 67
Kuratsuji H., Iida S. (1986) *Phys. Rev. Lett.* **56**, 1003
Kuroiwa J., Kasai M., Futamase T. (1993) *Phys. Lett.* **A182**, 330
Kurz H., Rauch H. (1969) *Z. Phys.* **220**, 419
Kwiat P.G., Mattle K., Weinfurter, H., Zeilinger A., Sergienko A.V., Shih Y. (1995) *Phys. Rev. Lett.* **75**, 4337
Lamb W.E. (1995) *Appl. Phys.* **B60**, 77
Lamehi-Rachti M., Mittig W. (1976) *Phys. Rev.* **D14**, 2543
Lamoreaux S.K. (1992) *Int. J. Mod. Phys.* **A7**, 6691
Lan S.-Y., Kuan P.C., Estey B., English D., Brown J.M. Hohensee M.A., Müller H. (2013) *Science* **339**, 554
Landau L.D., Lifshitz E.M. (1969) *Mechanics.* Pergamon Press, New York
Landkammer F.J. (1966) *Z. Phys.* **210**, 113
Lauterborn W., Kurz T., Wiesenfeldt M. (1995) *Coherent Optics.* Springer, Berlin
Lawson-Daku B.J., Asimov R., Gorceix O., Miniatura Ch., Robert J., Baudon J. (1996) *Phys. Rev.* **A54**, 5042
Lax M. (1951) *Rev. Mod. Phys.* **23**, 287
Lax M. (1952) *Phys. Rev.* **85**, 621
Layer H.P., Greene G.L. (1991) *Phys. Lett.* **A155**, 450
Ledinegg E., Schachinger E. (1983) *Phys. Rev.* **A27**, 2555
Lee H.W. (1995) *Phys. Rep.* **259**, 147
Lee W.T., Motrunich O., Allman B.E., Werner S.A. (1998) *Phys. Rev. Lett.* **80**, 3165
Leeb H., Eder G., Rauch H. (1984) *J. Phys.* **45**, C3–47
Leeb H., Schmiedmayer J., (1992) *Phys. Rev. Lett.* **68**, 1472
Leeb H., Teichtmeister C. (1993) *Phys. Rev.* **C48**, 1719
Leek P.J., Fink J.M., Blais A., Bianchetti R., Göppl M., Gambetta J.M., Schuster D.I., Frunzio L., Schoelkopf R.J., Wallraff A. (2007) *Science* **318**, 1889
Leggett A.J. (1984) *Proc. Found. Quantum Mech.* (ed. S. Kamefuchi), *Phys. Soc. Japan*, 74
Leggett A.J. (1986) *The Lesson of Quantum Mechanics* (eds. J. deBoer, E. Dal, O. Ulfbeck). Elsevier, Amsterdam
Leggett A.J. (1994) *Curr. Sci.* **67**, 785
Leggett A.J. (2008) *Rep. Prog. Phys.* **71**, 022001
Lekner J. (1987) *Theory of reflection of electromagnetic and particle waves* (ed. M. Nijhoff), App. A-5. Kluwer, Dordrecht
Lemmel H., (2007) *Phys. Rev.* **B76**, 144305
Lemmel H. (2010) Private communication, ILL
Lemmel H. (2013) *Found. Cryst.* **A69**, 1
Lemmel H., Wagh A.G. (2010) *Phys. Rev.* **A 82**, 033626
Lenef A., Hammond T.O., Smith E.T., Chapman M.S., Rubenstein R.A., Pritschard D.E. (1997) *Phys. Rev. Lett.* **78**, 760
Lenz F. (1972) *Z. Phys.* **249**, 462

Lenz H., Wohland G. (1984) *Optik* **67**, 315
Leonhardt U. (1997) *Measuring the quantum state of light*. Cambridge University Press, Cambridge
Lerche I. (1977) *Am. J. Phys.* **45**, 1154
Levy-Leblond J.-M. (1987) *Phys. Lett.* **125**, 441
Levy-Leblond J.-M., Balibar F. (1990) *Quantics*. North-Holland, Amsterdam
Lichte H. (1988) in *New Techniques and Ideas in Quantum Measurement Theory* (ed. D.M. Greenberger), p. 175. New York Academy of Science
Lindblad G. (1976) *Comm. Math. Phys.* **48**, 119
Lipperheide R., Reiss G., Leeb H., Fiedeldey H., Sofianos S.A. (1995) *Phys. Rev.* **B51**, 11032
Littrell K. (1997), PhD thesis, University of Missouri, Columbia
Littrell K., Werner S.A., Allman B.E. (1996) *J. Phys. Soc. Japan* **65**, 98
Littrell K.C. (2007) *Physica* **C385–6**, 1371
Littrell K.C., Allman B.E., Motrunich O.I., Werner S.A. (1998) *Acta Cryst.* **A54**, 562
Littrell K.C., Allman B.E., Werner S.A. (1997) *Phys. Rev.* **A56**, 1767
Lloyd H. (1831) *Transaction Royal Irish Academy* **17**, 171
Lorentz H.A. (1927) *Theorie der Strahlung*. Akad. Verlagsges., Leipzig
Loschak G. (1984) in *The Wave–Particle Dualism* (eds. S. Diner, F. Fargue, G. Loschak, F. Selleri). Reidel, Dordrecht
Loudon R. (1983) *The quantum theory of light*. Clarendon Press, Oxford
Loudon R., Knight P. L. (1987) *J. Mod. Opt.* **34**, 709
Lukas P., Mikula P., Kulda J., Eichhorn F. (1987) *Czech J. Phys.* **B37**, 993
Lushchikov V.I., Frank A.I. (1978) *JETP Lett.* **28**, 559
Lushchikov V.I., Taran Y.U., Shapiro F.L. (1970) *Sov. J. Nucl. Phys.* 10, 669
Maaza M., Hamidi D. (2012) *Phys. Rep.* **514**, 177
Maaza M., Pardo B., Chauvineau J.P., Raynal A., Menelle A., Bridou F. (1996) *Phys. Lett.* **A223**, 145
Macek W.M., Schneider J.R., Salamon R.M. (1964) *Appl. Phys.* **35**, 1556
Mach J. (1892) *Z. Instrumentenkd.* **12**, 89
Machida S., Namiki M. (1980) *Progr. Theor. Phys.* **63**, 1457, 1833
Machida S., Namiki M. (1986) *Proc. 2nd Int. Symp. Found. Quantum Mechanics*, Tokyo, p. 255. Phys. Soc. Japan
Machleidt R., Slaus I. (2001) *J.Phys.G: Nucl.Part.Phys.* **27**, R69
McReynolds A.W. (1951) *Phys. Rev.* **83**, 233
Madelung E. (1926) *Z. Phys.* **40**, 322
Maier-Leibnitz H. (1966a) *Nukleonik* **8**, 5
Maier-Leibnitz H. (1966b) *Sitzungsb. Bayr. Akad. Wiss.* 16
Maier-Leibnitz H., Springer T. (1962) *Z. Physik* **167**, 386
Maier-Leibnitz H., Springer T. (1963) *J. Nucl. Energy* **A/B17**, 217
Majkrzak C.F., Berk N.F. (1995) *Phys. Rev.* **B52**, 10827
Majkrzak C.F., Berk N.F., Dura J.A., Satija S.K., Karim A., Pedulla J., Deslattes R.D. (1998) *Physica* **B241–3**, 1101
Majkrzak C.F., Passel L. (1985) *Acta Cryst.* **A41**, 41
Mandel L. (1962) *J. Opt. Soc. Am.* **52**, 1335
Mandel L. (1963) *Progress in optics*, Vol. 2 (ed. E.Wolf), p. 183. North-Holland, Amsterdam
Mandel L. (1991) *Opt. Lett.* **16**, 1882
Mandel L., Wolf E. (1965) *Rev. Mod. Phys.* **37**, 231
Mandel L., Wolf E. (1995) *Optical coherence and quantum optics*. Cambridge University Press, Cambridge
Mannheim P.D. (1998) *Phys. Rev.* **A57**, 1260
Manoukin E.B. (2006) *Quantum theory: a wide spectrum*. Springer, Dordrecht
Marshall A.G., Verdun F.R. (1990) *Fourier transform in MNR, optics and mass spectroscopy*. Elsevier, Amsterdam

Marshall W., Lovesey S.W. (1971) *Theory of thermal neutron scattering*. Clarendon Press, Oxford
Martin D.M. (1967) *Magnetism in solids*. MIT Press, Cambridge, MA
Martin P.J., Oidaker B.G., Miklich A.H., Pritschard D.E. (1988) *Phys. Rev. Lett.* **60**, 515
Mashhoon B. (1988) *Phys. Rev. Lett.* **61**, 2639
Mashhoon B. (1999) *Gen. Rel. a. Grav.* **31**, 681
Mashhoon B., Neutze R., Hannan M., Stedman G.E. (1998) *Phys. Lett.* **A249**, 161
Massa E., Mana G., Kuetgens U., Ferroglio L. (2010) *J. Appl. Cryst.* **43**, 293
Matteucci G., Pozzi G. (1985) *Phys. Rev. Lett.* **54**, 2469
Matull R., Eschkoetter P., Rupp R.A., Ibel K. (1991) *Europhys. Lett.* **15**, 133
Matull R., Rupp R.A., Hehmann J., Ibel K. (1990) *Z. Phys.* **B81**, 365
Matzkin A., Pan A.K. (2013) *J. Phys. A: Meth. Theor.* **46**, 315307
Mayer S., Rauch H., Geltenbort P., Schmidt-Wellenburg P., Allenspach P., Zsigmond G. (2009) *Nucl. Instr. Meth.* **A608**, 434
Mehring M., Hoefer P., Grupp A., Seidel H. (1984) *Phys. Lett.* **106A**, 146
Mensky M.B. (2010) *Consciousness and quantum mechanics: live in parallel worlds*. World Scientific, Singapore
Mermin D. (1987) in *Hilisophical Consequences of Quantum Theory* (eds. J.T. Cushing and E. McMullin), p. 49. Notre Dame University Press, Notre Dame, IN
Mermin D. (1990) *Phys. Rev. Lett.* **65**, 1838
Mermin N.D. (1990) *Phys. Rev. Lett.* **65**, 3372
Mermin N.D. (1993) *Rev. Mod. Phys.* **65**, 803
Meschede D.H., Walther H., Müller G. (1985) *Phys. Rev. Lett.* **54**, 89
Messiah A. (1965) *Quantum mechanics*. North-Holland, Amsterdam
Meyerhoff M., Eyl D., Frey A., Andresen H.G., Annand J.R.M., Aulenbacher K., Becker J., Blume-Werry J., Dombo Th., Drescher P., Ducret J.E., Fischer H., Grabmayer P., Hall S., Hartmann P., Hehl T., Heil W., Hoffmann J., Kellie J.D., Klein F., Leduc M., Möller H., Nactigall Ch., Ostrick M., Otten E.W., Owens R.O., Plützer S., Reichert E., Rohe D., Schäfer M., Schearer L.D., Schmieden H., Steffens K.-H., Surkau R., Walcher Th. (1994) *Phys. Lett.* **B327**, 201
Mezei F. (1972) *Z. Phys.* **255**, 146
Mezei F. (1979) in *Neutron Interferometry* (eds. U. Bonse and H. Rauch), p. 265. Clarendon Press, Oxford
Mezei F. (1980b) in *Imaging Processes and Coherence in Physics* (eds. M.Schlenker et al.), p. 282. Springer, Heidelberg
Mezei F. (1986) *Physica* **B137**, 295
Mezei F. (1988) *Physica* **B151**, 34
Mezei F. (ed.) (1980a) *Neutron spin echo*. Springer, Berlin
Mezei F., Dagleish P.A. (1977) *Comm. Phys.* **2**, 4l
Mezei F., Ioffe A., Fischer P., Arif M., Jacobson D.L. (2000) *Physica* **B276–8**, 979
Michel F.C. (1964) *Phys. Rev.* **B133**, 329
Michelson A.A., Gale H.G., Pearson J. (1925) *Astrophys. J.* **61**, 140
Michelson J. (1881) *Am. J. Sci.* **122**, 120
Miniatura Ch., Robert J., LeBoiteux., Reinhard J., Baudon J. (1992) *Appl. Phys.* **B54**, 347
Misra B., Sudarshan E.C.G. (1977) *J. Math. Phys.* **18**, 756
Missiroli G.F., Pozzi G., Valdre U. (1981), J. Phys. **E14**, 649
Mitchell D.D., Powers P.N. (1936) *Phys. Rev.* **50**, 486
Mittelstedt P. (1989) in *Proc. 3rd Int. Symp. Found. Quantum Physics*, Tokyo, p. 153
Mittelstedt P., Prieur A., Schieder R. (1987) *Found. Phys.* **17**, 891
Moehring D.L., Madson M.J., Blinov B.B., Monroe C. (2004) *Phys. Rev. Lett.* **93**, 090410
Moellenstedt G., Lichte H. (1978) in *9th Int. Congr. Electron Microscopy*, Toronto, Vol. 1, 178
Möllenstedt G., Bayh W. (1962) *Phys. Blätter* **18**, 299
Möllenstedt G., Dueker H. (1954) *Naturwissenschaften* **42**, 41

Möllenstedt G., Wohland G. (1980) in *Electron Microscopy* (eds. P. Bredoro and G. Boom), Vol. I, p. 28. Leiden
Momose A. (1995) *Nucl. Instr. Meth.* **A352**, 622
Momose A., Takeda T., Itai Y. (1995) *Rev. Sci. Instr.* **66**, 1434
Momose A., Takeda T., Itai Y., Hirano K. (1996) *Nat. Med.* **2**, 473
Montgomery R. (1990) *Math. Phys.* **128**, 565
Morikawa Y., Otake Y. (1990) *Il Nuovo Cim.* **105B**, 507
Moshinsky M. (1952) *Phys. Rev.* **88**, 625
Mueckenheim W., Lokai P., Burghardt B. (1988) *Phys. Lett.* **A127**, 387
Mughabghab S.F. (1984) *Neutron cross sections*, Vol. 1, Part B. Academic Press, New York
Mughabghab S.F., Divadeeman M., Holden N.E. (1981) *Neutron cross sections*, Vol. 1. Academic Press, New York
Mukunda N., Simon R. (1993) *Ann. Phys. (N.Y.)* **228**, 205
Mulder F.M., Grigoriev S.V., Kraan W.H., Rekveldt M.Th. (2000) *Europhys. Lett.* **51**, 13
Müller H., Peters A., Chu S. (2010) *Nature* **463**, 926
Müller J.M., Bettermann D., Rieger V., Sengstock U., Sterr U., Ertmer W. (1995) *Appl. Phys.* **B60**, 199
Murayama Y. (1990a) *Found. Phys. Lett.* **3**, 103
Murayama Y. (1990b) *Phys. Lett.* **147**, 334
Muskat E., Dubbers D., Schaerpf O. (1987) *Phys. Rev. Lett.* **58**, 2047
Muynck de W.M., Martens H. (1990) *Phys. Rev.* **A42**, 5079
Nakatani S., Hasegawa Y., Tomimitsu H., Tahahashi T., Kikuta S. (1991) *Jap. J. Appl. Phys.* **30**, L867
Nakatani S., Takahashi T., Tomimitsu H., Kikuta S. (1996) *J. Phys. Soc. Japan* **65**, 77
Nakatani S., Tomimitsu H., Takahashi T., Kikuta S. (1992) *J. Appl. Phys.* **31**, L1137
Nakazato H., Namiki M., Pascazio S., Rauch H. (1995) *Phys. Lett.* **A199**, 27
Namiki M. (1988) *Found. Phys.* **18**, 29
Namiki M. (1992) *Stochastic quantization*. Springer, Heidelberg
Namiki M., Kanenaga M. (1998) *Phys. Lett.* **A249**, 13
Namiki M., Otake Y., Soshi H. (1987) *Prog. Theor. Phys.* 77, 508
Namiki M., Pascazio S., Nakazato H. (1997) *Decoherence and quantum measurements*. World Scientific, Singapore
Namiki M., Pascazio S. (1990) *Phys. Lett.* **147**, 430
Namiki M., Pascazio S. (1991) *Phys. Rev.* **A44**, 39
Namiki M., Pascazio S. (1992) *Found. Phys.* **22**, 451
Namiki M., Pascazio S. (1993) *Phys. Rep.* **232**, 301
Nazarenko V.M., Ryabov V.A., Serebrov V.L., Taldaev A.P. (1981), Phys. Rev. **102B**, 13
Nesvishevsky V.V. (1998) ILL, 96NE14T
Nesvizhevsky V.V., Pignol G., Protasov K.V. (2008) *Phys. Rev.* **D77**, 034020
Nesvizhevsky V.V., Protasov K.V. (2004) *Class. Quant. Grav.* **21**, 4557
Neumann von J. (1932) *Mathematische Grundlagen der Quantenmechanik*. Springer, Berlin
Newton I. (1686) *Mathematical principles of natural philosophy—the principia*, Vol. I/II (ed. F. Cajori). Berkeley, University of California Press
Newton I. (1730) *Optics: a treatise of reflections, inflections and colour of light*. Dover, New York (1952), based on the 4th edn, London
Nicklaus M., Hasselbach F. (1993) *Phys. Rev.* **A48**, 152
Nieto M.M. (1977) *Phys. Lett.* **60A**, 401
Nieto M.M. (1984) *Phys. Rev.* **A29**, 3413
Nieuwenhuizen Th.M., Spicka V., Mehmani B., Aghdami M.J., Khrennikov A. Yu. (eds.) (2007) *Beyond the Quantum*. World Scientific, Singapore
Noh J.W., Fougeres A., Mandel L. (1992) *Phys. Rev.* **A45**, 424
Noh J.W., Fougeres A., Mandel L. (1993) *Phys. Rev. Lett.* **71**, 2579
Nosov V.G., Frank A.I. (1991) *J. Moscow Phys. Soc.* **1**, 1

Nowak E. (1982) *Z. Phys.* **B45**, 265
Obermair G. (1967) *Z. Phys.* **204**, 215
Oberthaler M.K., Abfalterer R., Bernet S., Keller C., Schmiedmayer J., Zeilinger A. (1999) *Phys. Rev.* **A60**, 456
Okorokov A.I., Persedov V.V., Runov V.V., Sedlakova L.N., Chalupa B. (1985) *Sov. Phys. JETP* **62**, 19
Oliver W.D., Kim J.I., Liu R.C., Yamamoto Y. (1999) *Science* **284**, 299
Omnès R. (1988) *J. Stat. Phys.* **53**, 893
Omnès R. (1994) *The interpretation of quantum mechanics*, p. 448. Princeton University Press, Princeton
Omnès R. (1999) *Quantum philosophy*. Princeton University Press, Princeton
Opat G.I. (1991) *Rev. Sci. Instr.* **62**, 1947
Opat G.I. (1995) in *Advances in Quantum Phenomena* (eds. E.G. Bertrametti and J.M. Levy-Leblond), p. 89. Plenum Press, New York
Ostrick M., Herberg C., Andersen H.G., Annand J.R.M., Aulenbacher K., Becker J., Drescher P., Eyl D., Frey A., Grabmayr P., Hartmann P., Hehl T., Heil W., Hoffmann J., Ireland D., Kellie J.D., Klein F., Livingston K., Nachtigall Ch., Natter A., Otten E.W., Owens R.O., Reichert E., Rohe D., Schmieden H., Spengart R., Steigerwald M., Steffens K.-H., Walcher Th., Watson R. (1999) *Phys. Rev. Lett.* **83**, 276
Otake Y., Zeyen C.M.E., Forte M. (1996) *J. Neutron Res.* **4**, 215
Ou Z.Y. (1997) *Phys. Rev.* **A55**, 2598
Overhauser A.W., Colella R. (1974), *Phys. Rev. Lett.* **33**, 1237
Ozawa M. (2003) *Phys. Rev.* **A 67**, 042105
Ozawa M. (2005) *J. Opt.* **B7**, 5672
Paganin D.M. (2006) *Coherent X-ray optics*. Oxford University Press, Oxford
Page L.A. (1975) *Phys. Rev. Lett.* **35**, 543
Pan A.K., Home D. (2009) *Phys. Lett.* **A373**, 3430
Pancharatnam S. (1956) *Proc. Indian Acad. Sci.* **A44**, 247
Parazzoli L.P., Hankin A.M., Biedermann G.W. (2012) *Phys. Rev. Lett.* **109**, 230401
Paris M.G.A. (1999) *Phys. Rev.* **A59**, 1615
Paterek T., Kofler J., Prevedel R., Klimek P., Aspelmayer M., Zeilinger A. (2010) *New J. Phys.* **12**, 013019
Paul W., Trinks V. (1978) in *Fundamental Physics with Reactor Neutrons and Neutrinos* (ed. T. v. Egidy), *Inst. Phys. Conf. Ser.* **42**, p. 18
Paz J.P., Habib S., Zurek W.H. (1993) *Phys. Rev.* **D47**, 488
Pearle P. (1976) *Phys. Rev.* **D13**, 857
Pearle P. (1984a) *The Wave–Particle Dualism* (eds. S. Diners, D. Fargue, G. Loschack, F. Selleri), p. 457. D. Reidel, Dordrecht
Pearle P. (1984b) *Phys. Rev.* **D29**, 235
Pegg D.T., Barnett S.M. (1988) *Europhys. Lett.* **6**, 483
Pehkin M. (1999) *Found. Phys.* **29**, 481
Pendlebury J.M., Hei W., Sobolev Yu., Harris P.G., Richardson J.D., Baskin R.J., Doyle D.D., Geltenbort P., Green K., van der Grinten M.G.D., Iaydjiev P.S., May D.J.R., Smith K.F. (2004) *Phys. Rev.* **A70**, 032102
Pendlebury J.M., Smitt K.F., Golub R., Byrne J., McComb L., Summer T.J., Burnett S.M., Taylor A.R., Heckel B., Ramsey N.F., Green K., Morse J., Kilvington A.I., Baker C.A., Clark S.A., Mampe W., Ageron P., Miranda P.C. (1984) *Phys. Lett.* **136B**, 327
Penrose R. (1986) in *Quantum Concepts in Space and Time* (eds. R. Penrose and C.Y. Ishan). Clarendon Press, Oxford
Penrose R. (1994) *Quantum Reflections* (eds. J. Ellis and A. Amati). Cambridge University Press, Cambridge
Penrose R. (2004) *Road to reality: a complete guide to the laws of the universe*. Jonathan Cape, London
Penrose R., Rindler W. (1984) *Spinors and space-time*. Cambridge University Press, Cambridge

Peres A. (1979) *Phys. Rev. Lett.* **42**, 683
Peres A. (1993) *Quantum theory: concepts and methods.* Kluwer, Dordrecht
Perina J. (1973) *Coherence of light.* Van Nostrand, London
Perreault J.D., Cronin A.C. (2005) *Phys. Rev. Lett.* **95**, 133201
Peshkin M. (1995), in *Fundamental Problems in Quantum Theory* (eds. D.M. Greenberger and A. Zeilinger), *Ann. N.Y. Acad. Sci.* **255**, p. 330
Peshkin M., Ringo G.R. (1971) *Am. J. Phys.* **39**, 324
Peters A., Chung K.Y., Chu S. (1999) *Nature* **400**, 849
Peters A., Chung K.Y., Chu S. (2001) *Metrologica* **38**, 25
Petrascheck D. (1976) *Acta Phys. Austr.* **45**, 217
Petrascheck D. (1986) *Acta. Cryst.* **A42**, 289
Petrascheck D. (1987) *Phys. Rev.* **B35**, 6549
Petrascheck D. (1988) *Physica* **B151**, 171
Petrascheck D., Folk R. (1976) *Phys. Stat. Sol. (a)* **36**, 147
Petrascheck D., Rauch H. (1976) "Theorie des Interferometers", AIAU internal report, Atominstitut Vienna
Petrascheck D., Rauch H. (1984) *Acta Cryst.* **A40**, 445
Pfeiffer F., Grünzweig C., Bunk O., Frei G., David C. (2006) *Phys. Rev. Lett.* **96**, 215505
Philippides C., Dewdney C., Hiley B.J. (1979) *Il Nuovo Cim.* **52B**, 15
Phillips D.R. (2007) *J. Phys.: Nucl. Part.Phys.* **36**, 104004
Pineo W.F.E., Divadeeman M., Bilpuch E.G., Seth K.K., Newson H.W. (1974) *Ann. Phys.* **84**, 165
Pinsker Z.G. (1977) *Dynamical scattering of x-rays in crystals.* Springer, Berlin
Piron C. (1964) *Helv. Phys. Acta* **37**, 439
Piron C. (1990) *Mécanique quantique basis et applications.* Presses Polytechnique et Universitaires Romandes, Lausanne
Pitowsky I. (1982) *Phys. Rev. Lett.* **48**, 1299
Pleshanov N. (1994) *Physica* **B198**, 70
Pleshanov N. (1996) *Z. Phys.* **B100**, 423
Podurets K.M., Somenkov V.A., Shil'shtein S.Sh. (1965) *Sov. Phys. Tech. Phys.* **34**, 654
Poeyry H., Hiismaki P., Virjo A. (1975) *Nucl. Instr. Meth.* **126**, 421
Pokotilovski Yu.N. (1998) *Phys. Lett.* **A248**, 114
Pokotilovski Yu N. (2013) *J. Exp. Theor. Phys.* **116**, 609
Polchinski J. (1998) *String Theories.* Cambridge University Press, Cambridge
Popper K.R. (1967) in *Quantum Theory and Reality* (ed. M. Bunge). Springer, New York
Popper K.R. (1973) *Objektive Erkenntnis.* Hoffmann und Campe, Hamburg
Popper K.R. (1994) *Vermutungen und Widerlegungen.* J.C.B. Mohr, Tübingen
Post E.J. (1967) *Rev. Mod. Phys.* **39**, 475
Pound R.V., Rebka G.A. (1960) *Phys. Rev. Lett.* **4**, 337
Prigogine I. (1991) *Proc. Ecol. Phys. Chem.*, p. 8. Elsevier, Amsterdam
Pruner C., Fally M., Rupp R.A., May R.P., Vollbrandt J. (2006) *Nucl.Instr.Meth.* **A560**, 598
Pushin D.A., Arif M., Cory D.G. (2009) *Phys. Rev* **A79**, 053635
Pushin D.A., Arif M., Huber M.G., Cory D.G. (2008) *Phys. Rev. Lett.* **100**, 250404
Pushin D.A., Huber M.G. Arif M., Cory D.G. (2011) *Phys. Rev. Lett.* **107**, 150401
Rabi I.I. (1937) *Phys. Rev.* **51**, 652
Rabi I.I., Ramsey N.F., Schwinger J. (1954) *Rev. Mod. Phys.* **26**, 167
Radon F. (1917) *Ber. Sächs. Akad. Wiss.* **29**, 262
Ramsey N.F. (1949) *Phys. Rev.* **76**, 996
Ramsey N.F. (1956) *Molecular beams.* Oxford University Press, Oxford
Ramsey N.F. (1982), *Inst. Phys. Conf. Ser.* **64**, 5
Ramsey N.F. (1993) *Phys. Rev.* **A48**, 80

Rao D.N., Kumar V.N. (1994) *J. Mod. Optics* **41**, 1757
Rauch H. (1979), in *Neutron Interferometry* (eds. U. Bonse and H. Rauch) Clarendon Press, Oxford, p. 161
Rauch H. (1980) *Nukleonika* **25**, 855
Rauch H. (1984a) *J. de Physique* **45**, p. C3-197
Rauch H. (1984b) *Proc. Int. Symp. Found. Quantum Mechanics*, Tokyo, Phys. Soc. Japan, p. 277
Rauch H. (1985), in *Neutron Scattering in the Nineties*, IAEA, Vienna, p. 35
Rauch H. (1986) *Contemp. Phys.* **27**, 345
Rauch H. (1989a) *Nucl Instr. Meth.* **A284**, 156
Rauch H. (1989b) *Proc. 3rd Int. Conf. Found. Quantum Mechanics*, Tokyo, p. 3
Rauch H. (1991) *Proc. Found. Mod. Phys. Joensuu* (eds. P. Lahti, P. Mittelstaedt), p. 347. World Scientific, Singapore
Rauch H. (1993a) *Phys. Lett.* **A173**, 240
Rauch H. (1993b), *Proc. Quantum Measurement and Control* (eds. E. Ezawa and Y. Murayama), p. 223. North Holland, Amsterdam
Rauch H. (1995) *Physica* **B213 & 214**, 830
Rauch H. (1995b) in *Fundamental Problems in Quantum Physics* (eds. M. Ferrero and A. van der Merve.), p. 279. Kluwer, Dortrecht
Rauch H. (1997) in *Neutron Radiography* (eds. C.O. Fischer et al.), p. 61. Deutsche Ges. zerstörungsfreie Prüfung, Berlin
Rauch H. (2004) in *Lehrbuch der Experimentalphysik*, Vol. 3 *Optik*. Walter de Gruyter, Berlin
Rauch H., Badurek G., Bauspiess W., Bonse U., Zeilinger A. (1976) *Proc. Int. Conf. Interaction of Neutrons with Nuclei*, Lowell/MA, Vol. II, p. 1027
Rauch H., Kischko U., Petrascheck D., Bonse U. (1983) *Z. Phys.* **B51**, 11
Rauch H., Lemmel H., Baron M. (2002) *Nature* **417**, 630
Rauch H., Petrascheck D. (1978) in *Neutron Diffraction* (ed. H. Dachs), Top. Curr. Phys. **6**, 303. Springer, Heidelberg,
Rauch H., Seidl E. (1987) *Nucl. Instr. Meth.* **A255**, 32
Rauch H., Seidl E., Zeilinger A., Bauspiess W., Bonse U. (1978b) *J. Appl. Phys.* **49**, 2731
Rauch H., Seidl E., Tuppinger D., Petrascheck D., Scherm R. (1987) *Z. Phys.* **B69**, 313
Rauch H., Suda M. (1974) *Phys. Stat. Sol. (a)* **25**, 495
Rauch H., Suda M. (1995) *Appl. Phys.* **B60**, 181
Rauch H., Suda M. (1998) *Physica* **B241-243**, 157
Rauch H., Suda M., Pascazio S. (1999) *Physica* **B267–8**, 277
Rauch H., Summhammer J. (1984) *Phys. Lett.* **104**, 44
Rauch H., Summhammer J. (1992) *Phys. Rev.* **A46**, 7284
Rauch H., Summhammer J., Zawisky M., Jericha E. (1990) *Phys. Rev.* **A42**, 3726
Rauch H., Treimer W., Bonse U. (1974) *Phys. Lett.* **A47**, 369
Rauch H., Tuppinger D. (1985) *Z. Physik* **A322**, 427
Rauch H., Tuppinger D., Woelwitsch H., Wroblewski T. (1985) *Phys. Lett.* **165B**, 39
Rauch H., Vigier J.P. (1990) *Phys. Lett.* **A151**, 269; *Phys. Lett.* (1991) **A157**, 311
Rauch H., Waschkowski W. (1999) *Landolt-Börnstein*, Vol. **I/16A**, Chapter 6
Rauch H., Wilfing A., Bauspiess W., Bonse U. (1978a) *Z. Phys.* **B29**, 281
Rauch H., Woelwitsch H., Clothier R., Kaiser H., Werner S.A. (1992) *Phys. Rev.* **A46**, 49
Rauch H., Woelwitsch H., Kaiser H., Clothier R., Werner S.A. (1996) *Phys. Rev.* **A53**, 902
Rauch H., Zeilinger A., Badurek G., Wilfing A., Bauspiess W., Bonse U. (1975) *Phys. Lett.* **54A**, 425
Raum K., Koeller M., Zeilinger A., Arif M., Gähler R. (1995) *Phys. Rev. Lett.* **74**, 2859
Raum K., Weber M., Gähler R., Zeilinger A. (1997) *J. Phys. Soc. Japan* **65**, 277
Rayleigh J. (1896) *Proc. Roy. Soc.* **59**, 198
Razavy M. (1989) *Phys. Rev.* **A40**, 1

Rehacek J., Hradil Z., Zawisky M., Bonse U., Dubus F. (2005) *Phys.Rev.* **A71**, 023608
Rehacek J., Hradil Z., Zawisky M., Pascazio S., Rauch H., Perina J. (1999) *Phys. Rev.* **A60**, 473
Rekveldt M.T., van Dijk N.H., Gregoriev S.V., Kraan W.H., Bowman W.G. (2006) *Rev. Sci .Instr.* 77, 073902
Rekveldt M.Th. (1971) *J. Phys.* **32**, p. C579
Rekveldt M.Th. (1973) *Z. Phys.* **259**, 391
Rekveldt M.Th. (1996) *Nucl. Instr. Meth.* **B114**, 366
Rekveldt M.Th., Kraan W.H. (1987) *Nucl. Instr. Meth.* **B28**, 117
Renard R. (1964) *J. Opt. Soc. Am.* **54**, 1190
Richardson D.J., Kilvington A.I., Green K., Lamoreaux S.K. (1988) *Phys. Rev. Lett.* **61**, 2030
Riehle F., Kisters Th., Witte A., Helmcke J., Borde Ch. (1991) *Phys. Rev. Lett.* **67**, 177
Riehle F., Witte A., Kisters Th., Pritschard D.E. (1992) *Appl. Phys.* **B54**, 333
Riehs P., Kopecky S., Harvey J.A., Hill N.W. (1994) *Proc. Int. Conf. Nuclear Data for Science and Techn.* (ed. J.K. Dickens) Gatlinburg, p. 236
Rimawi K., Benenson R.E. (1975) *Nucl. Instr. Meth.* **123**, 195
Rode G., Jex H. (1999) *Phys. Lett.* **A251**, 236
Rogalski M.S., Palmer S.B. (1999) *Quantum physics*. Gordon & Beach Science, New York
Roll P.G., Krotkov R., Dicke R.H. (1964) *Ann. Phys.* **26**, 442
Rosman R., Rekveldt M.Th. (1991) *Phys. Rev.* **B43**, 8437
Rostamyan P.H., Bezirganyan P.A., Rostamyan A.M. (1989) *Phys. Stat. Sol.* **A116**, 489
Rostamyan P.H., Rostamyan A.M. (1989) *Phys. Stat. Sol.* **A126**, 29
Roy S.M., Sing V. (1999) *Phys. Lett.* **A255**, 201
Roy S.M., Singh V. (1993) *Phys.Rev.* **A48**, 3379 T
Rupp R.A., Hehmann J., Matull R., Ibel K. (1990) *Phys. Rev. Lett.* **64**, 301
Sachs R.G. (1962) *Phys. Rev.* **126**, 2256
Sagnac M.G. (1913) *C. R. Acad. Sci.* **157**, 708, 1410
Sakurai J.J. (1980) *Phys. Rev.* **D21**, 2993
Sakurai J.J. (1994) *Modern quantum mechanics*. Addison-Wesley, Reading, MA
Samuel J., Bhandari R. (1988) *Phys. Rev. Lett.* **60**, 2339
Sangster K., Hinds E.A., Barnett S.M., Riis E. (1993) *Phys. Rev. Lett.* **71**, 3641
Sangster K., Hinds E.A., Barnett S.M., Riis E., Sinclair A. (1995) *Phys. Rev.* **A46**, 1776
Sanz A.S., Miret-Artés S. (2002) *J. Phys. Cond. Matter* **14**, 6109
Sato Y., Packard R. (2012) *Phys. Today*, Oct., 31
Scheckenhofer H., Steyerl A. (1977) *Phys. Rev. Lett.* **39**, 1310
Scheckenhofer H., Steyerl A.(1981), Nucl. Instr. Meth. **179**, 393
Scheidl T., Ursin R., Kofler J., Ramelow S., Ma X.-S., Herbst T., Ratschbacher L., Fedrizzi A., Langford N.K., Jennewein T., Zeilinger A. (2009) *Proc. Nat. Acad. Sci. USA* **107**, 19708
Schellhorn U., Rupp R.A., Breer S., May R.P. (1997) *Physica* **B234–6**, 1068
Scherm R.(1981) pers. comm., PTB-Braunschweig
Schiff L.I. (1955) *Quantum mechanics*, p. 289. McGraw-Hill, New York
Schilpp P.A. (ed.) (1949) *Albert Einstein: Philosopher-Scientist*. Harper and Row, New York
Schleich W., Greenberger D.M., Rasel E.M. (2013a) *Phys. Rev. Lett.* **116**, 010401
Schleich W., Greenberger D.M., Rasel E.M. (2013b) *New J. Phys.* **15**, 013007
Schleich W., Pernigo M., Fam Le Kien (1991) *Phys. Rev.* **A44**, 2172
Schleich W., Walls D.F., Wheeler J.A. (1988) *Phys. Rev.* **A38**, 1177
Schleich W., Wheeler J.A. (1987) *Nature* **326**, 574
Schlenker M., Baruchel J. (1986) *Physica* **137B**, 309
Schlenker M., Bauspiess W., Graeff W., Bonse U., Rauch H. (1980) *J. Magn. Magn. Mater.* **15–18**, 1507
Schlenker M., Shull C.G. (1973) *J. Appl. Phys.* **44**, 4181
Schmidt E. (1907) *Math. Ann.* **63**, 433

Schmidt U., Baum G., Dubbers D. (1993) *Phys. Rev. Lett.* **70**, 3396
Schmiedmayer J., Chapman M.S., Ekstrom C.R., Hammond T.D., Wehinger S., Pritschard D.E. (1995) *Phys. Rev. Lett.* 74, 1043
Schmiedmayer J., Rauch H., Riehs P. (1988) *Phys. Rev. Lett.* **61**, 1065
Schmiedmayer J., Riehs P., Harvey J.A., Hill N.W. (1991) *Phys. Rev. Lett.* **66**, 1015
Schneider C.S., Shull C.G. (1971) *Phys. Rev.* **B3**, 830
Schoeberl F., Leeb H. (1986) *Phys. Lett.* **166B**, 355
Schoellkopf W., Toennis J.P. (1994) *Science* **266**, 1345
Schoellkopf W., Toennis J.P. (1996) *J. Chem. Phys.* **104**, 1155
Schoen K., Jacobson D.L., Arif M., Huffman P.R., Black T.C., Snow W.M., Lamoreaux S.K., Kaiser H., Werner S.A. (2003) *Phys.Rev.* **C67**, 044005
Schrödinger E. (1926) *Ann. Physik* **79**, 361, 81, 109
Schroeder U.E. (1980) *Nucl. Phys.* **B166**, 103
Schroedinger E. (1935) *Naturwissenschaften* **23**, 807, 825 and 844
Schuster M., Rauch H., Seidl E., Jericha E., Carlile C.J. (1990) *Phys. Lett.* **A144**, 297
Schütz G., Steyerl A., Mampe W. (1980) *Phys. Rev. Lett.* **44**, 1400
Schwahn D., Miksovsky A., Rauch H., Seidl E., Zugarek G. (1982) *Nucl. Instr. Meth.* **A239**, 229
Schwarz J.H. (1982) *Phys. Rep.* **89**, 223
Schwinger J. (1948) *Phys. Rev.* **73**, 407
Schwinger J., Scully M.O., Englert B.-G. (1988) *Z. Phys.* **D10**, 135
Scully M.O., Dowling J.P. (1993) *Phys. Rev.* **A48**, 3186
Scully M.O., Englert B.G., Schwinger J. (1989) *Phys. Rev.* **A40**, 1775
Scully M.O., Englerth B.G., Walther H. (1991) *Nature* **351**, 111
Scully M.O., Walther H. (1989) *Phys. Rev.* **A39**, 5229
Scully M.O., Zubainy M.S. (1997) *Quantum physics*. Cambridge University Press, Cambridge
Sears V. (1988), *Physica* **B151**, 146
Sears V.F. (1978) *Can. J. Phys.* **56**, 1261
Sears V.F. (1982) *Phys. Rep.* 82, 1
Sears V.F. (1982b) *Phys. Rev.* **D25**, 2023
Sears V.F. (1985) *Phys. Rev.* **A32**, 2524
Sears V.F. (1985) *Z. Phys.* **A321**, 443
Sears V.F. (1986a) *Phys. Rep.* **141**, 281
Sears V.F. (1986b) *Meth. Exp. Phys.* **23A**, 521
Sears V.F. (1989) *Neutron optics*. Oxford University Press, Oxford
Sears V.F. (1992) *Neutron News* **3**, 26
Sears V.F. (1996) *J. Phys. Soc. Japan* **65**, 1
Sears V.F., Shelley S.A. (1991) *Acta Cryst.* **A47**, 441
Selleri F. (1990) *Quantum paradoxes and physical reality*. Kluwer, Dordrecht
Seth K.K., Hughes D.J., Zimmermann R.L., Garth R.L. (1958) *Phys. Rev.* **110**, 692
Shapere A., Wilczek F. (eds.) (1988) *Geometric phases in physics*. World Scientific, Singapore
Shilstein S.Sh., Somenkov V.A., Dokashenko V.P. (1971) *JETP Lett.* **13**, 214
Shimizu F., Shimizu K., Takuma H. (1992) *Phys. Rev.* **A46**, R17
Shimony A. (1979) *Phys. Rev.* **A20**, 394
Shimony A. (1987) in *Philisophical Consequences of Quantum Theory* (eds. J.T. Cushing and E. McMullin), p. 25. Notre Dame University Press, Notre Dame, IN
Shin Y., Saba M., Pasquini T.A., Ketterle W., Pritchard D.E., Leanhardt A.E. (2004) *Phys. Rev. Lett.* **92**, 050405
Shirley J.H. (1965) *Phys. Rev.* **138**, B979
Shitnev V.I. (1989) *Z. Phys.* **B74**, 321
Shull C.G. (1963) *Phys. Rev. Lett.* **10**, 297
Shull C.G. (1967) *Trans. Amer. Cryst. Ass.* **3**, 1

Shull C.G. (1968) *Phys. Rev. Lett.* **21**, 1585
Shull C.G. (1969) *Phys. Rev.* **179**, 752
Shull C.G. (1973) *J. Appl. Cryst.* **6**, 61
Shull C.G. (1982) *Inst. Phys. Conf. Ser.* **64**, 157
Shull C.G., Atwood D.K., Arthur J., Horne M.A. (1980) *Phys. Rev. Lett.* **44**, 765
Shull C.G., Ferrier R.P. (1963) *Phys. Rev. Lett.* **10**, 295
Shull C.G., Wedgwood F.W. (1966) *Phys. Rev. Lett.* **16**, 513
Shull C.G., Wollan E.O., Strauser W.A. (1951) *Phys. Rev.* **81**, 481
Shull C.G., Zeilinger A., Squires G.L., Horne M.A., Atwood D.K., Arthur J. (1980a) *Phys. Rev. Lett.* **44**, 1715
Sillitto R.M., Wykes K. (1972) *Phys. Lett.* **A39**, 333
Silverman M.P. (1987) *Phys. Lett.* **A122**, 226
Silverman M.P. (1988 *Phys. Lett.* **A132**, 154
Silverman M.P. (1997) *Waves and grains: studies of light and learning.* Princeton University Press, Princeton
Simon B. (1983) *Phys. Rev. Lett.* **51**, 2167
Simon C., Zukovski M., Weinfurter H., Zeilinger A. (2000) *Phys. Rev. Lett.* **85**, 1783
Simon G.G., Schmitt Ch., Borkowski F., Walther V.H. (1980) *Nucl. Phys.* **A333**, 381
Sinha S. (1997) *Phys. Lett.* **A228**, 1
Sinha S., Samuel J. (2011) *Class. Quantum Grav.* **28**, 145018
Sinha S.K., Tolan M., Gibaud A. (1998) *Phys. Rev.* **B57**, 2740
Sippel D., Kleinstück K., Schulze G.E.R. (1962) *Phys. Stat. Sol.* **2**, K104
Sippel D., Kleinstück K., Schulze G.E.R. (1965) *Phys. Lett.* **14**, 174
Sjoeqvist E., Brown H.R., Carlsen H. (1997) *Phys. Lett.* **A229**, 273
Sjöqvist E. (2001) *Phys. Rev.* **A63**, 035602
Skalicky P., Oppolzer H. (1972) *Z. Metallk.* **63**, 73
Slaus I., Akaishi Y., Tanaka H. (1989) *Phys. Rep.* **173**, 257
Smith K.F., Crampin N., Pendlebury J.M., Richardson D.J., Shiers D., Green K., Kilvington A.I., Moir J., Prosper H.B., Thompson D., Ramsay N.F., Heckel B.R., Lamoreaux S.K., Ageron P., Mampe W., Steyerl A. (1990) *Phys. Lett.* **B234**, 4
Smith T.P. (2010) *Amer. Sci.* **98**, 478
Smithey D.T., Beck M., Cooper J., Raymer M.G. (1993) *Phys. Rev.* **A48**, 3159
Smolin L. (2011) *Emergent quantum mechanics*, lecture. *University of Vienna*
Spavieri G. (1999) *Phys. Rev.* **A59**, 3184
Spiridonov V. (1995) *Phys. Rev.* **A52**, 1909
Spohn H. (1980) *Rev. Mod. Phys.* **53**, 569
Sponar S., Klepp J., Loidl R., Filipp S., Badurek G., Rauch H. (2008) *Phys.Rev.* **A78**, 061604
Sponar S., Klepp J., Loidl R., Filipp S., Durstberger-Rennhofer K., Bertlmann R.A., Badurek G., Hasegawa Y., Rauch H. (2010) *Phys. Rev.* **A81**, 042113
Sprague G., Tomboulian D.H., Bedo D.E. (1955) *J. Opt. Soc. Am.* **45**, 756
Springer J., Zawisky M., Farthofer R., Lemmel H., Suda M., Kuetgens U. (2010a) *Nucl. Instr. Meth.* **A615**, 307
Springer J., Zawisky M., Lemmel H., Suda M. (2010b) *Found. Cryst.* **A66**, 17
Springer T. (1972) *Springer Tracts Mod. Phys.* **64**
Squires E. (1990) *Conscious mind in the physical world.* IOP, Bristol
Squires E.J. (1985) *Eur. J. Phys.* **A18**, 171
Squires G. L. (1978) *Thermal Neutron Scattering.* Cambridge University Press, Cambridge
Stapp H.P. (1972) *Am. J. Phys.* **40**, 1098
Stassis C. (1970) *Phys. Rev. Lett.* **24**, 1415
Stassis C., Oberteuffer J.A. (1974) *Phys. Rev.* **B10**, 303
Staudenmann J.L., Werner S.A., Colella R., Overhauser A.W. (1980) *Phys. Rev.* **A21**, 1419

Steane A., Szriffigiser P., Desbiolles P., Dalibard J. (1995) *Phys. Rev. Lett.* **74**, 4972
Stedman G.E. (1997) *Rep. Prog. Phys.* **60**, 615
Steinhauser K.A., Steyerl A., Scheckenhofer H. (1980) *Phys. Rev. Lett.* **44**, 1306
Stepanov S.A., Pietsch U., Baumbach G.T. (1995) *Z. Phys.* **B96**, 341
Stern A., Aharonov Y., Imry Y. (1990) *Phys. Rev.* **A41**, 3436
Steyerl A. (1975) *Nucl. Instr. Meth.* **125**, 461
Steyerl A. (1977) in *Neutron Physics*, Springer Tracts Mod. Phys. **80**, p. 57. Springer, Berlin
Steyerl A., Drexel W., Malik S.S., Gutsmiedl E. (1988) *Physica* **B151**, 36
Steyerl A., Ebisawa T., Steinhauser K.A., Utsuro M. (1981) *Z. Phys.* **B41**, 282
Steyerl A., Malik S.S., Steinhauser K.A., Berger L. (1979) *Z. Phys.* **B36**, 109
Stodolsky L. (1974) *Phys. Lett.* **B50**, 352
Stodolsky L. (1979) *Gen. Relat. Gravit.* **11**, 391
Stodolsky L. (1979a) in *Neutron Interferometry* (eds. U. Bonse and H. Rauch), p. 313. Clarendon Press, Oxford
Stodolsky L. (1982) *Nucl. Phys.* **B197**, 213
Stoll M.E., Wolff E.K., Mehring M. (1978) *Phys. Rev.* **A17**, 1561
Storey P., Tan S., Collett M., Walls D. (1994) *Nature* **367**, 626
Strobl M., Treimer W., Hilger A. (2003) *Nucl. Instr. Meth.* **B222**, 653
Suda M. (1995) *Quantum Semiclass. Opt.* 7, 901
Suda M. (2005) *Quantum interferometry in phase space.* Springer, Berlin
Suda M., Rauch H. (1996) *Acta Phys. Slov.* **46**, 499
Suda M., Rauch H., Peev M. (2004) *J. Optcs B: Qantum Semicclass. Opt.* **6**, 345
Suderashan E.C.G. (1963) *Phys. Rev. Lett.* **10**, 277
Sulyok G., Hasegawa Y., Klepp J., Lemmel H., Rauch H. (2010) *Phys. Rev.* **A81**, 053609
Sulyok G., Lemmel H., Rauch H. (2012) *Phys. Rev.* **A85**, 033624
Sumbaev O.I. (1981), Leningrad Inst. Phys., LINP-report no. 676
Summhammer J. (1989) *Il Nuovo Cim.* **Bl03**, 265
Summhammer J. (1993) *Phys. Rev.* **A47**, 556
Summhammer J. (1996) *Phys. Rev.* **A54**, 3155
Summhammer J. (1997) *Phys. Rev.* **A56**, 4324
Summhammer J., Badurek G., Rauch H., Kischko U. (1982) *Phys. Lett.* **90A**, 110
Summhammer J., Badurek G., Rauch H., Kischko U. A. Zeilinger (1983) *Phys. Rev.* **A27**, 2523
Summhammer J., Hamacher K.A., Kaiser H, Weinfurter H, Jacobson D.L., Werner S.A. (1995) *Phys. Rev. Lett.* 75, 3206
Summhammer J., Rauch H., Tuppinger D. (1987) *Phys. Rev.* **A36**, 4447
Sur B., Rogge R.B., Hammond R.P., Anghel V.N.P., Katsaras J. (2001) *Nature* **414**, 525
Szabo S., Adam P., Jansky J., Domokos P. (1996) *Phys. Rev.* **A53**, 2698
Szriftgiser J., Guèry-Odelin D., Arndt M., Dalibald J. (1996) *Phys. Rev. Lett.* 77, 4
Takagi S. (1962) *Acta Cryst.* **15**, 1311
Takagi S. (1969) *J. Phys. Soc. Japan* **26**, 1239
Takahashi Y. (1987) *Phys. Lett.* **A121**, 381
Tan S.M., Walls D.F. (1993) *Phys. Rev.* **A47**, 4663
Tanaka K. (1983) *Phys. Rev. Lett.* **51**, 378
Tang J.-S., Li Yu-L., Li C.F., Guo G.-C. (2013) *Phys. Rev.* **A88**, 014103
Tasaki S., Kawai T., Ebisawa T. (1995) *J. Appl. Phys.* **78**, 2398
Tasset F., Chupp T.E., Pique J.P., Steinhof A., Thompson A., Wassermann E., Ziade M. (1992) *Physica* **B180 & 181**, 896
Taupin D. (1961) *Bull. Soc. Fr. Mineral. Crist.* **84**, 51
Teague M.R. (1983) *J. Opt. Soc. Am.* **73**, 1434
Tegmark M. (1993) *Found. Phys.* **6**, 571
Tegmark M., Shapiro H.S. (1994) *Phys. Rev.* **E50**, 2538

Teo Y.S., Stoklasa B., Englert B.-G., Rehacek J., Hradil Z. (2912) *Phys. Rev.* **A85**, 042317
Terburg B.P., Verkerk P., Jericha E., Zawisky M. (1993) *J. Neutron Res.* **1**, 37
Teworte R., Bonse U. (1984) *Phys. Rev.* **B29**, 2102
Teworte R., Bonse U. (1987) *Z. Phys.* **B65**, 275
Thaler R.M. (1959) *Phys. Rev.* **114**, 827
Teo Y.S., Zhu H., Englert B.G., Rehacek J., Hradil Z. (2012) *Phys. Rev. Lett.* **107**, 020404
Thomas A.W., Theberge S., Miller G.A. (1981) *Phys. Rev.* **D24**, 216
Thomas L.H. (1926) *Nature* **117**, 514
Tollaksen J. (2007) *J. Phys. A: Math. Theor.* **40**, 9033
Tomimitsu H., Hasegawa Y., Aizawa K., Kikuta S. (1995) *Physica* **B213 & 214**, 836
Tomimitsu H., Hasegawa Y., Aizawa K., Kikuta S. (1999) *Nucl. Instr. Meth.* **A420**, 453
Tomimitsu H., Hasegawa Yu., Aizawa K. (2000) *Phys. Lett.* **A274**, 175
Tomita A., Chiao R.Y. (1986) *Phys. Rev. Lett.* **57**, 937
Tonomura A. (1987) *Rev. Mod. Phys.* **59**, 639
Tonomura A. (1998) *The quantum world unveiled by electron waves.* World Scientific, Singapore
Tonomura A., Endo J., Matsuda T., Kawasaki T. (1989) *Am. J. Phys.* **57**, 117
Tonomura A., Osakabe N., Matsuda T., Kawasaki T., Endo J. (1986) *Phys. Rev. Lett.* **56**, 762
Treimer W. (1979) in *Neutron Interferometry* (eds. U.Bonse and H.Rauch). Clarendon Press, Oxford
Treimer W., Feye-Treimer U. (1998) *Physica* **B 241–3**, 1228
Treimer W., Strobl M., Hilger A., Seifert C., Feye-Treimer U. (2003) *Appl. Phys. Lett.* **83**, 398
Trinker M., Jericha E., Bouman W.G., Loidl R., Rauch H. (2007) *Nucl. Instr. Meth.* **A579**, 1081
Trubnikov S.V. (1981) *Sov. J. Nucl. Phys.* **34**, 550
Tumulka R., Viale A., Zanghi N, (2007) *Phys. Rev.* **A75**, 055602
Tuppinger D. (1987) thesis, Techn. Univ. Vienna
Tuppinger D., Rauch H., Seidl E. (1988) *Physica* **B151**, 96
Unnerstall T. (1990) *Phys. Lett.* **A151**, 263
Utsuro M, Ignatovich V.K. (2010) *Handbook of Neutron Optics.* Wiley-VCH, Weinheim
Vacchini B. (2005) *Phys. Rev. Lett.* **95**, 55
Vaidman L. (2014) *Quantum Stud.: Math. Found.* DOI 10.1007/s40509-14-0008-4; arXiv: 1405.4222
Van Cittert P.H. (1934) *Physica* **1**, 201
Van der Merve S.(1985), *Rev. Mod. Phys.* **57**, 689
Van Hove L. (1954) *Phys. Rev.* **95**, 249
Varju K., Ryder L.H. (2000) *Am. J. Phys.* **68**, 404
Varro S. (2008) *Prog. Phys.* **56**, 91
Varro S. (2011) *Prog. Phys.* **59**, 296
Vedral V. (2011) *Decoding Reality: the universe as quantum information.* Oxford University Press, Oxford
Venugopalan A. (1997) *Phys. Rev.* **A56**, 4307
Venugopalan A., Ghosh R. (1995) *Phys. Lett.* **A204**, 11
Vigier J.P. (1985) *Pramana, J. Phys.* **25**, 397
Vigier J.P. (1988) *Physica* **B151**, 386
Viola L., Onofrio R. (1997) *Phys. Rev.* **D55**, 455
Vogel K., Risken H. (1989) *Phys. Rev.* **A40**, 2847
Von Laue M. (1931) *Ergebn. Exakt. Naturwiss.* **10**, 133
Von Neumann J. (1932) *Mathematische Grundlagen der Quantentheorie.* Springer, Berlin
Voronin V.V., Lapin E.P., Semenikhin S.Yu., Fedorov V. V. (2000) *JETP Letters* **71**, 76
Wagh A.G. (1999a) *Phys. Lett.* **A251**, 86
Wagh A.G. (1999b) *Phys. Rev.* **A59**, 1715
Wagh A.G. (1999c) *Phys. Lett.* **A259**, 81
Wagh A.G., Abbas S., Treimer W. (2011) *Nucl. Instr. Meth.* **A634**, S41
Wagh A.G., Rakhecha V.C. (1990) *Phys. Lett.* **A148**, 17
Wagh A.G., Rakhecha V.C. (1996) *Prog. Part. Nucl. Phys.* **37**, 485

Wagh A.G., Rakhecha V.C. (1997) *Phys. Rev. Lett.* **78**, 1399

Wagh A.G., Rakhecha V.C., Fischer P., Ioffe A. (1998) *Phys. Rev. Lett.* **81**, 1992

Wagh A.G., Rakhecha V.C., Summhammer J., Badurek G., Weinfurter H., Allman B.E., Kaiser H., Hamacher K., Jacobson D.L., Werner S.A. (1997) *Phys. Rev. Lett.* **78**, 755

Wahl H. (1970) *Optik* **30**, 508

Waler I. (1923) *Z. Phys.* **17**, 398

Wallace W.E., Jacobson D.L., Arif M., Ioffe A. (1999) *J. Appl. Phys.* **74**, 469

Walls D.F. (1966) *Am. J. Phys.* **45**, 952

Walls D.F. (1983) *Nature* **306**, 141

Walls D.F., Milburn G.J. (1985) *Phys. Rev.* **A31**, 2403

Walls D.F., Milburn G.J. (1994) *Quantum Optics*. Springer, Berlin

Walther H. (1998) *Proc. R. Soc. London* **A454**, 431

Wang L.J., Zou X.Y., Mandel L. (1991) *Phys. Rev. Lett.* **66**, 1111

Wang Y.-J., Anderson D., Bright V., Cornell E., Diot Q., Kishimoto T., Prentiss R., Saravanan S., Segal S., Wu S. (2005) *Phys. Rev. Lett.* **94**, 090405

Warner M., Gubernatis J.E. (1985) *Phys. Rev.* **B32**, 6347

Webb C.L., Godun R.M., Summy G.S., Oberthaler M.K., Faetonby P.D., Foot C.J., Burnett K. (1999) *Phys.Rev.* **A60**, R1783

Weber H.W., Schelten J., Lippmann G. (1973) *Phys. Stat. Sol. (a)* **57**, 515

Weber M., Zeilinger A. (1997) Universität Innsbruck, private communication

Weihs G. (2007), *Nature* **445**, 723

Weihs G., Jennewein T., Simon C., Weinfurter H., and Zeilinger A. (1998) *Phys. Rev. Lett.* **81**, 5039

Weiner R., Weise W. (1985) *Phys. Lett.* **159B**, 85

Weinberg S. (1978) *The first three minutes: a modern view of the origin of the universe*. Fontana, Glasgow

Weinfurter H., Badurek G. (1990) *Phys. Rev. Lett.* **64**, 1318

Weinfurter H., Badurek G., Rauch H., Schwahn D. (1988) *Z. Phys.* **B72**, 195

Werner S.A. (1976) *Proc. Int. Conf. Neutron Scattering*, Gatlinburg, p. 1060

Werner S.A. (1980a) *Phys. Today*, Dec., 2

Werner S.A. (1980b) *Phys. Rev.* **B21**, 1774

Werner S.A. (1996) *J. Phys. Soc. Japan* **65**, Suppl. A, 51

Werner S.A. (2012) *Found. Phys.* **42**, 122

Werner S.A., Berliner R.R., Arif M. (1986) *Physica* **137B**, 245

Werner S.A., Clothier R., Kaiser H., Rauch H., Woelwitsch H. (1991) *Phys. Rev. Lett.* **67**, 683

Werner S.A., Colella R., Overhauser A.W., Eagen C.F. (1975) *Phys. Rev. Lett.* **35**, 1053

Werner S.A., Kaiser H., Arif M., Clothier R. (1988) *Physica* **B151**, 22

Werner S.A., Kaiser H., Arif M., Hu H.C., Berliner R. (1986) *Physica* **136B**, 137

Werner S.A., Klein A.G. (1986) *Meth. Exp. Phys.* **23A**, 259

Werner S.A., Klein A.G. (2010) *J. Phys. A: Math. Theor.* **43**, 354006

Werner S.A., Staudenmann J. L., Colella R. (1979) *Phys. Rev. Lett.* **42**, 1103

Weyrl A. (1978) *Rev. Mod. Phys.* **50**, 221

Wheeler J.A. (1978) in *Mathematical fundations of quantum mechanics* (ed. A.R. Marlow). Academic Press, New York

Wheeler J.A. (1983) in *Quantum Theory and Measurement* (eds. J.A. Wheeler and W.H. Zurek), p. 182. Princeton University Press, Princeton

Wheeler J.A., Zurek W.H. (1983) *Quantum theory and measurement*. Princeton University Press, Princeton

Whitaker A. (1995) *Einstein, Bohr and the quantum dilemma*. Cambridge University Press, Cambridge

Whitaker A. (2012) *The new quantum age*. Oxford University Press, Oxford

Wietfeldt F.E., Huber M., Black T.C., Kaiser H., Arif D.L., Jacobson D.L., Werner S.A. (2005) *Physica*, **B385–6**, 1374

Wigner E.P. (1932) *Phys. Rev.* **40**, 749

Wigner E.P. (1963) *Am. J. Phys.* **31**, 6
Wigner E.P. (1970) *Am. J. Phys.* **38**, 1005
Williams W.G. (1988) *Polarized neutrons*. Clarendon Press, Oxford
Willis B.T.M., Carlile C.J. (2009) *Experimental neutron scattering*. Oxford University Press, Oxford
Wineland D.J., Itano W.M., Berquist J.C., Bollinger J.J., Prestage J.D. (1984) *At. Phys.* **9**, 3
Wipf H. (1997) *Hydrogen in Metals III*, Top. Appl. Phys. 73. Springer, Berlin
Wiringa R.B., Stoks V.G.J., Schiavilla R. (1995) *Phys. Rev.* **C51**, 38
Wirtz L., Tang J.-Z., Burgdoerfer J. (1997) *Phys. Rev.* **B56**, 1
Wiseman H.M. (1998) *Phys. Rev.* **A58**, 1740
Wiseman H.M., Harrison F. (1995) *Nature* **377**, 584
Wiseman H.M., Harrison F.E., Collett M.J., Tan S.M., Walls D.F., Killip K.B. (1997) *Phys. Rev.* **A56**, 55
Wodkiewicz K. (1984) *Phys. Rev. Lett.* **52**, 1064
Wodkiewicz K. (1988) *Phys. Lett.* **A115**, 304
Wolf E. (1989) *Phys. Rev. Lett.* **63**, 2220
Wolf P., Blanchet L., Bordè C.J., Raymond S., Salomon C., Cohen-Tanoudji C. (2011) *Class. Quantum Grav.* 28, 145017
Wolf P., Tourrenc P. (1999) *Phys. Lett.* **A251**, 241
Wood R.W. (1910) *Phil. Mag.* **20**, 770
Wood R.W. (1912) *Phil. Mag.* **23**, 310
Wooters W.K., Zurek W.H. (1979) *Phys. Rev.* **D17**, 473
Word R.E., Werner S.A. (1982) *Phys. Rev.* **B26**, 4190
Wu T.T., Yang C.N. (1975) *Phys. Rev.* **D12**, 3845
Yamazaki D., Ebisawa T., Kawai T., Tasaki S., Hino M., Akayoshi T., Achiwa H. (1998) *Physica* **B141–3**, 186
Young T. (1802) *Trans. R. Soc.* **92**, 387
Yuasa K., Facchi P., Nakazato H., Ohba I., Pascazio S., Tasaki S. (2008) *Phys. Rev.* **A77**, 043623
Yu Sixia, Oh C.H. (2014) *ArXiv.org: 1408.2477*
Yuen H.P. (1976) *Phys. Rev.* **A13**, 2226
Yurke B. (1986) *Phys. Rev. Lett.* **56**, 1515
Yurke B., McCall S.L., Klauder J.R. (1986) *Phys. Rev.* **A33**, 4033
Yurke B., Schleich W., Walls D.F. (1990) *Phys. Rev.* **A42**, 1703
Yurke B., Stoler D. (1986) *Phys. Rev. Lett.* **57**, 13
Zachariasen W. H. (1967) *The theory of X-ray diffraction in crystals*. Dover, New York
Zawisky M, Baron M., Loidl R. (2002) *Phys. Rev.* **A66**, 063608
Zawisky M. (2004) *Found. Phys. Lett.* **17**, 561
Zawisky M., Bonse U., Dubus F., Hradil Z., Rehacek J. (2004) *Europhys. Lett.* **68**, 215505
Zawisky M., Hasegawa Y., Rauch H., Hradil Z, Myska R., Perina J. (1998) *J. Phys.* **A31**, 551
Zawisky M., Rauch H., Hasegawa Y. (1994) *Phys. Rev.* **A50**, 5000
Zawisky M., Springer J., Farthofer R., Kuetgens U. (2010) *Nucl. Instr. Meth.* **A612**, 338
Zawisky M., Springer J., Lemmel H. (2011) *Nucl. Instr. Meth.* **A634**. S46
Zeeman P. (1914) *Acad. Sci. Amsterdam* **17**, 445
Zeeman P. (1915) *Acad. Sci. Amsterdam* **18**, 398, 1240
Zeeman P. (1927) *Arch. Nederland Sci. Exactes Nat., Ser. 3A* **10**, 131
Zeh H.D. (1970) *Found. Phys.* **1**, 69
Zeh H.D. (2001) *The physical basis of the direction of time*, 4th edn. Springer, Berlin
Zehnder L. (1891) *Z. Instrumentenkd.* **14**, 887
Zeidler A., Salmon P.S., Fischer H.E., Neuefeind J.C., Simonson J.M., Lemmel H., Rauch H., Markland T.E. (2011) *Phys. Rev. Lett.* **107**, 145501
Zeilinger A. (1979) in *Neutron Interferometry* (eds. U. Bonse and H. Rauch), p. 241. Clarendon Press, Oxford
Zeilinger A. (1981a) *Nature* **294**, 544

Zeilinger A. (1981b) *Am. J. Phys.* **49**, 882
Zeilinger A. (1984) *J. Phys.* **45**, C3-213
Zeilinger A. (1986a) in *Fundamental Aspects of Quantum Theory* (eds. V. Gorini and A. Frigero), NATO ASI Series B, Vol. 144, p. 311. Plenum Press, New York
Zeilinger A. (1986b) *Physica* **137B**, 235
Zeilinger A. (1999a) *Rev. Mod. Phys.* **71**, S288
Zeilinger A. (1999b) *Found.of Phys.* **29**, 631
Zeilinger A., Gaehler R.. Horne M.A. (1991) *Phys. Lett.* **A154**, 93
Zeilinger A., Gähler R., Shull C.G., Treimer W. (1982) *AIP Conf. Proc. Neutron Scattering* (ed. J. Faber), p. 93
Zeilinger A., Gähler R., Shull C.G., Treimer W. (1981) *AIP Conf. Proc.*(ed. H.C. Wolfe) **89**, 93
Zeilinger A., Gähler R., Shull C.G., Treimer W., Mampe K. (1988) *Rev. Mod. Phys.* **60**, 1067
Zeilinger A., Horne M.A. (1986) *Physica* **136B**, 14
Zeilinger A., Horne M.A., Shull C.G. (1984) *Proc. Int. Symp. Found. Quantum Mechanics*, Tokyo, Phys. Soc. Japan, p. 289
Zeilinger A., Shull C.G. (1979) *Phys. Rev.* **B19**, 3957
Zeilinger A., Shull C.G., Arthur J., Horne M.A. (1983) *Phys. Rev.* **A28**, 487
Zeilinger A., Shull C.G., Horne M.A., Finkelstein K.D. (1986) *Phys. Rev. Lett.* **57**, 3089
Zeilinger A., Shull C.G., Horne M.A., Squires G.L. (1979) in *Neutron Interferometry* (eds. U. Bonse and H. Rauch), p. 48. Clarendon Press, Oxford
Zernike F. (1938) *Physica* **5**, 785
Zernike F. (1950) *Opt. Soc. Am.* **40**, 326
Zettili N. (2001) *Quantum mechanics: concepts and applications*. Wiley, Chichester
Zeyen C.M.E., Otake Y., Tabaru T., Toperverg B. (1996) *J. Phys. Soc. Japan* **A65**, 177
Zhang H., Gallagher P.D., Satija S.K., Lindstrom R.M., Paul R.L., Russel T.P., Lambooy P., Kramer E.J. (1994) *Phys. Rev. Lett.* **72**, 3044
Zhang W.M., Feng D.H., Gilmore R. (1990) *Rev. Mod. Phys.* **62**, 867
Zhang Y.D., Badurek G., Rauch H., Summhammer J. (1994) *Phys. Lett.* **A188**, 225
Zhang X., Um M., Zhang J., An S., Wang Y., Deng D.-L., Shen C., Duan L.-M., Kim K. (2013) *Phys. Rev. Lett.* **110**, 070401
Zhao S., Yuan S., De Raedt H., Michielsen K. (2008) *Europhys. Lett.* **82**, 40004
Zimmer O., Ehlers G., Farago B., Humbolt H., Ketter W., Scherm R. (2002) *Eur. Phys. J.* **C4**, 1
Zimmer O., Felber J., Schärpf O. (2001) *Europhys. Lett.* **53**, 183
Zimmerman J.E., Mercereau J.E. (1965) *Phys. Rev. Lett.* **14**, 887
Zou X.Y., Grayson T.P., Mandel L. (1992) *Phys. Rev. Lett.* **69**, 3041
Zouw van der G., Weber M., Felber J., Gaehler R., Geltenbort P., Zeilinger A. (2000) *Nucl. Instr. Meth.* **A 440**, 568
Zurek W.H. (1981) *Phys. Rev.* **D24**, 15516
Zurek W.H. (1991) *Phys. Today*, **Oct.**, 36
Zurek W.H. (1993) *Progr. Theor. Phys.* **89**, 281
Zurek W.H. (1998a) *Phys. Scripta* **T76**, 186
Zurek W.H. (1998b) *Phil. Trans. R. Soc. Lond.* **A356**, 1793
Zurek W.H. (2003) *Rev. Mod. Phys.* **75**, 715
Zurek W.H., Habib S., Paz J.P. (1993) *Phys. Rev. Lett.* **70**, 1187
Zych M., Costa F., Pikovski I., Ralph T.C. Brukner C. (2012) *Class. Quantum Grav.* **29**, 224010
Zych M., Costa F., Pikovski I. Brukner C. (2011) *Nat. Commun.* **2**, 505

Index

V

W

X

Y

Z